Leveraging Microbial Technology for Food Security

Microorganisms play a vital role in addressing global food security challenges. *Leveraging Microbial Technology for Food Security: Solutions for Zero Hunger* examines their diverse applications in achieving zero hunger with a focus on practical solutions that transform food production, improve nutrition, and reduce environmental impact. Topics include biofertilization, genetic engineering, waste management, and microbial fermentation for alternative protein sources.

Through global case studies, the book demonstrates how microbial innovations have improved food security and provided scalable, sustainable approaches to reducing hunger. It also explores the potential of microorganisms to diversify diets and lessen dependence on traditional agricultural systems.

Written for scholars, policymakers, educators, farmers, and students, this book combines scientific analysis with real-world examples. Featuring case studies, data visualizations, and actionable insights, it is a valuable resource for those interested in biotechnology, food systems, and sustainable development.

Leveraging Microbial Technology for Food Security

Solutions for Zero Hunger

Edited by

Babafemi Raphael Babaniyi,
Barbara Sawicka, Mukul Machhindra Barwant,
and Sesan Abiodun Aransiola

CRC Press
Taylor & Francis Group
Boca Raton London New York

CRC Press is an imprint of the
Taylor & Francis Group, an **informa** business

Designed cover image: Shutterstock

First edition published 2026
by CRC Press
2385 NW Executive Center Drive, Suite 320, Boca Raton FL 33431

and by CRC Press
4 Park Square, Milton Park, Abingdon, Oxon, OX14 4RN

CRC Press is an imprint of Taylor & Francis Group, LLC

ISBN: 978-1-041-09589-7 (hbk)
ISBN: 978-1-041-09064-9 (pbk)
ISBN: 978-1-003-65088-1 (ebk)

DOI: 10.1201/9781003650881

Typeset in Times
by KnowledgeWorks Global Ltd.

Contents

Foreword

Food is both a basic necessity and a universal right, yet hunger continues to affect vast populations across every continent. In many regions, people face not only the absence of food but also poor nutritional quality, which undermines health, education, and economic opportunity. Traditional agricultural methods alone have not been sufficient to meet the demands of a growing population under the stresses of climate change, land degradation, and limited natural resources. The search for new, effective approaches is no longer optional, it is essential.

This book, *Leveraging Microbial Technology for Food Security: Solutions for Zero Hunger*, takes on this challenge by placing microorganisms at the center of the conversation. Microbes, though often overlooked, provide extraordinary opportunities for innovation in food production and distribution. They can enrich soil, protect crops, enhance the nutritional value of food, preserve harvests, recycle waste, and even restore damaged ecosystems. Their potential stretches far beyond supporting agriculture; it extends to reshaping the way societies secure and sustain food supplies in an increasingly uncertain world.

The chapters presented here combine scientific depth with practical relevance. They trace the causes of hunger, assess current interventions, and then present microbial biotechnology as a transformative tool for the future. Readers are guided through a wide range of applications—from microbial fertilizers and biopesticides to genetic engineering, biopreservation, and climate adaptation strategies. In doing so, the book does not merely describe technology but frames it as a practical response to one of humanity's oldest and most enduring struggles.

What sets this work apart is its forward-looking perspective. It acknowledges the technical and policy challenges that lie ahead, while also emphasizing the enormous opportunities for collaboration among scientists, farmers, industry leaders, and governments. In highlighting microbes as key partners in agriculture and nutrition, the book opens new pathways for innovation that could redefine global food security.

This is not just a book for academics or specialists. It is intended for anyone concerned with the future of food, whether as a researcher, policymaker, student, or practitioner. By presenting a clear vision of how microbial technology can be mobilized against hunger, it inspires not only a sense of urgency, but also of possibility.

At a time when the gap between food need and food availability continues to widen, *Leveraging Microbial Technology for Food Security: Solutions for Zero Hunger* offers both knowledge and hope. It reminds us that some of the smallest organisms on Earth may hold the key to solving one of humanity's greatest challenges.

Prof. Olabisi Peter Abioye
Head of Department, Public Health
Federal University of Technology Minna, Nigeria

Preface

Food is fundamental to life, yet hunger continues to afflict millions of people across the globe. Despite decades of progress, global hunger remains a persistent challenge, intensified by poverty, conflict, climate change, and unsustainable agricultural practices. The pursuit of the United Nations Sustainable Development Goal 2 (Zero Hunger) requires not only strengthening existing systems but also embracing innovative solutions capable of transforming the way food is produced, managed, and consumed.

This book, *Leveraging Microbial Technology Food Security: Solutions for Zero Hunger*, was conceived to present the immense potential of microorganisms in shaping sustainable food systems. Microbes, though microscopic, play macroscopic roles in agriculture, nutrition, and environmental health. Their ability to improve soil fertility, enhance crop resilience, fortify food with essential nutrients, and support waste management positions them as critical allies in addressing both the quantity and quality aspects of food security.

The structure of the book reflects a progression from understanding the global hunger crisis to exploring microbial innovations and their integration into sustainable agricultural practices. Section I lays the groundwork by examining the realities, causes, and consequences of hunger. Section II highlights practical applications of microbial technologies, from biofertilizers to food fortification. Section III delves into the synergies between microbiomes, soil, water, and agro-ecology in securing sustainable food production. Section IV emphasizes microbial strategies for climate-resilient agriculture, while Section V provides a forward-looking perspective on research, policies, and innovations that will shape the future of food security.

This book aims to serve as a comprehensive resource for researchers, educators, students, policymakers, and practitioners. It provides both scientific insight and practical guidance, bridging the gap between microbiological innovation and real-world challenges in agriculture and food systems. By uniting knowledge from diverse disciplines, it stresses the urgency and feasibility of leveraging microbial technology to advance global food security.

The preparation of this volume was guided by a simple yet profound belief that the solutions to some of humanity's greatest challenges can be found in the smallest of organisms. It is our hope that this book will not only inform but also inspire action, encouraging collaboration, innovation, and commitment in the shared journey toward a hunger-free world.

Babafemi Raphael Babaniyi, Ph.D.
Bioresources Development Centre, National Biotechnology Research and Development Agency (NBRDA), Ogbomoso, Nigeria

Barbara Sawicka, Professor
Department of Plant Production Technology and Commodities Science, Faculty of Agrobioengineering, University of Life Sciences, Lublin, Poland

Mukul Machhindra Barwant, Ph.D.
Department of Botany, Sanjivani Rural Education Society's, Sanjivani Arts, Commerce and Science College, Kopargaon, Ahmednagar, Maharashtra, India

Sesan Abiodun Aransiola, Ph.D.
Department of Microbiology, Faculty of Science, University of Abuja, Abuja, Nigeria

Acknowledgments

We sincerely thank the contributors of this book for their invaluable insights and timely responses throughout the development of this project. Their enthusiasm, dedication, and collaborative spirit were key to completing this volume within the planned schedule. We are truly grateful for their unwavering support. Our heartfelt appreciation also goes to the anonymous reviewers, whose constructive feedback greatly enhanced the quality of this book. A special note of thanks is extended to CRC Press and the production team for their outstanding guidance and steadfast support. Their professionalism and commitment were instrumental in bringing this project to completion in such a short timeframe. It has been both an honor and a privilege to work with them.

Babafemi Raphael Babaniyi, Ph.D.
Bioresources Development Centre, National Biotechnology Research and Development Agency (NBRDA), Ogbomoso, Nigeria

Barbara Sawicka, Professor
Department of Plant Production Technology and Commodities Science, Faculty of Agrobioengineering, University of Life Sciences Lublin, Poland

Mukul Machhindra Barwant, Ph.D.
Department of Botany, Sanjivani Rural Education Society's, Sanjivani Arts, Commerce and Science College, Kopargaon, Ahmednagar, Maharashtra, India

Sesan Abiodun Aransiola, Ph.D.
Department of Microbiology, Faculty of Science, University of Abuja, Abuja, Nigeria

About the Editors

Babafemi Raphael Babaniyi is a prominent researcher and academic in the field of Environmental Chemistry. He earned a B.Sc. in Industrial Chemistry from Caritas University, Enugu (2008), and obtained both an M.Tech. (2021) and a Ph.D. (2025) in Environmental Chemistry from the Federal University of Technology, Akure. His interdisciplinary expertise spans Environmental and Green Chemistry, Phytoremediation, Chemical Biology, Bioremediation, Environmental Impact Assessment, and Biodegradable Plastics. With over 40 publications in peer-reviewed journals and book chapters, Babafemi has made impactful contributions to global scientific dialogue. He currently serves as an Assistant Chief Research Officer at the National Biotechnology Development Agency, where he leads research initiatives focused on biotechnology and environmental sustainability. His commitment to addressing environmental challenges through science positions him as a key advocate for sustainable development and environmental health.

Barbara Sawicka is a distinguished full professor at the University of Life Sciences in Lublin, Poland, where she has been a faculty member in the Department of Plant Production Technology and Commodities Science since 2004. She has established herself as a leading scholar with an exceptional body of work comprising over 1,020 publications, including 152 monographs, 390 original research papers, and 27 books. Her research interests cover agronomy, bioengineering, food sciences, and environmental protection. Professor Sawicka has guided 11 Ph.D. candidates, 190 master's students, and 250 engineers, reflecting her deep commitment to academic mentorship. She also plays an active role in the editorial sphere, serving on the advisory boards of several journals and holding positions as deputy editor-in-chief and co-editor. Beyond academia, she has contributed significantly to civil and social initiatives and remains an active member of professional bodies such as the Polish Agronomic Society and the Polish Academy of Sciences. Her international presence is reinforced through research collaborations and scholarships across Europe, including Belgium, Lithuania, and Turkey. Fluent in multiple languages, Professor Sawicka continues to shape her field with a truly global impact.

Mukul Machhindra Barwant is the Dean of Research and Innovation and an Assistant Professor of Botany at Sanjivani Arts, Commerce and Science College, Kopargaon, Ahilyanagar, Maharashtra, India. His expertise includes phycology, nutraceuticals, herbal biotechnology, and environmental sciences. An accomplished researcher and editor, he serves on the editorial boards of 25 international journals, has reviewed 49 research articles, edited 41 books, authored 24 book chapters, and published 55 research papers in top journals published by Springer, Elsevier, and Taylor & Francis. Barwant holds 25 patents across India, the United Kingdom, and Germany, focusing on seaweed-based nutraceuticals and plant-derived solutions for healthcare and agriculture. He is also an active mentor, academic trainer, and organizer of national and international workshops on plant sciences and sustainability. His contributions have earned him multiple prestigious awards, including Best Researcher, Young Scientist, Young Phycologist, and Distinguished Researcher Awards (2021–2025), along with recognition for excellence in research publications and plant sciences.

Sesan Abiodun Aransiola holds a Ph.D. in Environmental Microbiology from the Federal University of Technology, Minna, Nigeria, where he also earned his B.Tech. (2009) and M.Tech. (2014) degrees. He is currently a Lecturer in the Department of Microbiology at the University of Abuja, Nigeria. His research centers on innovative strategies for soil and water remediation, with a particular focus on vermitechnology, phytoremediation, and microbial biotechnology. Aransiola has extensively explored the application of vermicasts and biofertilizers such as vinasse to support soil recovery from heavy metal and plastic pollution. His expertise spans environmental health, marine resources, and sustainable waste management. An award-winning scientist, he has authored over 90 scholarly publications, including high-impact journal articles, reviews, and book chapters. He has also co-edited several internationally recognized scientific volumes with publishers such as Elsevier, Springer, and CRC Press, covering topics including microbial biotechnology, soil microbiomes, marine bioprospecting, phytoremediation, and food safety. Previously, he served as an Assistant Chief Scientific Officer at the National Biotechnology Development Agency. Beyond academia, Aransiola actively contributes to community service and sustainability advocacy. He is a member of several professional bodies, including the Nigerian Society for Microbiology and the American Society for Microbiology.

Contributors

Bolaji Hamzat Abdulmujeeb
Nigerian Institute for Oceanography and Marine Research
Victoria Island, Lagos, Nigeria

Aisha Eniola Adegbite
Department of Microbiology
University of Abuja
Abuja, Nigeria

Akintunde Adewole Adegbite
Nigerian Institute for Oceanography and Marine Research
Victoria Island, Lagos, Nigeria

Maria M. Adeyemi
Department of Chemistry and Biochemistry
and
Department of Biological Sciences and Biotechnology
Caleb University Lagos
Lagos, Nigeria

Olarewaju John Adunbi
Bioresources Development Centre
National Biotechnology Development Agency
Abuja, Nigeria

Enaji Kanu Agabi
Department of Biochemistry & Biotechnology
Nile University of Nigeria
Abuja, Nigeria

Shireen Adeeb Mujtaba Ali
Salar-E-Millat Sultan Salah Uddin Owaisi Central Research Laboratory for Cellular and Molecular Medicine
Princess Esra Hospital
Deccan College of Medical Sciences
Kanchan Bagh, Telangana, India

Muhammad Anas
Department of Soil Science & Plant Nutrition
University of Selcuk Campus
Konya, Turkey

Moses Aniefiok
Anison Environmental Services Ltd.
Abuja, Nigeria

Ponnusamy Thillai Arasu
Department of Chemistry
Wollega University
Nekemte, Oromia, Ethiopia

Abraham Olufemi Asuku
Bioresources Development Centre
National Biotechnology Research and Development Agency
Ogbomoso, Nigeria

Piotr Barbaś
Department of Potato Agronomy
Institute of Plant Breeding and Acclimatization – National Research Institute
Branch Jadwisin, Serock, Poland

Victoria Bassey
Department of Chemistry and Biochemistry
Caleb University Lagos
Lagos, Nigeria
and
Department of Biological Sciences and Biotechnology
Caleb University Lagos
Lagos, Nigeria

Jennifer Chisom
National Biotechnology Research and Development Agency
Abuja, Nigeria

Mudasir A. Dar
School of Environment and Safety Engineering
Jiangsu University
Zhenjiang, China

Oluwatosin Olaoluwa Daramola
Department of Chemistry
Federal University of Technology Akure
Akure, Nigeria

Mohit Deswal
Department of Zoology
Maharshi Dayanand University
Rohtak, Haryana, India

Elayaraja Dhandapani
Northern Research and Development
MIGAL Galilee Research Institute
Kiryat Shmona, Israel

Namla Djadjitti
Department of Biological Sciences
King Abdulaziz University
Jeddah, Saudi Arabia

Kingsley Erhons Enerijiofi
Department of Biological Sciences
College of Basic Applied and Health Sciences
Glorious Vision University
Ogwa Edo State, Nigeria

Gloria Ginikanwa Ezeagu
Department of Biology, Microbiology & Science Laboratory Technology
Nile University of Nigeria
Abuja, Nigeria

Ayobami Grace Fajemidagba
Department of Physiology
Redeemer's University
Ede, Osun State, Nigeria

Mayowa Sikirat Giwa
Nigerian Institute for Oceanography and Marine Research
Lagos, Nigeria

Roseline Jebapriya Gunaseelan
Department of Marine Science
Bharathidasan University
Tiruchirappalli, Tamil Nadu, India

Himanshi Gupta
Council of Scientific and Industrial Research (CSIR)-Institute of Himalayan Bioresource Technology (IHBT)
Palampur, Himachal Pradesh, India

Pooja Kadyan
Department of Zoology
Maharshi Dayanand University
Rohtak, Haryana, India

Alex Kamgain Tchuenchieu
Food Evolution Research Laboratory
School of Tourism and Hospitality
College of Business and Economics
University of Johannesburg
Johannesburg, South Africa
and
Centre for Food and Nutrition Research
Institute of Medical Research and Medicinal Plants Studies
Yaoundé, Cameroon

Urvashi Kashyap
Department of Zoology
Government P. G. College New Tehri
Sri Dev Suman Uttarakhand University
Uttarakhand, India

Sudhir Kumar Kataria
Department of Zoology
Maharshi Dayanand University
Rohtak, Haryana, India

Rupinder Kaur
Academy of Scientific and Innovative Research (AcSIR)
Ghaziabad, India

Amjid Khan
UNESCO UNISA Africa Chair in Nanoscience and Nanotechnology
College of Graduate Studies
University of South Africa
Pretoria, South Africa

Bunmi C. Kotun
Department of Biological Sciences and Biotechnology
Caleb University
Lagos, Nigeria

Nur Koç Koyun
Department of Field Crops
Faculty of Agriculture
Selcuk University
Konya, Türkiye

Rahul Kumar
Department of Zoology
College of Agricultural
Bawal Chaudhary Charan Singh Haryana Agricultural University
Hisar, Haryana, India

Shailendra Kumar
Department of Microbiology
Dr. Ram Manohar Lohia Avadh University
Ayodhya, Uttar Pradesh, India

Naga Raju Maddela
Departamento de Ciencias Biológicas
Facultad la Ciencias de la Salud
Universidad Técnica de Manabí
Portoviejo, Manabí, Ecuador

Collins Madejomu
Department of Biology
Microbiology & Science Laboratory Technology
Nile University of Nigeria
Abuja, Nigeria

Vivek Manyapu
Department of Tropical Agriculture and International Cooperation
National Pingtung University of Science and Technology
Neipu, Pingtung County, Taiwan

Samuel Abasiama Moses
National Veterinary Research Institute (NVRI)
Vom, Nigeria

Shaikh Shagufta Mujiburrehman
P. T. Arts and Science College
Godhra, Gujarat, India

Annamalai Neelamegam
Thayer School of Engineering
Dartmouth College
Hannover, New Hampshire, United States

Sarah Chioma Nwakanma
Department of Microbiology
Faculty of Science
University of Abuja
Abuja, Nigeria

Odangowei Inetiminebi Ogidi
Department of Biochemistry
Bayelsa Medical University
Yenagoa, Bayelsa State, Nigeria

Itoroabasi Joseph Okon
National Biotechnology Research and Development Agency
Abuja, Nigeria

Deborah Oni-Orisan
Department of Chemistry and Biochemistry
and
Department of Biological Sciences and Biotechnology
Caleb University
Lagos, Nigeria

Beatrice Ovute Onodagu
Department of Microbiology
University of Nigeria Teaching Hospital
Ituku/Ozalla, Enugu, Nigeria

Eridiong Onyenweaku
College of Business and Economics
University of Johannesburg
Johannesburg, South Africa
and
Department of Human Nutrition and Dietetics
Faculty of Allied Health Sciences
College of Medical Sciences
University of Calabar
Cross River State, Nigeria

Oluwademilade Opesanwo
Department of Chemistry and Biochemistry
Caleb University
Lagos, Nigeria

Adewumi Yetunde Oriji
Department of Chemistry
Federal University of Technology
Akure, Nigeria

Gideon Taiwo Otolorin
Department of Chemistry
University of Ibadan
Ibadan, Nigeria

Santhanam Perumal
Department of Marine Science
Bharathidasan University
Tiruchirappalli, Tamil Nadu, India

Mallikarjun Chanchi Prashanthkumar
International Center of Excellence in Seafood Science and Innovation
Faculty of Agro-Industry
Prince of Songkla University
Hat Yai, Songkhla, Thailand

Piotr Pszczółkowski
Research Centre for Cultivar Testing
Słupia Wielka, Poland

Anburaj Raj
Department of Microbiology
Thanthai Hans Roever College (Autonomous)
Perambalur, Tamil Nadu, India

Geetanjli Sahni
Department of Zoology
Arya P. G. College
Panipat, Haryana, India

Jaya Saini
Department of Zoology
Maharshi Dayanand University
Rohtak, Haryana, India

Vanmathi Vidhyashini Selvaraj
Department of Biotechnology
Bishop Heber College (Autonomous)
Tiruchirappalli, Tamil Nadu, India

Balwant Singh
Department of Botany
K. S. Saket PG College
Dr. Ram Manohar Lohia Avadh University
Ayodhya, India

Stella Suanu
National Biotechnology Research and Development Agency
Abuja, Nigeria

Raja Subramaniyan
Atal Centre for Ocean Science and Technology
National Institute of Ocean Technology
Ministry of Earth Sciences
Andaman and Nicobar, India

Nonso Ugwuja
Department of Chemistry and Biochemistry
and
Department of Biological Sciences and Biotechnology
Caleb University
Lagos, Nigeria

Atabo Hilda Uyo
National Biotechnology Research and Development Agency
Abuja, Nigeria

Amir Vazirizadeh
Persian Gulf Research Institute (PGRI)
Fishery and Marine Biology Department
Persian Gulf University
Bushehr, Iran

Vanessa Williams
Department of Chemistry and Biochemistry
and
Department of Biological Sciences and Biotechnology
Caleb University
Lagos, Nigeria

Baikunth Kumar Yadav
Department of Zoology
Mahatma Gandhi Central University
Motihari, Bihar, India

Amina Yahaya
Department of Microbiology
Faculty of Science
University of Abuja
Abuja, Nigeria

Leh-Togi Zobeashia
National Biotechnology Research and Development Agency
Abuja, Nigeria

1 The Global Hunger Definitions, Statistics, and Solutions Overview

Sesan Abiodun Aransiola, Sarah Chioma Nwakanma, Aisha Eniola Adegbite, Amina Yahaya, and Naga Raju Maddela

1.1 INTRODUCTION

Hunger is simply the feeling one gets when their body is in need of food. It is that empty, painful, sometimes biting sensation one gets in their stomach when they haven't eaten in a while. But hunger is more than just a physical feeling, it also reflects deeper social and economic challenges that affect individuals, families, and as a whole, the communities (Khofi et al., 2024).

Based on what the World Food Programme (WFP, 2024) said, hunger is the discomfort brought on by a shortage of food. It is a condition that happens when people don't get sufficient food to satisfy their daily energy requirements. The Food and Agriculture Organization (FAO, 2024) defines hunger more formally as undernourishment, when a person's dietary intake is frequently below the minimum required to keep an active and healthy lifestyle.

Worldwide hunger can be better understood when one is able to differentiate the related terms associated with hunger:

1. *Hunger*: This is the discomfort or pain in the body brought on by inadequate food intake. It can be temporary or chronic depending on one's access to food (FAO, 2024).
2. *Undernourishment*: As defined by the FAO (2024), is when someone doesn't eat enough calories to maintain a healthy way of living. It's a long-term issue that affects health, energy, and productivity.
3. *Malnutrition*: This is a more extensive term that includes both undernutrition (too little food or nutrients) and overnutrition (too many nutrients or unhealthy foods). It can lead to problems like weakened immunity, stunted growth, and increased risk of disease (Ahmed et al., 2022).
4. *Food Insecurity*: This happens when people lack consistent access to enough reasonably priced, wholesome food. Food insecurity can range from going days without eating to worrying about where one's next meal will come from, according to the World Health Organization (WHO, 2024).

1.2 TYPES OF HUNGER

Hunger is not a one-size-fits-all condition. It comes in different forms, each having its own causes and effects. The three main types of hunger include chronic hunger, acute hunger, and hidden hunger (Fasano, 2025).

1.2.1 Chronic Hunger

Chronic hunger is a long-term, consistent lack of access to food. It happens when people go months or even years without getting the nutrients they need to lead healthy, active lives (Tack et al., 2021).

DOI: 10.1201/9781003650881-1

This type of hunger is often linked to poverty, poor infrastructure, and even weak agricultural systems.

For example, many rural communities in Africa's Sub-Saharan region and parts of South Asia experience chronic hunger because of ongoing poverty and food system challenges. In countries like Ethiopia or Bangladesh, millions of people rely on subsistence farming and often face seasonal food shortages (Tack et al., 2021).

1.2.2 Acute Hunger

Acute hunger is more severe and sudden. It happens in short bursts and is usually triggered by crises like conflict, natural disasters, or economic collapse. Acute hunger can lead to malnutrition and even death if not addressed quickly (Lenaerts and Demont, 2021).

For example, in Yemen, South Sudan, and parts of northern Nigeria, ongoing conflicts and displacement have left millions in dire need of food assistance. Additionally, in the Horn of Africa, droughts and floods have led to spikes in acute hunger, particularly among children and pregnant women (Lenaerts and Demont, 2021).

Another example includes the people of Palestine, who have lack of access to food and clean water due to the ongoing war with Israel, leading to the death of many Palestinians, especially children (Hardman, 2024).

1.2.3 Hidden Hunger

Hidden hunger occurs when people eat enough calories but don't get essential vitamins and minerals like iron, vitamin A, or iodine. This type of hunger is not as visible but it is very dangerous. It can lead to weakened immunity, poor mental development, and complications during pregnancy (Lowe, 2021).

For example, hidden hunger affects people in both developing and developed countries. In places like India, Nigeria, Palestine, and even parts of the United States, diets that are heavy in processed foods but low in nutrients can cause deficiencies—especially among women and children (Lenaerts and Demont, 2021).

1.3 KEY GLOBAL STATISTICS

Global hunger continues to be a major concern, affecting hundreds of millions of people. In 2022, there were about 735 million suffered from malnutrition (FAO, 2024). This means nearly 1 in 10 people worldwide lacked consistent access to enough food for a healthy lifestyle.

There is a regional breakdown of these statistics:

1. The region with the highest rate of hunger is still Sub-Saharan Africa, with about 20% of the population undernourished (Otekunrin, 2025).
2. South Asia, including countries like India, Pakistan, and Bangladesh, also has high hunger levels, especially among women and children (Otekunrin, 2025).
3. Hunger is increasing in areas impacted by war, including Yemen, Palestine, Sudan, and Afghanistan, where food systems are disrupted and humanitarian access is limited (Hardman, 2024).

1.4 THE GLOBAL HUNGER INDEX: MEASURING HUNGER, CAUSES, AND EFFECTS

Global Hunger Index (GHI) is a measure of hunger that calculates hunger at global, regional, and national levels. It measures hunger on a scale of 0–100 where zero signifies no hunger and 100, extremely alarming hunger. It is classified into; low (0–9.9), moderate (10 –19.9), serious (20–34.9), alarming (35–49.9), and extremely alarming (50 and above) (Okunlola and Aregbeshola, 2025). According to FAO, IFAD, UNICEF, WFP, and WHO (2025), GHI is calculated based on four indicators:

1. *Undernourishment*: The population under the age of five with insufficient caloric intake. This measures the percentage of people whose caloric intake is below the minimum dietary energy requirement.
2. *Child Stunting*: The number of children under the age of five who have low height for their age due to acute undernutrition often due to acute food shortage or disease.
3. *Child Wasting*: The share of children under the age of five who are low weight for their height due to chronic undernutrition.
4. *Child Mortality*: The mortality rate of children under the age of five, reflecting the fatal synergy of inadequate nutrition and unhealthy environments. This indicator is a sensitive measure of the overall health and nutritional status of a population, particularly its most vulnerable members.

The 2023 GHI, released in October 2023, painted a concerning picture of global hunger, indicating that progress against hunger worldwide has been stagnant since 2015. Large regional gaps were clear, as both South Asia and Sub-Saharan Africa scored 27.0, showing serious hunger levels that have remained for over 20 years. The world's 2023 GHI level was 18.3, seen as moderate. This was a mere 0.8 points below the world's 2015 GHI level of 19.1, signaling no improvement in the last eight years (Global Hunger Index, 2023). This stagnation is particularly alarming given the significant strides made in reducing hunger between 2000 and 2015. This negligible improvement is opposite compared to the big progress seen between the year 2000 and 2015, where focus programs like the Millennium Development Goals (MDGs) halved undernutrition in many parts of the world (von Grebmer et al., 2023). Critically, this plateau signals systemic failures in addressing overlapping crises, undermining Sustainable Development Goal (SDG) 2 and related goals like SDG 13 (Climate Action) and SDG 16.

Regional disparities were stark, with South Asia and Africa South of the Sahara both scoring 27.0, indicating serious hunger levels—a pattern persistent over two decades. Nine countries were provisionally designated with extremely alarming hunger: Angola, Somalia, Sierra Leone, Niger, Ethiopia, and Zambia (Global Hunger Index, 2023). Many other countries with alarming hunger worsened since 2015; for example, the Central African Republic's score of 42.3 reflected nearly half its population unable to meet minimum energy needs, while Niger faced high child stunting, wasting, and mortality amid conflict, climate crises, and price rises. Somalia endured successive droughts followed by floods, rendering crop production nearly impossible (Global Hunger Index, 2023). The report projected that 58 countries would fail to reach low hunger by 2030, necessitating accelerated, transformative food system changes.

Due to the stagnation in the improvement of global hunger, Zero Hunger by 2030 would be difficult to achieve at current pace. Millions remain undernourished and billions cannot afford a healthy diet. Several countries show serious or alarming hunger. The 2024 report shows global GHI scores in the moderate range while documenting large regional and country variation and noting that recent stalling largely reflects interacting crises since 2016. Acute food insecurity and famine risks are rising, with starvation weaponized in conflicts (Drishti, 2024). Underlying factors include large-scale armed conflicts (e.g., Gaza and Sudan, causing exceptional crises and displacement); escalating climate change impacts, reducing Sub-Saharan African agricultural productivity by 34% since 1961 (IPCC, 2022); economic challenges and debt crises limiting investments; and inequality within and between countries, disproportionately affecting fragile, conflict-prone states (Weldemichel, 42024). Regional trends persist, with Africa South of the Sahara (26.8) and South Asia (26.2) (Figure 1.1) exhibiting the severest hunger among other regions due to high undernourishment and child mortality. Additionally, 36 countries faced serious hunger, with hunger

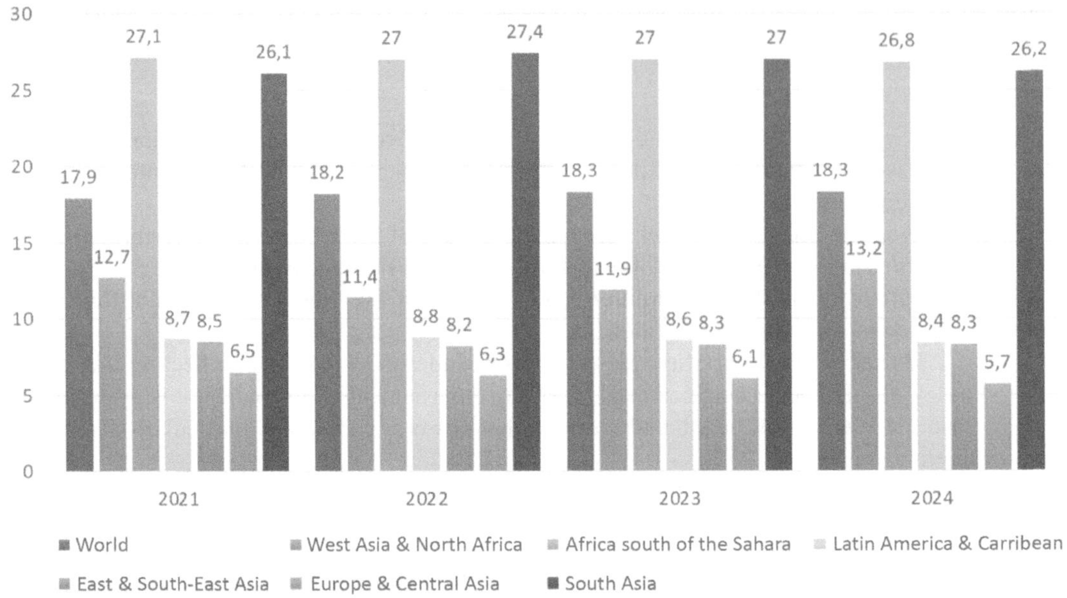

FIGURE 1.1 Graphical representation of global hunger index (GHI) from 2021 to 2024 in various regions.

increasing since 2016 in 22 countries and stalling (less than 5% decline) in another 20 (Global Hunger Index, 2024).

Climate change, population growth, economic and social inequalities, and a range of interconnected crises have greatly contributed to global hunger. Climate variability, from the rising temperatures, erratic rainfall, and extreme weather patterns, significantly disrupts the growth of crops and food availability (Nasser et al., 2021; Aransiola and Maddela, 2024b). Without action, this will ultimately increase undernourishment in over 73 million people globally. Africa's fast-growing population is expected to double by 2050, exacerbating pressure on food systems, increasing malnourishment risks due to insufficient food supply and nutrient-poor diets (Mumuni and Aleer, 2023). Economic and social inequalities further increase food insecurity, as access to food is heavily shaped by income gaps, poor infrastructure, and reliance on subsistence farming (Aransiola et al., 2024a). Gender inequities which limit marginalized groups and women to have access to land, credit, and ability to make decisions, deepen household vulnerability (Rao et al., 2024). Additionally, inadequate food production, land access issues, and the lingering effects of disease outbreaks as seen with COVID-19 (Okunlola and Aregbeshola, 2025). Also, armed conflicts, and displacement, such as those caused by the Ukraine war and insurgencies (Weldemichel, 2024). These factors, combined with high food prices, trade disruptions, unstable markets, rising external debt, and underfunded social protection programs like food aid or welfare, amplifies hunger risks (Hendrix and Brinkman, 2013). The combination of factors like climate change, droughts, floods (environmental), pollution outbursts, poverty, and insurgencies shapes the multifaceted nature of global hunger.

Hunger has the ability to exert profound and far-reaching consequences across health, development, and economic aspects of a nation, perpetuating vicious cycles that undermine human well-being and global progress. Malnutrition significantly increases child mortality, with undernourishment elevating infant death risks by 0.0119 percentage points for every 1% increase in prevalence (Gribble and Hausmann, 2023). It also causes stunting and wasting that can therefore result in the poor development of physical growth, cognitive development, and immune system health, contributing to 45% mortality among children below the age of five (Black et al., 2013). These health impacts are setbacks for SDG 2 (Zero Hunger), and even though child mortality has been decreasing over the years, environmental challenges may stall or reverse that progress (Lucas et al., 2019). Hunger and malnutrition don't just affect individuals' health, they also drain billions from the economy, reducing productivity, and keeping countries from growing. For instance, Sub-Saharan Africa loses about US$25 billion every year perpetuating poverty cycles which increases hunger levels (Gallup and Sachs, 2001; Global Nutrition Report, 2017). In Africa, where population growth does not align with food production, they are easily affected by hunger which makes it difficult for them to bounce back and stay food secure (Mumuni and Aleer, 2023). Collectively, these effects erode human capital, amplify inequalities, and impede sustainable development, necessitating integrated policy interventions to break the cycle of hunger and its cascading consequences (Horton and Steckel, 2013).

1.5 EMERGING TRENDS OF GLOBAL HUNGER

It is projected that around 733 million individuals were confronting starvation in 2023, which means that roughly one out of every eleven people on the planet, and an overwhelming one out of every five people in Africa, were affected (The Lancet Gastroenterology & Hepatology, 2022). The comeback of hunger is mostly caused by a combination of things, such as brutal wars, shifts in climate, slumps in the economy, and unfairness in how food is given out. These things all add up to a "food crisis created by humans" in different parts of the world (Rahaman et al., 2021).

Today, the situation of hunger around the world shows large differences between regions, with Africa having the highest levels of food insecurity. In contrast, Asia and Latin America have more stable or improving conditions (Flores, 2020). If current situations continue, it is expected that by 2030, 582 million individuals will suffer from ongoing hunger, with many of them living in Africa. This situation requires immediate action to achieve SDG 2, which seeks to end hunger completely by that year (Onyango et al., 2019).

Disagreements about global hunger often focus on how well present solutions work and how resources are shared fairly. Detractors claim that current systems mainly benefit specific areas and ignore the challenges that disadvantaged communities face, complicating the process of tackling the main causes of hunger (Upasana, 2023). Furthermore, conversations about funding for food security show serious deficiencies, with around 63% of low- and middle-income nations lacking enough financial support to carry out effective strategies against hunger (Freiria et al., 2022).

The problem of worldwide hunger has deep historical origins and has changed through many important events and crises. Major famines and food shortages have impacted the hunger situation and have often been worsened by outside influences like climate change, wars, and economic troubles (Rahaman et al., 2021). One of the most significant famines in history occurred during the Biafran War in Nigeria (1966–1967), when a blockade caused about 2 million civilians to starve. This event was crucial as it was the first famine widely reported in Western media, bringing global attention to the suffering of those facing hunger amid conflict (Gul et al., 2022).

Another important moment in the history of hunger was the Sahel Drought that occurred from the 1970s to the 1980s. This drought had a significant effect on the area between the Sahara Desert and the wetter Savanna, starting in the late 1960s and continuing through the 1980s. It led to very poor grazing conditions for livestock, causing a famine that resulted in over 1 million deaths in nations such as Mali, Chad, Mauritania, Niger, and Burkina Faso (Anderson et al., 2021).

Sub-Saharan Africa is known for having the highest rates of food insecurity in the world, which has serious effects on health. In 2017, this area had very high levels of food insecurity, with 55% of its people facing food shortages and 28% experiencing severe food shortages (Anderson et al., 2021). This region has consistently had the worst health outcomes, with the lowest life expectancy rates and the highest infant death rates. The ongoing food insecurity, worsened by poverty, political unrest, and illness, remains a significant issue for many countries in this area, leading to focused research to effectively tackle these challenges (Rahaman et al., 2021).

For a study by Beyene (2023), 31 countries in Sub-Saharan Africa were intentionally chosen based on their access to information that helps create models related to food security and health. The existing research on young people's nutrition in Sub-Saharan Africa shows notable gaps, particularly in countries located south of the Sahara Desert. Only a limited number of studies have been done in this area, indicating a need for more thorough research to grasp the specific nutritional difficulties adolescents face (Ross et al., 2021). The complexity of health issues in Sub-Saharan Africa, combined with its varied cultural settings, highlights the importance of creating context-aware strategies that consider the differences in food security and nutrition across the region (Owino, 2019).

In South America, nations like Brazil, Peru, and Colombia have implemented programs aimed at tackling food insecurity among young people. Studies show that there is increasing worry about food access and nutrition education in these nations, especially in poor communities (Vega Luz et al., 2018; Owino, 2019). Although research has indicated that intervention programs have had mixed success, they often encounter obstacles regarding stakeholder participation and how well the strategies can be adjusted to fit local situations (Owino, 2019). The varied geography and economic conditions across South America require customized approaches to effectively address food insecurity and malnutrition in young populations (Júnior et al., 2022).

Oceania, which includes nations such as Fiji and Tonga, has not been well represented in research on food security. Although the population is relatively small, food insecurity issues are significant, influenced by economic weaknesses and environmental problems. The few studies conducted in this area point to a chance for more investigation into the distinct nutritional challenges that teenagers face and how tailored solutions can be created to effectively tackle these issues (Savage et al., 2020).

European nations, along with those in North America and the Caribbean, also need more detailed research on the nutrition and food insecurity of adolescents. Only a limited number of studies have been carried out in these places, indicating a potential lack of understanding regarding how cultural and financial aspects affect food availability and eating habits among young people (Owino, 2019; Fram et al., 2022). The difficulties experienced by teenagers in these areas may be different from those in other regions, but it is equally important to address them as part of the global problem of hunger (Rahaman et al., 2021). The absence of thorough studies in various regions highlights the pressing need for more research to tackle the diverse and complicated nutrition issues that adolescents face around the world. Increasing research activities in neglected areas can offer valuable insights and guide the creation of effective, culturally suitable solutions (Fram et al., 2022; Lelijveld et al., 2023).

There is a notable difference in food insecurity trends across regions. Currently, Africa has the highest rates of hunger, with 20.4% of its population experiencing food insecurity (Mumuni and Aleer, 2023). On the other hand, Asia has a steady rate of 8.1%, while Latin America has made some progress, reducing its rate to 6.2% (Cuadrado-Roura and Aroca, 2013; International Monetary Fund, Asia and Pacific Dept, 2023). These differences highlight the critical need for targeted actions and increased funding to tackle the unique challenges that different regions face. Addressing these issues is vital for creating new solutions that not only resolve immediate food shortages but also foster long-term sustainability in food systems worldwide (Praseptiangga and Fauza, 2023).

1.6 FOOD SECURITY

According to Gallegos et al. (2023), food security is more than just eating, it's about having steady, dependable access to the appropriate type of food that one needs to stay healthy. A condition in which everyone, everywhere, has physical, social, and financial access to enough safe, nourishing food that satisfies their dietary needs and food choices for an active and healthy life is another way to describe food security (Peng and Berry, 2019).

According to Afrennetwork (2024), food security is made up of four essential elements: stability, utilization, availability, and access.

1. *Availability*: This component refers to food's actual physical existence in enough quantities. It involves food production, distribution, and storage (Afrennetwork, 2024). In terms of availability, a question like "Is the food physically present in markets or grocery stores?" should be asked.
2. *Access*: This simply means people have the ability to afford and buy the food (Afrennetwork, 2024). When talking about access to food, one should determine whether people have the financial and physical means to get the food. Even if it is available in markets, can people afford to buy it or even reach those markets?
3. *Utilization*: This means that the food being consumed provides the necessary nutrients for good health (Afrennetwork, 2024). This includes proper food preparation, clean water, and maybe basic knowledge of healthy diets.
4. *Stability*: This focuses on the availability, accessibility, and use of food that stays consistent and reliable over time (Afrennetwork, 2024). This is when the food is always available, despite the situation. Factors like conflict, inflation, climate disasters, or pandemics can disrupt stability which is more reason why food stability should be maintained.

If even one of these components is missing, a person or community can fall into food insecurity. For instance, a country may have abundant food production, but if inflation makes food unaffordable, the population may still go hungry.

1.7 NUTRITION SECURITY

Nutrition security focuses on the quality of food rather than just the quantity. It ensures that people are not just eating enough, but also eating right (García-Díez et al., 2021). Based on the work of Kesari and Noel (2023), it is possible for someone to eat three meals a day and still suffer from malnutrition if those meals lack the right balance of proteins, minerals, and vitamins.

Nutrition security is about sustained access to foods that promote health and well-being (Grosso et al., 2020). It also includes other factors that affect how the body uses food, like clean water, sanitation, healthcare, and nutrition education.

According to WHO (2022), achieving nutrition security needs more than just food—it also relies on systems that ensure good health, proper child care, and safe environments. For example, a child who eats well but constantly suffers from diarrhea due to dirty water may not absorb the nutrients they need, which can lead to malnutrition. All in all, nutrition security is the basis for a healthy life, and without it, there is no food security.

1.8 DISTINCTIONS BETWEEN FOOD AND NUTRITION SECURITY

Food security is not the same as nutrition security, despite their close relationship. Among their distinctions are:

Food Security	Nutrition Security
Focuses on access to enough food.	Focuses on access to nutritious food.
Focuses on the quantity of food.	Focuses on the quality of food.
The goal is to eliminate hunger.	The goal is to ensure health and well-being through proper nutrition.
This is measured by food availability and access.	This is measured by nutrient intake and health outcomes.
This can be met with repetitive diets.	This requires a varied, balanced diet.

To understand this better, imagine a family that eats mostly white rice because it's affordable and available. They may be food secure, but if they rarely consume vegetables, fruits, or proteins, they are not nutrition secure. This can lead to hidden hunger—micronutrient deficiencies that harm health even when people don't feel hungry (Galiè, 2019).

1.9 IMPORTANCE OF FOOD AND NUTRITION SECURITY

Both national development and individual well-being depend on food and nutrition security for the following reasons:

1. *Health*: This is the most important use of having both food and nutrition security. Proper nutrition helps children grow, strengthens immune systems, and reduces the chance of developing long-term conditions including diabetes or heart disease (Galiè, 2019).
2. *Education*: Malnourished—in this case lacking food or nutrient-deficient food—children struggle to concentrate and perform well in school due to stomach pains or dizzy spells (Grosso et al., 2020).
3. *Economic Productivity*: Healthy workers are more productive, reducing poverty and supporting the economy (Afrennetwork, 2024).
4. *Resilience*: Communities with good nutrition and food security are better equipped to withstand crises, like pandemics or climate disasters (Afrennetwork, 2024).

In regions where food insecurity is common, people often resort to the cheapest, most filling options—such as heavily processed foods that lack nutrients. Over time, this contributes to a double burden: undernutrition on the one hand and rising obesity or non-communicable diseases on the other hand.

1.10 REASONS FOR FOOD AND NUTRITION INSECURITY

Afrennetwork (2024) states that there are several factors that contribute to food and nutrition insecurity, but some of the major ones include:

1. *Poverty*: This restricts access to enough nourishing food because it is out of reach.
2. *Conflict and Displacement*: This can disrupt production and access of food due to insecurity.
3. *Shift in Climate Patterns*: This affects crop yields through droughts and floods.
4. *Poor Infrastructure*: This limits food transport and storage.
5. *Economic Instability*: This is a major cause which leads to high food prices and reduced purchasing power.

1.10.1 Implications of Nutrition and Food Insecurity

The repercussions of nutrition and food insecurity are severe because they lead to health deficiencies and even death. Some of these consequences according to Afrennetwork (2024) include:

1. *Malnutrition*: This is a major consequence which may manifest as undernutrition, micronutrient deficiencies, or even obesity in cases of poor-quality diet.
2. Stunted growth and delayed development in children.
3. Weak immune systems and increased illness.
4. Reduced productivity, weakened immune systems, and higher risk of chronic diseases in adults.
5. Hunger, starvation, and mortality in extreme cases.

1.11 POPULATIONS AT RISK

Certain groups are at a greater risk of experiencing food and nutrition insecurity due to biological, social or economic factors. According to Devine and Lawlis (2019), these groups include:

1. *Children under Five*: They are vulnerable to stunted growth and malnutrition.
2. *Pregnant and Breastfeeding Women*: This is because they have higher nutritional needs compared to non-pregnant/lactating women.
3. *Elderly People*: They have limited mobility and income.
4. *Low-Income Families*: Especially in rural or conflict zones.
5. *Refugees and Displaced Persons*: There is limited and unstable food access.

1.12 POSSIBLE SOLUTIONS AND STRATEGIES

Addressing food and nutrition insecurity requires an approach by different sectors, including both private and public.

1. *School Feeding Programs*: NGOs can provide at least one nutritious meal to children daily, which in turn supports both health and education.
2. *Biofortification and Food Fortification*: This improves the nutrient content in staple foods, such as adding vitamin A to oil or iron to flour (Osendarp et al., 2018).
3. *Nutrition Education*: This helps families make better and healthy food choices (Devine and Lawlis, 2019).
4. *Improved Agriculture and Infrastructure*: By investing in infrastructure, there is an increase in food availability and access.
5. *Social Protection Initiatives*: Initiatives like cash transfers and food subsidies can also strengthen food access.
6. *Sustainable Agriculture*: This enhances agricultural sustainability and resilient food systems, which enable long-term food and nutrition security (Elgar et al., 2020).

Security of food and nutrition is a key component of human health, development, and societal stability. While food security ensures access to enough food, nutrition security emphasizes on the quality and balance of nutrients needed for healthy living. Understanding their differences, causes, and consequences shows the urgent need for effective interventions. Children, women, the elderly, and displaced people continue to be the most vulnerable groups. Therefore, sustainable solutions—including education, food fortification, agricultural improvements, and social protection—are very important. Improving food and nutrition security is a step toward long-term national growth, resilience, and well-being, not just a way to combat hunger.

1.13 MICROBIAL AND BIOTECHNOLOGICAL APPROACHES FOR SUSTAINABLE AGRICULTURE AND GLOBAL HUNGER MITIGATION

With the global population estimated to reach about 9.7 billion by 2050, the need for sustainable, nutrient-rich food sources is intensifying, driven by constraints such as land scarcity, water limitations, climate change, and the environmental impact of livestock farming (United Nations, 2019). Traditional agriculture alone cannot meet these demands, necessitating innovative solutions like microbial-based nutrients and biotechnological advancements. Microbial proteins, including single-cell proteins (SCPs) from bacteria, fungi, yeasts, and algae, offer a high-protein, resource-efficient alternative that aligns with multiple United Nations SDGs, including SDG 2 (Zero Hunger), SDG 12 (Sustainable Consumption and Production), and SDG 13 (Climate Action) (Muazzam et al., 2025).

Microbial proteins, derived from bacteria (*Methylophilus methylotrophus*), fungi (*Fusarium venenatum*, *Aspergillus oryzae*), yeasts (*Saccharomyces cerevisiae*, *Candida utilis*), and algae (*Spirulina*, *Chlorella vulgaris*), provide a protein content of 40–80% dry weight, containing all nine essential proteins that matches the nutritional value of animal proteins (Muazzam et al., 2025). Bacteria are rich in lysine and threonine, fungi in threonine and glutamic acid, yeasts in B-vitamins and methionine, and algae in glutamic acid, lysine, valine, and antioxidants like phycocyanin (Figure 1.2), which offer anti-inflammatory benefits (Dimopoulou et al., 2025). Beyond protein, these microbes supply B-complex vitamins, iron, zinc, omega-3 fatty acids (from algae), and dietary fiber (beta-glucans from fungi), addressing nutritional deficiencies, particularly in plant-based diets (Graham and Ledesma-Amaro, 2023). Fermented foods produced by *Lactobacillus* and *Bifidobacterium* reduce disease risks, such as diabetes and heart disease, by lowering inflammation and improving metabolic health (Santacruz et al., 2018).

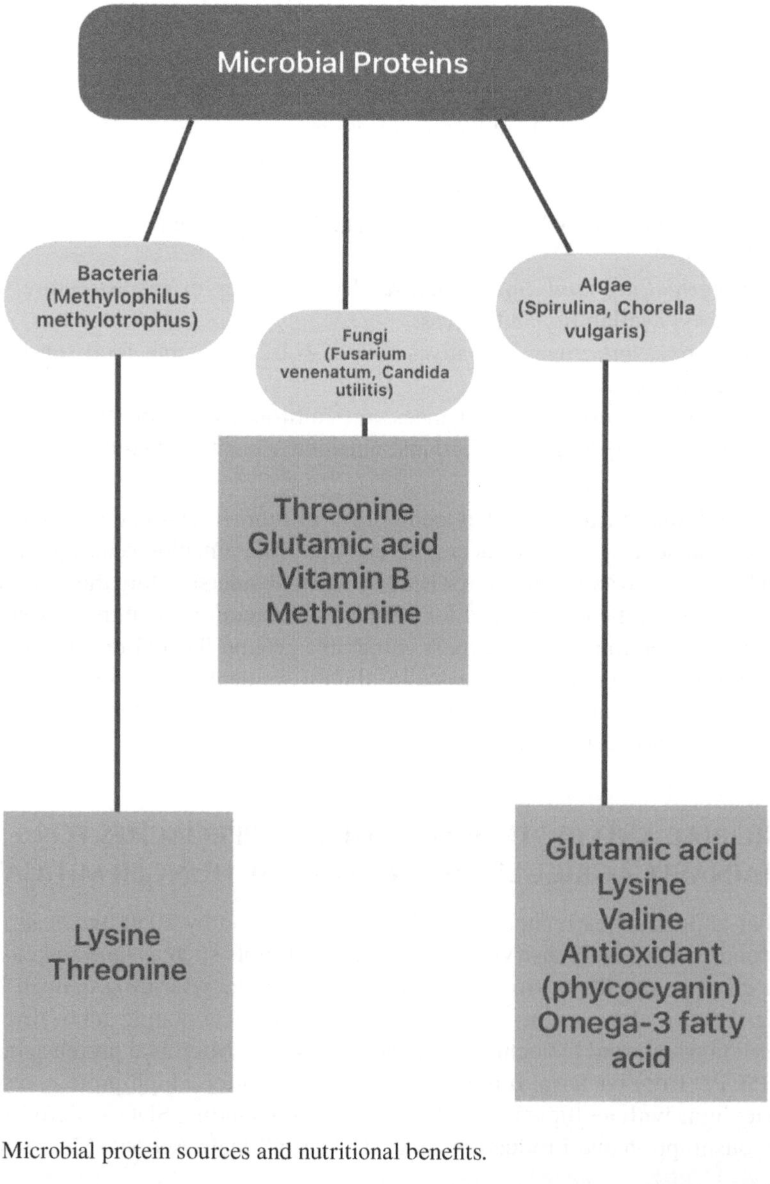

FIGURE 1.2 Microbial protein sources and nutritional benefits.

Microbial proteins are highly versatile in food applications due to their ability to mimic meat-like textures and flavors. *F. venenatum*-derived mycoprotein (Quorn) exhibits a fibrous texture that can be tailored through extrusion or structuring, making it an ideal plant-based meat alternative (Xian et al., 2025). Yeasts and fungal proteins, such as those from *S. cerevisiae* and *M. methylotrophus*, contribute umami flavors through glutamic acid and bioactive peptides, enhancing sensory appeal in products like protein powders, snacks, and meat analogs (Tamang et al., 2020). These properties make microbial proteins attractive to consumers, facilitating their integration into mainstream diets as sustainable, palatable alternatives to traditional meat (Anupama et al., 2024).

Microbial interventions address food security through four key approaches. First, *Lactobacillus* species, *Bacillus subtilis*, and *S. cerevisiae* enhance food preservation and safety through fermentation, producing compounds like lactic acid and ethanol that extend shelf life and improve nutritional profiles in products like yogurt and soy-based products (Ramírez Rojas et al., 2022). Second, SCP production using *Torulopsis* spp., *F. venenatum*, and *Spirulina* spp. provides nutrient-dense protein to combat malnutrition, leveraging waste streams to support SDG 2 (Azam et al., 2014; Ugbogu and Ugbogu, 2016). Third, algae like *Dunaliella tertiolecta* and *Chlorella* spp. produce biofuels, reducing fossil fuel reliance and supporting SDG 13 (Saad et al., 2019). Fourth, bio-control agents (*Trichoderma* spp., *Pseudomonas* spp.) and bio-fertilizers (*Lactobacillus casei, B. subtilis*) enhance crop yields and pest resistance, minimizing chemical inputs and supporting SDG 15 (Petrasch et al., 2019; Jouzani et al., 2017).

Microbial protein production leverages advanced biotechnological methods, including submerged fermentation (SmF) for *F. venenatum* and *S. cerevisiae*, solid-state fermentation (SSF) using agro-industrial waste (*A. oryzae*), and gas fermentation converting CO_2 and CH_4 into protein (*Cupriavidus necator*) (Muazzam et al., 2025). Genetic engineering enhances yields, such as *Aspergillus niger* producing chymosin for cheese-making or *S. cerevisiae* modified for increased vitamin B production (Graham and Ledesma-Amaro, 2023). The integration of CRISPR/Cas9 technology is proposed to optimize microbial genomes, enhancing their efficiency in food production (Bankefa et al., 2020). These include nutritional supplements (e.g., nutritional oil, food pigment), food additives (e.g., vanillin, lycopene), and alternative protein sources (e.g., protein for animal-free milk). Examples include *A. niger* for citric acid and *S. cerevisiae* for vanillin (Singh et al., 2020; Wang et al., 2021). They do not require large spaces or sunlight, making them suitable for decentralized production with consistent nutritional profiles. The document notes that microbes from traditional fermentation cultures, such as those used in dairy or brewing, can be adapted for future food production. Co-cultures of *Lactobacillus* and *Bacillus* improve efficiency by 20–30%, while artificial intelligence optimizes processes, increasing yields by 15–20% (Kumar et al., 2024). These innovations enable scalable, sustainable production of SCPs for food and animal feed, utilizing non-arable land and waste feedstocks (Azam et al., 2014).

Genome editing technologies, particularly CRISPR/Cas9, enable precise modifications to enhance crop traits without introducing foreign DNA and CRISPR can deactivate genes causing disease susceptibility, activate stress-tolerant genes, or improve nutritional pathways, accelerating crop improvement (Wilson and Harrison, 2021). Applications include developing disease-resistant crops (e.g., cassava resistant to brown streak disease) and drought-tolerant varieties, addressing global crop losses of ~40% due to pests and diseases and herbicide-tolerant crops. These advancements enhance climate resilience and support SDG 2 by ensuring stable food production.

Microbial protein production offers significant environmental advantages over livestock farming, including lower water and land use, reduced greenhouse gas emissions (CO_2 and CH_4), and decreased deforestation (Matassa et al., 2016). By utilizing waste streams like lignocellulose and industrial gases, microbes contribute to a circular bio-economy, with *Clostridium thermocellum* reducing landfill waste by 15% in pilot projects (Lynd et al., 2021). Algae fix CO_2 efficiently, supporting carbon sequestration, while biofuel production from *Chlorella* spp. aligns with SDG 13 (Saad et al., 2019). These benefits reduce the environmental footprint of food production, supporting SDGs 9, 12, and 15 by promoting green industrial processes and responsible resource use.

Biofertilizers, including rhizobia, mycorrhizae, and plant-growth-promoting rhizobacteria (PGPR) like *B. subtilis* and *Pseudomonas fluorescens*, enhance nutrient uptake, stress tolerance, and pest resistance, reducing reliance on synthetic fertilizers and restoring soil health (Verma et al., 2021). Systematic reviews report yield gains and improved resilience, though outcomes vary by strain and environment, requiring quality control and local production for scalability (Petrasch et al., 2019). Bio-control agents like *Trichoderma* spp. and *Bacillus thuringiensis* further reduce chemical pesticide use, enhancing agricultural sustainability and supporting SDG 15 (Jouzani et al., 2017).

Biofortification, through conventional breeding or genetic engineering, increases the nutritional value of staple crops, addressing "hidden hunger" affecting billions. Crops like vitamin A-enriched sweet potatoes and high-zinc rice improve micronutrient status, with studies showing health and cognitive benefits, especially in deficient populations (Verma et al., 2021). Integration into school feeding and seed systems enhances policy uptake, supporting SDG 2. Microbial contributions, such as *S. cerevisiae* producing vitamins, complement biofortification efforts, ensuring nutrient-dense diets (Anupama et al., 2024).

Precision fermentation uses microbes to produce targeted proteins, lipids, and micronutrients, such as whey proteins, heme, and vanillin (*A. niger*, *S. cerevisiae*). This approach supports fortified and therapeutic foods, addressing fragile supply chains and improving affordability for low-income contexts and by the end of the fermentation process, the desired molecules are taken out and purified (Graham and Ledesma-Amaro, 2023). By requiring minimal space and sunlight, precision fermentation enables decentralized production with consistent nutritional profiles, aligning with SDG 9 (Wallace, 2021).

Insects provide complete amino acids, polyunsaturated fats (up to 33%), and micronutrients, require 30% less feed conversion than poultry and emit fewer greenhouse gases (Nirmal et al., 2024) making them good protein alternatives. Duckweeds as well with up to 45% protein and high bioavailability, doubles biomass every 48 hours and fixes nitrogen, reducing fertilizer needs offering a significant sustainable protein alternatives, though contamination risks for duckweed require management (Habeeb et al., 2025).

Microbial and biotechnological approaches, from SCP production to genome editing, offer scalable, sustainable solutions to global hunger and environmental challenges. By leveraging the nutritional, environmental, and production advantages of microbes, alongside biofortification, precision fermentation, and herbicide-tolerant crops, these strategies address SDGs while complementing policy efforts. Continued research, regulatory support, and consumer acceptance are essential to realize their full potential, ensuring a resilient, equitable food future (Akinsemolu et al., 2023).

1.14 ADVANCES AND FURTHER DIRECTIONS IN COMBATING GLOBAL HUNGER

Global hunger continues to be a persistent and difficult issue in today's world, impacting countless individuals in spite of advancements in farming practices, technological innovations, and economic progress on a global scale (Shebanina et al., 2024). The study by the FAO reported approximately 733 million individuals across the planet experienced malnourishment back in 2023, which equates to almost 10% of the world's population (FAO, IFAD, UNICEF, WFP, and WHO, 2023). The fact that starvation continues even though there is enough food being produced worldwide demonstrates the multifaceted aspect of this issue, which is not only impacted by how much food is being produced, but by societal, governmental, ecological, and financial factors as well (Sethi et al., 2020). The battle against hunger has been ongoing for many years; the Agricultural Movement of the 1960s led to significant increases in food production. Nevertheless, differences in food access, harm to nature, and unstable social and political situations have kept food shortages persisting. Currently, innovations in scientific and technological fields, newly created systems of governance, and cooperative approaches present new potentials to eradicate starvation, but considerable obstacles are still present (Zreik and Iqbal, 2024; Semchuk et al., 2025).

One of the biggest improvements in fighting hunger has been using new technology in farming. Precision agriculture, which involves tools like satellite photos, geographic information systems, and data analysis, has changed how farming is done. It helps farmers use water, fertilizers, and pesticides more effectively. By adjusting farming inputs to match the particular conditions of their soil and crops, precision farming boosts productivity while also lessening harm to the environment (Shebanina et al., 2024; Yilmaz and Yilmaz, 2025). For instance, variable rate irrigation systems apply water only where it is necessary, minimizing waste and increasing crop production in areas with limited water. In nations with small farmers, like Kenya and India, mobile apps have been created to give farmers up-to-date information on weather, soil quality, and market prices, helping them make better choices and reduce losses (Sethi et al., 2020; Onyeaka et al., 2024).

Biotechnology has been essential in improving food safety. Beginning in the 1990s with the launch of genetically modified plants, the field has grown to include crops that can withstand drought, like certain corn, cotton that resist pests, and enriched crops such as Golden Rice, which has more vitamin A (Feng and Squires, 2025). These plants can help fight hunger and malnutrition by increasing production in tough conditions and boosting the nutritional value of basic foods. In Africa, introducing cassava that fights diseases and sweet potatoes with added nutrients has helped address the lack of calories and essential vitamins (Samad et al., 2025). Still, not everyone has adopted these advancements because of worries about safety, ownership rights, and how consumers feel. To overcome these challenges, it is important to communicate clearly, avoid conflicting rules, and create policies that make it fair for everyone to access these biotechnological advancements (Nair and Joy, 2024).

Besides biotechnology and advanced farming techniques, the mix of artificial intelligence, machine learning, and the Internet of Things (IoT) in agriculture, often referred to as smart farming, has expanded the chances for food production. Smart farming equipment includes drones that monitor crop health, AI systems that predict how much will be harvested, and IoT sensors that manage soil and water resources (Sethi et al., 2020; Onyeaka et al., 2024). These innovations can significantly enhance food system efficiency while reducing waste and damage to the environment. For example, AI systems are increasingly being utilized to predict pest invasions and the spread of diseases, allowing farmers to take preventive measures before significant crop losses happen (Feng and Squires, 2025). However, while these technologies show great potential, small farmers in low-income nations often cannot access them due to financial, infrastructure, and knowledge challenges. Therefore, plans must focus on fair access, skill development, and funding methods that encourage broad use of these technologies (Agboklou et al., 2024; Semchuk et al., 2025).

Moreover, improvements in managing supply chains have been crucial in tackling global hunger. Losses after harvest, especially in developing nations, can make up as much as one-third of all food produced (De Andrade Júnior, 2020). New developments in logistics, cold storage systems, and digital tracking tools are helping to cut down these losses and make sure more food gets to markets and consumers (Von Braun et al., 2024). For example, blockchain technology is being tested to improve openness in food supply chains, which reduces waste and ensures farmers receive fair payments. In rural areas of Africa and South Asia, solar-powered cold storage units have allowed farmers to keep perishable goods, such as vegetables and milk, fresh for longer, leading to less waste and higher incomes for families (Onyeaka et al., 2024).

Despite advancements in technology, addressing hunger cannot happen without strong policies and good governance. International guidelines like the United Nations Sustainable Development Goals for Zero Hunger and the Global Food Security Strategy offer crucial plans for action (Adkins, 2018; Shebanina et al., 2024; Zreik and Iqbal, 2024). These guidelines stress the importance of connecting food production, nutrition, water use, and climate change adaptation. However, the issue of policy separation continues to be a problem (Sethi et al., 2020; Agboklou et al., 2024). For instance, the different treatment of food and water policies in US laws shows how separate institutions can weaken comprehensive efforts to fight hunger. Therefore, future efforts must focus on integrated governance that connects food security with environmental protection, trade laws, and social safety nets (Zreik and Iqbal, 2024; Yilmaz and Yilmaz, 2025).

Global cooperation is also a vital aspect of fighting hunger. Agencies such as the FAO, the WFP, and the International Fund for Agricultural Development work together with governments, non-profit organizations, and businesses to fund, carry out, and grow various initiatives (Von Braun et al., 2024). These collaborations not only offer financial help and expertise but also promote the exchange of effective practices across different areas. Particularly, public-private partnerships have shown success in creating environments that attract investment, leading to innovation and greater market access for small farmers (Nair and Joy, 2024). For instance, the initiatives in sub-Saharan Africa have increased the output of small-scale farmers twofold by offering them loans, guidance, and improved seed selections with help from public-private partnerships (Samad et al., 2025).

Economic factors continue to influence the ongoing issue of hunger. Even in places where food is plentiful, poverty and inequality stop many people from getting enough nutrition. Increasing food costs, unpredictable markets, and a lack of job options make food insecurity worse (Semchuk et al., 2025). Tackling these ongoing issues needs economic development that involves the community. Groups like cooperatives, microfinance organizations, and farmers' associations have shown they can help communities become stronger and more self-sufficient in food production (Agboklou et al., 2024). Additionally, charitable organizations have been important in bridging funding gaps, especially through blended finance methods that mix grants with business investments to make lending safer for small farmers (Feng and Squires, 2025).

Environmental problems are strongly connected to the fight against hunger. Changes in the climate may be the most significant long-term threat to global food systems since severe weather events like storms, dry spells, and floods frequently ruin crops and ways of making a living (Samad et al., 2025). Farming itself also adds to greenhouse gas emissions, water pollution, and land damage, creating a harmful cycle between environmental harm and food shortages. To address these problems, environmentally friendly farming methods such as minimal ploughing, planting trees alongside crops, and holistically managing pests are being promoted worldwide. Smart watering practices, including drip irrigation and collecting rainwater, are crucial in regions with scarce water, where agriculture consumes nearly 70% of available freshwater. Moreover, projects that restore land and regenerate soil are crucial for reversing degradation, which currently impacts a third of the world's farmland (Shebanina et al., 2024).

Conflict makes the global hunger crisis even worse. Wars and civil disturbances hinder farming, damage infrastructure, and force people to move, leading many to rely on emergency support (De Andrade Júnior, 2020). According to De Waal (2024), conflict is the main cause of hunger in many of the worst-affected areas today. Therefore, creating peace and providing humanitarian aid are key parts of long-term strategies for food security. Future actions should focus on both the underlying issues of conflict and the immediate needs of those impacted, making sure that rebuilding agriculture and boosting resilience are prioritized along with humanitarian support (Agboklou et al., 2024; Okoh and Oladokun, 2024).

Looking into the future, various effective methods may assist in ending hunger. Merging food production with healthcare, education, and social assistance is crucial. For instance, school meal programs not only provide meals but also enhance education and community growth (Sethi et al., 2020; Shebanina et al., 2024). Additionally, local and regional strategies designed to address specific problems are important. While some places, such as South America, have made progress in lowering hunger through successful policies, regions like the Sahel and the Caribbean need focused actions to improve resilience against climate change (Yilmaz and Yilmaz, 2025). Encouraging local efforts like farmer cooperatives, seed exchanges, and community markets can spark new ideas and promote sustainability, and this is another important approach (Samad et al., 2025). Similarly, enhancing digital farming with AI, mobile applications, and online markets presents a major chance to advance access to agricultural information and marketplaces (Okoh and Oladokun, 2024). Lastly, ongoing funding for agricultural research and development is essential to maintain a balance between productivity and ecological sustainability, ensuring that food systems remain reliable for future generations (Onyeaka et al., 2024).

In conclusion, fighting global hunger requires more than just producing more food; it involves tackling social inequalities, environmental harm, and political chaos (Adkins, 2018). Progress in technology, policy, and global partnerships has created a solid foundation, yet much more work is needed. The way forward for global food security lies in connected approaches that link farming innovation with climate resilience, fair policies, and community-led efforts (Nair and Joy, 2024). If global partners work together, invest sustainably, and distribute resources equitably, achieving a world without hunger can shift from being a dream to becoming a reality.

1.15 CONCLUSION

Global hunger remains one of the biggest issues facing people today. Even though there have been significant advancements in producing food and technology, millions of individuals worldwide still experience hunger and poor nutrition. The numbers relating to hunger not only show the size of the problem but also its ongoing nature, indicating that hunger is not just about having enough food, but is also closely linked to poverty, inequality, conflict, and environmental issues. Therefore, understanding hunger, undernourishment, and food insecurity is crucial for grasping the full extent of the issue and for developing effective solutions. Examining global hunger shows that solving it requires more than just increasing agricultural output; it also needs fair access, sustainable methods, and coordinated efforts from countries around the world. While statistics alone highlight the urgent need for action, they also remind us of the real people represented in these numbers, children suffering from stunted growth, families who cannot get healthy food, and communities that are at risk from various hardships. Moreover, these figures establish a basis for accountability, creating markers to measure progress and identify areas that need attention. In the end, addressing hunger is both a moral responsibility and a practical necessity. As the world works toward achieving Zero Hunger, having clear definitions and trustworthy data is extremely important. These provide an overview of current situations as well as a roadmap for future actions. By using these understandings to shape policies, enhance resilience, and foster inclusive solutions, the global community can get closer to a time when hunger is no longer a major issue in human society.

REFERENCES

Adkins, M. (2018). The global food security act: America's strategic approach to combating world hunger. Journal of Food Law and Policy, 14(2), 345–363. https://doi.org/10.54119/jflp.qysl1733

Afrennetwork. (2024). What Are the Four Pillars of Food Security? Foundation of Food. Pretoria, Gauteng, South Africa: Afrenet—African Food Research Network. https://afrenet.org/what-are-the-four-pillars-of-food-security/

Agboklou, K. E., Özkan, B., & Gujrati, R. (2024). Challenges to achieving zero hunger by 2030: The impact of armed conflict on global food security with a focus on the Russo-Ukrainian war. Journal of Lifestyle and SDGs Review, 4, e01595. https://doi.org/10.37497/2965-730x.sdgsreview.v4.n00.pe01595

Ahmed, M. H., Vasas, D., Hassan, A., & Molnár, J. (2022). The impact of functional food in prevention of malnutrition. PharmaNutrition, 19(100288), 100288. https://doi.org/10.1016/j.phanu.2022.100288

Akinsemolu, A., Onyeaka, H., Fagunwa, O., Adenuga, A. H. (2023). Toward a resilient future: The promise of microbial bioeconomy. Sustainability, 15, 7251. https://doi.org/10.3390/su15097251

Anderson, W., Taylor, C., McDermid, S., Ilboudo-Nébié, E., Seager, R., Schlenker, W., Cottier, F., De Sherbinin, A., Mendeloff, D., & Markey, K (2021). Violent conflict exacerbated drought-related food insecurity between 2009 and 2019 in sub-Saharan Africa. Nature Food, 2(8), 603–615. https://doi.org/10.1038/s43016-021-00327-4

Anupama, A., Kumar, S., & Singh, R (2024). Microbial proteins as sustainable alternatives for food security: Current status and future perspectives. Journal of Agricultural and Food Chemistry, 72(4), 1801–1814. https://doi.org/10.1021/acs.jafc.3c07654

Aransiola, S. A., Akinsola, R. O., Abioye, O. P., & Maddela, N. R. (Eds.). (2024a). Emerging Contaminants in Food and Food Products (1st ed.). Boca Raton, FL: CRC Press. Taylor and Francis Group. https://doi.org/10.1201/9781003438236

Aransiola, S. A., & Maddela, N. R. (Eds.). (2024b). Phytoremediation in Food Safety: Risks and Prospects (1st ed.). Boca Raton, FL: CRC Press. Taylor and Francis Group. https://doi.org/10.1201/9781032683751

Azam, S., Zeeshan, K., Bashir, A., Khan, I., & Ali, J. (2014). Production of single cell protein from orange peels using Aspergillus niger and Saccharomyces cerevisiae. Global Journal of Biotechnology and Biochemistry, 9(1), 14–18.

Bankefa, O. E., Oladeji, S. J., Ayilara-Akande, S. O., & Lasisi, M. M. (2020). Microbial redemption of "evil" days: A global appraisal to food security. Journal of Food Science and Technology, 58(6), 2041–2053. https://doi.org/10.1007/s13197-020-04725-7

Beyene, S. D. (2023). The impact of food insecurity on health outcomes: Empirical evidence from sub-Saharan African countries. BMC Public Health, 23(1). https://doi.org/10.1186/s12889-023-15244-3

Black, R. E., Victora, C. G., Walker, S.P., Bhutta, Z.A., Christian, P., de Onis, M., Ezzati, M., Grantham-McGregor, S., Katz, J., Martorell, R., & Uauy, R. (2013). Maternal and child undernutrition and overweight in low-income and middle-income countries. The Lancet, 382(9890), 427–451. https://doi.org/10.1016/S0140-6736(13)60937-X

Cuadrado-Roura, J. R., & Aroca, P. (2013). Facing the need for regional policies in Latin America. In Cuadrado-Roura, J., & Aroca, P. (Eds), Regional Problems and Policies in Latin America. Advances in Spatial Science (pp. 21–42). Berlin, Heidelberg: Springer. https://doi.org/10.1007/978-3-642-39674-8_2

De Andrade Júnior, H. (2020). Combating global hunger and forging a path to food sovereignty. Cadernos De Saúde Pública, 36(1), 212. https://doi.org/10.1590/0102-311x00208719

Devine, A., & Lawlis, T. (2019). Nutrition and vulnerable groups. Nutrients, 11(5), 1066. https://doi.org/10.3390/nu11051066

De Waal, A. (2024). Hunger in global war economies: Understanding the decline and return of famines. Disasters, 49(1), e12661. https://doi.org/10.1111/disa.12661

Dimopoulou, M., Kolonas, A., Stagos, D., & Gortzi, O. (2025). A review of the sustainability, chemical composition, bioactive compounds, antioxidant and antidiabetic activity, neuroprotective properties, and health benefits of microalgae. Biomass, 5, 11.

Drishti, I. A. S. (2024). Global Hunger Index 2024. https://www.drishtiias.com/summary-of-important-reports/global-hunger-index-2024-1

Elgar, F. J., Pickett, W., Pförtner, T. K., Gariépy, G., Gordon, D., Georgiades, K., Davison, C., Hammami, N., MacNeil, A. H., Da Silva, M. A., & Melgar-Quiñonez, H. R. (2020). Relative food insecurity, mental health and wellbeing in 160 countries. Social Science & Medicine, 268, 113556.

FAO, IFAD, UNICEF, WFP, & WHO. (2023). The State of Food Security and Nutrition in the World 2023: Urbanization, Agrifood Systems Transformation and Healthy Diets Across the Rural–Urban continuum. eBooks. Rome: FAO. https://doi.org/10.4060/cc3017en

FAO, IFAD, UNICEF, WFP, & WHO. (2025). The State of Food Security and Nutrition in the World 2025: Financing to End Hunger, Food Insecurity and Malnutrition in All Its Forms. Rome: Food and Agriculture Organization of the United Nations.

Fasano, A. (2025). The physiology of hunger. New England Journal of Medicine, 392(4), 372–381. https://doi.org/10.1056/nejmra2402679

Feng, H., & Squires, V. R. (2025). Combatting hunger, malnutrition and food insecurity - A scourge of global drylands: A narrative review. Discoveries in Agriculture and Food Science, 13(1), 29–41. https://doi.org/10.14738/dafs.131.18273

Flores, C. S. (2020). To overcome the plight of hunger. Textos & Contextos (Porto Alegre), 19(2), e39636. https://doi.org/10.15448/1677-9509.2020.2.39636

Food and Agriculture Organization of the United Nations (FAO). (2024). Hunger and Food Insecurity. https://www.fao.org/hunger/en

Fram, M. S., Nguyen, H. T., & Frongillo, E. A. (2022). Food insecurity among adolescent students from 95 countries is associated with diet, behavior, and health, and associations differ by student age and sex. Current Developments in Nutrition, 6(3), nzac024. https://doi.org/10.1093/cdn/nzac024

Freiria, C. N., Arikawa, A., Van Horn, L. T., Corona, L. P., & Wright, L. Y. (2022). Food insecurity among older adults living in low- and middle-income countries: A scoping review. The Gerontologist, 64(1). https://doi.org/10.1093/geront/gnac161

Galiè, A. (2019). Leveraging gender for food and nutrition security through agriculture. Encyclopedia of Food Security and Sustainability, 3, 426–431. https://doi.org/10.1016/b978-0-08-100596-5.21568-0

Gallegos, D., Booth, S., Pollard, C. M., Chilton, M., & Kleve, S. (2023). Food security definition, measures and advocacy priorities in high-income countries: A Delphi consensus study. Public Health Nutrition, 26(10), 1986–1996.

Gallup, J. L., & Sachs, J. D. (2001). The economic burden of malaria. American Journal of Tropical Medicine and Hygiene, 64(1–2 Suppl), 85–96.

García-Díez, J., Gonçalves, C., Grispoldi, L., Cenci-Goga, B., & Saraiva, C. (2021). Determining food stability to achieve food security. Sustainability, 13(13), 7222.

Global Hunger Index. (2023). 2023 Global Hunger Index Report. Bonn/Dublin: Concern Worldwide and Welthungerhilfe.

Global Hunger Index. (2024). 2024 Global Hunger Index: How Gender Justice Can Advance Climate Resilience and Zero Hunger. Bonn/Dublin: Concern Worldwide and Welthungerhilfe.

Global Nutrition Report. (2017). Nourishing the SDGs. Bristol, UK: Development Initiatives.

Graham, A. E., & Ledesma-Amaro, R. (2023). The microbial food revolution. Nature Communications, 14, 2231. https://doi.org/10.1038/s41467-023-37891-1

Gribble, J. N., & Hausmann, R (2023). The impact of food insecurity on health outcomes. BMC Public Health, 23(1), 1–12.

Grosso, G., Mateo, A., Rangelov, N., Buzeti, T., & Birt, C. (2020). Nutrition in the context of the sustainable development goals. European Journal of Public Health, 30(Supplement_1), i19–i23. https://doi.org/10.1093/eurpub/ckaa034

Gul, N. A., Shabeer, N. M. G., Abbasi, N. R. A., & Khan, N. W. (2022). Africa's poverty and famines: Developmental projects of China on Africa. Perennial Journal of History, 3(1), 165–194. https://doi.org/10.52700/pjh.v3i1.109

Habeeb, F., Majid, D., Makroo, H. A., Castagnini, J. M., Barba, F. J., & Dar, B. N. (2025). Duckweed as a sustainable protein source: Extraction methods, functional properties, and applications in food systems. Food Chemistry, 493(Part 2), 145854. https://doi.org/10.1016/j.foodchem.2025.145854

Hardman, N. (2024). Hopeless, starving, and besieged. Human Rights Watch. https://www.hrw.org/report/2024/11/14/hopeless-starving-and-besieged/israels-forced-displacement-palestinians-gaza

Hendrix, C. S., & Brinkman, H. J. (2013). Food insecurity and conflict dynamics. International Interactions, 39(4), 805–829.

Horton, S., & Steckel, R. H. (2013). Global Economic Losses Attributable to Malnutrition 1900–2000 and Projections to 2050. Cambridge University Press.

International Monetary Fund. Asia and Pacific Dept. (2023). Regional economic outlook: Asia and Pacific, May 2023 [Ebooks]. In International Monetary Fund eBooks. USA: International Monetary Fund. https://doi.org/10.5089/9798400238062.086

IPCC. (2022). Climate Change 2022: Impacts, Adaptation, and Vulnerability. Intergovernmental Panel on Climate Change.

Jouzani, G. S., Valijanian, E., & Sharafi, R. (2017). Bacillus thuringiensis: A successful insecticide with new environmental features and tidings. Applied Microbiology and Biotechnology, 101, 2691–2711. https://doi.org/10.1007/s00253-017-8175-y

Júnior, J. G., De Amorim, L. M., Lima, N. N. R., & Neto, M. L. R. (2022). Challenges to food safety for children and adolescents in Brazil. Journal of Pediatric Nursing, 65, e7–e8. https://doi.org/10.1016/j.pedn.2021.11.033

Kesari, A., & Noel, J. Y. (2023). Nutritional Assessment. StatPearls Publishing. https://www.ncbi.nlm.nih.gov/books/NBK580496/

Khofi, L., Manderson, L., & Moyer, E. (2024). Speaking of hunger: Food shortages, poverty and community assistance in urban South Africa. Ecology of Food and Nutrition, 63(4), 323–342. https://doi.org/10.1080/03670244.2024.2361249

Kumar, S., Anupama, A., & Singh, R. (2024). Artificial intelligence in optimizing microbial protein production: A review. Biotechnology Advances, 62, 108087.

Lelijveld, N., Wrottesley, S. V., Aburmishan, D., Hemler, E. C., Fentahun, N., Ali, Z., Menezes, R., Roche, M., Sharma, D., Wazny, K., Ross, D. A., Sethi, V., & Mates, E. (2023). Research priorities for nutrition of school-aged children and adolescents in low- and middle-income countries. PLOS One, 18(1), e0280510. https://doi.org/10.1371/journal.pone.0280510

Lenaerts, B., & Demont, M. (2021). The global burden of chronic and hidden hunger revisited: New panel data evidence spanning 1990–2017. Global Food Security, 28, 100480. https://doi.org/10.1016/j.gfs.2020.100480

Lowe, N. M. (2021). The global challenge of hidden hunger: Perspectives from the field. Proceedings of the Nutrition Society, 80(3), 283–289.

Lucas, P., Hilderink, H. B. M., Janssen, P. H. M., Samir, K. C., van Vuuren, D. P., & Niessen, L. (2019). Future impacts of environmental factors on achieving the SDG target on child mortality—A synergistic assessment. Global Environmental Change, 57, 101925. https://doi.org/10.1016/j.gloenvcha.2019.05.009

Lynd, L. R., Beckham, G. T., & Guss, A. M. (2021). Clostridium thermocellum for lignocellulosic biomass conversion: Advances and applications. Biotechnology for Biofuels, 14(1), 85.

Matassa, S., Boon, N., Pikaar, I., & Verstraete, W. (2016). Microbial protein: Future sustainable food supply route with low environmental footprint. Microbial Biotechnology, 9, 568–575.

Muazzam, A., Samad, A., Alam, A. N., Hwang, Y. H., & Joo, S. T. (2025). Microbial proteins: A green approach towards zero hunger. Foods, 14, 2636.

Mumuni, S., & Aleer, M. J. (2023). Zero Hunger by 2030 – Are we on track? Climate variability and change impacts on food security in Africa. Cogent Food & Agriculture, 9(1). https://doi.org/10.1080/23311932.2023.2171830

Nair, R., & Joy, J. (2024). Eradication of global hunger at UN initiative: Holacracy process enriched by human will and virtue. In I. S. Dunmade, M. O. Daramola, & S. A. Iwarere (Eds.), Green Energy and Technology (pp. 477–494). Cham: Springer. https://doi.org/10.1007/978-3-031-47215-2_27

Nasser, H. A., Mahmoud, M., Tolba, M. M., Radwan, R. A., Gabr, N. M., ElShamy, A. A., Yehya, M. S., Ziemke, A., & Hashem, M. Y. (2021). Pros and cons of using green biotechnology to solve food insecurity and achieve sustainable development goals. Euro-Mediterranean Journal for Environmental Integration, 6(29). https://doi.org/10.1007/s41207-020-00240-5

Nirmal, N., Anyimadu, C. F., Khanashyam, A. C., Bekhit, A. Ed. A., & Dhar, B. K. (2024). Alternative Protein Sources: Addressing Global Food Security and Environmental Sustainability. Sustainable Development, 33(3), 3958–3969. https://doi.org/10.1002/sd.3338

Okoh, P. A., & Oladokun, A. (2024). Using digital resources in agriculture as a strategy for curbing the hunger crisis in Nigeria. Nigeria Journal of Home Economics (NIG-JHEC), 12(10), 116–123. https://doi.org/10.61868/njhe.v12i10.320

Okunlola, A. F., & Aregbeshola, A. R. (2025). Evaluating food production shocks and their effects on population growth and hunger in Africa. Journal of Population and Social Studies, 33, 336–360. https://doi.org/10.25133/JPSSv332025.018

Onyango, A. W., Jean-Baptiste, J., Samburu, B., & Mahlangu, T. L. M. (2019). Regional overview on the double burden of malnutrition and examples of program and policy responses: African Region. Annals of Nutrition and Metabolism, 75(2), 127–130. https://doi.org/10.1159/000503671

Onyeaka, H., Adeboye, A. S., Bamidele, O. P., Onyeoziri, I., Adebo, O. A., Adeyemi, M. M., & Thera-Sekgweng, S. N. (2024). Beyond hunger: Unveiling the rights to food in sub-Saharan Africa. Food and Energy Security, 13(1), e530. https://doi.org/10.1002/fes3.530

Osendarp, S. J. M., Martinez, H., Garrett, G. S., Neufeld, L. M., De-Regil, L. M., Vossenaar, M., & Darnton-Hill, I. (2018). Large-scale food fortification and biofortification in low- and middle-income countries: A review of programs, trends, challenges, and evidence gaps. Food and Nutrition Bulletin, 39(2), 315–331. https://doi.org/10.1177/0379572118774229

Otekunrin, O. A. (2025). A critical assessment of the interplay of conflict, hunger, poverty, and food insecurity in Africa. Food and Humanity, 4, 100544. https://doi.org/10.1016/j.foohum.2025.100544

Owino, V. O. (2019). Challenges and opportunities to tackle the rising prevalence of diet-related non-communicable diseases in Africa. Proceedings of the Nutrition Society, 78(4), 506–512. https://doi.org/10.1017/s0029665118002823

Peng, W., & Berry, E. (2019). Food Security - an overview | ScienceDirect Topics. https://www.sciencedirect.com/topics/food-science/food-security

Petrasch, S., Knapp, S. J., Van kan, J. A. L., & Blanco-ulate, B. (2019). Grey mould of strawberry, a devastating disease caused by the ubiquitous necrotrophic fungal pathogen Botrytis cinerea. Molecular Plant Pathology, 20(6):877–892. https://doi.org/10.1111/mpp.12794

Praseptiangga, D., & Fauza, G. (2023). Sustainable development of food supply chains toward food security, managing and reducing losses. In R. Accorsi & R. Bhat (Eds.), Sustainable Development and Pathways for Food Ecosystems, Integration and Synergies (pp. 51–68). Academic Press. https://doi.org/10.1016/b978-0-323-90885-6.00003-x

Rahaman, A., Kumari, A., Zeng, X., Khalifa, I., Farooq, M. A., Singh, N., Ali, S., Alee, M., & Aadil, R. M. (2021). The increasing hunger concern and current need in the development of sustainable food security in the developing countries. Trends in Food Science & Technology, 113, 423–429. https://doi.org/10.1016/j.tifs.2021.04.048

Ramírez Rojas, A. A., Swidah, R., & Schindler, D. (2022). Microbes of traditional fermentation processes as synthetic biology chassis to tackle future food challenges. Frontiers in Bioengineering and Biotechnology, 10, 982975. https://doi.org/10.3389/fbioe.2022.982975

Rao, N., Vercillo, S., & Torvikey, G. D. (2024). Gender justice, climate resilience, and food and nutrition security. In N. Rao, S. Vercillo, & G. D. Torvikey (Eds.), Global Hunger Index 2024. Bonn: Welthungerhilfe and Concern Worldwide.

Ross, D. A., Mshana, G., & Guthold, R. (2021). Adolescent health series: The health of adolescents in Sub-Saharan Africa: Challenges and opportunities. Tropical Medicine & International Health, 26(11), 1326–1332. https://doi.org/10.1111/tmi.13655

Saad, M. G., Dosoky, N. S., Zoromba, M. S., & Shak, H. M. (2019). Algal biofuels: current status and key challenges. Energies, 12, 1920. https://doi.org/10.3390/en12101920

Samad, A., Naz, E., Batool, S., & Amjad, A (2025). The New Geopolitics of Hunger: When Food Becomes the Ultimate Weapon. Review Journal of Social Psychology & Social Works, 3(2), 311–327.

Santacruz, A., Meza, M., & Torres, R. (2018). Health benefits of Lactobacillus species in fermented foods. Food Microbiology, 76, 345–354. https://doi.org/10.1016/j.fm.2018.06.005

Savage, A., Bambrick, H., & Gallegos, D. (2020). From garden to store: Local perspectives of changing food and nutrition security in a Pacific Island country. Food Security, 12(6), 1331–1348. https://doi.org/10.1007/s12571-020-01053-8

Semchuk, I., Mishchenko, V., Petrukha, N., Strochenko, N., & Kapelista, I. (2025). Agricultural practices' impact on achieving the Sustainable Development Goal: Eliminating world hunger. Grassroots Journal of Natural Resources, 8(1), 737–758. https://doi.org/10.33002/nr2581.6853.080131

Sethi, G., Avila Bedregal, L. P., Cassou, E., Constantino, L. F., Hou, X., Jain, C. M., Messent, F. J., Morales, X. Z., Mostafa, I., Pascual, J. C. G., Martinant de Preneuf, F., Thapa, D., Trinidad, R. Q., Yndriago, R., & Youssefi-Vash, F. (2020). Addressing Food Loss and Waste: A Global Problem with Local Solutions (Vol. 1 of 2, p. 128): Main Report (English). Washington, DC: World Bank Group. http://documents.worldbank.org/curated/en/549221601305124823

Shebanina, O., Poltorak, A., & Chorniy, D. (2024). Global food security: Challenges in achieving the sustainable development goals. Ukrainian Black Sea Region Agrarian Science, 28(4), 9–20. https://doi.org/10.56407/bs.agrarian/4.2024.09

Singh, R., Kumar, M., & Shukla, P. (2020). Microbial production of food additives: Vanillin and lycopene. Biotechnology Letters, 42(7), 1097–1106. https://doi.org/10.1007/s10529-020-02856-7

Tack, J., Verbeure, W., Mori, H., Schol, J., den Houte, K., Huang, I., Balsiger, L., Broeders, B., Colomier, E., Scarpellini, E., & Carbone, F. (2021). The gastrointestinal tract in hunger and satiety signalling. United European Gastroenterology Journal, 9(6), 727–734. https://doi.org/10.1002/ueg2.12097

Tamang, J. P., Cotter, P. D., Endo, A., Han, N. S., Kort, R., Liu, S. Q., Mayo, B., Westerik, N., & Hutkins, R. (2020). Fermented foods in a global age: East meets West. Comprehensive Reviews in Food Science and Food Safety, 19, 184–217.

The Lancet Gastroenterology & Hepatology. (2022). The world is failing on hunger. The Lancet. Gastroenterology & Hepatology, 7(10), 895. https://doi.org/10.1016/s2468-1253(22)00280-1

Ugbogu, E. A., & Ugbogu, O. C. (2016). A review of microbial protein production: Prospects and challenges. FUW Trends in Science & Technology, 1(1), 182–185.

UN World Food Programme (WFP). (2024). WFP's Strategy to improve diets and address malnutrition 2024 – 2030. https://www.wfp.org/publications/wfps-strategy-improve-diets-and-address-malnutrition-2024-2030

United Nations. (2019). 17 Goals to Transform the World for Persons With Disabilities. New York: United Nations.

Upasana, S. (2023). The impacts of poverty on hunger: An examination of the relationship between socioeconomic status and food insecurity. International Journal of Agricultural Science and Food Technology, 9(2), 041–043. https://doi.org/10.17352/2455-815x.000190

Vega Luz, Y. N., Mora Ana, P. V., & Espitia Paula, J. P. (2018). Case-Study: Nutritional status and contextual Factors in the food security of teenager students in Cartagena (Colombia). Advance Journal of Food Science and Technology, 16(SPL), 47–54. https://doi.org/10.19026/ajfst.16.5936

Verma, S., Chakdar, H., Kumar, M., Varma, A., & Saxena, A. K. (2021). Microorganisms as a sustainable alternative to traditional biofortification of iron and zinc: Status and prospect to combat hidden hunger. Journal of Soil Science and Plant Nutrition, 21, 2179–2195. https://doi.org/10.1007/s42729-021-00473-5

Von Braun, J., Chichaibelu, B. B., Laborde, D., & Torero Cullen, M. Cost of ending hunger – Consequences of complacency and financial needs for SDG2 achievement, ZEF. Discussion Papers on Development Policy (No. 347, p. 103). Bonn: Center for Development Research, May 2024.

von Grebmer, K., Bernstein, J., Wiemers, M., Reiner, L., Bachmeier, M., Hanano, A., Chéilleachair, R.N., Foley, C., Sheehan, T., Gitter, S., Larocque, G., and Fritschel, H. (2023). 2023 Global Hunger Index. Concern Worldwide and Welthungerhilfe.

Wallace, R. (2021). Microbial biotechnology for future food systems: Adapting to climate change. Trends in Biotechnology, 39(10), 987–990. https://doi.org/10.1016/j.tibtech.2021.03.004

Wang, Y., Zhang, H., & Li, Z. (2021). Precision fermentation for sustainable food production: Microbial synthesis of vanillin. Synthetic and Systems Biotechnology, 6(4), 291–298. https://doi.org/10.1016/j.synbio.2021.09.004

Weldemichel, T. G. (2024). Hunger and famine. In B. Warf (Ed.), The Encyclopedia of Human Geography. Cham: Springer. https://doi.org/10.1007/978-3-031-25900-5_173-1

Wilson, S. A., & Harrison, M. J. (2021). CRISPR/Cas9 in microbial biotechnology: Applications in food production. Nature Biotechnology, 39(6), 678–686.

World Health Organization (WHO). (2022). Nutrition. https://www.who.int/health-topics/nutrition

World Health Organization (WHO). (2024). Hunger Numbers Stubbornly High for Three Consecutive Years as Global Crises Deepen: UN Report.

Xian, Y., Yin, L., Wu, Z., Chen, D., Sun, Z., Jiang, X., & Tian, L. (2025). Characterization and Preliminary Safety Evaluation of Mycoprotein from Fusarium compactum MM-135. Frontiers in Sustainable Food Systems, 9, 1583642.

Yilmaz, H., & Yilmaz, A. (2025). Hidden hunger in the age of abundance: The nutritional pitfalls of modern staple crops. Food Science & Nutrition, 13(2), e4610. https://doi.org/10.1002/fsn3.4610

Zreik, M., & Iqbal, B. A. (2024). Evaluating China's strategies for achieving Zero Hunger and SDG 2: From policy to plate. Academia Nutrition and Dietetics, 1(1), 1–8. https://doi.org/10.20935/acadnutr7387

2 Global Action to End Hunger
Initiatives and Challenges

Moses Aniefiok and Samuel Abasiama Moses

2.1 INTRODUCTION

In recent time, food as one of the basic needs of man has become a difficult thing to come by in most households leading to wide spread hunger and starvation in most part of the world. The challenges being faced by most countries in combatting food insecurity have been well stated with obvious scary pictures of the impact of global food shortage everywhere in the world. While some countries are responding marginally well, most countries are overwhelmed by the enormous catastrophic consequences of food insecurity that is likely to be liken to the seven *lean* years of famine of 2700 BC era in Egypt, that led to widespread dislocation of families into regions where food could be found as recorded in the Bible (Genesis Chapters 41 and 47).

Food insecurity everywhere has elicited several initiatives in many part of the world aimed at increase food production to meet the growing hungry population, dietary needs, and food preferences for an active and healthy life (FAO, 2015; Bernal, 2024). Bernal (2024) pointed out that feeding a population that was expected to exceed 9 billion by 2050 was extremely complex and requires not only adequate and sustainable food supply but also a collective effort from different sectors of the society. Some countries have adopted multifaceted approaches to tackle food insecurity, but some of these initiatives have not abated the effects on the growing number of people who cannot afford basic food requirements in several parts of the world, especially in third-world countries particularly in the sub-Saharan region.

The major causes of food insecurity in recent times have been wide-spread conflicts, wars and political instabilities, climate crisis, poverty, natural and man-made disasters, forced migration as a cause and consequence of hunger, gender inequality and discrimination, bad policies and poor economies, food wastage as well as other cross-cutting issues like poor or inadequate adaptation of technologies as well as the inability to use mechanized and modern farming methods. To reverse this trend requires a major shift in all spheres of current and imaginal initiatives. This call for concerted efforts by all stakeholders, and government readiness to play the leading role in galvanizing the nexus between indigenous initiatives and modern technological innovations to ensure food availability for all. One of such initiatives is the application of microbial technology to improve food security. There is need for all stakeholders in the food value change to synergize, adopt implementable policies with clear long- and short-term plan of action to tackle food insecurity. It is envisaged that a unified approach will result in the reversal of the dooming catastrophe facing mankind. The time to act is now, if human race is not on the path of extinction in the face of natural abundance of availability of what it takes to feed mankind. Therefore, this chapter examines steps taken by all stakeholders to ensure food security in terms of availability and accessibility as well as utilization of food and stability of food system.

2.2 CAUSES AND IMPACT OF GLOBAL HUNGER – REGIONAL DIFFERENCES AND SEVERITY

We have mentioned some of the causes of global hunger. We briefly explain the impact of some of them with regional differences in severity.

DOI: 10.1201/9781003650881-2

As at today, wars and conflicts are major causes of food insecurity and hunger. In many parts of the world today, huge farming population is being wasted in senseless conflicts and wars. Many people are being forced to relocate from their ancestral homes to other places of relative safety, often in internally displaced persons (IDP) camps, where they have to depend on food aids. The impact of the Russian-Ukraine war on global food security has been highlighted by many scholars and economist. Lin et al. (2023) found that the Russia-Ukraine war would lead to a trade drop of 60%, increase price of wheat by 50%, hence leading to severe food insecurity with decreased purchasing power for wheat above 30%, especially for countries that heavily rely on wheat imports from Ukraine, such as Egypt, Turkey, Mongolia, Georgia, and Azerbaijan. There have been reported conflicts of different magnitude in some other parts of the world.

As at 2024, there were reports of serious conflicts in Myanmar, Ukraine, Palestine (Gaza), Syria, Yemen, Haiti, Afghanistan, Mexico, Armenia-Azerbaijan border, the Sahel (Burkina Faso and Niger), Ethiopia, Sudan, South Sudan, Democratic Republic of Congo, Somalia, Central African Republic, the Lake Chad Basin, etc. Most of the conflict regions are obviously in Africa. These conflicts have had very devastating impact on food production, thereby increasing the problems of food insecurity in these regions, with spill-over effects in the entire world.

In Nigeria, the effects of banditry, farmers-herders conflicts, rise in cases of kidnapping and the activities of Boko Haram have crippled food production in every part of the country leading to soaring food prices, hunger, and starvation. Ani et al. (2022) noted the negative effects of continuous armed confrontations over natural resources on food security in Nigeria. According to them, these conflicts were undermining human security, human dignity, economic and social well-being as well as the general prosperity of the nation. The recurring herders-farmers conflict, according to Ani et al. (2022) was unknown in Nigeria before 2000, but the episode has reached a dangerous dimension in recent times, affecting food production adversely. They reported that between 2009 and 2020, over 4,606 lives were lost to these conflicts in several states including Adamawa, Anambra, Benue, Borno, Ebonyi, Enugu, Kaduna, Kogi, Nasarawa, Plateau, Sokoto, Taraba, and Zamfara. Within the period also, over 300,000 persons were displaced from their ancestral home with huge loss of means of livelihood. These dislocations have impacted negatively on food production, leading to widespread high cost of food and hunger in the land.

Another major cause of food insecurity is the impact of climate change. Climate change is a global phenomenon with catastrophic consequences in almost all sectors of human endeavor. It impacts agriculture and food security in most profound ways, altering temperature and precipitation patterns, which lead to decreased crop yields, water scarcity, and increased risks of extreme weather events. Most regions of the world are extremely effected, but the impact of climate on food security is widespread. Extreme weather events due to climate volatility like flooding, drought, and wild fire are common occurrence in most part of the world today. Climate change leads to rise in sea level, coastal erosion, and excessive rain fall continue to damage farmland and destroy crops in many areas, leaving farmers stranded without anything to show for their labor. It also leads to increase prevalence of parasites and diseases that affect livestock and aquatic lives. Conversely, drought due to increase temperature has destroyed farmland, crops, and livestock in many parts of the world especially in the sub-Saharan region. Ani, et al. (2022) noted that climate change has negatively affected food security in Nigeria, leading to continuous armed confrontations over natural resources thereby undermining human security in the country. They observed that climate change was increasingly becoming a major threat to agricultural productivity in Nigeria. For instance, they noted that abnormal onset and cessation of rainfall, rising temperature and alteration in relative humidity have negative consequences on agricultural activities and food systems in Nigeria. They maintained that climate change was the key driver of food insecurity in Nigeria.

The effects of climate change are severe, leading to some changes in agricultural production that threatens food availability, food quality, accessibility, and stability, which ultimately contribute to food insecurity. The immediate impact is food scarcity, which leads to increased food prices and rising poverty, while the long-term effects include hunger, malnutrition, starvation, and death. FAO,

IFAD, UNICEF, WFP, and WHO (2021) confirmed that 720 to 811 million people were undernourished in 2021. They further projected that about 200,000 people were at a *catastrophic level* of food insecurity due to the effects of climate change, while additional 8 to 80 million people would be at risk of hunger by 2050.

2.3 GLOBAL ACTIONS TO END HUNGER

There is obvious palpable hunger in many parts of the world today and hunger is one of the leading cause of death in the world. The effects of hunger as a result of lack of food in adequate quantity and quality is felt everywhere. This challenge requires urgent actions at local, sub-national, national, and international levels. Expected actions should include research, policy realignment, and the formulation of long-term and short-term plans of action, adaptation of technology, and the commitment of all stakeholders in holistic manner to solve the problem. The general framework for global actions to end hunger is presented in this sub-section.

2.3.1 International Initiatives

The impact of hunger on humanity came to international front-burner in 2012, when the United Nation's Secretary launched the *Zero Hunger* initiative during the Rio + 20 Conference on Sustainable Development. The end hunger initiative further gathered momentum in 2021, when the UN convened the *Food System Summit* as part of the Decade of Action to achieve the Sustainable Development Goals (SDGs) by 2030. The Zero Hunger initiative hung on SDG 2, with eight specific targets to be achieved in 2030. These are as follows:

- that all people to have access to safe and nutritious food;
- end all forms of malnutrition;
- double the productivity and incomes of small-scale food producers;
- ensure sustainable food production and resilient agricultural practices;
- maintain the genetic diversity in food production;
- increase investment in rural infrastructure, agricultural research, technology, and livestock gene banks;
- prevent agricultural trade restrictions, market distortions and export subsidies and ensure stable food commodity markets and timely access to information.

Since the setting up of these targets, many international and national agencies, organizations, and governments have been working assiduously to address food insecurity and the growing impact of hunger and malnutrition. Some countries have made significant move toward attainment of these set goal and targets, while many others have not made any meaning progress. As at 2024, 42 countries, including Nigeria, were ranked as facing serious hunger on the *Global Hunger Index,* while situation was said to be alarming in 7 countries of Burundi, Chad, Lesotho, Madagascar, Somalia, South Sudan, and Yemen. Nigeria was ranked 110th and Somalia 127th out of the 127 countries assessed and ranked (Wiemers et al., 2024). WHO (2024), reported that in 2023, about 733 million (9%) of the world population was facing serious hunger and 20% of these number was in Africa, while UNICEF also reported that 295 million people in 53 countries and territories were at acute level of hunger in 2024 (UNICEF, 2025).

The following UN agencies have been working to end or reduce food insecurity worldwide. They are World Food Programme (WFP), aims to bring food assistance to millions of people and to help prevent hunger; the World Bank (WB), investing in agriculture and rural development to boost food production and nutrition; Food and Agriculture Organization (FAO), aims to make sure people have regular access to enough high-quality food to lead active and healthy lives; UNICEF, dedicating to helping children in need through – the *Scaling Up Nutrition*; International Fund for Agricultural

Development (IFAD), focusing on rural poverty reduction, and elimination of hunger and malnutrition. Other organizations working to end hunger include but not limited to Care, the Hunger Project, the Bread for the World, Clean Cooking Alliance, Action against Hunger, Rise against Hunger, Heifer International, Project Concern International, Alliance to End Hunger, Feed the Children, Caritas International, Association for International Agriculture and Rural Development (AIARD), World Central Kitchen, Food for the Poor, etc. The work of most of these UN agencies and international organizations has been hampered by inadequate funding (WHO, 2024) and other resources to reach the ever increasing hungry population, as well as those affected by conflicts and wars.

Other global initiatives aim at ending hunger include efforts at tackling the effects of climate change, steps to end conflicts and war in different parts of the globe, research, innovations and application of technology. According to Ani et al. (2022), the Kyoto Protocol to the United Nations Framework Convention on Climate Change (KP-UNFCCC), was aimed to increase food production and end hunger. The KP-UNFCCC of 1998 aimed to reduce emissions of greenhouse gases, to combat global warming; requiring industrialized nations and economies in transition to set legally binding targets to limit and reduce their emissions. This was among others to promote sustainable agriculture in light of climate change considerations. As its primary purpose, the Kyoto Protocol was an agreement among developed nations to reduce carbon dioxide emissions and greenhouse gases in an effort to minimize the impacts of climate change (Tardi, 2025). The Protocol further put responsibilities upon Parties to implement adequate measures to reduce the impact of Climate Change in their countries or limit the effects of such impacts from extending to other countries, and the world's governments to agree to try to stabilize emissions of the gases that can change the climate by 2000 (Anderson, 1998). However, not much has been achieved since most countries with high emission refused to take the expected measures to end emission even after 2000.

The UN is overwhelmed with the number of countries and territories currently experiencing conflicts and war situation. The Security Council has been described as a "toothless bulldog" because while it has the power to make decisions on international peace and security, it lacks the ability to enforce them, principally as it has no military formation (UKEssays, 2018 and Mugovera, 2023). The inability of UN Security Council to intervene in all the global conflict situations has led to its relevance being queried (Heath, 2021). Similarly, the African Union (AU) has been described as a toothless bulldog and a talking club of African leaders as it claimed non-interference in internal affairs of member states even when each is at the throat of other (Michael, 2023). This inactive or inability of the global and regional bodies to intervene in conflict and war situation has further undermined their existence and encourage escalation of turmoil and conflict in different parts of the world, which has adversely affected food production and increases food insecurity worldwide.

A lot have been done at the global stage in research, innovation, and technology to increase food production to end hunger. Organizations like WFP, FAO, IFAD, International Food Policy Research Institute (IFPRI), and International Institute of Tropical Agriculture (IITA) have been involved in various researches leading to innovative agricultural practices, improved crop varieties, and developing resilient crops adaptive to climate change and increase yield, using data-driven approaches and leveraging on technology for efficient food storage and distribution to reduce food wastage. These efforts are all geared toward improving food security and addressing malnutrition.

Globally, there have been ongoing researches and innovation aimed at improving and increasing food production. For instance, FAO and Gate Foundation as well as other international organizations have invested in research and innovation to increase food production, improve food security, and address challenges of climate change. FAO and Gates Foundation in particular, have initiated the utilization of digital technologies to enhance food productivity and sustainability. Most of their efforts aimed primarily at developing new crop varieties and improving existing farming practices. Klerkx and Villalobos (2024) confirmed that there have been many technological advances in agriculture and food systems over the past 20 years. Such technologies according to them include pervasive digitalization, gene editing technologies, biotechnology, precision fermentation and molecular technologies. Others include the digital transformation pathways that encompass digital agriculture

(DA) and vertical farming and, more broadly, "smart farming" in which all supply chain actors are connected through technologies such as sensors, the Internet of Things, platforms, and blockchains. Klerkx and Villalobos (2024) explained the terms *AgriFoodTech start-ups*, to mean technological initiatives such as *Agricultural robotics, Pixel Farming Robotics, SwarmFarm, Hello Tractor,* etc., being spearheaded by some private and public organizations in many countries, some of which have grown to large-scale application.

Most of the researches and innovations have been focused on the application of microbial technology in food production. This technology leverages microorganisms to produce various products, and genetic engineering enables enhancement of microorganisms for specific tasks, like producing food ingredients, enzymes, or pharmaceuticals. Genetically modified organisms (GMOs), including bacteria, yeast, and fungi, are central to microbial technology. GMOs, particularly genetically engineered microorganisms (GEMs), are essential tools in biotechnology, and are becoming more broad approach to achieve food security. Advances in this technology are gaining traction daily and more organisms are being manipulated to achieve obvious purpose in several areas of human endeavor including food production. As the days go by, we are likely to see more advances and intervention in application of microbial technology to achieve food security. Ultimately, there is need for sustain investment in the agriculture sector to reduce hunger and poverty, improving food security, creating employment, and building resilience to disasters and shocks. Global Alliance against Hunger and Poverty (2024) stated that Governments, Development Banks, International Organizations, and Philanthropies partners should bring socioeconomic inclusion programs to scale up food production so as to help lift 100 million more people out of poverty by 2030. If the world would promote sustainable agriculture with modern technologies and fair distribution systems, we can sustain the whole world's population and achieve zero hunger.

2.3.2 Regional Initiatives

Hunger and its impacts cut across every region of the world: from Africa to the Americas, and from Europe to Asia, the story is the same. However, the severity of the impacts has not been evenly distributed and some regions are more impacted than others. For instance, the North Americas appear to be coping well than most countries of the South. Argentina, Ecuador, Peru, and Suriname of the South America were said to be facing serious hunger as at 2024 (FAO, 2025). Conversely, the North Americas of United States and Canada are embracing technologies and innovations in food production and processing more than their southern counterpart, hence are less impacted. Yet, in regional comparison, Africa is the worst hit, as one in every five African were said to be facing hunger in 2023 (Signe, 2025), and most African countries are just trying to adopt some form of technological innovation in food production. Indeed, Brickstone (2024) observed that agricultural sector in Africa was faced with serious challenges as agricultural productivity in the continent was lagging behind other regions because of poor utilization of technology.

It is worthy of note that in Africa as at today, agricultural technology is beginning to emerge. But still, most African countries are still grabbling with seasonal farming, post-harvest losses, traditional land holding system, subsistent farming practices, open-grazing of livestock, backyard farming, and the use of crude farming implements. Therefore, the use of technology to increase food production varied and is uneven across the entire Africa region (Signe, 2025). Obviously, Africa needs technologies to improve food production to meet their food need. This need is even more amplified now than ever, with more African countries currently facing severe food insecurity in recent time.

The need for African to move up the technological ladder in food production is increasing daily. Africans cannot afford to maintain the traditional farming practices. They should deploy technological mix with indigenous initiatives like the traditional irrigation system in some parts of Africa – Swarm irrigation in Gambia, Burundi, Serra Leone, Malawi, Zambia, Tanzania and Ghana; Spate irrigation in Ethiopia, and so on. Traditional irrigation system in the inland valley of Bida region in

Nigeria has been used for improved rice cultivation (Fu et al., 2010). Other areas of improved food production and value change are begging for attention and need to be addressed. Earlier, Moussa (2002) had noted that Africa needed the use of high-yielding varieties and improved seeds along with the application of appropriate technology and fertilizers to improve food production.

High-yielding and improved varieties of seeds are the products of technology innovation. The use of microbial technology has proved to be the game changer in improved agricultural production everywhere. In this regard, Moussa (2002) noted that attention was being given to the African capacity of agricultural research and technology for effective use biotechnology as the only effective technique for future food supply growth in Africa. Indeed, the IITA has been playing vital role in research in application of biotechnology and genetic engineering to develop improved crops varieties like cassava, cowpea, maize, banana, and plantain and disseminating same across sub-Saharan Africa to enhance food security (IITA, 2025).

Coker et al. (2015) have also examined the impact of science and technology, particularly the contribution of biodiversity, biotechnology, and biosafety on sustainable food security in some countries of Africa including South Africa, Zimbabwe, Uganda, Kenya, Cameroon, Tanzania, and Nigeria. He concluded that for Africa to achieve sustainable food security, she must lay a solid foundation for science and technology by investing more on education, research and development (R&D) as well as initiate public policies aimed at sharing adaptive and practical innovations and breakthroughs in science, technology, and sustainability of food in the African continent among African States. AUDA-NEPAD (2023) also supported this position, saying that to tackle the challenges surrounding low agricultural yield in Africa, African governments have to embrace emerging digital technologies. Therefore, it is imperative that Africa must adopt and use available technology suitable to her climate condition and capacity to adapt and to enhance food production so as to change the narrative as regard food insecurity and impact of hunger in the continent. The Caribbean countries are already ahead and a country like Brazil has embraced technology, hence improving her food production capabilities. Abiri et al. 2023), have pointed out that DA as a pioneering technology was helping to meet the growing global demand for sustainable food production. Hence, according to him, integrating different sub-branches of DA technologies such as artificial intelligence, automation and robotics, sensors, and data analytics into agriculture practices would help to reduce waste, optimize farming inputs, and enhance crop production.

2.3.3 National Initiatives (Nigeria as a Case)

Nigeria has been grabbling with food insecurity that has worsened with emergence of every new government since the beginning of the Fourth Republic in 1999. The country has not been able to adequately feed her teeming population for a couple of years now. Many families are unable to afford food rich in adequate nutrients because of economic hardship, which WFP (2025) affirmed that 37% Nigerians were living below poverty line of US$1.8 per day. Therefore some women and most children who are more vulnerable are malnourished and stunted (Ani et al., 2022; Osagie-Jacobs, 2024; Ogunlade et al., 2024; Ajetunmobi, 2024; WFP, 2025). Osagie-Jacobs (2024) reported that 31.8 million Nigerians were suffering from acute food insecurity compounded with malnutrition. WFP (2025) projected that over 26.5 million Nigerians were going to face acute hunger by August 2024, a staggering increase from 18.6 million in 2023. According to the Program, the deplorable situation was largely caused by conflict and insecurity, rising inflation, and the impact of the climate crisis. UNICEF (2023) had earlier predicted that nearly 25 million Nigerians were at risk of facing hunger by August 2023 due to food insecurity. As a result, Nigeria has relied heavily on food import and any policy toward reducing food importation often creates ripple effects in the polity.

Nigeria has been facing a huge food import bills for a long time now. Most of the country's foreign earnings have been spent on food importation. In fact, according to Nigeria Custom Service (NCS), the country's food import bills have been steadily rising, skyrocketed to ₦920.54 billion in the first quarter of 2024, representing a quantum leap of 95.28% for the same period in 2023 (Amao,

2024). The military coups in West Africa countries of Mali, in August 2020 and May 2021; Burkina Faso in January 2022 and September 2022; and Niger Republic in July 2023 led to some sanctions by Economic Community of West African States (ECOWAS). The closure of Nigeria's border with its neighbors and particularly that of Niger Republic at the wake of the coup there, witnessed sharp increase in the prices of essential basic food items especially rice and frozen food. This policy saw the price of a bag of 50 kilogram of rice jumped to about ₦120,000 from ₦28,000 before May 2023. The ripple effect of that policy though of great national interest is still being felt till today.

The challenge of food insecurity in Nigeria has been exacerbated by climate change, economic hardship, and most devastating internal conflicts and insecurity since 2010. Banditry, kidnapping for ransom, farmers-herder's crashes, Boko Haram, and even communal conflicts have had serious negative implications on food production in the country. The situation has been compounded by the audacity of terrorists demanding down payment of large sums of money and some form of taxes by community members in some states before they are allowed relative access to their farms. Because of large-scale and wide spread insecurity in Nigeria, people can no longer go to farm. Lawal (2024) confirmed that wide and sustained insecurity was impacting on food security in Nigeria. FAO, European Union, and CIRAD (2022) noted that internal conflicts, high unemployment, and the effects of climate change were contributing to high food insecurity in Nigeria. They confirmed that the incidents of insecurity and violence have greatly disrupted agrifood system activities in the country. The attackers are reported to be wielding heavy sophisticated weapons with superior firing power often overwhelming that of state actors.

Natural disasters like flood and drought have added to the challenge in a very significant way. For instance, when flood occurs, large farmland and crops are often destroyed, leaving farmer with nothing to show for their labor. FAO (2024) raised the alarm that devastating flood was worsening the already deteriorating food security situation in Nigeria as at 2024. Since 2012, flood has been wreaking havoc in most communities in almost all the states of the federation, destroying farmlands, infrastructure, and rendering large proportion of the population homeless.

Another serious challenge to food security in Nigeria is the continued use of crude farming implement like hoes and cutlasses, which obviously hindered large-scale food production. Efforts aimed at expanding mechanized farming has been very slow, as Nigerian elites will rather prefer buying exotic cars rather than farming equipment like tractors to assist in food production. Similarly, leveraging microbial *biotechnology* is engulfed in controversies bothering on safety to the point that some more concerned Nigerians will verify whether food items were organic or GMOs derived before purchase. These multi-faceted problems are hindering food production, challenging food security, and thereby exposing majority of Nigerian population to hunger, starvation, and malnutrition.

Nigeria needs to adopt and implement some adaptive strategies that would enable her cope with the challenges of climate change to ensure food availability through sustainable agriculture system. Such measures should also address some challenges that lead to farmers-herders crises in parts of the country that have worsened human security situation as well (Ani et al., 2022). There is also the need to tackle the impact of climate change holistically so as to address the worsening food insecurity in Nigeria. Indeed, Ani et al. (2022), also affirmed that to address the emerging risks posed by climate change on food and human security, approaches to climate change vulnerabilities must seek to deploy new strategies to agricultural practices, adaptability of agricultural practices to changing environment and regulation of anthropogenic factors that trigger climate change.

Over the years, Nigeria has adopted many innovative approaches to address food insecurity and reduce the impact of hunger in the country. Some of these approaches have been well-intended, while most have been shrouded in controversy and corrupt practices. Indeed, FAO, European Union, and CIRAD (2022) observed that underdeveloped agrifood value chains were one of the key challenges facing sustainable food system in Nigeria. For instance, the 2011–2015 Agricultural Transformation Agenda (ATA) of the Nigeria, according to them failed to meet the set targets. Obviously, such failure has led to a situation where the country has continued to import USD 3 billion to 5 billion worth of food annually; indicating weak implementation of relevant policies and plans in the agriculture sector. To date, agriculture and food production do not seem to have the required impetus and catalyst to stimulate national consciousness to meet the nation's food requirement. Nonetheless, some laudable plans and programs have been initiated by Nigerian government aimed at addressing food security challenges, improve agricultural productivity, and enhance the livelihoods of the general population. Some of the initiatives, approaches, and programs include:-

1. **ATA** – aimed to transform the agricultural sector for food security, income growth, and employment generation.
2. **Growth Enhancement Support Scheme (GESS)** – aimed to provide subsidized inputs like fertilizers and seeds to farmers.
3. **Anchor Borrowers' Programme (ABP)** – aimed to link smallholder farmers with reputable large-scale companies (anchors) for financing, technical support, and guaranteed pricing.
4. **Youth Empowerment in Agriculture Programme (YEAP)** – aimed to encourage youth participation in agriculture.
5. **Livelihood Improvement Family Enterprises (LIFE) Programme** – focused on improving livelihoods through agricultural interventions.
6. **National Agricultural Seeds Council (NASC)** – meant to regulate and promote the quality of seeds in Nigeria.
7. **Commercial Agriculture Credit Scheme (CACS)** – aimed to provide credit to commercial agricultural enterprises.
8. **National Accelerated Food Production Programme (NAFPP)** – designed to enhance farmers' productivity in selected crops.
9. **Operation Feed the Nation (OFN)** – a mass mobilization program to encourage public participation in agriculture.
10. **Green Revolution Programme** – focused on increasing food production and self-sufficiency.
11. **Agriculture Promotion Policy (APP)** – focused on various aspects of the agriculture sector.
12. **River Basin Development Authorities (RBDAs)** – focused on developing water resources for irrigation, providing water for agricultural production, and improving rural infrastructure.

From the above listed initiatives and programs, it is a fact that attempts have been made and laudable programs initiated aimed at improving the agriculture system so as to increase food production. However, due to poor implementation and continued policy changes, Nigeria is yet to achieve food sufficiency. Sometimes, most policy somersault appears as if the country is in a panic mode. For instance, the current President Bola Ahmed Tinubu administration recently demerged livestock from Agriculture to create Federal Ministry of Livestock Development (FMLD) on July 9, 2024. According to the reason behind its creation, the new ministry is to modernize livestock sector, ensure its sustainability as well as bring to an end the perennial farmers-herders conflicts and increase livestock production. Nigerians will have to wait for the expected outcome of this new initiative. Then, on June 24, 2025, the administration launched 2,000 tractors, in a bid to revolutionize the agriculture sector and achieve food sovereignty. Though inputting 2,000 tractors definitely is a far cry from the reality (translating to 54 tractors per each state, including the federal capital territory (FCT), Abuja); mechanize farming in Nigeria appears to have come too late.

While trying to mechanize Nigeria's agriculture system in the 21st century, there are yet some thorny issues to be addressed. The traditional land holding system would not allow for expansion of agricultural holdings beyond subsistence level. FAO, European Union, and CIRAD (2022) have pointed out that the agriculture sector in Nigeria was heavily relying on rain-fed subsistence farming, with smallholder farmers constituting of approximately 88% of the farming population, with an average farm size of 0.5 hectare. This agriculture system, according to them is dominated by small family farms, managed and tended to mixed system of crops and livestock, and sometimes engage in fish farming, which is often characterized by low yields, post-harvest losses, and food safety issues such as food-borne diseases. The farmers in this system lack adequate access to inputs, credit and technology, and only about 2% of them have access to irrigation. Additionally, only 26% of them sell their agricultural products, indicating a high level of subsistence farming.

To address the challenge of traditional land holding system, there is need for deliberate efforts to be made to develop the vast undeveloped and uncultivated land mass across the country to expand agriculture system, if the mechanize farming system is to impact the food system in Nigeria. Easy access to land for farming purpose should be a top priority for every community and government. Each tier of government should initiate policy and strategy to ensure easy access to land for farming. Ranching must be encouraged as a way to stem the tide of herders-farmers conflict and to improve the quality of animal products from our livestock so as to improve protein supply in our food supply chain. As regard credit facility to farmers, the Bank of Agriculture (BOA) should be strengthened to provide more credit facilities for agricultural and non-agricultural value chain activities, promote capacity of cooperatives and extension services, and offer financial inclusion services to farmers.

Water resource management is another issue and posing serious challenge to agriculture practices in Nigeria. As stated earlier, large percentage of the farming population rely on rain for agricultural activities. With the unpredictability of the weather pattern due to climate crisis, it is apparently clear that Nigerians cannot depend on rain-fed agriculture to meet their food need. In March 2025, the Federal Government reconstituted the management team/boards for the 12 River RBDAs. These authorities were established in 1976 to manage water resources for irrigation and control flood. However, for years now, they were moribund with little to show for their existence. It is hope that they would breathe life into agriculture-water management system for improved food production to reverse the reliance on rain-fed agriculture.

It is time for Nigeria agriculture system to fully embrace technology for improve food production. The country is already making progress in GMOs for the production of improved varieties of seeds for cotton, cowpea, maize, soybean and wheat. Researches are ongoing to develop improved seeds for potato, cassava, and sorghum. There is need to expand researches into other crops and educationally address intractable controversies bothering on safety of GMOs in Nigeria. The National Biotechnology Research and Development Agency (NBRDA) should think out a more innovative way to address this issue. Nigeria should also adopt smart agriculture to improve food supply.

2.3.4 Local Initiatives

Achieving zero hunger is possible, but requires concerted efforts by all, including individuals, communities, local leaders, and governments. Everyone must stand up firmly to tackle food insecurity to preserve humanity from extinction due to hunger. There are some local initiatives aimed at food production in many communities in Nigeria today. Two that are becoming more attractive and widespread are *concrete-based planting and backyard poultry.*

As the food situation become more severe, some people who hitherto floored their large open space and setback in their residential compound have now adopted innovative way of turning those spaces into *farmlands*. In this *concrete-based planting,* empty bags of rice, cement, or animal feeds are deployed. The person ordered for humus/top soil to be brought to his compound. The soil may be mixed with appropriate organic manure and neatly packed into the empty bags to about ¾ full. The intended seed is introduced and watered intermittently at regular interval.

Backyard farming has been a traditional agriculture system for producing small quantity of food, especially vegetables that are easily accessible for family immediate food need. However, planting on concrete floor in bags or other containers is innovative that can help improve families' food requirement even in towns and city. The practice removed labor cost for weeding and other logistics for inspection and nurturing and can be carried out any season of the year as watering can be done during dry season to maintain plant growth and vitality.

Similarly, backyard poultry and keeping other animals like goats, sheep, pigs, rabbits, and cattle is a wide spread animal husbandry in most African traditional setting including Nigeria. Most families rely on this for their protein need and some even sell surplus stocks to meet other family needs like payment of school fees for ward.

Backyard poultry farming is very common in most cities in Nigeria. A study carried out in Kubwa in FCT indicated that 88% of the poultry farms were located within 100 meter radius from residential buildings and in fact, 44% were located within 10 meter radius from residential buildings (Moses and Reign, 2025). The authors noticed that none of the poultry farms in the study area met the minimum allowable 500 meters distance from residential buildings. Though this practice poses significant health risk, it was noted that backyard poultry farming was contributing to the food supply system in most parts of Nigeria.

To achieve zero hunger in most rural communities in Nigeria, governments must expand the social safety nets through ensuring that the social register is updated to capture all those that should be there and ensure that all people, regardless of their location, gender, or social status, have access to food and adequate nutrition. As stated by WFP (2022), effort should be made to reduce food waste by helping rural farmers connect to the market and ensuring that all bottlenecks along the food logistic supply chain, particularly the "toll gate-keepers" who extort sundry fees from food trucks are eliminated. Rural farmers should have access to credit facility, agriculture extension services and training, improved seeds and other farm inputs as well as guaranteed security to enable them cultivate more food for Nigeria growing population.

2.4 THE APPLICATION OF TECHNOLOGY IN FOOD PRODUCTION AND PRESERVATION – THE ROLES OF MICROBES

The application of technology in improved food production is gaining traction by the day. Santos-Beneit (2024) stated that the concept of genetic engineering, which involves the artificial manipulation, modification, and recombination of deoxyribonucleic acid (DNA) or other nucleic acid molecules to alter organisms, has garnered significant interest over the past few decades. He affirmed that recent advancements in genetic and molecular biology have propelled genetic engineering into the forefront of scientific and technological disciplines of microbial biotechnology and genetic engineering. According to him, microbial biotechnology was playing a pivotal role in shaping the field of genetic engineering, contributing significantly to the development of microbial biotechnology being leveraged in food security today.

Despite some controversies surrounding concerns about potential risks to human health and the environment, GMOs as derivative of microbial technology are increasingly being used in food production to enhance crop traits and increase yields. Indeed, Hanlon and Sewalt (2021) and Santos-Beneit (2024) have stated that microorganisms have been used throughout human history and in many centuries in aid of food production, even before humans knew that these organisms were responsible for the fermentation processes in food. According to them, the use of GEMs, often referred to as Genetically Modified Microorganisms (GMMs), has become widespread in the production of food processing aids and other food ingredients over the past several decades. They said that GEMs were advancing food production by increasing efficiency, reducing waste and resource requirements, and ultimately enabling beneficial innovations such as the cost-effective fortification of food with essential nutrients to achieve unique food processing capabilities and enhance food nutrients requirement for growth and development.

GMOs as offshoot of GEMs have demonstrated great potential for reducing food insecurity in not too long distant time if all doubts surrounding its safety are scientifically cleared. Aside from increased crop yields, GMOs are said to be resistant to pests and diseases and able to tolerate adverse environmental conditions like drought and cold. They also have the capability of improved shelf lives and better storage quality of food, hence reducing food waste, and increasing the levels of certain nutrients in stored food. These products are known to enhance nutritional values of food as well as reduce production cost associated with pest, disease, and herbicide use (Paparini and Romano-Spica, 2004; WHO, 2014; Diaz and Fridovich-Keil, 2025).

Some common foods genetically modified and approved in many countries include apple, corn/maize, common beans, canola, cowpea, rice, soybeans, sugar beets, alfalfa, cotton, papaya, potatoes, pineapple, tomatoes, and squash (National Biosafety Authority, 2023; Singapore Food Agency, 2025). For livestock and aquaculture: chicken, pigs, goats, salmon, and tilapia have been successfully produced through genetically modification. More researches are on-going to add more crops, aquatic organisms, and livestock on the list. For instance, as at 2024, catfish, medaka, and zebrafish were being studied to be genetically modified.

GMOs, including bacteria, yeast, and fungi, are central to microbial technology. Microbial technology leverages microorganisms to produce various products, and genetic engineering enables the enhancement of these microorganisms for specific tasks, like producing food ingredients, enzymes, or pharmaceuticals. GMOs, particularly GEMs, are essential tools in biotechnology, providing a platform for precise genetic manipulation to achieve desired outcomes. They should definitely perfect tools to zero hunger by 2030.

2.5 CONCLUSION

Global food insecurity is posing serious challenge to human existence as a large proportion of the world population is facing devastating hunger due to intractable conflicts and wars, impact of climate change, economic hardship, and poor governmental policies everywhere as well as non-utilization of technologies in food production. Many regions and countries of the world have taken steps to mitigate the impact of food shortages but achieving zero hunger is still a far dream, but a possible reality if all stakeholders decide to review current strategies and adopt more pragmatic approaches and smart agricultural system, back-up with emerging technologies to increase food production. Transforming the agriculture systems and its value chain to achieve a more resilient, inclusive, and sustainable food system requires honest commitments, sound policies, and renewed sustainable investments in agriculture in rural and urban areas as well as provision of social safety nets protection so that poor and vulnerable people can have access to food and able to improve their livelihoods. This is urgent, it is imperative, and the world leaders owe it a duty to ensure zero hunger for continued human existence.

REFERENCES

Abiri, R., Rizan, N., Balasundram, S. K., Shahbazi, A. B. and Abdul-Hamid, H. (2023). Application of digital technologies for ensuring agricultural productivity. Heliyon, 9 (12). www.doi.org/10.1016/j.heliyon.2023.e22601.

African Union Development Agency (AUDA-NEPAD) (2023). Fertile Ground for Digitalisation: Adopting Digital Technologies to Improve Farming and Food Security in Africa. www.nepad.org/blog/fertile-ground-digitalisation-adopting-digital-technologies-improve-farming-and-food-security

Ajetunmobi, O. A. (2024). Socio-economic impact of food insecurity on vulnerable population in Nigeria [Commentary]. Journal of Agriculture, Food Systems, and Community Development, 13(4), 13–18. https://doi.org/10.5304/jafscd.2024.134.010

Amao, O. (2024). Nigeria's Food Import Bill Skyrockets to 920.54 Billion in Q1 of 2024 Ships & Ports. WWW.shipsandports.com.ng/nigerias-food-import-bill-skyrockets-to-n920-54-billion-in-q1-2024/

Anderson, J. W. (1998). The Kyoto Protocol on Climate Change BACKGROUND, UNRESOLVED ISSUES AND NEXT STEP. Washington, DC: Resource for the Future.

Ani, K. J., Anyika, V. O. and Mutambara, E. (2022). The impact of climate change on food and human security in Nigeria. International Journal of Climate Change Strategies and Management, 14(2), 148–167. https://doi.org/10.1108/IJCCSM-11-2020-0119.

Bernal, P. (2024). How are microbes helping end hunger? Microbial Biotechnology. https://doi.org/10.1111/1751-7915.14432.

Brickstone (2024). Digital Technology in Transforming Africa's Food System. www.brickstone.africa/digital-technology-in-africas-food-system/

Coker, M. A., George-Genyi, E. M., Agishi, T. V. and Abumbe, G. T. (2015). Science, technology and sustainable food production in Africa: An assessment. Mediterranean Journal of Social Sciences, 6 (6 S4). https://doi.org/10.5901/mjss.2015.v6n6s4p285.

Diaz, M. and Fridovich-Keil, J. L. (2025). Genetically Modified Organism. www.britannica.com/science/genetically-modified-organism (Accessed 03/06/25).

FAO (2015). Climate change and food security: risks and responses. www.openknowledge.fao.org/server/api/core/bitstreams/a4fd8ac5-4582-4a66-91b0-55abf642a400/content

FAO (2024). FAO Warns of Worsening Food Insecurity in Nigeria as Devastating Floods Continue. FAO in Nigeria. www.fao.org/nigeria/news/detail-events/en/c/1710177/ (Accessed 25/06/25).

FAO (2025). New UN Report: 74 percent of Latin American and Caribbean Countries Are Highly Exposed to Extreme Weather Events, Affecting Food Security. www.fao.org/newsroom/detail/new-un-report–74-percent-of-latin-american-and-caribbean-countries-are-highly-exposed-to-extreme-weather-events–affecting-food-security/en (Accessed 09/06/25).

FAO, European Union and CIRAD (2022). Food Systems Profile – Nigeria. Catalysing the Sustainable and Inclusive Transformation of Food Systems. Rome, Brussels and Montpellier, France. https://doi.org/10.4060/cc3380en

FAO, IFAD, UNICEF, WFP, and WHO (2021). In Brief to the State of Food Security and Nutrition in the World 2021. Transforming Food Systems for Food Security, Improved Nutrition and Affordable Healthy Diets for All. FAO. https://doi.org/10.4060/cb5409en.

Fu, R. H. Y., Abe, S., Wakatsuki, T. and Maruyama, M. (2010). Traditional farmer-managed irrigation system in Central Nigeria. Japan Agricultural Research Quarterly: JARQ, 44(1). https://doi.org/10.6090/jarq.44.53

Global Alliance against Hunger and Poverty (2024). Global Alliance. www.globalallianceagainsthungerandpoverty.org

Hanlon, P. and Sewalt, V. (2021). GEMs: Genetically engineered microorganisms and the regulatory oversight of their uses in modern food production. Critical Reviews in Food Science and Nutrition, 61(6), 959–970. https://doi.org/10.1080/10408398.2020.1749026

Heath, R. (2021). Is a Toothless United Nations Still Relevant? *Politico*. www.politico.com/newsletters/global-insider/2021/09/20/is-a-toothless-united-nations-still-Frirelevant-494401 (Accessed 30/05/25).

IITA (2025). Improving Crops. www.iita.org/research/our-research-themes/improving-crops/

Klerkx, L. and Villalobos, P. (2024). Are AgriFoodTech start-ups the new drivers of food systems transformation? An overview of the state of the art and a research agenda. Global Food Security, Vol. 40. https://doi.org/10.1016/j.gfs.2023.100726

Lawal, A. A. (2024). Insecurity and Its Impact on Food Security in Nigeria. AgroNigeria. www.agronigeria.ng/insecurity-and-its-impact-on-food-security-in-nigeria.

Lin, F., Li, X., Jia, N., Feng, F., Huang, H., Huang, J., Fan, S., Ciais, P. and Song, X. (2023). The impact of Russia-Ukraine conflict on global food security. Global Food Security, 36. https://doi.org/10.1016/j.gfs.2022.100661

Michael, A. (2023). An appraisal of OAU as a toothless bulldog and its transformation to African Union, 1963-2002. Innovare Journal of Social Science, 11(6). https://doi.org/10.22159/ijss.2023v11i6.49648.

Moses, A. and Reign, A. (2025). Poultry production and the risk of zoonotic diseases in Kubwa District of Bwari Area Council, Federal Capital Territory, Abuja Nigeria. International Journal of Research and Scientific Innovation, XII(II):448–459. https://doi.org/10.51244/IJRSI.2025.12020039

Moussa, S. Z. (2002). Technology Transfer for Agriculture Growth in Africa. AfDB: Economic Research Paper No. 72. www.afdb.org/fileadmin/uploads/afdb/Documents/Publications/00157678-EN-ERP-72.PDF.

Mugovera, G. (2023). Is UN Arbiter of Peace or Toothless Bulldog? *The Patriot*. www.thepatriot.co.zw/feature/is-un-arbiter-of-peace-or-toothless-bulldog/ (Accessed 30/05/25).

National Biosafety Authority (2025). List of Globally Commercialized Genetically Modified Organisms (GMOs) (Plants, animals, microorganisms, and vaccines), and derived products as at June 2025. Kenya: NBA. www.biosafetykenya.go.ke/commercialized-gmo-status (Accessed 27/12/25).

Nigeria Biosafety Clearing House GM Crop Events Approved in Nigeria. www.isaaa.org/gmapprovaldatabase/approvedeventsin/default.asp?CountryID=NG

Ogunlade, T. A., Esohe, O. O., Amarachi, N. C., Kwadwo, O. J., Nwafor, M. V. and Ifeoluwa, O. B. (2024). Impact of food insecurity among household with under 5 children in Nigeria: A systematic review. Asian Journal of Food Research and Nutrition, 3(4), 983–996. https://www.journalajfrn.com/index.php/AJFRN/article/view/191.

Osagie-Jacobs, J. (2024). 31.8m Nigerians Are Suffering from Acute Food Insecurity. www.web.facebook.com/PlanningNigeria/posts/318m-nigerians-suffering-from-acute-food-insecurity-over-318m-nigerians-are-suff. (Accessed 27/12/25).

Paparini, A. and Romano-Spica, V. (2004). Public health issues related with the consumption of food obtained from genetically modified organisms. Biotechnology Annual Review, 10, 85–122. https://doi.org/10.1016/S1387-2656(04)10004-5

Santos-Beneit, F. (2024). What is the role of microbial biotechnology and genetic engineering in medicine? MicrobiologyOpen, 13(2), e1406. https://doi.org/10.1002/mbo3.1406.

Signe, L. (2025). How Emerging Technologies Can Boost Africa's Agriculture and Agribusiness. Global Economy and Development. www.brookings.edu/articles/how-emerging-technologies-can-boost-africas-agriculture-and-agribusiness/ (Accessed 09/06/25).

Singapore Food Agency (2025). List of Approved Genetically Modified Crops for (i) Use as food for Direct Consumption, (ii) Use as Food Ingredients and (iii) Further Processing to become Ingredients for Other Food, as at February 2025. www.sfa.gov.sg/docs/default-source/regulatory-standards-frameworks-and-guidelines/list-of-approved-gm-crops-for-use-as-food-or-as-food-ingredients.pdf

Tardi, C. (2025). What Is the Kyoto Protocol? Definition, History, Timeline, and Status. Investopedia. www.investopedia.com/terms/k/kyoto.asp.

UKEssays (2018). The UN as a Toothless Bull Dog Politics Essay. https://www.ukessays.com/essays/politics/the-un-as-a-toothless-bull-dog-politics-essay.php?vref=1

UNICEF (2023). 25 million Nigerians at High Risk of Food Insecurity in 2023. www.unicef.org/press-releases/25-million-nigerians-high-risk-food-insecurity-2023

UNICEF (2025). Acute Food Insecurity and Malnutrition Rise for Sixth Consecutive Year in World's Most Fragile Regions. www.unicef.org/press-releases/acute-food-insecurity-and-malnutrition-rise-sixth-consecutive-year-worlds-most (Accessed 29/05/25).

WFP (2022). How to End World Hunger: 6 Solutions for Zero Hunger. www.wfpusa.org/articles/how-to-end-world-hunger-6-zero-hunger-solutions/

WFP (2025). WFP Nigeria Country Brief, February 2025. www.ecoi.net/en/document/2129684.html or www.reliefweb.int/attachments/b4415d9e-0a80-4651-ba5b-756bcb308d30/WFP%20Nigeria%20Country%20Brief%20February%202025.pdf (Accessed 27/12/25)

WHO (2014). Food, Genetically Modified. www.who.int/news-room/questions-and-answers/item/food-genetically-modified#:~:text=GM%20foods%20are%20developed%20–%20and,where%20pest%20pressure%20is%20high.

WHO (2024). Hunger Numbers Stubbornly High for Three Consecutive Years as Global Crises Deepen: UN Report. www.who.int/news/item/24-07-2024-hunger-numbers-stubbornly-high-for-three-consecutive-years-as-global-crises-deepen--un-report (Accessed 29/05/25).

Wiemers, M., Bachmeier, M., Hanano, A., Chéilleachair, R. N., Vaughan, A., Foley, C., Mann, H., Weller, D., Radtke, K. and Fritschel, H. (2024). Global Hunger Index How Gender Justice Can Advance Climate Resilience and Zero Hunger. Bonn: Welt Hunger with HILFE, Concern Worldwide & IFHV. www.globalhungerindex.org (Accessed 29/05/25).

3 Impacts of Climate Change and Environmental Degradation on Food Security

Bolaji Hamzat Abdulmujeeb, Gideon Taiwo Otolorin, Akintunde Adewole Adegbite, Mayowa Sikirat Giwa, and Adewumi Yetunde Oriji

3.1 INTRODUCTION

Climate change and environmental degradation present increasingly severe threats to global food security undermining the stability, availability, accessibility, and utilization of food systems. Rising global temperatures, altered precipitation patterns, increased frequency and severity of extreme weather events, and progressive environmental degradation have affected the agricultural productivity and the livelihoods of millions of smallholder farmers negatively, particularly in low-income and climate-vulnerable regions such as sub-Saharan Africa and South Asia. Climate variability has been found to significantly reduce food security by affecting crop, livestock, and fishery sectors, with long-term shifts in climatic conditions increasing the risk of hunger and malnutrition.

Environmental degradation such as soil erosion, declining soil fertility, water scarcity, and biodiversity loss further magnifies the negative impacts of climate change on food systems. For instance, extreme weather events like droughts and floods not only reduce agricultural yields but also damage critical infrastructure and increase post-harvest losses, thereby reducing food availability and increasing food prices.

Recent studies and systematic reviews demonstrate that climate change can alter the suitability of agricultural land, disrupt planting and harvesting cycles, and intensify pest and disease pressures on crops, all of which contribute to declining food production and heightened food insecurity. These stresses are compounded in regions with limited adaptive capacity and institutional support, where economic and social vulnerabilities interact with environmental change to increase the risk of food crises. Moreover, climate-driven impacts such as water scarcity and glacier retreat threaten not just agricultural production but also water resources critical for irrigation and livestock, intensifying competition for finite natural resources, and undermining long-term food security. Given these intersecting pressures, there is an urgent need to integrate climate adaptation measures, sustainable environmental management, and resilient agricultural practices into food security frameworks to safeguard food systems against ongoing and future environmental change.

3.2 CLIMATE CHANGE AND ITS DIRECT IMPACTS ON FOOD SECURITY

The relationship between climate change and food security represents a critical global concern with far-reaching implications. Understanding the individual concepts and the ways in which they interact is essential for developing resilient agricultural systems and safeguarding livelihoods, especially in vulnerable regions. This section provides an overview of the two key concepts climate change

 DOI: 10.1201/9781003650881-3

and food security and examines the fundamental pillars that support food security: availability, access, utilization, and stability.

Climate Change: Climate change refers to the long-term alteration of global or regional climate patterns, particularly shifts in temperature, precipitation, and the frequency of extreme weather events. While the Earth has experienced natural climate fluctuations throughout its history, the current phase of rapid climate change is largely driven by human activities. Key contributors include the burning of fossil fuels, deforestation, industrial emissions, and intensive agricultural practices (IPCC, 2022; NASA, 2022). According to EPA (2021), these activities have significantly increased the concentration of greenhouse gases (GHGs) notably carbon dioxide (CO_2), methane (CH_4), and nitrous oxide (N_2O) which trap heat in the atmosphere and contribute to global warming.

The impacts of climate change are diverse and far-reaching, affecting both natural ecosystems and human systems. Key observed and projected consequences include:

- Rising global temperatures, leading to more frequent heatwaves and altered growing seasons.
- Changes in precipitation patterns, resulting in increased occurrences of droughts and floods.
- Sea level rise, threatening coastal agriculture and freshwater supplies.
- Greater intensity and frequency of extreme weather events, such as hurricanes and cyclones.
- Loss of biodiversity, including essential crop varieties and pollinators critical to food production.

These changes pose serious risks to ecological stability and agricultural systems worldwide. Importantly, the effects of climate change are not evenly distributed. Developing countries particularly those that depend on rain-fed agriculture are disproportionately affected due to limited adaptive capacity, weaker infrastructure, and constrained access to technological and financial resources (Adger et al., 2003; FAO, 2016).

Food Security: The Food and Agriculture Organization (FAO) defines food security as a comprehensive and multidimensional concept, stating that "all people, at all times, have physical, social, and economic access to sufficient, safe, and nutritious food to meet their dietary needs and food preferences for an active and healthy life" (FAO, 1996, 2020). This definition emphasizes not only the availability of food but also the importance of consistent access, cultural acceptability, and nutritional adequacy. Food security is essential to individual well-being, national stability, and sustainable development.

To operationalize this concept, the FAO outlines four interrelated and mutually reinforcing pillars that must be satisfied simultaneously to achieve food security:

1. *Food Availability*: This refers to the consistent presence of sufficient quantities of food through domestic production, imports, food aid, or stockpiles. It depends on factors such as agricultural productivity, land use, water availability, and the resilience of food systems to shocks like climate change and conflict.
2. *Food Access*: Access includes both physical and economic means by which individuals obtain food. Even when food is available, people may not be food secure if they cannot afford it or reach it due to poverty, lack of infrastructure, or social inequalities. Access also encompasses the influence of market systems, purchasing power, and social safety nets.
3. *Food Utilization*: This pillar focuses on the proper biological use of food, requiring not only sufficient caloric intake but also balanced nutrition, safe food handling, clean water, and adequate health care. Poor utilization often linked to malnutrition and disease can result from limited knowledge of nutrition, inadequate food storage, and poor sanitation.

4. *Food Stability*: Stability refers to the reliability and consistency of the other three pillars over time. A population is considered food secure only if it has access to food at all times and is not vulnerable to periodic disruptions caused by environmental, economic, or political factors such as climate shocks, inflation, or conflict.

These four pillars is crucial in assessing and understanding food security through the impact of external stressors particularly climate change and environmental degradation on agricultural systems and human well-being. Only when all four dimensions are robust can food systems be considered secure and sustainable.

3.3 CLIMATE CHANGE AND ITS DIRECT EFFECTS ON FOOD SECURITY

Climate change has become one of the most significant threats to global food security, directly affecting the core pillars of food availability, access, utilization, and stability. The impacts are particularly severe in low- and middle-income countries, where agriculture is often rain-fed, climate-sensitive, and central to livelihoods. One of the most immediate and visible effects of climate change is the alteration of **agricultural productivity**. Rising global temperatures can shorten growing seasons, reduce crop yields, and alter the geographic suitability of key food crops. Heat stress negatively affects plant physiology and growth, while also increasing evapotranspiration rates, leading to greater water demand in areas already facing scarcity.

- **Temperature Changes**

  Global warming is one of the most evident and well-documented consequences of climate change, with significant implications for agricultural productivity. Rising temperatures directly affect both crop and livestock systems, particularly in regions already operating near critical temperature thresholds (Lobell et al., 2011). Many staple crops including wheat, maize, and rice are highly sensitive to temperature variations. Even moderate increases in temperature can accelerate crop maturation, reduce grain-filling periods, and ultimately result in lower yields and diminished nutritional quality (Ortiz et al., 2008). Crops exposed to prolonged heat stress may also experience reductions in photosynthetic efficiency, poor pollination, and increased sterility, especially during critical growth stages. As a result, warming trends threaten not only the quantity of food produced but also its quality and reliability. Higher temperatures can also lead to geographic shifts in crop suitability, forcing the relocation of certain crops to cooler regions. However, such shifts are not always feasible due to land limitations, infrastructure constraints, and socioeconomic factors, leaving many farming communities at heightened risk. Livestock production is similarly affected. Animals such as cattle, sheep, goats, and poultry experience **heat stress** when ambient temperatures exceed their thermal comfort zones. This stress can lead to reduced feed intake, lower weight gain, decreased reproductive efficiency, reduced milk and egg production, and, in severe cases, increased mortality (Schlenker and Roberts, 2009). Heat stress also weakens immune systems, making livestock more vulnerable to diseases and infections. In both crop and animal production systems, the rising frequency of heatwaves compounds these impacts, increasing the unpredictability of yields and incomes. These temperature-induced effects threaten food availability and income security, especially for smallholder farmers who lack access to irrigation, cooling systems, or climate-resilient technologies.
- **Extreme Weather Events**

  Climate change has significantly increased the frequency, duration, and intensity of extreme weather events such as hurricanes, storms, floods, and droughts (IPCC, 2021). These events pose direct and severe threats to food security by disrupting agricultural production, damaging infrastructure, and destabilizing food supply chains. Their growing unpredictability and magnitude make food systems increasingly vulnerable, particularly in regions already exposed to climatic stressors. Extreme rainfall and flooding can destroy standing crops, erode topsoil, and render farmland unusable for subsequent planting seasons. Floodwaters may also carry contaminants, affecting soil fertility and posing health risks to communities reliant on local food production (Brown and Funk, 2008). Conversely, prolonged droughts reduce soil moisture and deplete surface and groundwater resources, severely limiting irrigation potential and leading to widespread crop failure (Cline, 2007). Livestock and fisheries are also impacted, as water scarcity affects grazing lands, fish habitats, and animal health. Beyond direct impacts on food production, extreme weather events often damage the critical infrastructure needed to store, transport, and distribute food. Roads, bridges, storage facilities, and markets are particularly vulnerable to storms and flooding. When this infrastructure is compromised, food distribution is delayed or halted altogether, limiting market access and disrupting local economies (FAO, 2018). These disruptions can trigger sudden spikes in food prices and reduce the availability of essential goods. Poor and marginalized communities are disproportionately affected, as they often lack the financial means to absorb price increases or relocate to unaffected areas. The result is reduced food access, increased reliance on food aid, and heightened food insecurity (Hossain and Islam, 2013).

- **Soil and Water Impacts**

  Water availability and soil fertility are foundational to agricultural productivity and food security. However, both are increasingly threatened by the effects of climate change (Elliott et al., 2015). Shifting rainfall patterns ranging from severe droughts to intense, unseasonal downpours have led to both water scarcity in some regions and excessive water accumulation (waterlogging) in others (Wheeler and Braun, 2013). This growing unpredictability disrupts irrigation planning, planting schedules, and crop management strategies, especially for smallholder farmers with limited adaptive capacity (Lobell et al., 2008). In drought-prone areas, reduced rainfall and declining groundwater levels limit the availability of water for irrigation, leading to reduced crop yields and, in some cases, complete harvest failure. Conversely, in areas experiencing increased precipitation, waterlogging can damage root systems, inhibit nutrient uptake, and foster the spread of plant diseases, all of which negatively affect crop performance.

  In tandem with water stress, **soil degradation** has emerged as a serious constraint on agricultural productivity. Climate-related factors such as heavy rainfall, wind, and high temperatures accelerate erosion, deplete soil nutrients, and contribute to salinization and desertification. These processes are often exacerbated by unsustainable land use practices, including overgrazing, deforestation, and improper tillage methods (FAO, 2016). Degraded soils lose their ability to retain moisture and nutrients, undermining crop growth and increasing dependency on costly fertilizers and irrigation. In areas already characterized by low soil fertility, these impacts can be especially devastating leading to persistent low yields, land abandonment, and increased food insecurity (Pretty et al., 2018).

  Soil degradation and loss of arable land due to erosion, salinization, and desertification processes accelerated by climate change further undermine food production. Rising sea levels also threaten coastal agricultural zones, contaminating freshwater and fertile lands with saltwater intrusion.

- **Pests and Diseases**

  Climate change has contributed to the increased proliferation of agricultural pests and diseases, posing a serious threat to crop yields and global food security. Rising temperatures, coupled with altered precipitation patterns, create favorable conditions for the survival, reproduction, and geographic expansion of numerous pest species and plant pathogens (Cameron et al., 2013). Insects such as locusts, armyworms, and aphids thrive in warmer environments, and their breeding cycles accelerate under high temperatures.

These pests can cause widespread destruction of crops, leading to significant reductions in food availability, particularly in regions that rely heavily on agriculture for both subsistence and income (FAO, 2017). Similarly, changes in humidity and temperature can intensify the occurrence of plant diseases, including bacterial blight, viral infections, and fungal outbreaks such as mildew and rust (Pimentel et al., 2005). One notable example is the increasing prevalence of wheat rust particularly stem rust (*Puccinia graminis f. sp. tritici*) which has been linked to shifting climate conditions and has severely impacted wheat yields in parts of Africa, the Middle East, and Asia (Singh et al., 2011). These outbreaks not only diminish harvests but also reduce the nutritional and economic value of crops. The impact of these biological stressors is most profound in low-income regions, where limited access to pest control technologies, resistant crop varieties, and effective monitoring systems leaves farmers particularly vulnerable (Fischer et al., 2002). As pests and diseases continue to expand into new areas, their unpredictability adds another layer of complexity to climate-induced agricultural risk.

Post-harvest losses are another direct consequence. Higher temperatures and increased humidity can promote the spread of pests, fungi, and diseases, leading to spoilage during storage and transport. These losses contribute to food scarcity, drive up prices, and reduce the nutritional quality of available food.

- **Economic and Social Impact**

The direct effects of climate change on agricultural productivity have profound economic and social implications, particularly in developing countries where a significant proportion of the population relies on agriculture for both income and food (Thornton et al., 2009). As climate-induced stresses such as droughts, floods, and temperature extremes lead to reduced crop and livestock yields, the livelihoods of farmers especially smallholders are severely impacted. Declining agricultural output translates to lower household incomes, increased poverty rates, and reduced access to food (Alderman and Yemtsov, 2016). In many low-income and agrarian economies, agriculture forms the backbone of the national economy. When climate events disrupt farming activities, the ripple effects extend beyond individual households to affect local markets, employment rates, and food availability. Financial instability among farmers often leads to decreased investments in inputs such as seeds, fertilizers, and irrigation, further lowering productivity and perpetuating a cycle of vulnerability (Binswanger-Mkhize, 2012).

Climate change also contributes to **food price volatility**, a critical social and economic challenge. As agricultural production becomes increasingly uncertain due to erratic weather patterns, supply shortages push prices higher and create unpredictable food markets. These fluctuations disproportionately affect the poor, who spend a larger share of their income on food. The inability to afford a nutritious diet increases the risk of malnutrition and other health problems, particularly among children, pregnant women, and other vulnerable groups (FAO, 2020).

In extreme cases, the economic stress induced by rising food prices and dwindling agricultural returns can contribute to **social unrest**. Historical evidence has shown that spikes in food prices can exacerbate tensions, especially in regions already grappling with inequality, unemployment, and weak governance. Food insecurity fueled by climate change can thus become a trigger for conflict, displacement, and broader societal instability (Maxwell et al., 2017).

Furthermore, **fisheries and aquaculture**, which are critical sources of protein for millions, are affected by warming waters, ocean acidification, and changing marine ecosystems. Shifts in fish distribution and declines in fish stocks threaten food supply in many coastal and island communities. These direct impacts of climate change not only reduce food availability but also undermine access and utilization by pushing up food prices, reducing incomes from farming, and compromising the nutritional quality of diets. The cumulative effects weaken food system **stability**, especially for vulnerable populations with limited capacity to adapt.

- **Vulnerable Populations**

  The impacts of climate change on food security are not experienced equally across all communities. Certain populations particularly those in developing countries are disproportionately affected due to their high dependence on climate-sensitive sectors such as rain-fed agriculture, coupled with limited adaptive capacity (Adger et al., 2003). These populations often lack access to critical resources, infrastructure, technology, and institutional support systems that are essential for coping with the adverse effects of climate variability (Sperling and Seligman, 2017). Rural communities in low-income regions face heightened exposure to food insecurity as their livelihoods and local food systems are closely tied to unpredictable climatic conditions. Crop failures, livestock losses, and water scarcity can rapidly undermine food availability and economic stability in such settings. The lack of diversification in income sources further exacerbates the vulnerability of these households. Gender inequality compounds the issue. Women play a central role in food production, processing, and household nutrition, particularly in sub-Saharan Africa, South Asia, and other developing regions (Meinzen-Dick et al., 2014). Despite their contributions, women often have limited access to land, credit, agricultural extension services, and decision-making platforms. As a result, the burden of climate-related food insecurity falls more heavily on them. When crops fail or livestock perish, it is frequently women who must secure alternative food sources and manage household needs under increasingly difficult conditions (Doss, 2018). Additionally, vulnerable population groups such as children, the elderly, and individuals with health challenges face heightened risks due to food insecurity. Malnutrition and undernutrition caused by reduced food availability and diversity can lead to long-term health and developmental problems, particularly in young children during critical growth stages (Vineis et al., 2011). In older populations, weakened immune systems increase the risk of illness when adequate nutrition is not maintained.

3.4 ENVIRONMENTAL DEGRADATION AND ITS IMPLICATIONS (ROLE IN FOOD SECURITY)

Environmental degradation encompassing deforestation, soil erosion, pollution, and biodiversity loss represents a profound and escalating threat to global food security. These interlinked processes compromise agricultural productivity, destabilize critical ecosystem services, and adversely impact human health and well-being (Richardson, 2010). Although traditional monoculture-based farming systems have historically driven gains in food production, their ecological inefficiencies and dependence on non-renewable inputs render them unsustainable over the long term (Çakmakçı et al., 2023). In light of mounting environmental challenges, there is an urgent need to transition toward diversified, ecologically sound farming systems that enhance both resilience and sustainability. Achieving a sustainable food system in the 21st century necessitates moving beyond conventional agricultural paradigms that degrade natural resources, toward integrated policy frameworks that holistically address environmental, economic, and social dimensions (Wahbeh et al., 2022). A nuanced understanding of the complex interdependencies between environmental degradation and food security is essential. Environmental stressors can disrupt crop and livestock systems through both direct and indirect mechanisms, thereby threatening the foundational pillars of food security availability, access, utilization, and stability. These cascading impacts not only reduce agricultural output but also compromise nutritional quality and public health outcomes (Ramp, 2014).

Ensuring global food security amid a rapidly expanding population is increasingly jeopardized by the accelerating pace of environmental degradation, which threatens the productivity, resilience, and stability of agricultural systems worldwide. The widespread reliance on unsustainable agricultural practices driven largely by the urgency to meet rising food demands is placing immense pressure on ecosystems, thereby compromising the long-term viability of food production (Díaz-Rodríguez et al., 2025). This intensifying crisis underscores the critical interdependence between environmental health and agricultural output, calling for a comprehensive understanding of how degradation processes affect the four key dimensions of food security: availability, accessibility, affordability, and stability (Çakmakçı et al., 2024). The challenge is particularly acute in developing nations, where food systems are highly dependent on natural resources and often lack the infrastructure and institutional capacity required for resilience. In these contexts, environmental degradation not only diminishes agricultural productivity but also obstructs broader poverty alleviation and sustainable development efforts (Paramesh et al., 2023). With the global population projected to surpass 10 billion by 2050, an estimated 70% increase in agricultural output will be required to meet future food needs placing even greater strain on land, water, and energy resources (Chakraborti et al., 2023). This scenario highlights the urgent need to transition toward sustainable agricultural models that prioritize environmental integrity and resource conservation. Such systems are essential not only for maintaining productivity but also for safeguarding ecosystem services, supporting climate adaptation, and securing long-term food security in an increasingly volatile climate (Zarei et al., 2021). The subsequent sections examine the multifaceted implications of environmental degradation on global food security.

3.4.1 Land Degradation (Desertification, Erosion, and Deforestation)

Land degradation, encompassing desertification, soil erosion, and deforestation poses a critical threat to global agricultural productivity and food security. Desertification, characterized by the transformation of fertile land into arid, unproductive terrain, significantly reduces the availability of arable land for crop cultivation and livestock grazing, directly undermining food availability and agricultural resilience (Ansari et al., 2023). Soil erosion, driven by wind and water, is particularly detrimental, as it strips away the nutrient-rich topsoil layer essential for crop growth. The loss of organic matter, soil nutrients, and water retention capacity leads to lower yields and heightened vulnerability to crop failure. Similarly, deforestation often resulting from agricultural expansion, logging, and urbanization exacerbates land degradation by eliminating forest cover that provides critical ecosystem services such as carbon sequestration, hydrological regulation, and soil stabilization (Kertolli et al., 2024). Driven by both natural processes and anthropogenic pressures, land degradation contributes to a persistent decline in soil health, agricultural output, and ecosystem functionality, thereby threatening the sustainability of food systems globally (Dhakal and Rai, 2020). It is closely intertwined with climate vulnerability and socioeconomic marginalization, disproportionately affecting smallholder farmers and low-income rural populations who depend directly on land resources for their livelihoods (Kieti et al., 2016; Singh et al., 2022). Current estimates indicate that land degradation affects the livelihoods of over 1.5 billion people and leads to the loss of approximately 15 billion tons of fertile soil annually (Abebaw, 2019; Arora, 2019). The intensification of agricultural production particularly through monoculture cropping, excessive chemical input use, and inefficient irrigation has accelerated degradation across vast landscapes (Wolde and Bade, 2020). The economic cost of land degradation in agriculture is projected to reach approximately US$500 billion per year, underscoring the urgent need for comprehensive strategies to combat desertification, rehabilitate degraded lands, and promote sustainable land management (SLM) practices (Pacheco et al., 2018).

3.4.2 Soil Quality and Fertility Loss

Soil degradation is not an unavoidable consequence of agricultural intensification and socioeconomic advancement; rather, it signals the failure to implement effective and SLM strategies (Pásztor, 2021). The primary challenge lies in balancing the imperative to increase food production with the necessity of conserving and restoring soil health (Abenu et al., 2023). Tackling soil degradation is vital not only for sustaining agricultural productivity but also for preserving biodiversity and maintaining essential ecological functions that underpin long-term ecosystem resilience (Bachri et al., 2021). Declining soil quality and fertility constitute a critical aspect of environmental degradation, with direct consequences for global food security. This degradation is frequently driven by unsustainable agricultural practices, including excessive tillage, overuse of chemical inputs, and monoculture systems, which lead to the depletion of essential macronutrients such as nitrogen, phosphorus, and potassium as well as trace micronutrients. The loss of these nutrients compromises crop yields and diminishes the nutritional value of agricultural produce (Irawan and Antriyandarti, 2021). Structural deterioration of soils, manifested in compaction, impaired water infiltration, and restricted root development, exacerbates fertility decline and significantly impairs productive capacity. Therefore, these challenges demands a transition toward integrated soil fertility management and conservation-oriented farming systems that enhance soil health and ensure agricultural sustainability.

Intensive farming systems characterized by monocultures, continuous cropping, and excessive tillage have been major contributors to the accelerated degradation of soils. These practices lead to the depletion of soil organic matter, disrupt the balance of microbial communities essential for nutrient cycling, and increase susceptibility to erosion (El-Naggar et al., 2019). In densely populated agrarian economies such as India, the implications of soil degradation are particularly acute. Despite supporting a significant portion of the global population and livestock, India accounts for only a small share of the world's arable land (Gomiero, 2016).

Efforts to sustain food self-sufficiency have driven a marked rise in chemical fertilizer use; however, while this strategy temporarily improves crop yields, it has also exacerbated environmental degradation through nutrient leaching, soil acidification, and biodiversity erosion.

The degradation of soil health and fertility directly undermines agricultural productivity and compromises food and nutritional security, especially in regions where agriculture remains the primary livelihood source.

Soil degradation, commonly characterized by the loss of organic carbon, essential nutrients, and soil structure, critically undermines agricultural productivity by impairing nutrient retention, water infiltration, and root penetration. These impacts are particularly pronounced in developing countries, where limited financial and institutional capacity restricts the adoption of adaptive and restorative land-use practices. In response to this growing crisis, SLM strategies have gained recognition as essential tools for reversing soil degradation and restoring ecosystem services. Conservation agriculture including no-till systems, crop rotation, cover cropping, integrated crop–livestock systems, and carbon sequestration interventions such as afforestation, reforestation, and improved grazing management offers promising pathways to enhance soil health and long-term resilience (Adão et al., 2025). Nevertheless, the widespread implementation of these practices remains hindered by socioeconomic barriers, inadequate institutional support, and policy gaps, highlighting the need for coordinated interventions that facilitate farmer engagement, provide economic incentives, and create enabling policy environments. The excessive and unregulated use of synthetic fertilizers once promoted to boost short-term crop yields has instead contributed to long-term soil degradation, nutrient imbalances, environmental contamination, and the erosion of food quality (Arora et al., 2024). Compounding this, the decline in soil biodiversity, especially the depletion of microbial communities essential for nutrient cycling, organic matter decomposition, and natural disease regulation, exacerbates soil fertility losses. Globally, approximately 40% of croplands suffer from erosion, nutrient depletion, and overgrazing, posing a significant threat to global food security (Gomiero, 2013). Reversing this trend necessitates a holistic and integrated approach that combines sustainable agricultural practices, ecological restoration strategies, and robust policy frameworks. Such efforts are crucial not only for reclaiming degraded lands and bolstering agricultural productivity but also for securing the resilience and sustainability of food systems in the face of ongoing environmental and climatic pressures (Paramesh et al., 2023).

3.4.3 Pollution (Air, Water, and Soil) and Its Effects on Food Safety

Pollution, encompassing air, water, and soil contamination, poses a significant threat to food safety and security, with cascading effects on human health and ecosystem integrity. Air pollutants, such as particulate matter, ozone, and nitrogen oxides, can directly contaminate crops, reduce photosynthetic activity, and alter plant growth patterns, leading to yield losses and diminished nutritional value (Wang, 2023). Water pollution, stemming from industrial discharges, agricultural runoff, and domestic sewage, introduces a myriad of contaminants into aquatic ecosystems, including heavy metals, pesticides, pathogens, and excess nutrients. These pollutants accumulate in aquatic organisms, posing risks to human health through the consumption of contaminated fish and seafood (Li et al., 2021). Soil pollution, arising from industrial activities, mining operations, and improper waste disposal, introduces toxic substances into the soil environment, including heavy metals, organic pollutants, and radioactive materials. These pollutants can be taken up by plants, contaminating food crops and posing risks to human health through dietary exposure. Most contaminants originate from human activities such as industrial processes and mining, poor waste management, unsustainable farming practices, accidents ranging from small chemical spills to accidents at nuclear power plants, and the many effects of armed conflicts (FAO and UNEP, 2021). Soil contamination, in particular, hampers the achievement of Sustainable Development Goals (SDGs), including achieving zero hunger, ending poverty, ensuring healthy lives and human well-being, halting and reversing land degradation and biodiversity loss, and making cities safe and resilient (FAO and UNEP, 2021). The combined effects of air, water, and soil pollution present a significant and escalating threat to food safety, with far-reaching implications for public health, ecosystem integrity, and agricultural sustainability. Exposure to environmental contaminants particularly heavy metals, persistent organic pollutants (POPs), and hazardous industrial byproducts has been linked to an increased risk of carcinogenic diseases, endocrine disruption, and other chronic health outcomes (Lu et al., 2015). These threats are particularly acute in regions where regulatory oversight is weak and environmental monitoring systems are underdeveloped. Addressing these multifaceted challenges requires integrated and adaptive strategies that encompass pollution prevention, source control, environmental remediation, and robust regulatory frameworks aimed at protecting both food systems and human health (FAO and UNEP, 2021; Ajibade et al., 2021).

The infiltration of pollutants into terrestrial and aquatic ecosystems disrupts critical ecological functions, degrades natural habitats, and interferes with essential environmental services. Anthropogenic activities such as industrial emissions, intensive agriculture, fossil fuel combustion, and inadequate waste management are primary drivers of air, water, and soil contamination, contributing to toxic exposures in both humans and wildlife (Bashir et al., 2020; Gamage et al., 2021). Water bodies, in particular, have become increasingly contaminated with inorganic and organic chemicals originating from petrochemical industries, mining operations, vehicle emissions, and urban runoff (Borah et al., 2020). In the case of soil pollution, hazardous substances from pesticides, fertilizers, heavy metals, and untreated industrial waste accumulate in agricultural soils, posing persistent risks to biodiversity, crop safety, and groundwater quality.

In response, there is an urgent need to scale up the deployment of sustainable remediation technologies. Biological remediation especially bioremediation and phytoremediation offers an ecologically sound alternative to conventional physico-chemical approaches, enabling the degradation, transformation, or immobilization of a wide range of organic and inorganic pollutants within the environment (Gupta and Gandhi, 2023; Mir et al., 2023). By prioritizing ecosystem-based management and integrating pollution control into agricultural and industrial policies, it becomes possible to mitigate the cumulative risks of environmental contamination and secure the foundations of food safety, public health, and environmental sustainability for future generations (Santos et al., 2025).

3.4.4 Biodiversity Loss and Agricultural Sustainability

Biodiversity loss, characterized by the decline in the variety of life at all levels of biological organization, poses a profound threat to agricultural sustainability and food security. The global decline in biodiversity diminishes the resilience of agricultural systems, making them more vulnerable to pests, diseases, and environmental stresses (Vikas and Ranjan, 2024). Agricultural activities contribute about 10% of GHG emissions and are one of the main causes of biodiversity loss (García-Huante et al., 2024). Monoculture farming practices, which involve the cultivation of a single crop species over large areas, reduce genetic diversity and create homogenous landscapes that are susceptible to widespread outbreaks of pests and diseases. The loss of pollinators, such as bees, butterflies, and other insects, disrupts pollination processes, leading to reduced crop yields and diminished fruit and seed production. The loss of natural habitats, such as forests, wetlands, and grasslands, reduces the availability of wild relatives of crops, which serve as important sources of genetic diversity for crop improvement and adaptation to changing environmental conditions. Traditional agricultural techniques, such as chemical fertilizers, pesticides, fungicides, and herbicides, enable the protection of crop plants against pathogens and ensure better yield (Antoszewski et al., 2022). However, these agricultural practices are associated with significant environmental side effects, including soil and water pollution and the loss of biodiversity. Conserving and promoting biodiversity in agricultural landscapes is essential for enhancing ecosystem services, improving crop productivity, and ensuring the long-term sustainability of food production systems (Kocira and Staniak, 2025). Agricultural landscapes can provide habitats for biodiversity, promote connectivity between protected areas, and increase the capacity of species to respond to environmental threats (Wanger et al., 2020). Supporting biodiversity in agriculture can promote more sustainable food production, improve nutrition and livelihoods, and strengthen the resilience of agricultural systems (Udawatta et al., 2019).

Harnessing biodiversity for sustainable agriculture requires more than a return to traditional practices it demands a nuanced understanding of how agrobiodiversity contributes to improved nutrition, enhanced food security, and ecological resilience (Pandey et al., 2022). Biodiversity-rich agroecosystems offer multiple benefits beyond agricultural output, including critical ecosystem services such as nutrient cycling, climate regulation, pest suppression, and pollination. By enriching biodiversity within production systems, farmers can enhance biological integrity and promote ecosystem health, ultimately contributing to more sustainable and resilient food systems (Maitra et al., 2021). Reducing the intensity of agricultural management practices particularly the excessive use of chemical fertilizers and pesticides is key to conserving biodiversity at the local scale (Gonthier et al., 2014). Ecological engineering approaches, such as planting pest-repellent crops or preserving habitat refuges for beneficial insects, have been increasingly promoted as nature-based solutions for sustainable pest control (Lu et al., 2014). This strategy reflects a broader agroecological perspective, which recognizes biodiversity not only as a passive component to be protected but also as an active driver of ecosystem functionality and agricultural sustainability (Altieri, 1999; Cozim-Melges et al., 2024). Furthermore, biodiversity conservation in agriculture requires spatial thinking that transcends the field level to include the configuration, quality, and connectivity of agricultural and semi-natural habitats across broader landscapes (Jeanneret et al., 2021). Agricultural landscapes, when properly managed, can support wild biodiversity without compromising and in some cases enhancing agricultural productivity and rural livelihoods (Scherr and McNeely, 2008). Traditional farming systems offer valuable lessons in this regard, as they often reflect centuries of co-evolution between farmers and local ecosystems, demonstrating how biodiversity can be effectively integrated into food production (Altieri, 2004).

However, barriers remain. Many farmers tend to optimize short-term profits without accounting for the long-term ecological impacts of their decisions, such as biodiversity loss and resource degradation (Brunetti et al., 2019). Addressing this challenge requires targeted policies, education, and incentives that align farmer behavior with biodiversity goals. Integrating biodiversity conservation into agricultural policy and practice is not only critical for environmental health but also for ensuring future food security and socioeconomic stability (Singh, 2024). The conservation of natural resources especially soil and water remains fundamental to the environmental and economic sustainability of both agriculture and forestry systems (Nalluri and Karri, 2023). Moving beyond the outdated paradigm of separating conservation from production areas, modern agricultural strategies increasingly embrace multifunctional landscapes that deliver both ecological and agronomic benefits. Ultimately, agrobiodiversity is a cornerstone of sustainable agriculture, and its preservation and enhancement must be at the heart of any meaningful agroecological transition (Maitra et al., 2021).

3.4.5 Water Scarcity and Pollution (Impact on Irrigation and Water Resources)

Water scarcity intensified by the combined pressures of climate change, pollution, and over-extraction poses a critical challenge to sustainable irrigation and effective water resource management, with direct implications for global food security. The growing competition for limited water resources among agriculture, industry, and domestic sectors is particularly acute in arid and semi-arid regions, where water availability is already constrained and agriculture heavily depends on irrigation. Unsustainable irrigation practices have led to the widespread depletion of groundwater aquifers, reducing long-term water availability for agriculture and contributing to ecological risks such as land subsidence and saltwater intrusion in coastal zones (Zarei et al., 2021). These hydrological stresses threaten not only food production but also the stability of ecosystems and human settlements. Inefficient irrigation techniques, such as flood irrigation, lead to water losses through evaporation, runoff, and deep percolation, reducing the amount of water available for crop production. Climate change is altering precipitation patterns, increasing the frequency and intensity of droughts and floods, and exacerbating water scarcity in many regions, with profound implications for agricultural production.

Water is a critical resource, and its sustainable management is essential for addressing the challenges posed by climate change, population growth, and the rising demands of agricultural and industrial activities (Lakhiar et al., 2024). Widespread water pollution and substandard water infrastructure exacerbate water scarcity and complicate government efforts to manage the water demand of the agricultural sector (Bahn et al., 2021). Addressing water scarcity requires a multifaceted approach that includes improving water use efficiency, promoting water conservation, and investing in water infrastructure. Furthermore, the use of non-conventional water resources, such as recycled water and saline groundwater, has been suggested as an alternative source to ensure the sustainable production of food crops (El-Hendawy et al., 2022). The challenge for the agricultural sector is to produce more food with less water, particularly as concerns about water scarcity and food security intensify (Soothar et al., 2021). Also, there is a need for major improvements in water resource use and irrigation technology and management (Hamdy et al., 2003). Agriculture will always be considered one of the major users of water resources and is still required to be more efficient especially in arid and semiarid climates (Ahmad et al., 2022). Ultimately, the sustainable management of water resources is crucial for ensuring food security, protecting ecosystems, and promoting sustainable development in the face of increasing water scarcity and pollution (Ortuani et al., 2019; Lakhiar et al., 2024). The agricultural sector utilizes a significant portion of the available freshwater globally (Abioye et al., 2022). Agriculture relies heavily on water, with approximately 70% of the world's freshwater used for irrigation to cultivate 25% of arable land (Lakhiar et al., 2024). Municipal and industrial water demands are expected to increase faster than those of agriculture, potentially crowding out agricultural allocations (Dubois, 2011). Water recycling and reuse technologies offer possibilities for more extensive use of water, depending on cost (Vaux Jr, 2011). To effectively tackle water scarcity and ensure sustainable agricultural production, an integrated approach is essential, combining technological innovations, policy changes, and community participation. This approach requires the coordinated effort and cooperation of governments, farmers, scientists, and local communities (Reinman, 2015). To ensure sufficient water availability for plant growth, excessive volumes of ground and river waters are routinely withdrawn, leading to critically depleted water supplies in many parts of the world and resulting in international concerns about food and water security (Gallo et al., 2021). Agriculture relies heavily on irrigation, so efficient water management is essential for meeting the rising demand for food and reducing poverty and hunger sustainably (Mekonnen and Gerbens-Leenes, 2020). The focus should shift from merely developing water resources to actively managing them, incorporating strategies for conservation and mindful utilization (Tewari and Tewari, 2003).

3.4.6 Over-Exploitation of Natural Resources (Fisheries, Forests, and Freshwater)

Over-exploitation of natural resources, encompassing fisheries, forests, and freshwater sources, poses significant threats to environmental sustainability and food security. Unsustainable fishing practices, deforestation, and excessive water extraction disrupt ecosystems, diminish biodiversity, and compromise the long-term availability of these vital resources. Over-exploitation of resources can lead to severe consequences, including resource depletion, habitat destruction, and ecosystem collapse. In fisheries, overfishing depletes fish stocks, disrupts marine ecosystems, and threatens the livelihoods of communities that depend on fishing for their sustenance and income (Tsolakis et al., 2019). Over-extraction of water from rivers, lakes, and aquifers can lead to water scarcity, reduced water quality, and ecological damage. Over-exploitation of forests leads to deforestation, which has far-reaching consequences for climate change, biodiversity loss, and soil erosion. The intricate interplay between resource over-exploitation and food security is particularly evident in small-scale fisheries, which contribute significantly to the diets and livelihoods of millions of people worldwide. However, unsustainable fishing practices and habitat degradation threaten the viability of these fisheries, jeopardizing food security and economic stability. Similarly, the dependence on forests for fuelwood, timber, and non-timber forest products can lead to deforestation and forest degradation, impacting biodiversity, soil fertility, and water resources, ultimately affecting agricultural productivity and food security. Addressing the over-exploitation of natural resources requires a holistic approach that incorporates sustainable management practices, effective regulations, and community participation. For example, implementing sustainable fishing practices, such as catch limits, gear restrictions, and marine protected areas, can help conserve fish stocks and protect marine ecosystems. Reforestation and afforestation initiatives, coupled with sustainable forest management practices, can help restore degraded forests and promote biodiversity.

Moreover, promoting alternative livelihood options for communities that depend on over-exploited resources can reduce pressure on these resources and enhance economic resilience. Sustainable use of natural resources is fundamental to the socioeconomic life of every nation (Hassan et al., 2025). Unsustainable exploitation has serious environmental, social, and economic consequences, affecting human well-being. The pursuit of ecological, economic, and social sustainability is vital to ensuring the long-term health and productivity of fisheries (Rice, 2017). Governments must design policies that offer incentives to protect resource systems and impose penalties for free-riding. Without a common framework to organize findings, isolated knowledge does not cumulate (Ostrom, 2009). Managing natural resources sustainably and safeguarding vital ecosystems requires understanding the complex interactions between the biological, environmental, political, managerial, and governance aspects involved (Nilsson et al., 2019). A sustainable livelihood can cope with and recover from stresses and shocks, maintain or enhance its capabilities and assets, and provide sustainable livelihood opportunities for the next generation (Agbeja et al., 2021). Rational utilization and proper management of forest resources are the most viable ways to prevent mass destruction of forests and large-scale species extinction (Kumari et al., 2019).

3.5 REGIONAL AND SOCIOECONOMIC VARIABILITY

The 21st century has ushered in an era of intensified environmental disruption, bringing into sharp focus the concept of vulnerability defined as the propensity to suffer harm coupled with limited capacity to anticipate, cope with, and recover from environmental stressors (Soomro et al., 2025). In the context of food security, vulnerability is deeply shaped by regional disparities and socioeconomic inequalities, which influence the degree to which communities can prepare for, absorb, and adapt to the multifaceted impacts of climate change and environmental degradation.

Environmental changes particularly those induced by climate variability do not affect all regions or populations uniformly. Instead, they exacerbate existing inequalities, disproportionately impacting low-income countries and marginalized populations that are heavily dependent on climate-sensitive livelihoods such as subsistence farming, livestock rearing, and small-scale fisheries (Kumar et al., 2025). These communities often reside in ecologically fragile regions such as arid lands, coastal zones, and mountainous terrains where environmental thresholds are already being pushed to their limits. The combined pressures of rising temperatures, altered rainfall patterns, increased frequency of droughts and floods, and ongoing land degradation have led to reduced agricultural productivity, food price volatility, and recurring food crises. Regions such as sub-Saharan Africa, Southeast Asia, and parts of Latin America face particularly acute vulnerabilities. In sub-Saharan Africa, for instance, the majority of the population depends on rain-fed agriculture, which is highly susceptible to erratic rainfall and prolonged dry spells. Likewise, densely populated delta regions in Southeast Asia are confronting increased salinity intrusion, sea-level rise, and displacement of agricultural zones all of which compromise food production and water security (IPCC, 2022; World Bank, 2023). In these areas, even minor climatic shocks can translate into major food insecurity events due to the absence of buffering infrastructure and adaptive safety nets.

In addition to environmental stressors, a confluence of socioeconomic factors further compounds vulnerability. Widespread poverty, underdeveloped rural infrastructure, limited access to education and health care, and weak governance mechanisms restrict the ability of communities to respond effectively to environmental change. In many developing nations, inadequate road networks, storage facilities, and irrigation systems hinder efficient food production, distribution, and market access (HLPE, 2020). Smallholder farmers often lack access to financial services, crop insurance, and agricultural extension services, making it difficult to recover from shocks or adopt improved practices. These systemic barriers perpetuate a cycle of poverty and ecological degradation, creating entrenched vulnerabilities that undermine sustainable food systems.

Moreover, gender inequality is a critical yet often overlooked factor that significantly affects food security outcomes. Women in rural communities frequently serve as key actors in food production, processing, and household nutrition. Despite their central role, they are routinely excluded from land ownership, credit schemes, decision-making forums, and training programs (UN Women, 2023b). As a result, their capacity to adapt to environmental and economic shocks is severely constrained. This disparity not only weakens household resilience but also has intergenerational consequences, particularly for children's nutrition and health. Addressing gender disparities through inclusive agricultural policies, targeted training, and access to resources is therefore fundamental to enhancing the adaptive capacity of vulnerable populations. Ultimately, reducing regional and socioeconomic disparities in vulnerability requires integrated strategies that couple climate adaptation with poverty alleviation, infrastructural development, and social protection. This includes expanding access to education, enhancing market linkages, investing in resilient agricultural technologies, and strengthening institutional frameworks that support marginalized populations. Tailoring interventions to local contexts while ensuring that policies are inclusive, participatory, and gender-responsive will be essential for building robust food systems that can withstand and recover from future environmental challenges.

3.6 ECONOMIC IMPACTS ON FOOD PRODUCTION AND DISTRIBUTION

The escalating impacts of climate change and environmental degradation present multifaceted challenges to global food security, exerting considerable economic pressure on food production and distribution systems (Haro et al., 2021). These pressures manifest through disruptions in both global and local food markets, increased costs associated with climate-induced disasters, the intricate dynamics of supply chains and trade, and the imperative need for financial strategies to manage climate impacts on food systems (Braun, 2020). Climate variability, marked by rising temperatures and altered precipitation patterns, has historically been a principal source of fluctuations in global food production, particularly in developing countries, and these fluctuations are expected to intensify with continued climate change (Tripathi et al., 2016). Ecological disruptions caused by climate change will likely create unprecedented challenges in the future that will put at risk the ability of people to access sufficient, safe, and nutritious foods locally (Ferguson et al., 2022). Agronomists face the dual challenge of enhancing food production sustainably amidst a changing climate, while simultaneously reducing climate-forcing factors (Khalid, 2020). The intersection of climate change and agriculture necessitates a comprehensive understanding of the vulnerabilities inherent in current food systems and the implementation of proactive measures to mitigate economic risks (Garnett et al., 2013). The following are the economic impacts of climate change on food production and distribution.

3.6.1 Disruptions in Global and Local Food Markets

Climate change-induced variability and extreme weather events are significantly disrupting global and local food markets, resulting in price volatility, reduced yields, and impaired access to food, especially for vulnerable populations (Wheeler and Braun, 2013). Alterations in temperature and precipitation patterns are already affecting the suitability of growing conditions for various crops, leading to geographical shifts in agricultural production zones (Pratap et al., 2024). Extreme weather events, such as droughts, floods, and cyclones, are becoming more frequent and intense, causing widespread crop failures and livestock losses, thereby destabilizing food supply chains (Arora, 2019; Dias et al., 2024). The increasing frequency of such extreme events are having a devastating effect on agriculture, causing unpredicted changes in precipitation and increases in GHG emissions (Shahzad et al., 2021). These disruptions trigger changes in price expectations with cascading effects beyond the region in crisis (Carriquiry et al., 2022). The impacts are particularly severe in regions highly dependent on rain-fed agriculture, where even slight deviations from normal rainfall patterns can lead to substantial reductions in crop yields. Moreover, the high frequency of flooding events can diminish the land's capacity to catch rainwater (Kingra et al., 2018). Climate change can compromise the integrity of agroecosystems and health, negatively impacting food security by modifying crop types, increasing weed invasion and pests, and altering plant physiological and biochemical processes (Semeraro et al., 2023). These effects reverberate through local markets, leading to food price inflation and reduced availability, disproportionately affecting low-income households that spend a larger portion of their income on food. Globally, these localized disruptions aggregate to create instability in international food markets, affecting trade flows and exacerbating food insecurity in import-dependent countries. Furthermore, climate change-induced shifts in agricultural productivity can lead to trade disputes and geopolitical tensions as countries compete for scarce resources, further destabilizing global food markets (Neyra et al., 2025).

3.6.2 Economic Impacts of Climate-Induced Disasters on Agricultural Systems

The escalating economic costs of climate-induced disasters and extreme weather events present serious challenges to global food production systems and demand urgent investments in adaptation and resilience-building. In 2018 alone, natural disasters accounted for approximately USD 225 billion in economic losses worldwide (Arora, 2019). These losses include both direct damages such as the destruction of crops, livestock, and agricultural infrastructure and indirect costs, such as supply chain disruptions, elevated insurance premiums, and increased spending on emergency relief and recovery efforts (Sivakumar, 2020). Climate extremes also degrade soil quality, reduce water availability, and promote the spread of pests and diseases, further eroding agricultural productivity. The cumulative impact of these disruptions is substantial, contributing to estimated losses of 30–50% in global agricultural productivity (Bibi and Rahman, 2023). Smallholder farmers and rural communities are disproportionately affected, as they often lack the financial and institutional buffers to absorb climate shocks. Rising production costs, falling farm incomes, and fluctuating food prices undermine food security and economic stability in these vulnerable regions. Simultaneously, the growing frequency and severity of extreme weather events place additional strain on national budgets, diverting public funds from long-term development programs toward disaster response and infrastructure repair.

Beyond agriculture, climate-related disasters disrupt broader systems including transportation, energy, and trade which compound their effects on food production and distribution networks. Tackling these interlinked challenges requires a multi-pronged strategy. Key priorities include investing in climate-resilient infrastructure, enhancing early warning systems, implementing risk transfer mechanisms (e.g., climate insurance), and promoting policies that support SLM and reduce GHG emissions.

Moreover, the compounding effects of multiple extreme events such as storms followed by landslides can dramatically worsen agricultural losses and diminish farmers' adaptive capacity (Manescu et al., 2025). Agrometeorological interventions offer a promising pathway to mitigate these risks. Strengthening national agrometeorological networks, developing efficient climate data systems, ensuring timely dissemination of early warnings, and fostering research on climate risk trends are essential strategies to reduce the vulnerability of agriculture to climate extremes (Sivakumar, 2020). A robust agrometeorological framework, coupled with sound policy and investment, will be indispensable in building climate-resilient food systems.

3.6.3 Climate Change and the Vulnerability of Food Supply Chains and Trade

Supply chains and trade play a critical role in ensuring food security by connecting producers to consumers and facilitating the movement of food from surplus to deficit regions, but are increasingly vulnerable to climate change impacts. Climate change induced disruptions to agricultural production can have cascading effects throughout the entire supply chain, leading to bottlenecks, delays, and increased transportation costs. For example, droughts can reduce crop yields, leading to shortages in raw materials for food processing industries, while floods can damage transportation infrastructure, disrupting the movement of goods to markets. Similarly, extreme weather events can disrupt port operations and shipping routes, affecting international trade flows and exacerbating food insecurity in import-dependent countries. Agricultural supply chains are particularly vulnerable to climate-related disruptions due to their reliance on natural resources and exposure to weather variability (Yuan et al., 2024). The intricate web of activities involved in producing, processing, distributing, and consuming food requires significant coordination and is vulnerable to climate change which can have an impact on the quantity and quality of food produced. Climate change also carries significant implications for trade policy and international agreements. As nations seek to protect their food supply amid increasing volatility, there is a growing risk of trade barriers, export restrictions, and protectionist measures. Such actions can destabilize global food markets and disproportionately impact low-income, import-dependent countries (Bajaj et al., 2025). Building resilient food supply chains in the face of climate change demands a complex approach. This includes strategic investments in climate-resilient infrastructure, diversification of sourcing and production strategies, enhancement of logistical networks, and improved coordination among stakeholders at local, national, and international levels. Only through such integrative efforts can food systems adapt to the challenges of a changing climate while ensuring equitable access to food worldwide.

3.6.4 Financial Strategies for Climate-Resilient Food Systems

Financial strategies are crucial for managing climate impacts on food systems, enabling stakeholders to mitigate risks, adapt to changing conditions, and build resilience against future shocks (Sulewski et al., 2020). This involves risk transfer mechanisms, insurance products, and innovative financing models to incentivize climate-smart agriculture (CSA) and promote SLM practices. Access to financial resources is essential for farmers to invest in climate-resilient technologies, such as drought-resistant seeds, efficient irrigation systems, and soil conservation measures (Bhandari, 2021). In addition, financial incentives can encourage the adoption of climate-smart agricultural practices, such as agroforestry, conservation tillage, and integrated crop-livestock systems, which enhance productivity while reducing GHG emissions. Government policies also play a crucial role in creating an enabling environment for climate-resilient agriculture, through measures such as carbon pricing, subsidies for sustainable practices, and regulations to protect natural resources (Niles and Salerno, 2018).

To effectively confront the escalating challenges posed by climate change and ensure the long-term sustainability of global food systems, increased investment in research and development (RandD) is imperative. Such investments are vital for driving innovation in technologies and adaptive practices that strengthen the resilience of agricultural systems while enhancing productivity, resource efficiency, and environmental sustainability (Yang et al., 2024c). Innovations emerging under the umbrella of CSA including biotechnology, precision farming, artificial intelligence, and remote sensing are being developed to strengthen food security, reduce GHG emissions, and bolster the adaptive capacity of vulnerable farming communities (Stock and Gardezi, 2021a). Beyond technological innovation, access to agricultural inputs remains a cornerstone of resilience. Collaborative financing mechanisms such as public-private partnerships, cooperative lending schemes, and value chain financing can enhance farmers' access to essential inputs like fertilizers, improved seeds, and pest control agents, thus reducing climate-related production risks (Darma et al., 2025). Strengthening the role of local financial institutions and cooperatives also promotes financial inclusion and empowers smallholder farmers to manage climate variability more effectively (Oostendorp et al., 2019). To further support adaptation, governments and international development agencies must facilitate access to climate finance. This includes the provision of concessional credit, the establishment of credit guarantee facilities, and the promotion of index-based insurance schemes tailored to agricultural risk profiles (Nzuma et al., 2010). Insurance products that cover climate-induced hazards such as droughts, floods, and extreme weather events are particularly vital in protecting farmers' livelihoods and reducing their exposure to climate shocks. Crucially, integrating climate-smart agricultural strategies into national adaptation and mitigation plans will help align public and private investments with climate-resilient development goals. Embedding CSA within broader policy frameworks ensures that financial incentives, regulatory environments, and institutional capacities work synergistically to support sustainable food production in the face of a changing climate (World Bank, 2011).

3.7 ADAPTATION STRATEGIES FOR ENHANCING FOOD SECURITY

While adaptation has historically been a fundamental human response to environmental change, the current pace and scale of global environmental change (GEC) demand more deliberate, proactive, and strategic forms of adaptation. Unchecked, these changes may lead to further negative feedback loops, undermining both ecosystem stability and food security (Burns and Machado Des Johansson, 2017). Paradoxically, food system activities themselves have significantly contributed to the environmental degradation that now threatens food production and distribution. In parallel, food insecurity is often rooted in persistent structural issues, including chronic poverty, underdeveloped markets, and sociocultural barriers that limit access to resources and services. Adaptation strategies aimed at improving food security or mitigating the degradation of ecosystem services are not inherently effective or universally applicable. Scholars advocating for "sustainable adaptation" (Inderberg et al., 2015) emphasize that climate adaptation must simultaneously address broader development challenges, including poverty alleviation and social equity. Similarly, the concept of resilience in ecological systems highlights the importance of enabling systems to absorb external shocks without compromising their functional integrity (Cumming et al., 2006; Folke, 2006).

In response to the growing impacts of climate change and environmental stress, a diverse and integrated set of adaptation strategies has become essential to ensure global food security. These strategies encompass technological innovation, sustainable agricultural practices, efficient water management, and supportive policy frameworks.

- Technological advancements play a crucial role in enhancing the resilience of food systems. Innovations such as drought-tolerant crop varieties, improved seed genetics, and climate-resilient breeding programs help sustain agricultural productivity in increasingly hostile environments (FAO, 2023b). In addition, precision agriculture utilizing tools like satellite imagery, remote sensing, and data analytics enable more efficient use of inputs such as water, fertilizers, and pesticides, thereby reducing environmental harm while maintaining output (World Bank, 2023).
- Sustainable farming practices offer long-term solutions that support both productivity and ecological integrity. Approaches such as agroecology, conservation agriculture, and integrated pest management contribute to soil health, biodiversity conservation, and the provision of ecosystem services. Agroecology, in particular, leverages local knowledge, organic inputs, and diversified cropping systems to promote environmentally sound and socially just food production (Altieri et al., 2024). Conservation agriculture, which involves minimal soil disturbance, crop rotation, and permanent soil cover, improves soil structure and moisture retention, enhancing resilience to climate extremes.
- Efficient water management is also essential, especially in areas facing increasing water scarcity and irregular rainfall. Techniques such as drip irrigation, rainwater harvesting, and the deployment of drought forecasting tools help optimize water use and reduce reliance on unpredictable rainfall patterns (Jain et al., 2024). Community-led water governance models further enhance local adaptability and ensure more equitable resource allocation in water-stressed regions.
- Finally, strong policy frameworks and international cooperation are vital to the success of adaptation efforts. National Adaptation Plans (NAPs), early warning systems, and comprehensive food security strategies must be supported by adequate financial investment, institutional capacity, and stakeholder engagement. Global initiatives, such as the United Nations Framework Convention on Climate Change (UNFCCC) and the Global Alliance for Climate-Smart Agriculture, play a key role in facilitating cross-border collaboration, resource mobilization, and the dissemination of best practices (IPCC, 2022; FAO, 2023b).

Achieving enduring food security amid climate-related challenges requires that these adaptation strategies be inclusive and equity-driven. Special attention must be given to the needs of vulnerable groups, including smallholder farmers and marginalized communities, ensuring that adaptation promotes sustainability, innovation, and social justice.

3.8 MITIGATION STRATEGIES FOR ENHANCING FOOD SECURITY

Ensuring food security in the face of escalating climate threats, a collection of integrated mitigation strategies must be implemented, combining technological innovation, sustainable resource management, inclusive development, and robust policy frameworks. These include the promotion of CSA, which simultaneously addresses productivity, adaptation, and mitigation needs. CSA approaches including drought-tolerant crop varieties, improved livestock breeds, conservation agriculture, and agroforestry help optimize water and nutrient use, sequester carbon, and reduce GHG emissions while enhancing productivity and resilience (Stock and Gardezi, 2021b; Yang et al., 2024b). Alongside CSA, sustainable land and soil management practices such as minimal tillage, crop rotation, cover cropping, and organic amendments are vital for preserving soil health. These techniques maintain soil fertility and microbial diversity, reverse degradation, and bolster agricultural output, especially in ecologically fragile regions (El-Naggar et al., 2019; Pásztor, 2021).

Water resource efficiency is another critical mitigation pillar. As water scarcity intensifies, efficient water management practices like drip irrigation, rainwater harvesting, and wastewater reuse can substantially reduce water consumption and enhance irrigation reliability. Equally important are policy reforms to curb unsustainable groundwater extraction and to regulate pollution from agricultural and industrial activities, protecting freshwater ecosystems (Zarei et al., 2021; FAO, 2023a). Ensuring the financial inclusion of smallholder farmers is also essential. Expanding access to microloans, concessional credit, and index-based insurance mechanisms can enhance farmers' ability to invest in climate-resilient technologies and buffer against production shocks induced by droughts, floods, and extreme weather (Nzuma et al., 2010; Darma et al., 2025).

Crucially, mitigation strategies must be gender-responsive and community-driven. Women, who play a significant role in food production, often face structural barriers to land ownership, financing, and inputs. Gender-sensitive policies, support for women-led cooperatives, and improved access to agricultural services are fundamental to achieving equitable adaptation outcomes and enhancing food system resilience (UN Women, 2023a). Involving communities in resource governance through participatory planning and local stewardship builds adaptive capacity and strengthens ecosystem resilience.

Finally, policy integration and international cooperation are indispensable. National policies must embed food system resilience within climate action plans, especially Nationally Determined Contributions (NDCs), while ensuring coherence across sectors such as agriculture, water, and health. International collaboration through technology transfer, climate finance, and knowledge exchange is critical for enabling developing countries to meet sustainability goals. Instruments like climate adaptation funds, global research partnerships, and multilateral initiatives play a pivotal role in closing capability and resource gaps (World Bank, 2011; IPCC, 2022). Together, these mitigation strategies lay the foundation for a climate-resilient, inclusive, and sustainable global food system.

3.9 THE ROLE OF GLOBAL GOVERNANCE AND POLICY

Effectively addressing the complex challenges that climate change and environmental degradation pose to food security necessitates coordinated action across local, national, and global levels. Global governance mechanisms and policy frameworks play a pivotal role in aligning stakeholder efforts, mobilizing resources, and ensuring accountability in the transition toward sustainable and equitable food systems. Key priorities within governance structures include the integration of climate change and food security policies. Due to the intrinsic link between the two, it is imperative that strategies for climate adaptation and mitigation are embedded within broader national and international food security agendas. Such integrated policy approaches can generate synergies among agricultural development, environmental sustainability, and climate resilience, ultimately leading to more coherent and effective interventions (FAO, 2023c).

International frameworks such as the *Paris Agreement* and the *United Nations Sustainable Development Goals (SDGs)* have established foundational principles for addressing food security in the context of climate change. Notably, SDG 2 ("Zero Hunger") and SDG 13 ("Climate Action") underscore the imperative to eliminate hunger while enhancing agricultural resilience to climate impacts. The *Paris Agreement*, through NDCs, encourages countries to incorporate agricultural adaptation and food system resilience into their climate action plans (IPCC, 2022). The successful implementation of these global commitments relies on the active engagement of multiple actors, including national governments, non-governmental organizations (NGOs), and local communities. Governments are responsible for creating enabling policy environments that support sustainable agricultural investment and climate risk management. NGOs often play a complementary role by facilitating capacity-building initiatives and delivering support to vulnerable populations. Local communities and Indigenous peoples contribute valuable traditional knowledge and resource stewardship practices, offering place-based insights into sustainable land and food system management (UNESCO, 2023). Despite these efforts, significant implementation challenges persist. Fragmented institutional frameworks, limited financial resources, political instability, and weak inter-agency coordination continue to constrain effective policy execution. In many low- and middle-income countries, insufficient institutional capacity further undermines the scaling and enforcement of successful adaptation measures (HLPE, 2020). Strengthening institutional capacity is therefore essential to improving governance outcomes. This includes increased investment in agricultural research, extension services, and transparent monitoring systems, as well as enhancing institutions' ability to respond rapidly and effectively to climate-related risks. Additionally, international cooperation and financing mechanisms such as the Green Climate Fund (GCF) and the Global Environment Facility (GEF) are crucial in supporting climate adaptation and resilience efforts in financially and environmentally vulnerable regions (World Bank, 2023). Finally, there is a growing recognition of the value of community-based and Indigenous approaches to food system governance. These approaches emphasize local ownership, cultural relevance, and ecological sustainability, offering alternative models that complement formal, top-down governance structures. Promoting inclusive, equitable governance that respects local rights and knowledge systems is central to building adaptive capacity and achieving long-term food security in the face of climate change.

3.10 THE FUTURE OF FOOD SECURITY IN A CHANGING CLIMATE

The future of food security is intricately tied to the escalating challenges posed by climate change, environmental degradation, population growth, and deep-rooted socioeconomic inequalities. As climate-induced disruptions become more frequent and severe, the ability of food systems to ensure the core pillars of food security availability, access, utilization, and stability is under growing threat. Addressing this crisis requires a transformation of agricultural and food systems toward sustainability, resilience, and equity. Central to this transformation is the adoption of climate-resilient strategies that move away from unsustainable, high-input agricultural models and embrace practices that restore ecosystems, reduce emissions, and strengthen adaptive capacities. Approaches such as climate-smart agriculture, agroecology, and regenerative farming offer promising pathways to balance productivity with ecological stewardship (Stock and Gardezi, 2021a). Technological innovation will also play a critical role in shaping the resilience of future food systems. Advances in biotechnology, artificial intelligence, precision irrigation, and digital monitoring tools offer opportunities to improve yields, reduce waste, and optimize resource use even in the face of unpredictable climatic conditions (Yang et al., 2024a). However, to ensure these innovations reach those who need them most particularly smallholder farmers in the Global South sustained investments in research, capacity-building, and equitable technology transfer are essential (World Bank, 2023). At the same time, the future of food security cannot be decoupled from social equity. Women, rural communities, and other marginalized groups often face disproportionate risks from climate change while lacking access to land, finance, and decision-making processes. Promoting gender-inclusive agricultural policies and supporting women-led cooperatives are vital for enhancing community resilience and ensuring that adaptation efforts are both fair and effective (UN Women, 2023b).

The global response must also include stronger policy alignment and financial mobilization. Integrating sustainable food systems into national development agendas and climate adaptation plans such as NDCs will help harmonize agricultural development with broader climate goals.

International cooperation is equally critical. Mechanisms such as climate finance, concessional loans, green bonds, and index-based insurance schemes must be expanded to support low-income countries in building resilient food systems (Nzuma et al., 2010; UNFCCC, 2023). Furthermore, future food systems must be prepared for emerging risks, including pest outbreaks, geopolitical disruptions, and supply chain shocks. Building robust early warning systems, fostering public-private partnerships, and diversifying food sources are necessary components of adaptive governance (IPCC, 2022; Verma et al., 2025).

Ultimately, the future of food security will depend on our collective capacity to address the root causes of vulnerability while embracing innovative, inclusive, and sustainable pathways. Ensuring food for a projected 10 billion people by 2050 without breaching planetary boundaries will require decisive action that redefines how we produce, distribute, and consume food. In this context, sustainability is not merely an option but a necessity for ensuring human well-being and ecological balance in a rapidly changing world.

3.11 CONCLUSION

The combined challenges of climate change and environmental degradation present significant and growing risks to global food security. This review has demonstrated that regional and socioeconomic disparities critically influence vulnerability; low-income, agrarian, and marginalized populations particularly women and children bear the greatest burden. Regions such as sub-Saharan Africa and Southeast Asia are of particular concern, as their limited adaptive capacity exacerbates the disruptive effects of climate-related disturbances on food systems. Adaptation measures, including agricultural innovation, sustainable farming practices, effective water management, and comprehensive policy frameworks, represent promising avenues toward enhanced resilience. However, their successful implementation depends on robust governance, strong institutional support, and the coherent integration of global climate and food security priorities. Strengthening the roles of governments, international organizations, NGOs, and Indigenous communities is essential to ensure that adaptation initiatives are equitable, contextually relevant, and sustainable over the long term. The future implications for food security in the context of ongoing climatic and environmental changes are deeply concerning. Without immediate and coordinated action, the frequency and intensity of food insecurity crises are likely to increase, thereby compromising progress toward global development objectives. There is an urgent need for cross-sectoral collaboration, increased investment in climate-resilient agriculture, and a firm commitment to addressing the fundamental causes of environmental degradation. This critical moment demands decisive leadership and concerted collaboration. Policymakers must integrate climate and food policies, researchers must produce and disseminate evidence-based solutions, and the global community must mobilize the necessary resources and political commitment. The future of food security depends on our collective ability to act immediately, equitably, and with resolute determination.

REFERENCES

Abebaw, W. A. (2019). Review on impacts of land degradation on agricultural production in Ethiopia. *Journal of Resources Development and Management*, *57*, 21–29.

Abenu, A., Yusufu, F. A., & Sahabo, M. M. (2023). Smallholder farmers' perceptions on drivers of land degradation and its impact on their livelihoods in the Kuje Area Council of Abuja, Nigeria. *Journal of Applied Sciences and Environmental Management*, *27*(6), 1203–1206.

Abioye, E. A., Hensel, O., Esau, T. J., Elijah, O., Abidin, M. S. Z., Ayobami, A. S., … Nasirahmadi, A. (2022). Precision irrigation management using machine learning and digital farming solutions. *AgriEngineering*, *4*(1), 70–103.

Adão, F., Pádua, L., & Sousa, J. J. (2025). Evaluating soil degradation in agricultural soil with ground-penetrating radar: A systematic review of applications and challenges. *Agriculture*, *15*(8), 852.

Adger, W. N., Huq, S., Brown, K., Conway, D., & Hulme, M. (2003). Adaptation to climate change in the developing world. *Progress in Development Studies, 3*(3), 179–195. https://doi.org/10.1191/1464993403ps060oa

Agbeja, Y., Oluyede, E., & Oyedepo, V. (2021). Socio-economic characteristics assessment and livelihood coping strategies of small-scale fisheries in Ogun waterside local government area, Nigeria. *International Journal of Fisheries and Aquatic Studies, 9*(2), 266–276.

Ahmad, U., Alvino, A., & Marino, S. (2022). Solar fertigation: A sustainable and smart IoT-based irrigation and fertilization system for efficient water and nutrient management. *Agronomy, 12*(5), 1012.

Ajibade, F. O., Adelodun, B., Lasisi, K. H., Fadare, O. O., Ajibade, T. F., Nwogwu, N. A., … Wang, A. (2021). Environmental pollution and their socioeconomic impacts. In A. Kumar, V. K. Singh, P. Singh, & V. K. Mishra (Eds.), *Microbe Mediated Remediation of Environmental Contaminants* (pp. 321–354). Woodhead Publishing.

Alderman, H., & Yemtsov, R. (2016). How can safety nets contribute to economic growth? *Social Protection and Labor Discussion Paper No. 1603.* World Bank Group.

Altieri, M. A. (1999). The ecological role of biodiversity in agroecosystems. In M. G. Paoletti (Ed.), *Invertebrate Biodiversity as Bioindicators of Sustainable Landscapes* (pp. 19–31). Elsevier.

Altieri, M. A. (2004). Linking ecologists and traditional farmers in the search for sustainable agriculture. *Frontiers in Ecology and the Environment, 2*(1), 35–42.

Altieri, M.A., Nicholls, C.I., Dinelli, G. and Negri, L. (2024). Towards an agroecological approach to crop health: Reducing pest incidence through synergies between plant diversity and soil microbial ecology. *npj Sustainable Agriculture, 2*(1), 6.

Ansari, M. A., Ravisankar, N., Ansari, M. H., Babu, S., Layek, J., & Panwar, A. S. (2023). Integrating conservation agriculture with intensive crop diversification in the maize-based organic system: Impact on sustaining food and nutritional security. *Frontiers in Nutrition, 10*, 1137247.

Antoszewski, M., Mierek-Adamska, A., & Dąbrowska, G. B. (2022). The importance of microorganisms for sustainable agriculture—A review. *Metabolites, 12*(11), 1100.

Arora, N. K. (2019). Impact of climate change on agriculture production and its sustainable solutions. *Environmental Sustainability, 2*(2), 95–96.

Arora, P. K., Tripathi, S., Omar, R. A., Chauhan, P., Sinhal, V. K., Singh, A., … Singh, V. P. (2024). Next-generation fertilizers: The impact of bionanofertilizers on sustainable agriculture. *Microbial Cell Factories, 23*(1), 254.

Bachri, S., Irawan, L. Y., Fathoni, M. N., Fawaid, A. M., Nuraini, S. G., Utomo, K. S. B., & Aldianto, Y. E. (2021). Developing land capability to reduce land degradation and disaster incident in Bendo Watershed, Banyuwangi. *IOP Conference Series: Earth and Environmental Science, 630*(1), 012004.

Bahn, R. A., Yehya, A. A. K., & Zurayk, R. (2021). Digitalization for sustainable agri-food systems: Potential, status, and risks for the MENA region. *Sustainability, 13*(6), 3223.

Bajaj, K., Mehrabi, Z., Kastner, T., Jägermeyr, J., Müller, C., Schwarzmüller, F., Hertel, T. W., & Ramankutty, N. (2025). Current food trade helps mitigate future climate change impacts in lower-income nations. *PLOS One, 20*(1), e0314722.

Bashir, I., Lone, F. A., Bhat, R. A., Mir, S. A., Dar, Z. A., & Dar, S. A. (2020). Concerns and threats of contamination on aquatic ecosystems. *Bioremediation and Biotechnology: Sustainable Approaches to Pollution Degradation*, 1–26.

Bhandari, M. P. (2021). Climate change impacts on agriculture, a case study of Bangladesh, India, Nepal, and Pakistan. *SocioEconomic Challenges, 5*(2), 35–48.

Bibi, F., & Rahman, A. (2023). An overview of climate change impacts on agriculture and their mitigation strategies. *Agriculture, 13*(8), 1508.

Binswanger-Mkhize, H. P. (2012). Is there too much hype about index-based agricultural insurance? *Journal of Development Studies, 48*(2), 187–200. https://doi.org/10.1080/00220388.2011.625411

Borah, P., Kumar, M., & Devi, P. (2020). Types of inorganic pollutants: Metals/metalloids, acids, and organic forms. In P. Devi, P. Singh, & S. K. Kansal (Eds.), *Inorganic Pollutants in Water* (pp. 17–31). Elsevier.

Braun, J. V. (2020). Climate change risks for agriculture, health, and nutrition. *Health of People, Health of Planet and Our Responsibility: Climate Change, Air Pollution and Health*, 135–148.

Brown, M. E., & Funk, C. C. (2008). Food security under climate change. *Science, 319*(5863), 580–581. https://doi.org/10.1126/science.1154102

Brunetti, I., Tidball, M., & Couvet, D. (2019). Relationship between biodiversity and agricultural production. *Natural Resource Modeling, 32*(2), e12204.

Burns, T.R. and Machado Des Johansson, N. (2017). Disaster risk reduction and climate change adaptation—A sustainable development systems perspective. *Sustainability, 9*(2), 293.

Çakmakçı, R., Salık, M. A., & Çakmakçı, S. (2023). Assessment and principles of environmentally sustainable food and agriculture systems. *Agriculture*, *13*(5), 1073.

Çakmakçı, S., Polatoğlu, B., & Çakmakçı, R. (2024). Foods of the future: Challenges, opportunities, trends, and expectations. *Foods*, *13*(17), 2663.

Cameron, D., Worthington, R., & Lonsdale, W. M. (2013). Climate change and invasive alien species. In R. P. Keller, M. W. Cadotte, & G. Sandiford (Eds.), *Bioinvasions and Globalization: Ecology, Economics, Management, and Policy* (pp. 290–309). Oxford University Press.

Carriquiry, M., Dumortier, J., & Elobeid, A. (2022). Trade scenarios compensating for halted wheat and maize exports from Russia and Ukraine increase carbon emissions without easing food insecurity. *Nature Food*, *3*(10), 847–850.

Chakraborti, R., Davis, K. F., DeFries, R., Rao, N. D., Joseph, J., & Ghosh, S. (2023). Crop switching for water sustainability in India's food bowl yields co-benefits for food security and farmers' profits. *Nature Water*, *1*(10), 864–878.

Cline, W. R. (2007). *Global Warming and Agriculture: Impact Estimates by Country*. Peterson Institute for International Economics.

Cozim-Melges, F., Ripoll-Bosch, R., Veen, G. F., Oggiano, P., Bianchi, F. J., Van Der Putten, W. H., & Van Zanten, H. H. (2024). Farming practices to enhance biodiversity across biomes: A systematic review. *npj Biodiversity*, *3*(1), 1.

Cumming, G. S., Cumming, D. H. M., & Redman, C. L. (2006). Scale mismatches in social ecological systems: Causes, consequences, and solutions. *Ecology and Society*, *11*(1), Article 14.

Darma, R., O'Connor, P., Akzar, R., Tenriawaru, A. N., & Amandaria, R. (2025). Enhancing sustainability in rice farming: Institutional responses to floods and droughts in pump-based irrigation systems in Wajo District, Indonesia. *Sustainability*, *17*(8), 3501.

Dhakal, A., & Rai, R. K. (2020). Who adopts agroforestry in a subsistence economy? *Preprints*. https://doi.org/10.20944/preprints202003.0146.v1

Dias, M. C., Silva, S., Galhano, C., & Lorenzo, P. (2024). Olive tree belowground microbiota: Plant growth-promoting bacteria and fungi. *Plants*, *13*(13), 1848.

Díaz-Rodríguez, A. M., Parra Cota, F. I., Cira Chávez, L. A., García Ortega, L. F., Estrada Alvarado, M. I., Santoyo, G., & de Los Santos-Villalobos, S. (2025). Microbial inoculants in sustainable agriculture: Advancements, challenges, and future directions. *Plants*, *14*(2), 191.

Doss, C. (2018). Women and agricultural productivity: Reframing the issues. *Development Policy Review*, *36*(1), 35–50. https://doi.org/10.1111/dpr.12243

Dubois, O. (2011). *The state of the world's land and water resources for food and agriculture: Managing systems at risk* (xxiii + 285 pp.). Food and Agriculture Organization of the United Nations and Earthscan.

El-Hendawy, S., Dewir, Y. H., Elsayed, S., Schmidhalter, U., Al-Gaadi, K., Tola, E., … Hassan, W. M. (2022). Combining hyperspectral reflectance indices and multivariate analysis to estimate different units of chlorophyll content of spring wheat under salinity conditions. *Plants*, *11*(3), 456.

Elliott, J., Deryng, D., Müller, C., Frieler, K., Konzmann, M., Gerten, D., & Rosenzweig, C. (2015). The global impacts of climate change on agriculture. *Nature Climate Change*, *4*, 287–291. https://doi.org/10.1038/nclimate2475

El-Naggar, A., Lee, S. S., Rinklebe, J., Farooq, M., Song, H., Sarmah, A. K., … Ok, Y. S. (2019). Biochar application to low fertility soils: A review of current status, and future prospects. *Geoderma*, *337*, 536–554.

Environmental Protection Agency (EPA). (2021). *Climate Change Indicators: Greenhouse Gases*. https://www.epa.gov/climate-indicators/greenhouse-gases

FAO. (2023a). *The State of Food and Agriculture 2023: Water Resources for Sustainable Agriculture*. Rome: Food and Agriculture Organization of the United Nations.

FAO. (2023b). *The State of Food Security and Nutrition in the World*. Rome: Food and Agriculture Organization of the United Nations.

FAO. (2023c). *Soil Health Monitoring and Data Systems: Enhancing Decision-Making*. Rome: Food and Agriculture Organization of the United Nations.

FAO and UNEP. (2021). *Global Assessment of Soil Pollution: Report*. Rome: Food and Agriculture Organization of the United Nations and United Nations Environment Programme.

Ferguson, C. E., Tuxson, T., Mangubhai, S., Jupiter, S., Govan, H., Bonito, V., … Waide, M. (2022). Local practices and production confer resilience to rural Pacific food systems during the COVID-19 pandemic. *Marine Policy*, *137*, 104954.

Fischer, G., Shah, M., Tubiello, F. N., & van Velthuizen, H. (2002). Climate change and agricultural vulnerability (152 pp.). *International Institute for Applied Systems Analysis (IIASA), Laxenburg*, Austria.

Folke, C. (2006). Resilience: The emergence of a perspective for social–ecological systems analyses. *Global Environmental Change, 16*(3), 253–267.

Food and Agriculture Organization (FAO). (1996). *Rome Declaration on World Food Security and World Food Summit Plan of Action*. FAO.

Food and Agriculture Organization (FAO). (2016). *The State of Food and Agriculture 2016: Climate Change, Agriculture and Food Security*. FAO.

Food and Agriculture Organization (FAO). (2017). Locust Watch. *Food and Agriculture Organization of the United Nations*. Retrieved from http://www.fao.org/ag/locusts/en/

Food and Agriculture Organization (FAO). (2018). *The Impact of Disasters and Crises on Agriculture and Food Security 2017*. FAO.

Food and Agriculture Organization (FAO). (2020). *The State of Food Security and Nutrition in the World 2020*. FAO. https://doi.org/10.4060/ca9692en

Gallo Jr, A., Odokonyero, K., Mousa, M. A., Reihmer, J., Al-Mashharawi, S., Marasco, R., … Mishra, H. (2021). Superhydrophobic sand mulches increase agricultural productivity in arid regions. arXiv preprint arXiv:2102.00495.

Gamage, A., Punniamoorthy, T., & Madhujith, T. (2021). Starch based hybrid nanomaterials for environmental remediation. In M. O. Emeje (Ed.), *Starch: Evolution and Recent Advances*. IntechOpen.

García-Huante, A., Gómez-Merino, F. C., Trejo-Téllez, L. I., & López-Herrera, A. (2024). Potential of Cold-Water Agriculture (ColdAg) in Mexico: Challenges and opportunities for sustainable food production. *Sustainability, 16*(10), 4298.

Garnett, T., Appleby, M. C., Balmford, A., Bateman, I. J., Benton, T. G., Bloomer, P., Burligame, B., Dawkins, M., Dolan, L., Fraser, D., Herrero, M., Hoffmann, I., Smith, P., Thornton, P. K. Toulmin, C., Vermeulen, S. J., & Godfray, H. C. J. (2013). Sustainable intensification in agriculture: Premises and policies. *Science, 341*(6141), 33–34.

Gomiero, T. (2013). Alternative land management strategies and their impact on soil conservation. *Agriculture, 3*(3), 464–483.

Gomiero, T. (2016). Soil degradation, land scarcity and food security: Reviewing a complex challenge. *Sustainability, 8*(3), 281.

Gonthier, D. J., Ennis, K. K., Farinas, S., Hsieh, H. Y., Iverson, A. L., Batáry, P., … Perfecto, I. (2014). Biodiversity conservation in agriculture requires a multi-scale approach. *Proceedings of the Royal Society B: Biological Sciences, 281*(1791), 20141358.

Gupta, P. K., & Gandhi, M. (2023). Bioremediation of organic pollutants in soil–water system: A review. *BioTech, 12*(2), 36.

Hamdy, A., Ragab, R., & Scarascia-Mugnozza, E. (2003). Coping with water scarcity: water saving and increasing water productivity. *Irrigation and Drainage: The Journal of the International Commission on Irrigation and Drainage, 52*(1), 3–20.

Haro, A., Mendoza-Ponce, A., Calderón-Bustamante, Ó., Velasco, J. A., & Estrada, F. (2021). Evaluating risk and possible adaptations to climate change under a socio-ecological system approach. *Frontiers in Climate, 3*, 674693.

Hassan, A., Ibrahim, R. L., Raimi, L., Omokanmi, O. J., & S Senathirajah, A. R. B. (2025). Balancing growth and sustainability: can green innovation curb the ecological impact of resource-rich economies? *Sustainability, 17*(10), 4579.

HLPE. (2020). *Food Security and Nutrition: Building a Global Narrative Towards 2030*. High Level Panel of Experts on Food Security and Nutrition.

Hossain, M., & Islam, M. S. (2013). Impacts of climate change on food security and livelihoods in Bangladesh. *International Journal of Science and Nature, 4*(3), 472–479.

Inderberg, T. H., Eriksen, S., O'Brien, K. and Sygna, L. (2015). *Climate Change Adaptation and Development*. London and New York: Routledge.

Intergovernmental Panel on Climate Change (IPCC). (2014). *Climate Change 2014: Impacts, Adaptation, and Vulnerability*. Cambridge University Press.

Intergovernmental Panel on Climate Change (IPCC). (2021). *Climate Change 2021: The Physical Science Basis*. Cambridge University Press.

Intergovernmental Panel on Climate Change (IPCC). (2022). *Sixth Assessment Report: Impacts, Adaptation, and Vulnerability*. Intergovernmental Panel on Climate Change.

IPCC. (2022). *Climate change 2022: Impacts, adaptation and vulnerability. Contribution of Working Group II to the Sixth Assessment Report of the Intergovernmental Panel on Climate Change*. In H.-O. Pörtner, D. C. Roberts, M. Tignor, E. S. Poloczanska, K. Mintenbeck, A. Alegría, … B. Rama (Eds.). Cambridge University Press.

Irawan, S., & Antriyandarti, E. (2021). Physical deterioration of soil and rice productivity in rural Java. In *Journal of Physics: Conference Series* (Vol. 1825, No. 1, p. 012103). IOP Publishing.

Jain, S., Srivastava, A., Khadke, L., Chatterjee, U., & Elbeltagi, A. (2024). Global-scale water security and desertification management amidst climate change. *Environmental Science and Pollution Research*, *31*(49), 58720–58744.

Jeanneret, P., Aviron, S., Alignier, A., Lavigne, C., Helfenstein, J., Herzog, F., … Petit, S. (2021). Agroecology landscapes. *Landscape Ecology*, *36*(8), 2235–2257.

Kertolli, E., Prosperi, P., Harbouze, R., Moussadek, R., Echchgadda, G., & Belhouchette, H. (2024). The water–energy–food–ecosystem nexus in North Africa dryland farming: a multi-criteria analysis of climate-resilient innovations in Morocco. *Agricultural and Food Economics*, *12*(1), 34.

Khalid, S. (2020). Agronomy food security climate change and the sustainable development goals. In Amanullah (Ed.), *Agronomy Climate Change & Food Security*. IntechOpen.

Kieti, R. N., Kauti, M. K., & Kisangau, D. P. (2016). Biophysical conditions and land use methods contributing to watershed degradation in Makueni County, Kenya. *Journal of Ecosystem & Ecography*, *6*(4), 1000216.

Kingra, P. K., Setia, R., Kaur, S., Kaur, J., Singh, S., Singh, S. P., Kukal, S. S., & Pateriya, B. (2018). Analysis and mapping of spatio-temporal climate variability in Punjab using classical statistics and geostatistics. *Mausam*, *69*(1), 147–155.

Kocira, A., & Staniak, M. (2025). Role of agriculture in implementing the concept of sustainable food system. *Agriculture*, *15*(10), 1041.

Kumar, P., Sarda, R., Yadav, A., Ashwani, Gonencgil, B., & Rai, A. (2025). Farmer's perception of climate change and factors determining the adaptation strategies to ensure sustainable agriculture in the cold desert region of Himachal Himalayas, India. *Sustainability*, *17*(6), 2548. https://doi.org/10.3390/su17062548

Kumari, R., Banerjee, A., Kumar, R., Kumar, A., Saikia, P., & Khan, M. L. (2019). Deforestation in India: Consequences and sustainable solutions. In M. N. Suratman, Z. A. Latif, G. de Oliveira, N. Brunsell, Y. Shimabukuro, & C. A. Costa Dos Santos (Eds.), *Forest Degradation around the World*, 1–18. IntechOpen.

Lakhiar, I. A., Yan, H., Zhang, C., Wang, G., He, B., Hao, B., … Rakibuzzaman, M. (2024). A review of precision irrigation water-saving technology under changing climate for enhancing water use efficiency, crop yield, and environmental footprints. *Agriculture*, *14*(7), 1141.

Li, Q., Zhu, K., Liu, L., & Sun, X. (2021). Pollution-induced food safety problem in China: trends and policies. *Frontiers in Nutrition*, *8*, 703832.

Lobell, D. B., Burke, M. B., Tebaldi, C., Mastrandrea, M. D., Falcon, W. P., & Naylor, R. L. (2008). Prioritizing climate change adaptation needs for food security in 2030. *Science*, *319*(5863), 607–610. https://doi.org/10.1126/science.1152339

Lobell, D. B., Schlenker, W., & Costa-Roberts, J. (2011). Climate trends and global crop production since 1980. *Science*, *333*(6042), 616–620. https://doi.org/10.1126/science.1204531

Lu, Y., Song, S., Wang, R., Liu, Z., Meng, J., Sweetman, A. J., … Wang, T. (2015). Impacts of soil and water pollution on food safety and health risks in China. *Environment International*, *77*, 5–15.

Lu, Z. X., Zhu, P. Y., Gurr, G. M., Zheng, X. S., Read, D. M., Heong, K. L., … Xu, H. X. (2014). Mechanisms for flowering plants to benefit arthropod natural enemies of insect pests: Prospects for enhanced use in agriculture. *Insect Science*, *21*(1), 1–12.

Maitra, S., Brestic, M., Bhadra, P., Shankar, T., Praharaj, S., Palai, J. B., … Hossain, A. (2021). Bioinoculants—Natural biological resources for sustainable plant production. *Microorganisms*, *10*(1), 51.

Manescu, A. C., Barna, F. M., Regep, H. D., Manescu, C. M., & Cerba, C. (2025). The Impact of Extreme Weather Events on Agricultural Insurance in Europe. *Agriculture*, *15*(9), 995.

Maxwell, D., Vaitla, B., Tesfay, G., & Abadi, N. (2017). Constraints to humanitarian response: A case study of Ethiopia's 2016 drought. *Disasters*, *41*(3), 516–538. https://doi.org/10.1111/disa.12217

Meinzen-Dick, R., Quisumbing, A., Doss, C., & Theis, S. (2014). Women's land rights as a pathway to empowerment: A conceptual framework. *IFPRI Discussion Paper 01347*.

Mekonnen, M. M., & Gerbens-Leenes, W. (2020). The water footprint of global food production. *Water*, *12*(10), 2696.

Mir, R. A., Mantoo, A. G., Sofi, Z. A., Bhat, D. A., Bashir, A., & Bashir, S. (2023). Types of environmental pollution and its effects on the environment and society. In *Geospatial analytics for environmental pollution modeling: Analysis, control and management* 1–31. Cham: Springer Nature Switzerland.

Nalluri, N., & Karri, V. (2023). Superior effect of nature based solutions in soil and water management for sustainable agriculture. *Plant Science Today*, *11*, 6–20.

NASA. (2022). *Global Climate Change: Vital Signs of the Planet.* https://climate.nasa.gov/

Neyra, J. M., Cequea, M. M., & Schmitt, V. G. H. (2025). Current practices and key challenges associated with the adoption of resilient, circular, and sustainable food supply chain for smallholder farmers to mitigate food loss. *Frontiers in Sustainable Food Systems, 9*, 1484933.

Niles, M. T., & Salerno, J. D. (2018). A cross-country analysis of climate shocks and smallholder food insecurity. *PLOS One, 13*(2), e0192928.

Nilsson, J. A., Fulton, E. A., Johnson, C. R., & Haward, M. (2019). How to sustain fisheries: Expert knowledge from 34 nations. *Water, 11*(2), 213.

Nzuma, J. M., Makau, J., & Mulwa, R. M. (2010). Agricultural adaptation to climate change in Kenya: Micro-level analysis of options and constraints. *IFPRI Discussion Paper 01089.*

Oostendorp, R., van Asseldonk, M., Gathiaka, J., Mulwa, R., Radeny, M., Recha, J., Wattel, C., & van Wesenbeeck, L. (2019). Inclusive agribusiness under climate change: A brief review of the role of finance. *Current Opinion in Environmental Sustainability, 41*, 18–22.

Ortiz, R., Sayre, K. D., Govaerts, B., Gupta, R., Subbarao, G. V., Ban, T., Hodson, D., Dixon, J. M., Iván Ortiz-Monasterio, J., & Reynolds, M. (2008). Climate change: Can wheat beat the heat? *Agriculture, Ecosystems & Environment, 126*(1–2), 46–58. https://doi.org/10.1016/j.agee.2008.01.019

Ortuani, B., Facchi, A., Mayer, A., Bianchi, D., Bianchi, A., & Brancadoro, L. (2019). Assessing the effectiveness of variable-rate drip irrigation on water use efficiency in a Vineyard in Northern Italy. *Water, 11*(10), 1964.

Ostrom, E. (2009). A general framework for analyzing sustainability of social-ecological systems. *Science, 325*(5939), 419–422.

Pacheco, F. A. L., Fernandes, L. F. S., Junior, R. F. V., Valera, C. A., & Pissarra, T. C. T. (2018). Land degradation: Multiple environmental consequences and routes to neutrality. *Current Opinion in Environmental Science & Health, 5*, 79–86.

Pandey, D. K., Adhiguru, P., Momin, K. C., & Kumar, P. (2022). Agrobiodiversity and agroecological practices in 'jhumscape' of the Eastern Himalayas: don't throw the baby out with the bathwater. *Biodiversity and Conservation, 31*(10), 2349–2372.

Paramesh, V., Mohan Kumar, R., Rajanna, G. A., Gowda, S., Nath, A. J., Madival, Y., … Toraskar, S. (2023). Integrated nutrient management for improving crop yields, soil properties, and reducing greenhouse gas emissions. *Frontiers in Sustainable Food Systems, 7*, 1173258.

Pásztor, L. (2021). Advanced GIS and RS applications for soil and land degradation assessment and mapping. *ISPRS International Journal of Geo-Information, 10*(3), 128.

Pimentel, D., Zuniga, R., & Morrison, D. (2005). Update on the environmental and economic costs associated with alien-invasive species in the United States. *Ecological Economics, 52*(3), 273–288. https://doi.org/10.1016/j.ecolecon.2004.10.002

Pratap, D., Tamuly, G., Ganavi, N. R., Anbarasan, S., Pandey, A. K., Singh, A., & Iberaheem, M. (2024). Climate change and global agriculture: Addressing challenges and adaptation strategies. *Journal of Experimental Agriculture International, 46*(6), 799–806.

Pretty, J., Benton, T. G., Bharucha, Z. P., Dicks, L. V., Flora, C. B., Godfray, H. C. J., … Wratten, S. D. (2018). Global assessment of agricultural system redesign for sustainable intensification. *Nature Sustainability, 1*, 441–446. https://doi.org/10.1038/s41893-018-0114-0

Ramp, W. (2014). Complicating food security: Definitions, discourses, commitments. *Canadian Studies in Population, 41*(4), 117–134.

Reinman, S. L. (2015). Open knowledge repository. *Reference Reviews, 29*(5), 21–22.

Rice, J. (2017). Achieving and maintaining sustainable fisheries. *UN Chronicle, 54*(2), 62–65.

Richardson, R. B. (2010). Ecosystem services and food security: Economic perspectives on environmental sustainability. *Sustainability, 2*(11), 3520–3548.

Santos, M., Rebola, S., & Evtuguin, D. V. (2025). Soil remediation: Current approaches and emerging bio-based trends. *Soil Systems, 9*(2), 35.

Scherr, S. J., & McNeely, J. A. (2008). Biodiversity conservation and agricultural sustainability: towards a new paradigm of 'ecoagriculture' landscapes. *Philosophical Transactions of the Royal Society B: Biological Sciences, 363*(1491), 477–494.

Schlenker, W., & Roberts, M. J. (2009). Nonlinear temperature effects indicate severe damages to U.S. crop yields under climate change. *Proceedings of the National Academy of Sciences, 106*(37), 15594–15598. https://doi.org/10.1073/pnas.0906865106

Semeraro, T., Scarano, A., Leggieri, A., Calisi, A., & De Caroli, M. (2023). Impact of climate change on agro-ecosystems and potential adaptation strategies. *Land, 12*(6), 1117.

Shahzad, M. F., Abdulai, A., & Issahaku, G. (2021). Adaptation implications of climate-smart agriculture in rural Pakistan. *Sustainability, 13*(21), 11702.

Singh, R. (2024). Assessing the impact of sustainable agriculture practices on biodiversity conservation. *Journal of Sustainable Solutions*, *1*(2), 1–5.

Singh, R. P., Hodson, D. P., Huerta-Espino, J., Jin, Y., Njau, P., Wanyera, R., Herrera-Foessel, S., & Ward, R. W. (2011). The emergence of Ug99 races of the stem rust fungus is a threat to world wheat production. *Annual Review of Phytopathology*, *49*, 465–481. https://doi.org/10.1146/annurev-phyto-072910-095423

Singh, M., Yousuf, A., Singh, H., Singh, S., Hartsch, K., von Werner, M., ... Ali, H. R. (2022). Simulation accuracy of EROSION-3D model for estimation of runoff and sediment yield from micro-watersheds. *Water*, *14*(3), 280.

Sivakumar, M. V. (2020). Climate extremes and impacts on agriculture. *Agroclimatology: Linking Agriculture to Climate*, *60*, 621–647.

Soomro, S., Sahito, J. G. M., & Gilal, F. (2025). The link between climate change and human health. In J. Bento & M. M. Torres (Eds.), *Global Perspectives on Climate Change, Inequality, and Multinational Corporations* (pp. 183–208). Springer Nature Switzerland.

Soothar, R. K., Singha, A., Soomro, S. A., Chachar, A. U. D., Kalhoro, F., & Rahaman, M. A. (2021). Effect of different soil moisture regimes on plant growth and water use efficiency of sunflower: Experimental study and modeling. *Bulletin of the National Research Centre*, *45*, 1–8.

Sperling, F., & Seligman, R. (2017). *Transitioning to Climate-Resilient Development: Perspectives from Communities in Peru*. World Bank Group.

Stock, R., & Gardezi, M. (2021a). Make bloom and let wither: Biopolitics of precision agriculture at the dawn of surveillance capitalism. *Geoforum*, *122*, 193–203.

Stock, R., & Gardezi, M. (2021b). Climate-smart agriculture and adaptive capacity. *Journal of Environmental Planning and Management*, *64*(5), 887–906.

Sulewski, P., Wąs, A., Kobus, P., Pogodzińska, K., Szymańska, M., & Sosulski, T. (2020). Farmers' attitudes towards risk—An empirical study from Poland. *Agronomy*, *10*(10), 1555.

Tewari, A., & Tewari, S. (2003). Water resource management in India: issues and options. *Indian Journal of Public Administration*, *49*(3), 411–420.

Thornton, P. K., van de Steeg, J., Notenbaert, A., & Herrero, M. (2009). The impacts of climate change on livestock and livestock systems in developing countries: A review of what we know and what we need to know. *Agricultural Systems*, *101*(3), 113–127. https://doi.org/10.1016/j.agsy.2009.05.002

Tripathi, A., Tripathi, D. K., Chauhan, D. K., Kumar, N., & Singh, G. S. (2016). Paradigms of climate change impacts on some major food sources of the world: A review on current knowledge and future prospects. *Agriculture, Ecosystems & Environment*, *216*, 356–373.

Tsolakis, N., Aivazidou, E., & Srai, J. S. (2019). Sensor applications in agrifood systems: Current trends and opportunities for water stewardship. *Climate*, *7*(3), 44.

Udawatta, R. P., Rankoth, L. M., & Jose, S. (2019). Agroforestry and biodiversity. *Sustainability*, *11*(10), 2879.

UN Women. (2023a). *Gender-Responsive Climate-Smart Agriculture Toolkit*. New York: United Nations Entity for Gender Equality and the Empowerment of Women.

UN Women. (2023b). *Progress on Gender Equality: The Critical Role of Rural Women in Food Systems*. UN Women report.

UNESCO. (2023). Indigenous Knowledge and Climate Change Adaptation. https://unesdoc.unesco.org/ark:/48223/pf0000381896

UNFCCC. (2023). *Climate Finance Overview: Biennial Assessment*. Bonn: United Nations Framework Convention on Climate Change.

Vaux Jr, H. (2011). Water conservation, efficiency, and reuse. *Elements*, *7*(3), 187–191.

Verma, R., Osei, M., & Tariq, A. (2025). *Food Systems under Threat: Climate Risk and Agricultural Vulnerability in the Global South*. Global Environmental Change, 78, 102658.

Vikas, & Ranjan, R. (2024). Agroecological approaches to sustainable development. *Frontiers in Sustainable Food Systems*, *8*, 1405409.

Vineis, P., Chan, Q., & Khan, A. (2011). Climate change impacts on water salinity and health. *Journal of Epidemiology and Global Health*, *1*(1), 5–10. https://doi.org/10.1016/j.jegh.2011.09.001

Wahbeh, S., Anastasiadis, F., Sundarakani, B., & Manikas, I. (2022). Exploration of food security challenges towards more sustainable food production: A systematic literature review of the major drivers and policies. *Foods*, *11*(23), 3804.

Wang, H. (2023). Research on environmental change and sustainability. *Highlights in Science, Engineering and Technology*, *50*, 226–232.

Wanger, T. C., DeClerck, F., Garibaldi, L. A., Ghazoul, J., Kleijn, D., Klein, A. M., ... Weisser, W. (2020). Integrating agroecological production in a robust post-2020 Global Biodiversity Framework. *Nature Ecology & Evolution*, *4*(9), 1150–1152.

Wheeler, T., & Braun, J. V. (2013). Climate change impacts on global food security. *Science, 341*(6145), 508–513.

Wolde, T., & Bade, A. (2020). Impact of selected soil and water conservation measures on sustainability of soil moisture content at Bambasi District, Benishangul Gumuz Region, Ethiopia. *Journal of Energy, Environmental & Chemical Engineering*, *5*(3), 42–47.

World Bank. (2011). *Increased Productivity and Food Security, Enhanced Resilience and Reduced Carbon Emissions for Sustainable Development: Opportunities and Challenges for a Converging Agenda-Country Examples*. World Bank.

World Bank. (2023). *Transforming Food Systems for Resilience and Sustainability*. Washington, DC: World Bank Group.

Yang, J., Li, C., & Mohammed, A. (2024a). Technological innovations in climate-resilient agriculture. *Agricultural Systems*, *227*, 103755.

Yang, X., Liu, Z., & Wang, M. (2024b). Smart agriculture and digital innovation in climate adaptation. *Technological Forecasting and Social Change*, *200*, 122077.

Yang, Y., Tilman, D., Jin, Z., Smith, P., Barrett, C. B., Zhu, Y. G., Burney, J., D'Odorico, P., Fantke, P., Fargione, J., Finlay, J. C., Rulli, M. C., Sloat, L., Groenigen, K. J. V., West, P. C., Ziska, L., Michalak, A. M., Team, T. C., & Lobell, D. B. (2024c). Climate change exacerbates the environmental impacts of agriculture. *Science*, *385*(6713), eadn3747.

Yuan, M., Hu, H., Xue, M., & Li, J. (2024). Framework for resilience strategies in agricultural supply chain: assessment in the era of climate change. *Frontiers in Sustainable Food Systems*, *8*, 1444910.

Zarei, S., Bozorg-Haddad, O., Singh, V. P., & Loáiciga, H. A. (2021). Developing water, energy, and food sustainability performance indicators for agricultural systems. *Scientific Reports*, *11*(1), 22831.

4 Food Hunger and Public Health

Long-Term Consequences

Eridiong Onyenweaku and Alex Kamgain Tchuenchieu

4.1 AN OVERVIEW OF GLOBAL FOOD HUNGER

Global food hunger remains one of the most pressing challenges of the 21st century. Despite significant progress in agricultural productivity and food distribution, millions of people around the world continue to suffer from chronic undernourishment also called food hunger. According to FAO, IFAD, UNICEF, WFP, and WHO (2024), as of 2023, almost 735 million people face hunger globally. The causes of food hunger are not only complex but multifaceted, involving socio-economic, political, and environmental aspects. By way of definition, food hunger refers to the physical sensation of desiring food due to insufficient dietary intake (WFP, 2022). In academic and policy contexts, it is often measured as food insecurity which is the lack of consistent access to enough nutritious food for an active and healthy life. The FAO classifies food insecurity into two: moderate and severe levels; while moderate food insecurity involves uncertainty over access to food and a reduction in quality and quantity of meals, severe food insecurity is characterized by skipping meals, going days without eating, or chronic undernourishment. According to FAO, IFAD, UNICEF, WFP, and WHO (2024), some major causes of global food hunger include poverty and inequality, conflict and displacement, climate change and environmental degradation, economic crises and food price volatility as well as poor governance and policy failures.

4.1.1 Regional Differences in the Distribution of Food Hunger

The World Food Programme (WFP) (2022) reports that food hunger is unevenly distributed across the globe. Sub-Saharan Africa continues to bear the highest burden, with over 25% of the population facing chronic hunger. South Asia also reports high levels, especially in rural and marginalized communities. In contrast, high-income countries tend to experience food insecurity in specific vulnerable groups, such as the homeless, unemployed, or migrants. Global hunger, measured by the prevalence of undernourishment (PoU), increased sharply from 2019 to 2021, and has persisted at nearly the same level for three consecutive years, still affecting up to 9.1% of the population in 2023 compared with 7.5% in 2019 (FAO, IFAD, UNICEF, WFP, & WHO, 2024). Figure 4.1 shows the regional distribution of chronic hunger – projecting that the number of undernourished people in the different continents may remain high by the year 2030.

The recent stagnation and progress reversal in combating hunger, food insecurity, and malnutrition have significantly derailed global efforts toward achieving Sustainable Development Goal (SDG) Targets 2.1 and 2.2, which aim to end hunger and all forms of malnutrition by 2030. Efforts to address the underlying drivers of these challenges have been insufficient, and the growing frequency and intensity of these contributing factors threaten to worsen existing conditions. Without targeted and coordinated actions to mitigate these risks, the trajectory will continue to decline, further exacerbating global food and nutrition security.

DOI: 10.1201/9781003650881-4

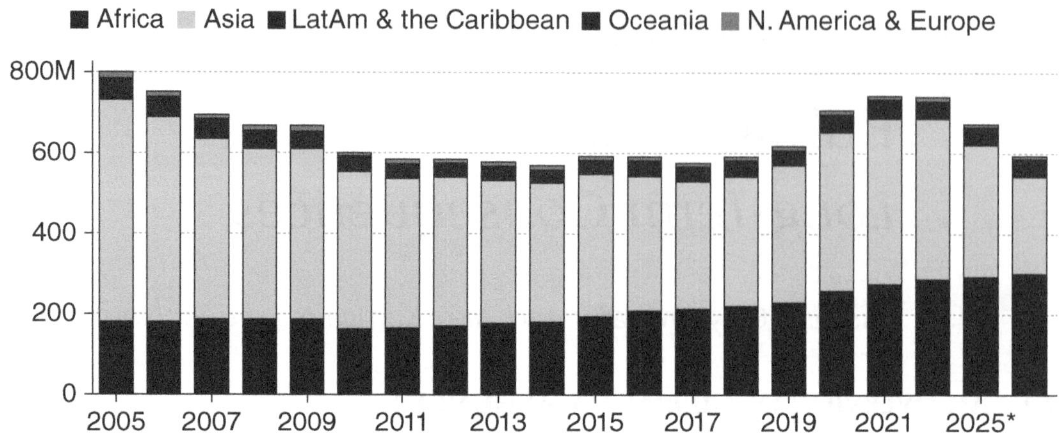

FIGURE 4.1 Regional distribution of chronic hunger.

Both external drivers of agrifood systems – such as armed conflict and climate extremes – and internal factors – including low agricultural productivity, insufficient supply of nutrient-dense foods (particularly fruits and vegetables), and the widespread availability of inexpensive, highly processed, energy-dense foods rich in fats, sugars, and/or salt – are contributing to rising costs of nutritious food (FAO, IFAD, UNICEF, WFP, & WHO, 2021). This escalation in cost has rendered healthy diets increasingly unaffordable for a significant portion of the global population. The resulting challenges to food security and nutrition are not confined to rural areas but are evident across the entire rural–urban continuum, as highlighted in the previous edition of this report (FAO, IFAD, UNICEF, WFP, & WHO, 2023).

Various international organizations are making useful and targeted efforts to combat food hunger globally; these include the United Nations – SDG 2 (Zero Hunger), World Food Programme (WFP), FAO, and UNICEF. These organizations work collaboratively with governments and civil society to address both immediate food needs and long-term sustainable food supply. The key strategies adopted include:

1. Supporting sustainable agriculture and smallholder farmers
2. Improving food distribution and storage systems
3. Strengthening social protection programs
4. Investing in nutrition and education
5. Mitigating the impact of climate change on food systems

Global food hunger is a serious challenge with far-reaching implications for health, development, and social stability. Tackling it requires a coordinated and multifaceted approach that handles both the symptoms and root causes of hunger. While progress has been made, achieving food security for all remains an urgent and ongoing challenge that demands sustained global commitment and innovation. According to WFP (2022), the IMF and World Bank are making historically large payments to nations, including those in vulnerable situations. Since COVID-19, WFP has provided technical assistance and expertise to more than 50 governments for government-to-person payments. In addition, WFP (2022) reports assisting 5 million people during the lean season in the Central Sahel of Africa, where over 9.6 million people experience food insecurity in Burkina Faso, Niger, and Mali.

In essence, while global levels of hunger and food insecurity have essentially not changed significantly for some years, many subregions of the world have recorded notable progress in this regard.

4.2 CAUSES AND INDICATORS OF FOOD HUNGER

Food hunger, manifested in the form of undernutrition, malnutrition, or outright starvation remains one of the most widespread global dilemmas. Despite enhanced levels of food production, hunger is a global issue, precipitated by a complex array of economic, environmental, political, and social factors. The following section elaborates on the complex causes of food hunger and identifies the metrics employed to measure and track it by scholars and global agencies.

4.2.1 Underlying Causes of Food Insecurity

4.2.1.1 Poverty and Economic Inequality

Poverty remains the most consistent and significant driver of food hunger globally. Low-income households are often unable to afford adequate and nutritious food, making them vulnerable to food insecurity (Saxena et al., 2025). In many developing nations, income disparity exacerbates this condition, as a small elite controls food-producing resources while large swaths of the population struggle to survive.

4.2.1.2 Conflict and Displacement

Armed conflicts displace persons, damage infrastructure, and cause serious disruption to agricultural production. The recent Tigray conflict in Ethiopia is a case in point, with areas that have witnessed conflict showing exceedingly high rates of food insecurity, particularly for vulnerable populations such as pregnant and lactating women (Woldegebriel et al., 2024). Food supply lines during periods of war are usually the targets of deliberate attacks, further exacerbating hunger situations (El Bilbeisi & Soboh, 2025).

4.2.1.3 Climate Change and Environmental Degradation

Climate stressors like droughts, floods, and desertification reduce the potential of food production, particularly for economies that rely heavily on agriculture. Research in Pakistan has demonstrated how climate variability overlaps with population growth to undermine food security (Abbas et al., 2025). The adverse impact of environmental change is highest in sub-Saharan Africa and South Asia, where the potential for adaptation is low.

4.2.1.4 Globalization and Trade Inequities

Global trade regimes are more attuned to export-driven, large-scale agribusiness rather than local food sovereignty. This displaces the local farmers and makes nations over-reliant on foreign foods that may be unaffordable in the event of global disruptions (Iacuone et al., 2024). The COVID-19 pandemic demonstrated how fragile globalized food systems can exacerbate hunger with supply chain disruptions.

4.2.1.5 Food Wastage and Inefficiency in Distribution

Ironically, global food production is more than sufficient to feed the world, but hunger persists due to massive food wastage and loss. It is estimated that one-third of all food produced globally is lost (Bostancı et al., 2024). With unequal distribution channels, it creates food abundance in some places while scarcity in others.

4.2.1.6 Gender Imbalances and Social Inequality

Women, especially in less developed countries, face institutional constraints in accessing land, credit, and agricultural equipment. This exclusion reduces their productive efficiency and leads to nutrition deficiencies in households that are reliant on them (Ruhi et al., 2025).

4.2.2 Indicators of Food Insecurity

Accurate measurement of hunger requires the use of in-depth and multifaceted indicators. The major tools employed globally are as follows:

1. **Global Hunger Index (GHI)**
   GHI is a comprehensive indicator that holds together four components: undernourishment, child wasting, child stunting, and child mortality. It gives a scalable indicator to compare the severity of hunger across countries. For example, the lowest-scoring countries in the GHI are typically torn apart by conflict or abject poverty (Iacuone et al., 2024).
2. **Household Food Insecurity Access Scale (HFIAS)**
   The HFIAS measures food access at the household level based on worry about food quantity, quality, and reduction in food intake. It is employed in both humanitarian emergency and development contexts (Aden & Mohammed, 2024).
3. **Anthropometric Measurements**
   Stunting (low height-for-age), wasting (low weight-for-height), and underweight (low weight-for-age) in children are key indicators that UNICEF and WHO employ to measure hunger and undernutrition among children. Maternal and child malnutrition, for instance, was widely measured in urban Kenya based on these indicators (Aden & Mohammed, 2024).
4. **Food Consumption Score (FCS)**
   This indicator combines dietary diversity, food frequency, and the relative nutritional importance of different food groups. Low FCS score is an indicator of poor dietary diversity and potential caloric or micronutrient deficits (Mohammadlo et al., 2024).
5. **Famine Early Warning Systems Network (FEWS NET)**
   The system detects early signs of food emergencies by utilizing satellite imaging, economic modeling, and ground observations. Its role in averting mass starvation events has been validated through its capacity to initiate international aid (Linares Quero et al., 2023).

4.3 EFFECTS OF FOOD INSECURITY ON POPULATION DYNAMICS

Food insecurity, whether transient, chronic, or intermittent, impacts populations in numerous physiological, psychological, social, and economic ways. As the global community aggressively works toward the vision of "Zero Hunger" (SDG 2), understanding the extensive effects of hunger is necessary for the development of meaningful policies and interventions.

4.3.1 Physiological and Health Effects

Hunger for food directly affects the health of human beings, particularly vulnerable groups like children, expectant women, and the elderly. Chronic malnourishment leads to stunted growth, weakened immune systems, greater susceptibility to illnesses, and elevated mortality rates.

Malnutrition in mothers in conflict zones, such as Northern Uganda, has been directly associated with child wasting and developmental delay. Dewan (2025) highlights that violence exposure and food insecurity among mothers worsens child malnutrition in low- and middle-income countries (LMICs), commonly precipitating intergenerational cycles of adverse health effects. Additionally, food insecurity situations are associated with a greater prevalence of micronutrient deficiency, often called "hidden hunger," that can lead to developmental disorders, anemia, and permanent cognitive impairment (Mishra et al., 2025).

4.3.2 Mental Health and Psychological Well-Being Implications

The mental effect of food insecurity is very serious. Chronic hunger leads to high rates of stress, anxiety, depression, and feelings of powerlessness. Mezieobi et al. (2025) reported a high incidence of eco-anxiety and depression among smallholder farmers who faced repeated food insecurity, especially those suffering the compounding consequences of climate change and displacement. Children who grow up under conditions of food deprivation exhibit long-term cognitive and emotional impairments. Georgiou et al. (2025) modeled how hunger disrupts behavioral coherence in group dynamics, revealing insights into how malnutrition may even impair collective community responses to crises.

4.3.3 Disruption of Education and Cognitive Development

Undernourished children perform poorly in school both due to their weaker physical state and their reduced levels of intellectual capacity. The effects begin early hungry children are more apt to miss school, have difficulty focusing, and ultimately drop out of school (Ramadugu, 2024). This forms a cycle of hunger and poverty since education is directly related to income and access to food in the future. Longitudinal studies also show that insufficient nutrition in childhood is linked with reduced productivity in adulthood and decreased lifetime incomes, pointing to the need for successful school feeding programs (Saavedra & Prentice, 2023).

4.3.4 Social Instability and Conflict

Hunger for food can both be a cause and a result of social instability. When food becomes limited, the struggle over available resources is heightened, which may develop into communal or international conflicts. This has been seen in Nigeria, where food scarcity caused by climate change has fueled mass migration and civil conflict (Ezenwajiobi, 2025).

In Sudan, Hemmeda (2025) describes that famine has devastated war-stricken and humanitarian withdrawal communities, causing a compounding crisis that risks regional destabilization.

4.3.5 Economic Implications

Food insecurity destabilizes economies by decreasing labor productivity, raising health expenses, and hindering long-term human capital development. Communities afflicted by hunger engage less in economic activities and tend to depend on subsidies or unofficial labor markets. Jangra and Tiwari (2025) observe that the combined effect of hunger with low-energy efficiency and trade restrictions severely dampens economic resilience in developing markets. Also, nations ravaged by long-term hunger fail to attract investment and establish proper public health and education systems, thereby reinforcing poverty (Wilts, 2025).

4.3.6 Cultural and Ethical Implications

Hunger upsets social norms, changes food cultures, and confronts ethical frameworks of equity and justice. During the COVID-19 pandemic, Mohamed-Saleem and Osman (2025) examined how Muslim families in Toronto tried to maintain dietary norms during food insecurity, showing how hunger can intersect with religious identity and social cohesion. The consequences of food insecurity are neither localized nor short-term. They spread through communities and nations, hindering development, health, and social equilibrium. Ending hunger involves not just the enhancement of food production but the remolding of global systems for equitable access, safe supply chains, and socio-economic inclusion.

4.4 LONG-TERM CONSEQUENCES OF FOOD HUNGER

The long-term manifestation of food insecurity, is a severe challenge with extensive implications for public health, especially if it occurs in the initial stages of life. Aside from the short-term risks of malnutrition and starvation, long-term hunger for food has extensive and long-lasting influences on the physiological, intellectual, psychological, and socio-economic aspects of health throughout an individual's life course. Over the past several years, a mounting volume of evidence has shown that food deprivation experiences, especially during developmentally critical time points create indelible marks on the body and brain, thus setting the stage for a lifetime of vulnerability to illness.

4.4.1 Mental Health and Cognitive Impairment

A seminal research study conducted by Li et al. (2025) using longitudinal data on more than 9,900 individuals demonstrated that child hunger is a robust predictor of poorer self-rated health, greater risks of depression, worse memory, and worse cognitive functioning in middle age and old age. Further, the effects did not diminish over time; rather, they manifested deteriorating trajectories over time, especially for mental health results like depression and cognitive impairment. Lifestyle determinants like poor sleep, smoking, and reduced physical activity have been identified as mediating variables, thus further ingraining health inequalities that originate from early deprivation of food.

4.4.2 Weakened Immune Function

Physiologically, hunger during growth phases disrupts the development of essential organ systems. Sharda (2024) stated that recurrent undernutrition compromises immune function, thus predisposing one to infectious diseases and increasing the risk for chronic conditions, including cardiovascular disease and type 2 diabetes. Micronutrient deficiency, especially of iron, zinc, and folic acid has been associated with long-term damage in immunity, metabolism, and brain development (Sharda, 2024). Mateus et al. (2025) believe that future risk-benefit evaluations should feature dose-response relationships of the nutrients in conjunction with more comprehensive health effects to guide recommendations and interventions regarding diet.

4.4.3 Child Health Effects

Furthermore, Gupta and Kaicker (2023) also indicate that protein-energy malnutrition leads to stunting, child mortality, and heightened susceptibility to infections such as pneumonia and malaria because of lowered resistance. The neurocognitive effects of hunger are also disturbing. Childhood nutritional deficiencies influence brain structure and function, undermining executive functions, attention, and learning capacities. Al Afnan (2024) highlights that hunger-induced cognitive delays which undermine school performance and reduce one's chances of getting employed in the future, hence generating cycles of poverty and food insecurity.

4.4.4 Maternal Health Effects and Emotional Health Effects

The findings are consistent with Mosca and Nolan (2025), who set those prenatal nutritional deficiencies, such as rubella-related food shortages, create lower educational levels, increased disability, and poorer adult health outcomes. Emotionally, hunger has also been associated with greater psychological stress, anxiety, and depression. There is evidence that food-insecure children and adults have high cortisol levels and symptoms of chronic stress that are associated with psychiatric disorders and emotional regulation impairment (Sharda, 2024; Park et al., 2024). The second study showed that food-insecure adults in the United States not only incurred greater healthcare expenses and had restricted access to medical care but were also far more prone to financial difficulty and emotional distress (Park et al., 2024).

Food insecurity also affects behavior and decision-making and has significant implications for dietary habits and general well-being. Fakhreddine et al. (2025) demonstrated that emotions related to food throughout crises like the COVID-19 pandemic were strongly associated with perceived food availability and nutritional stress, thus generating affect-driven food behaviors that encouraged calorically dense but nutritionally poor foods. Hunger impairs the ability to regulate attention, making individuals prioritize taste and immediate gratification over nutritional value. This is referred to by Neal et al. (2024) as cognitive discounting under hunger conditions. Socially, food hunger erodes trust and social bonds, alienating and exposing individuals. Jaspars (2023) contends that prolonged hunger destroys social bonds, induces violence, and promotes fragmentation of communities, hence rendering populations vulnerable to control and authoritarian governance. In war-torn areas such as Gaza, hunger has accelerated the breakdown of health systems, caused increases in infectious diseases, and induced long-term psychological trauma, especially in children (Kooli & Kooli, 2025). The humanitarian emergency underscores how hunger intersects with infrastructure decay and social breakdown to compound public health emergencies (Fang et al., 2024).

4.4.5 Economic Implications

Economically, hunger limits human capital formation and degrades productivity. Reuveni (2025) argues that high prices and poor infrastructure disproportionately affect low-income households, which either reduce caloric intake or consume less nutritious alternatives. Such decisions lead to higher spending on long-term health, decrease economic productivity, and exacerbate intergenerational poverty. Similarly, Sharma et al. (2024) argue that the use of sustainable farming techniques is essential in averting future food shortages and consolidating national resilience. In the absence of concerted policies toward addressing the structural determinants of food insecurity, hunger across the world is likely to continue until 2030, with severe implications for public health.

Summarily, these studies highlight that the chronic nature of food insecurity has consequences that extend far beyond short-term nutritional afflictions. These consequences include, but are not limited to, weakened immune systems, decreased educational and economic prospects, mental illness, and societal disintegration.

4.5 STRATEGIES/APPROACHES TO TACKLE FOOD HUNGER

The issue of hunger, as a complex problem with interconnected socio-economic, environmental, and public health dimensions, calls for an integrated response in the framework of both relief response and resilience-building in the long term. At the global level, there is greater convergence between policy innovation, agricultural transformation, nutrition-sensitive intervention, and technological assistance to create comprehensive systems that will combat hunger and provide sustainable food security. A key approach includes the utilization of non-food social safety net programs, which aim primarily at removing economic burdens from households as well as increasing food availability. A systematic scoping review conducted by Duffy et al. (2025) establishes that the receipt of income and housing support in the United States is generally associated with reductions in food insecurity, thereby underscoring the importance of comprehensive welfare systems in mitigating hunger. Furthermore, Scott et al. (2025) demonstrate that joint provision of maternal cash transfers with health-nutrition services in South Asia, though underused, significantly improves both reach and effectiveness of food assistance programs. From an agricultural perspective, peri-urban and urban agriculture has become an essential avenue to achieving localized food production and economic self-sustainability. According to Eigenbrod and Gruda (2015), the use of such urban farming methods as vertical gardens and communal green areas has the potential to enhance availability of fresh foods, even in high-density locations with limited land cover, where food imports are the primary source of sustenance. These strategies also strengthen micro-entrepreneurial endeavors, especially when supplemented by facilitating policies and the strengthening of technical competencies. The advent of technology innovations, particularly decentralized ledger technologies (DLTs), creates new opportunities for promoting transparency and equity in food distribution. Ismail and Adane, in their 2024 study, suggest the implementation of a blockchain-based hunger management system with the goal of addressing corruption, inefficiency, and ineffective targeting common in centralized relief systems (Ismail & Adane, 2024). With traceability and accountability in the food supply chain, DLTs have the potential to make sure that aid reaches the intended beneficiaries in addition to reducing wastage in the system. One of the key evolutions in dietetic interventions that is progressively gaining traction is the "Food is Medicine" model, which integrates nutritional aspects with paradigms in healthcare. According to Rahman et al. (2024), produce prescriptions and medically tailored meals are the foundation blocks of this model, thereby closing the disparity between diet quality and the prevention and control of chronic diseases. When paired with regenerative agriculture, these interventions do not only guarantee individual health but also facilitate ecological sustainability. The shape of economic policy to enhance both food access and the affordability of food prices is of significant relevance. Reuveni (2025) demonstrates that access is constrained by food price volatility and weak infrastructure in low-income areas, especially in sub-Saharan Africa and Latin America. Evidence demonstrates investments in logistics, subsidies, and price stabilization as policy choices for making food access equitable. Parallel arguments elucidated by Kar (2025) emphasize food availability, affordability, and access as foundations of food security aimed at redressing structural disparities.

Additionally, neglected grain legumes, that is, cowpeas and Bambara groundnuts, have been identified as priority crops in order to improve food and nutritional security, especially under climatic stress conditions. According to Vilakazi et al. (2025), these legumes are imbued with high nutritional value in addition to supporting environmental resilience; the objective is to develop targeted policies, improve market development, and increase funding for research in order to facilitate the inclusion of such legumes in food systems. At the macro-policy level, the Post-Malabo Agenda and the US National Strategy on Hunger and Health are both guiding documents for governance of the food system across sectors. Phungwayo (2020) highlights the function of inter-ministerial coordination between the agricultural, nutrition, and social protection sectors in propelling systemic change in African settings.

Fleischhacker et al. (2025) also highlight the necessity of federal involvement in coordinating food system changes with US public health outcomes. Growing emphasis on nutrition-sensitive social protection, as outlined by Yuan et al. (2024), highlights the importance of integrated programming that harmonizes food production with health and education outcomes. Their research on COVID-19's effects on food security emphasizes that resilience planning, local food production, and high-tech innovation must be made central pillars of strategies that are designed to survive future shocks.

Overall, ending hunger for food entails multisectoral approaches that involve emergency response, structural transformation, agricultural transformation, and social justice. Because the approaches overlap, they must be localized and guided by the principles of data, equity, and sustainability to ensure they have a lasting effect.

4.6 APPLICATION OF TECHNOLOGICAL ADVANCEMENTS IN THE FIGHT AGAINST FOOD HUNGER

The integration of technological innovation into food systems has been increasingly significant in reducing global hunger. Contemporary technological advances such as genetically modified organisms (GMOs), sophisticated greenhouse technologies, virtual and augmented reality (VR/AR) being utilized for nutritional education, and sophisticated waste management systems are being leveraged to address the intricate problem of food insecurity. These technologies enhance not only the availability and accessibility of food but also sustainability, learning, and efficiency in the utilization of resources in food chains and agricultural systems. GMOs are a modern technique in agricultural biotechnology, with the objective of improving crop yield and tolerance to pests and climatic stress. A recent research by Sakhong et al. (2024) highlights the contribution of GMOs to enhancing soil fertility and facilitating recycling of nutrients, particularly in areas under the threat of soil degradation from climate change. Their research advocates for the application of GMOs, biofertilizers, and conservation tillage as a method for developing more sustainable food systems. Likewise, Mešić et al. (2024) propose the incorporation of genetically modified crops into vertical farming systems to promote the resilience of crops while, in turn, lowering land and water use qualities that are vital in urban locations and arid regions. Greenhouse and controlled environment agriculture (CEA) systems are also gaining prominence due to their capacity to lengthen growing seasons and optimize conditions for the production of high-yielding crops. They utilize hydroponic and aquaponic techniques to increase food yield per square meter, showing great advantages in urban food deserts. As stated by Wibisono et al. (2025), when CEA practices are integrated with precision agriculture technologies such as IoT sensors and AI analytics, they assist in forecasting plant health and nutrient requirements, consequently reducing waste and enhancing efficiency. Simultaneously, waste management technologies are essential in alleviating food insecurity by transforming surplus and byproduct food. Öztürk et al. (2025) present a comprehensive model that merges machine learning, Internet of Things (IoT), and blockchain technology to predict demand, oversee food inventory, and automate the detection of spoilage. This collaboration enhances proactive management of inventory and food redistribution to at-risk communities. In the same vein, Amrutkar et al. (2024) present "ECOFEAST," a platform that enables food donations via real-time geolocation and inventory analysis, emphasizing how digital platforms can connect food surplus with hunger. Educational technologies can significantly impact the fight against hunger by enhancing nutrition literacy.

Capecchi et al. (2025) created ARFood, a game utilizing AR to educate children about healthy eating habits and environmental sustainability. The game employs AI-driven virtual assistants to offer insights on food selections, showcasing how immersive technologies can improve public awareness and promote lasting behavioral shifts. The possibilities of VR and AR in education have been enhanced by Uçar (2024), who highlights their role in experiential learning settings that connect the knowledge gap in food sustainability and health for various communities. Sustainable packaging and circular economy frameworks are synergistic technological strategies for alleviating hunger. Yoğun & Yıldırım (2025) carried out an extensive bibliometric analysis showing that circular food systems, intelligent urban recycling, and digital upcycling are crucial in reducing food waste and enhancing resource recovery. Their research indicates that incorporating circular design concepts in food retail and logistics can significantly lessen reliance on landfills and ensure surplus food is redirected effectively. Clark et al. (2025) examined how AI-powered food inventory management could alleviate hunger at the household level, revealing that AI technologies from the hospitality sector could be tailored for home kitchens to monitor expiration dates, offer meal ideas, and enhance purchasing habits. Moreover, digital technologies such as intelligent bins and IoT-enabled composting systems are more frequently being utilized in community environments to promote sustainable food utilization and nutrient cycling, as noted by Saraswathi et al. (2024).

In total, the use of frontier technologies in genetic engineering, e-learning, and artificial intelligence-based food systems presents a comprehensive approach to food shortage prevention. The prospects of these technologies are dependent on conducive policies, multisectoral coordination, and active public involvement. Investment in these technologies not only satisfies current food demands but also enables the building of more resilient and inclusive food systems in the future.

4.6.1 The Way Forward

Mitigation strategies and policy options to combat food insecurity should be effective, policymakers must invest in climate-resilient agriculture, implement drought-tolerant crop varieties, enhance irrigation infrastructure, and encourage agroecological practices. The strategies must also strengthen Social Safety Nets, expand food subsidy programs, conditional cash transfers, and school feeding programs. Local communities' ownership of their food systems should be encouraged by empowering smallholder farmers and decreasing reliance on imports. The journey to "Zero Hunger" (SDG 2) is complex but can be achieved. It demands concerted efforts on underlying causes and enhancing the validity of measurements of food hunger.

To treat food hunger as a public health crisis, we need to take collective action across sectors. This includes implementing nutrition-sensitive social protection programs, universal school feeding programs, climate-resilient agriculture, and food price stabilization policies to reduce hunger exposure. Food hunger, in brief, is both a result and a symptom of system inequities, with long-term and compounding effects on individuals and societies. To mitigate its effects in the long term, a system response must be brought to bear that consolidates food policy, health systems, education, and social equity. Only through sustained and multisectoral interventions are intergenerational food hunger and its attendant traumas substantially mitigated.

REFERENCES

Abbas, S., Haider, A., Kousar, S., Lu, H., Lu, S., Liu, F., Li, H., Miao, C., Feng, W., Ahamad, M.I. and Mehmood, M.S. (2025). Climate variability, population growth, and globalization impacting food security in Pakistan. *Scientific Reports*, *15*(1), 1–20. https://doi.org/10.1038/s41598-025-88916-2

Aden, E.H. and Mohammed, S.A. (2024). Household food insecurity and its association with nutritional status of children 6–59 months of age in Kebridehar. *Indonesian Food and Nutrition Progress*, *21*(2), 81–94.

Al Afnan, L. (2024). Hunger in America: Examining the far-reaching effects of food insecurity in the United States of America. *International Journal of Translational Medical Research and Public Health*, *8*, (e014). https://doi.org/10.25259/IJTMRPH_43_2024

Amrutkar, P., Bondarde, S., Kale, A., Patil, N. and Habbu, S. Dr. (2024). ECOFEAST – A food redistribution platform for sustainable food waste management and community empowerment. *International Journal for Research in Applied Science and Engineering Technology*, *12*(11), 2133–2140. https://doi.org/10.22214/ijraset.2024.65493

Bostancı, S., Vasilev, V. and Yıldırım, S. (2024). Strategies and policies in recycling food waste: The intermediary role of municipalities. *Journal of Recycling Economy & Sustainability Policy*, *3*(2), 160–167.

Capecchi, I., Borghini, T., Bellotti, M. and Bernetti, I. (2025). Enhancing education outcomes integrating augmented reality and artificial intelligence for education in nutrition and food sustainability. *Sustainability*, *17*(5), 1–29. https://doi.org/10.3390/su17052113

Clark, Q.M., Kanavikar, D.B., Clark, J. and Donnelly, P.J. (2025). Exploring the potential of AI-driven food waste management strategies used in the hospitality industry for application in household settings. *Frontiers in Artificial Intelligence*, *7*, 1–14. https://doi.org/10.3389/frai.2024.1429477

Dewan, S.M. (2025). *Maternal Exposure to Violence and Child Malnutrition in LMICs: A Cross-Sectional, Population-Based Study in Conflict-Affected Northern Uganda* (Doctoral dissertation). University of Cambridge Repository. https://doi.org/10.17863/CAM.117207

Duffy, E.W., Poole, M.K., Gonzalez, D., Petimar, J., Kinsey, E.W., Shafer, P.R., Baldwin-SoRelle, C. and Austin, A.E. (2025). Beyond food assistance: A scoping review examining associations of nonfood social safety net programs in the United States with food insecurity and nutrition outcomes. *Nutrition Reviews*, *83*(7), 1344–1357. http://doi.org/10.1093/nutrit/nuae216

Eigenbrod, C. and Gruda, N. (2015). Urban vegetable for food security in cities. A review. *Agronomy for Sustainable Development*, *35*, 483–498. http://doi.org/10.1007/s13593-014-0273-y

El Bilbeisi, A. and Soboh, R. (2025). From crisis to recovery: How Gaza can overcome its post-war malnutrition challenges. *Frontiers in Public Health*. https://doi.org/10.3389/fpubh.2025.1612823

Ezenwajiobi, C.C. (2025). *Tackling Food Insecurity and Boosting Agricultural Productivity and Food Access in Nigeria: Areas for Legislative Intervention*. NILDS-Department of Democracy and Governance – Information brief.

Fakhreddine, L.B., Sánchez, M. and Pindado, E. (2025). Health crises, community health habits, and pandemic-driven food sentiments: A twitter-based study in the United States. *Agribusiness*, 1–16. https://doi.org/10.1002/agr.22024

Fang, Z., Xiao, Q. and Gao, Y. (2024). Impact of major emergencies on zero hunger, achieving food security, and improving nutrition. *Frontiers in Sustainable Food Systems*, *8*, 1–18. https://doi.org/10.3389/fsufs.2024.1408454

FAO, IFAD, UNICEF, WFP and WHO. (2021). The State of Food Security and Nutrition in the World 2021. Transforming Food Systems for Food Security, Improved Nutrition and Affordable Healthy Diets for All. Rome, FAO. https://doi.org/10.4060/cb4474en

FAO, IFAD, UNICEF, WFP and WHO. (2023). *The State of Food Security and Nutrition in the World 2023. Urbanization, Agrifood Systems Transformation and Healthy Diets across the Rural–Urban continuum*. Rome, FAO. https://doi.org/10.4060/cc3017en

FAO, IFAD, UNICEF, WFP and WHO. (2024). *The State of Food Security and Nutrition in the World 2024 – Financing to End Hunger, Food Insecurity and Malnutrition in All Its Forms*. Rome. https://doi.org/10.4060/cd1254en

Fleischhacker, S., Chandran, K., Cotwright, C., Winters, M., Abley, M., Chester, D., Jacobs-Young, C. and Long, C. (2025). Progress to implement the national strategy on hunger, nutrition, and health at the US department of agriculture. *Nutrition Today*, *60*(1), 10–19. https://doi.org/10.1097/NT.0000000000000726

Georgiou, F., Buhl, C., Green, J.E.F., Lamichhane, B. and Thamwattana, N. (2025). Including population and environmental dynamic heterogeneities in continuum models of collective behaviour with applications to locust foraging and group structure. *PLOS Computational Biology*, *21*(4), 1–21. https://doi.org/10.1371/JOURNAL.PCBI.1011469

Gupta, A. and Kaicker, A. (2023). Food security and human health. *Journal of Ecophysiology and Occupational Health*, 99–104. https://doi.org/10.18311/jeoh/2023/34445

Hemmeda, L. (2025). The humanitarian crisis in Sudan: The devastating effects of decreased aid. *Medicine, Conflict and Survival*, 1–5. https://doi.org/10.1080/13623699.2025.2491970

Iacuone, S., Ferrari, F. and Fuschi, M. (2024). Food as a critical metaphor for globalization: Spazialization of territorial imbalances. Evidence from the Global South. *Romanian Review on Political Geography/Revista Română de Geografie Politică*, *26*(2). https://doi.org/10.30892/rrgp.262108-384

Ismail, F.S.M. and Adane, D. (2024). Leveraging decentralized ledger technology to enable equitable distribution of food resources to fight hunger. *ESS Open Archive Eprints*, *408*, 1–8. https://doi.org/10.22541/AU.171220964.40848880/V1

Jangra, A., and Tiwari, S. (2025). Empowering nutrition: biofortification to combat hidden hunger. In H. Pathak, W.S. Lakra, A. Gopalakrishnan, and K.C. Bansal (Eds.), *Advances in agri-food systems*. Springer, Singapore. https://doi.org/10.1007/978-981-96-0763-1_17

Jaspars, S. (2023). Food and power in protracted crisis: How systems and institutions influence livelihoods, food security, and nutrition. *Food and Nutrition Bulletin*, *44*(2S), S23–S31. https://doi.org/10.1177/03795721231202236

Kar, J. (2025). A study on the food security, hunger and poverty: A social menace of the nation. *International Journal of Science and Research Archive*, *14*(1), 898–901. https://doi.org/10.30574/ijsra.2025.14.1.0130

Kooli, C. and Kooli, E. (2025). The humanitarian and public health crisis in Gaza: Impact, challenges, and international responses. *Avicenna*, *2025*(1), 1–12. https://doi.org/10.5339/avi.2025.1

Li, D., Guo, X., Zhang, W., Li, W., Zhang, T., Liu, Z., Su, M. and Li, Z. (2025). The association between childhood hunger experiences and health in middle and old age: A longitudinal study over 10 years. *BMC Public Health*, *25*(1), 1–11. https://doi.org/10.1186/s12889-025-21345-y

Linares Quero, A., Pérez de Armiño, K. and Sánchez Montero, M. (2023). Improving famine early warning systems: A conflict-sensitive approach. *Conflict, Security & Development*, *23*(1), 23–42. https://doi.org/10.1080/14678802.2023.2188570

Mateus, G., Ferreira-Pêgo, C. and Assunção, R. (2025). Mapping the quantitative dose–response relationships between nutrients and health outcomes to inform food risk–benefit assessment. *Foods*, *14*(8), 1–23. https://doi.org/10.3390/foods14081420

Mešić, A., Jurić, M., Donsì, F., Maslov Bandić, L. and Jurić, S. (2024). Advancing climate resilience: Technological innovations in plant-based, alternative and sustainable food production systems. *Discover Sustainability*, *5*(1), 1–30. https://doi.org/10.1007/s43621-024-00581-z

Mezieobi, K.C., Alum, E.U., Egwu, C.K., Uti, D.E., Alum, B.N., Christine, A. and Daniel, O. (2025). Strategies for resilience: Mitigating the effects of climate change on hunger and mental health. *Journal of Agriculture and Food Research*, 1–11. https://doi.org/10.1016/j.jafr.2025.102023

Mishra, S., Jha, C.K., Goswami, D. and Saraf, M. (2025). *Pseudomonas fluorescens*: An influencer for food crop biofortification by siderophore-mediated iron acquisition. In P. Pandey, S. Dheeman, and D. K. Maheshwari (Eds.), *Microorganisms resilience to climate change. Microorganisms for Sustainability*, vol 56. Springer, Singapore. https://doi.org/10.1007/978-981-96-3748-5_10

Mohamed-Saleem, A., and Osman, I.I. (2025). Food insecurity among Toronto Muslim households during COVID and the role of key Muslim charitable institutions. In R. Jureidini, S.F. Hassan, and D. Aassouli (Eds.), *Food security and Islamic ethics* (pp. 266–290). Edward Elgar Publishing.

Mohammadlo, M., Fazli, A.R., Sajjadi, J. and Cheraghi, M. (2024). Assessing the resilience of rural settlements in Mahneshan Township against food insecurity in drought conditions. *Desert Management, 12*(3), pp. 73–98.

Mosca, I. and Nolan, A. (2025). The long-term effects of in utero exposure to Rubella. *Oxford Bulletin of Economics and Statistics*, pp. 1–18. https://doi.org/10.1111/obes.12663

Neal, C., Pepper, G.V., Allen, C., Shannon, O.M. and Nettle, D. (2024). No effect of hunger on the memory of food images and prices. *Adaptive Human Behavior and Physiology, 10*(3), 303–323. https://doi.org/10.1007/s40750-024-00247-0

Öztürk, A.B., Heathcote-Fumador, I.E., McSey, I.A., M'Nkubitu, E., Thomi, D., Wainaina, S. and Taherzadeh, M.J. (2025). Food waste management through machine learning, IoT, and blockchain. In R. Sirohi, A. Tarafdar, L. P. de S. Vandenberghe, M. J. Taherzadeh, and A. Pandey (Eds.), *Sustainable Technologies for Food Waste Management* (pp. 233–262). https://doi.org/10.1201/9781032706030-12

Park, S., Chen, J. and Bustamante, A.V. (2024). Adverse consequences of food insecurity among US adults beyond health outcomes. *American Journal of Preventive Medicine, 66*(1), 146–153. https://doi.org/10.1016/j.amepre.2023.09.003

Phungwayo, T. (2020). *Assessing multi-sectoral policy coordination for achievement food and nutrition security outcomes in Eswatini* (Doctoral dissertation, Stellenbosch University).

Rahman, M.S., Wu, O.Y., Battaglia, K., Blackstone, N.T., Economos, C.D. and Mozaffarian, D. (2024). Integrating food is medicine and regenerative agriculture for planetary health. *Frontiers in Nutrition, 11*, 1–8. https://doi.org/10.3389/fnut.2024.1508530

Ramadugu, D. (2024). Increasing Rural Nigeria's Food Security by Bolstering Climate Resilience in Agriculture. [Global Youth Institute paper]. World Food Prize Foundation.

Reuveni, S. (2025). The role of food prices and market access in contributing to hunger in low-income populations. *Eximia, 14*, 69–96. https://doi.org/10.47577/eximia.v14i1.531

Ruhi, T.N., Muhumuza, R., Namanya, G. and Mayhew, S.H. (2025). Interconnections between health, environment, demographic, and livelihood factors: Shaping food security in Rukiga, Uganda. In *The Elgar Companion to Health and the Sustainable Development Goals* (pp. 226–247). Edward Elgar Publishing. https://doi.org/10.4337/9781803927244.00022

Saavedra, J.M. and Prentice, A.M. (2023). Nutrition in school-age children: A rationale for revisiting priorities. *Nutrition Reviews, 81*(7), 823–843. https://doi.org/10.1093/nutrit/nuac089

Sakhong, R., Jaiswal, P., Yosung, L., Shukla, Y.K. and Kumari, S. (2024). Technology innovation: A green approach to soil health. *International Journal of Plant & Soil Science, 36*(12), 613–622. https://doi.org/10.9734/ijpss/2024/v36i125235

Saraswathi, P., Loghitha, K.Y., Poorvaja, J.V., Prabha, M. and Dhiyanesh, B. (2024). Optimizing Waste Management Systems: A Technological Approach. In *2024 IEEE International Conference on Blockchain and Distributed Systems Security (ICBDS)* (pp. 1–5). IEEE.

Saxena, D.R., Saxena, A.D., Tupkar, N.J., Karim, F.A. and Irving, A.L. (2025). Understanding sustainable development goal 2: Zero hunger. In *Smart Technologies for Sustainable Development Goals* (pp. 20–41). CRC Press. https://doi.org/10.1201/9781003569664-2

Scott, S., Neupane, S., Tasic, H., Akseer, N., Kim, S.S., Alderman, H., Carducci, B. and Heidkamp, R. (2025). Co-coverage of social protection programs and maternal and child nutrition interventions in Bangladesh, India and Nepal. *medRxiv*, pp.2025–01. https://doi.org/10.1101/2025.01.22.25320966

Sharda, K.L.S. (2024). The impact of hunger on physical and mental health. *International Journal for Multidisciplinary Research*, 6(4). https://doi.org/10.36948/IJFMR.2024.V06I04.24860

Sharma, T., Mrabet, R., Singh, A., Kumar, N., Baghla, D., Kumar, S., Rana, B.B. and Chauhan, G. (2024). Global hunger: The need for smart and sustainable agriculture. In Y.K. Mohanta, I. Chakrabartty, and S. Muthupandian (Eds.), *Sustainable Green Nanotechnology* (pp. 129–151). CRC Press. https://doi.org/10.1201/9781003389408-8

Uçar, E.D. (2024). Exploring the potential of augmented reality (AR) and virtual reality (VR) in developing immersive experiences for education, entertainment, and training: Innovations, applications, and future prospects. *Human Computer Interaction, 8*(1), 47. http://dx.doi.org/10.62802/1n99md93

Vilakazi, B., Mafongoya, P.L., Odindo, A.O. and Phophi, M.M. (2025). The role of neglected grain legumes in food and nutrition security and human health. *Sustainability, 17*(1), 1–28. https://doi.org/10.3390/su17010350

Wibisono, D.A.S., Saw, C.Y., Wu, T.Y. and Chau, C.F. (2025). Advancing industrial food byproduct management: Strategies, technologies, and progress in waste reduction. *Processes, 13*(1), 1–27. https://doi.org/10.3390/pr13010084

Wilts, R. (2025). *Computable General Equilibrium assessment of Sustainable Development Goals in low-income countries with household detail* (Doctoral dissertation, Universitäts-und Landesbibliothek Bonn).

Woldegebriel, A.G., Gebrehiwot, G.G., Teka, A.G., Gebreyesus, H.A., Weldeslassie, T.T., Desta, A.A., Yigzaw, H.K., Bezabih, N.M., Baryau, H.K., Haileselassie, M., and Gebregziabiher, M.B. (2024). High manifestations of food insecurity and hunger among pregnant and lactating women during post-conflict in Tigray, Ethiopia: A community-based study. *BMC Pregnancy and Childbirth, 24*(1), 385. https://doi.org/10.1186/s12884-024-06550-8

World Food Programme (WFP). (2022). Global Food Crisis 2022. Update on the World's Unprecedented Needs 2022. https://www.wfp.org/publications/update-global-food-crisis-2022

Yuan, X., Zhong, M., Huang, X., Hussain, Z., Ren, M., and Xie, X. (2024). Industrial production of functional foods for human health and sustainability. *Foods (Basel, Switzerland), 13*(22), 3546. https://doi.org/10.3390/foods13223546

Yoğun, K. and Yıldırım, N. (2025). Research and conceptual landscape of food waste and business models in the context of the circular economy: A literature review at the intersection of SDG 2–Zero hunger and SDG 12–Responsible consumption and production. *Journal of Lifestyle and SDGs Review, 5*(1), 1–28. https://doi.org/10.47172/2965-730X.SDGsReview.v5.n01.pe04757

5 Unraveling the Nexus Between Food Insecurity and Mental Health

Abraham Olufemi Asuku and Ayobami Grace Fajemidagba

5.1 INTRODUCTION

Food security (FS) is the assured availability of safe, nutrients-containing food which meets cultural requirements and is gotten in a way that is acceptable socially. However, food insecurity (FI), which affects many households worldwide and is brought on by a lack of resources, leads to malnutrition (Shen et al., 2015). Despite tremendous improvements in global food production for the past 2 decades and 5 years, close to 795 million people still FI (Jones, 2017). Additionally, overweight (Moradi et al., 2019a), anemia (Moradi et al., 2018), diabetes (Seligman et al., 2007), sustained high blood pressure (Pérez-Escamilla et al., 2014), asthma (Kirkpatrick et al., 2010), adverse development (Moradi et al., 2019b), and cancer (Gany et al., 2014) are all more common in people with FI.

FI may affect not only nutritional and physical health but it also affects mental health (MH). People with FI also experience psychological issues, such as worry and despair, as a result of their inability to support themselves and their children and their inaccessibility to culturally appropriate and affordable food (Carter et al., 2011). Whitaker et al. (2006) claim that FI causes uncertainty regarding the ability of someone to save and get adequate food, which triggers a stress reaction that feeds depressive and anxious feelings. Furthermore, eating in socially unacceptable ways may lead to depressive symptoms as isolation, shame, guilt, and helplessness (Nanama & Frongillo, 2012; Bernal et al., 2016). According to Weaver and Hadley (2009), it also makes income disparities in families and communities worse, which might increase cultural sensitivity and worsen MH. A worldwide FI assessment demonstrated that it was connected with different psychosocial stressors and poorer MH in different regions (Jones, 2017). Even after adjusting for socioeconomic characteristics and confounding demographic, these relationships remained (Carter et al., 2011). The different ways that FI could be involved in common mental disorders as well as the wider social consequences of FI related to self-efficacy and cultural norms could be the reason for the consistency of findings from one culture to another from past study (Nanama & Frongillo, 2012; Bernal et al., 2016).

According to Ponda et al. (2017), anxiety, stress, and depression all have a major role in the global burden of disease and in raising public concerns about socioeconomic differences and how they affect society's financial cost. Stress in early period of life has such a great effects on the brain development; which eventually impact behavior and MH for the rest of affected person's life (Asuku et al., 2025). Depressive disorders manifest as sadness, apathy, low self-esteem or guilt, difficulty eating or sleeping, exhaustion, and difficulties focusing. Accordingly, stress is defined as a particular connection between someone and his environment that is appraised by the person as taxing or exceeding his or her resources and putting his or her well-being in to danger (Araiza & Lobel, 2018), and anxiety-related disorders are a category of mental illnesses that are typified by feelings of fear and anxiety (Dew et al., 2015).

According to WHO estimates, the global numbers of people that suffer from anxiety disorder is 264 million, while over 322 million people worldwide suffer from depression (WHO, 2017). Since observational studies have not yet shown the consistent FI impacts on stress, anxiety, and depression,

 DOI: 10.1201/9781003650881-5

the aforementioned data could be inadequate to conclude from. As a result, this chapter highlights the connections that exist between FI and the probability of anxiety, stress, and depression.

5.2 UNDERSTANDING FOOD SECURITY

Since more than a billion individuals are inaccessible to enough nutrient-dense food in a way that is acceptable socially for them to live active lifestyles, FS is a worldwide issue (Bernal et al., 2014). Based on the Food and Agricultural Organization (FAO) definition, FS is achieved when the people around the world continuously has socioeconomic and physical access to safe, sufficient, and nourishing food that can meet their preferences and dietary requirements for healthy and active lifestyle (Bernal et al., 2014). This thorough definition of FS included significant elements of food accessibility, availability, as well as utilization. In the past, more terms were introduced, including food system sustainability and stability (Berry et al., 2015).

Leroy et al. (2015) define "food availability" as the steady and unbroken supply of enough food in a suitable quality and quantity to provide a healthy living condition. Also, availability of food takes in to consideration food safety and sociocultural acceptability to guarantee that food is served in a way that honors human dignity and is free of hazardous or microbial components that may have adverse effects on the health of humans (Leroy et al., 2015). A vital part of food security, food availability, has been the subject of various ideas to enhance the production of food worldwide (Synthesis of Technical Background Documents, 2018). Also, it entails creating adequate transportation infrastructure to ensure the availability of food even in rural regions from its source (Mittal et al., 2018).

Although availability and access to food are closely related, the former does not always determine the latter. The resources and policies needed to ensure that everyone is accessible to enough supply of wholesome, secure, and socially acceptable food are referred to as food access. These tools address a variety of subjects, including economic status (finances), physical (reliable transportation networks), and social (traditional eating patterns) (Ashby et al., 2016). Since 2014, there has been an upsurge in world hunger (FAO, 2018). The malnourished persons global percentage rose to 10.6% in 2015 and subsequently to 11% in 2016 (FAO, 2018). The FAO reported that 821 million individuals, or around 1 in 9 persons, were undernourished in 2017 (FAO, UNICEF, WFP, WHO, 2023). Up to 60% of the world's hungry are women, and around 5 million children under five years mortality cases were recorded yearly from causes associated to malnutrition, according to the FAO's 2018 figures on FI (FAO, UNICEF, WFP, WHO, 2023). The increase in FS raises the possibility that the Sustainable Development Goal (SDG) of putting an end to hunger by 2030 would be unachievable (FAO, UNICEF, WFP, WHO, 2023). FI is especially pronounced in developing countries, which have low per capita income (Pawlak, 2016). The prevalence of severe FI in Sub-Saharan Africa was as high as 33.8% and the prevalence of FS in severe, moderate, or mild, was as high as 76.1%, according to a 2016 study by Jones that looked at the 2014 Gallup World Poll cross-sectional data in 149 nations (Jones, 2017). Food insecure individuals are mostly in Asia. According to estimates, 460.2 million individuals in Pacific and Asia had extreme FI in the year 2021. In 2021, 49.8% of the world's 923.7 million inhabitants resided in the region. In comparison with 17.4 million in Southeastern Asia, 28.0 million in Eastern Asia, and 2.0 million in Oceania, 412.9 million people in Southern Asia experienced extreme FS (FAO, UNICEF, WFP, WHO, 2023). According to a 2015 study, FS disproportionately affects women, children, and racial minorities, impacting around 16 million families in the United States, with a prevalence of up to 12.7% (Bruening et al., 2017; Myers & Painter, 2017).

5.3 THE PSYCHOLOGICAL IMPACT OF HUNGER

According to Weaver and Hadley (2009), hunger is a stressful state caused by an involuntary food absence and an insufficient calorie intake. It is most common in Sub-Saharan Africa. Compared to East Asia and the Americas, South Asia has substantially higher prevalence rates (Ritchie et al., 2023). Incidence rates remain below 2.5% in both North America and Europe. The prevalence of

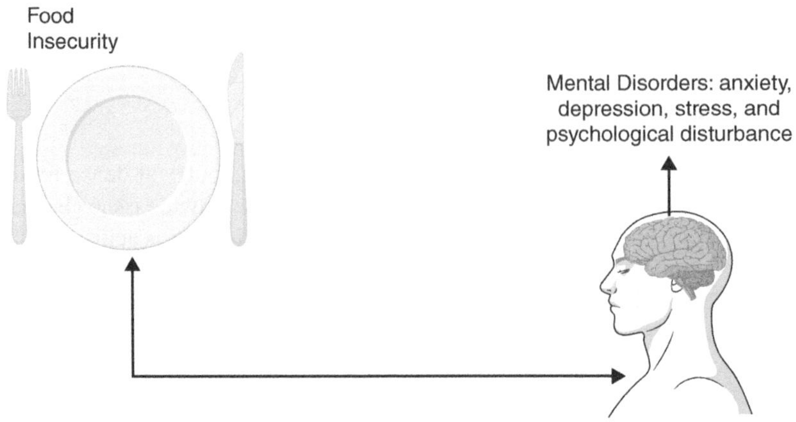

FIGURE 5.1 The connection between FI and MH.

hunger, which is characterized as a state of tension caused by an imposed absence of food and an inadequate calorie intake, makes these numbers concerning (Weaver & Hadley, 2009). Hunger is most prevalent in Sub-Saharan Africa. On the other hand, compared to East Asia and the Americas, South Asia has substantially higher prevalence rates (Ritchie et al., 2023). Incidence rates remain below 2.5% in both North America and Europe. The larger number of the world's old population lives in Sub-Saharan Africa, and over 0.8 billion people remain hungry worldwide (FAO, IFAD, UNICEF, WFP, WHO, 2023). These are concerned figures considering the great impact of malnutrition and FI on MH.

The WHO defines MH as a condition of being healthy in which the person is aware that his abilities, can tolerate the life stresses, work fruitfully and productively, and can impact his community positively (FAO, IFAD, UNICEF, WFP, WHO, 2023). FS is recognized to impact people's MH in addition to their body weight (Morales & Berkowitz, 2016; Weaver et al., 2021). Psychological acceptability issues, which lead to feelings of deprivation and anxiety about food availability, are among the conditions associated with FS. Thus, MH could suffer due to FI (Myers, 2020) as shown in Figure 5.1.

5.4 IMPACTS OF FI AND MALNUTRITION ON MH

Foods influence our physical well-being, but research demonstrates that they could as well have effect on our MH (Asuku et al., 2024a), as food intake has been reported to be regulated by the central nervous system (Asuku et al., 2024b), which leads to the field of psychiatry and nutrition. Jones carried out a global study that looked at the nexus between FI and poor MH across 149 nations discovered a substantial dose-response connection between the two (Jones, 2017). The study discovered a nexus between the degree of FI and a slow deterioration in MH (Jones, 2017). Weaver and Hadley evaluated 27 quantitative and qualitative researches from developing countries and found that food poverty is linked to melancholy, anxiety, tension, and guilt (Weaver & Hadley, 2009). FI and MH are correlated in both directions, according to a 2009 systematic study, with hunger causing poor emotional conditions and vice versa (Bruening et al., 2017).

Other psychological symptoms like worry, sleep issues, and suicidal thoughts have also been connected with FI and hunger (Bruening et al., 2017). Proper nutrition is crucial for the healthy maturation of the brain of fetus at pregnancy and the neonatal period, based on the study on MH and cognitive development (Cook & Frank, 2008). Due primarily to its function in shortages of micronutrient, FI is associated with retardation of early growth, impaired development of cognition, and MH indices poorer scores (Black et al., 2017). In food-insecure families, children have higher chances of displaying negative behaviors, as well as experiencing psychosocial stress and parental antagonism (Gee & Asim, 2019; Leung et al., 2020).

Grineski et al. (2018) demonstrated how food hardship impacts executive functions, children's academic performances, and social skills. The authors discovered that mathematical ability, self-control, and working memory were remarkably impacted by FI. Interpersonal and executive functions are also worse in kids from families that experienced FI. According to their findings, an increase in FI was connected to worse math scores and working memory (Grineski et al., 2018). A significant link of some sort was identified in most of cross-sectional researches that investigated the relationship that exists between depression and FI (Tarasuk et al., 2013; McCurdy et al., 2015). A plethora of researches have also found a connection between FI and depression, with FI elevating the chances of being diagnosed with depression or experiencing the symptoms of depression (Siefert et al., 2007; Melchior et al., 2009). Bronte-Tinkew et al. (2007) reported that mothers who passed through FI have greater susceptibility of displaying sadness symptoms than those who were not. Some authors reported that the connection was either reciprocal (Huddleston-Casas et al., 2009) or inverse, with FI caused by depression (Garg et al., 2015; Laraia et al., 2015). It has been reported that depressive mothers have higher susceptibility to be food insecure than those without depression (Garg et al., 2015). Other maternal deprivation variables were also assessed in relation to FI and depression. For instance, Corman et al. (2016) reported that females who have passed through main depressive episode at baseline demonstrated higher susceptibility of experiencing FI and inadequate housing at follow-up. A plethora of researches involving women who were pregnant have associated FI with prenatal and postpartum depression (Hromi-Fiedler et al., 2011; Sidebottom et al., 2014).

Food secure, pregnant women were less susceptible to have prenatal symptoms of depression than their colleagues who were food insecure (Hromi-Fiedler et al., 2011; Sidebottom et al., 2014). A plethora of researches have unraveled the nexus between FI, stress, and anxiety (Tarasuk et al., 2013; Martin et al., 2016).

Pregnant women who undergone FI reported greater levels of stress than those who did not, and those who pass through any level of FI at pregnancy period or three months after giving birth were more susceptible to report feeling stressed a year later, according to research Laraia et al. (2015). Martin et al. (2016) examined the degree of stress experienced by Canadians and found that when FS decreased, the probability of experiencing high levels of stress increased. Nevertheless, Trapp et al. (2015), in a cross-sectional study examined FI among 222 low-income women and their children and found no difference in the subjective stress levels of the food secure and insecure groups. Women who were food insecure were studied for emotional or disordered eating habits (Dressler & Smith, 2015; Sharpe et al., 2016). Laraia et al. (2015) and Sharpe et al. (2016) discovered bivariate relationships between FI and emotional or disturbed eating habits. Nevertheless, in models that involved sociodemographic factors, Laraia et al. (2015) did not discover any meaningful correlations between FI and abnormal eating behaviors. Emotional eating may serve as a mediator between MH, FI, and other food-related outcomes like weight and dietary consumption, according to Dressler and Smith (2015), who examined the connection between emotional eating and depression.

Furthermore, FI raised the susceptibility of having poor mental and physical health (Sharkey et al., 2011), and comorbid mental and physical health conditions increased susceptibility to FI (Tarasuk et al., 2013). This was the conclusion of a research that investigated a variety of mental and physical health conditions. There was significant emphasis on the implications for other family members, especially children. The authors documented that females in homes where there are FI had higher depressive levels that were linked to their offspring's poor well-being (Bronte-Tinkew et al., 2007).

5.5 GLOBAL PERSPECTIVES ON THE NEXUS BETWEEN FI AND MH: CASE STUDIES

FI is an intricate issue that affects persons in many diverse nations and regions. It remains a remarkable global concern. FI and MH are associated, and this offers vast opportunities as well as obstacles for the sustainable development of the world. Table 5.1 shows the case studies on the connection between FI and MH across various parts of the world.

TABLE 5.1
Case Studies on How FI Affects MH

S/No	Regions and Countries	Affected Groups	Outcomes	References
1.	149 countries across the globe	General	The state of MH deteriorated as FI increases	Jones (2017)
2.	138 countries across the globe	Individuals who is 15 years and above	They discovered a substantial correlation between lower subjective well-being and food insecurity.	Frongillo et al. (2017)
3.	147 countries, more- and less-developed countries	General	Poor subjective well-being was linked to food insecurity, and the correlation was larger in more-developed than in less-developed nations.	Frongillo et al. (2018)
4.	Korea	19 years and above adults	Higher likelihood of reporting perceived stress and depressive symptoms.	Chung et al. (2016)
5.	Denmark	18 years old or more	Greater risks of psychological distress	Lund et al. (2017)
6.	Canada	Aged 19 years and older	Lower levels of psychological well-being.	Hossain and Lamb (2019)
7.	United States	Adults	Mental distress and increased frequency of insufficient sleep	Liu et al. (2014)
8.	United States	African-American adults	Higher and serious psychological stress	Allen et al. (2018)
9.	Australia	Adults	The connection between FI and mental problems, as indicated by generalized depressive and anxiety symptoms was strengthened by drought exposure.	Friel et al. (2014)
10.	Uganda	18 years and above adults	Greater depression symptom severity	Perkins et al. (2018)
11.	United States	Above 50 years	More depressive symptoms	Pak and Kim (2019)
12.	United States	12–17 years adolescents	Adolescent girls report higher levels of anxiety than do boys.	Maynard et al. (2019)
13.	India	13–19 years teenage girls	High degrees of psychological anguish, anxiety, depression, and lack of behavioral control	Rani et al. (2018)
14.	Ethiopia	17–21 years youths	Lack of sleep or memory problems, as well as psychological problems.	Jebena et al. (2016)
15.	United States	Freshmen college students	Greater depressive symptoms	Darling et al. (2017)
16.	South India	People living with HIV (PLHIV)	PLHIV with food insecurity who were male but not female also reported higher levels of depression.	Heylen et al. (2015)
17.	Ecuador	Women with babies	Mental health issues like stress and sadness, as well as low mental health scores	Weigel et al. (2016)
18.	Rio de Janeiro, Brazil,	Women	Psychological and physical violence	Moraes et al. (2016)
19.	United States	Fathers and mothers who are Nationals	serious psychological distress among both fathers and mothers	Tseng et al. (2017)
20.	United States	Men and women	Greater depressive symptoms. Low psychological benefits of biological fruitfulness in women	Ciciurkaite and Brown (2018)

5.6 CONCLUSION

FI assessment and its linkage to MH in this chapter showed how challenging and urgent it is to tackle these worldwide issues. Various locations, innovative methods, and responses thorough case studies assist to have a profound comprehension of the intricate nature of FI. The documented impacts on MH are vital because they underscore the urgent need for comprehensive interventions that include social, diet, and healthcare support. There are glimpses of hope in spite of the challenges, thanks to global success stories and innovative solutions. The solution comprises interventions that are technology-based and social network programs which focus on the needed adaptability to take into cognizance the diverse sociocultural, environmental, and economic factors that lead to FI.

The future requires a deliberate implementation of economic, healthcare, social policies tailored toward worldwide collaboration and governance that are evidence-based. A future where people are both emotionally balanced and food secure is made possible by the focus on research and assessment, MH assistance integration into community activities, and the understanding of the MH and FI connections. A holistic approach provides the template for development, which is sustainable and equitable, promoting a community where accessibility to enough wholesome food and health of the people are given first priority. This chapter highlights the nexus between FI and MH, with emphasis on diverse case studies. In order to effectively address these interrelated concerns, thorough plans and compelling policy recommendations are essential. There is need for further studies, cooperation among nations, and the incorporation of MH services into community activities to advance sustainable growth that prioritizes mental and nutritional health. Although this chapter on FI and MH provides useful information, there are few limitations. The study's main focus areas are responses and case studies from specific locations, and initiatives. This can provide limitation to the generalization of the finding in other geographic and socioeconomic circumstances. Different research and geographical locations may employ different approaches to measure the outcomes of FI and MH, which could affect the findings' robustness and comparability. However, this chapter does set out a great path for further studies, pointing out places where more literature, policy, or methods trials might enhance understanding and solutions to the issues of FI and MH connection around the world.

REFERENCES

Allen, N.L., Becerra, B.J., & Becerra, M.B. (2018). Associations between food insecurity and the severity of psychological distress among African-Americans. Ethnicity & Health. 23(5):511–520. https://doi.org/10.1080/13557858.2017.1280139.

Araiza, A.M., & Lobel, M. (2018). Stress and eating: Definitions, findings, explanations, and implications. Social and Personality Psychology Compass. 12, e12378.

Ashby, S., Kleve, S., McKechnie, R., & Palermo, C. (2016). Measurement of the Dimensions of food insecurity in developed countries: A systematic literature review. Public Health Nutrition. 19, 2887–2896.

Asuku, A.O., Adebayo, P.O., & Ogungbangbe, G.O. (2025). Stress and gender differences in brain development. In Moradikor, N., Orish, C. (eds) Stress and the Brain: Evidence from Molecular Mechanisms to Interventions and Therapies, Elsevier. https://doi.org/10.1016/bs.pbr.2025.01.013

Asuku, A.O., Ayinla, M.T., & Ajibare, A.J. (2024a). Measuring mood in nutritional research. In: Mohamed, W., Kobeissy, F. (eds) Nutrition and Psychiatric Disorders. Nutritional Neurosciences. Springer, Singapore. https://doi.org/10.1007/978-981-97-2681-3_10

Asuku, A.O., Ayinla, M.T., & Ajibare, A.J. (2024b). Central nervous system regulation of eating and brain functions. In: Mohamed, W., Kobeissy, F. (eds) Nutrition and Psychiatric Disorders. Nutritional Neurosciences. Springer, Singapore. https://doi.org/10.1007/978-981-97-2681-3_4

Bernal, J., Frongillo, E.A., Herrera, H.A., & Rivera, J.A. (2014). Food insecurity in children but not in their mothers is associated with altered activities, school absenteeism, and stunting. Journal of Nutrition. 144, 1619–1626.

Bernal, J., Frongillo, E.A., & Jaffe, K. (2016). Food insecurity of children and shame of others knowing they are without food. Journal of Hunger & Environmental Nutrition. 11, 180–194.

Berry, E.M., Dernini, S., Burlingame, B., Meybeck, A., & Conforti, P. (2015). Food security and sustainability: Can one exist without the other? Public Health Nutrition. 18, 2293–2302.

Black, M.M., Walker, S.P., Fernald, L.C.H., Andersen, C.T., DiGirolamo, A.M., Lu, C., McCoy, D.C., Fink, G., Shawar, Y.R., & Shiffman, J. (2017). Early childhood development coming of age: Science through the life course. Lancet. 389, 77–90.

Bronte-Tinkew, J., Zaslow, M., Capps, R., Horowitz, A., & McNamara, M. (2007). Food insecurity works through depression, parenting, and infant feeding to influence overweight and health in toddlers. Journal of Nutrition. 137, 2160–2165.

Bruening, M., Dinour, L.M., & Chavez, J.B.R. (2017). Food insecurity and emotional health in the USA: A systematic narrative review of longitudinal research. Public Health Nutrition. 20, 3200–3208.

Carter, K.N., Kruse, K., & Blakely, T. (2011). The association of food security with psychological distress in New Zealand and any gender differences. Social Science & Medicine. 72, 1463–1471.

Chung, H.K., Kim, O.Y., Kwak, S.Y., Cho, Y., Lee, K.W., & Shin, M.J. (2016). Household food insecurity is associated with adverse mental health indicators and lower quality of life among Koreans: Results from the Korea National Health and Nutrition Examination Survey 2012–2013. Nutrients. 8(12). https://doi.org/10.3390/nu8120819.

Ciciurkaite, G., & Brown, R.L. (2018). Food insecurity, psychological distress and alcohol use: understanding the salience of family roles for gender disparities. Health Sociology Review. 27(3):294–311. https://doi.org/10.1080/14461242.2018.1461574.

Cook, J.T., & Frank, D.A. (2008). Food security, poverty, and human development in the United States. Annals of the New York Academy of Sciences. 1136, 193–209.

Corman, H., Curtis, M.A., Noonan, K., & Reichman, N.E. (2016). Maternal depression as a risk factor for children's inadequate housing conditions. Social Science & Medicine. 149, 76–83.

Darling, K.E., Fahrenkamp, A.J., Wilson, S.M., D'Auria, A.L., & Sato, A.F. (2017). Physical and mental health outcomes associated with prior food insecurity among young adults. Journal of Health Psychology. 22(5):572–581. https://doi.org/10.1177/1359105315609087.

Dew, M.A., Rosenberger, E.M., & Myaskovsky, L. (2015). Depression and anxiety as risk factors for morbidity and mortality after organ transplantation: a systematic review and meta-analysis. Transplant. 100, 988.

Dressler, H., & Smith, C. (2015). Depression affects emotional eating and dietary intake and is related to food insecurity in a group of multiethnic, low-income women. Journal of Hunger & Environmental Nutrition. 10, 496–510.

FAO, IFAD, UNICEF, WFP, WHO. (2023). The state of food security and nutrition in the world 2023: Urbanization. Agrifood Systems Transformation and Healthy Diets across the Rural–Urban Continuum, Food and Agriculture Organization, Rome, Italy.

FAO, UNICEF, WFP, WHO. (2023). Asia and the Pacific—Regional Overview of Food Security and Nutrition 2022; Urban Food Security and Nutrition. FAO, Bangkok, Thailand.

Food and Agricultural Organization (FAO). (2018). Building Climate Resilience for Food Security and Nutrition; The State of Food Security and Nutrition in the World. FAO, Rome, Italy.

Friel, S., Berry, H., Dinh, H., O'Brien, L., & Walls, H.L. (2014). The impact of drought on the association between food security and mental health in a nationally representative Australian sample. BMC Public Health. 14:1102. https://doi.org/10.1186/1471-2458-14-1102.

Frongillo, E.A., Nguyen, H.T., Smith, M.D., & Coleman-Jensen, A. (2017). Food insecurity is associated with subjective well-being among individuals from 138 countries in the 2014 Gallup World Poll. Journal of Nutrition. 147(4):680–687. https://doi.org/10.3945/jn.116.243642.

Frongillo, E.A., Nguyen, H.T., Smith, M.D., & Coleman-Jensen, A. (2018). Food insecurity is more strongly associated with poor subjective well-being in more-developed countries than in less developed countries. Journal of Nutrition. 149(2):330–335. https://doi.org/10.1093/jn/nxy261.

Gany, F., Lee, T., & Ramirez, J. (2014). Do our patients have enough to eat? Food insecurity among urban low-income cancer patients. Journal of Health Care for the Poor and Underserved. 25, 1153.

Garg, A., Toy, S., Tripodis, Y., Cook, J., & Cordella, N. (2015). Influence of maternal depression on household food insecurity for low-income families. Academic Pediatrics. 15, 305–310.

Gee, K.A., & Asim, M. (2019). Parenting while food insecure: Links between adult food insecurity, parenting aggravation, and children's behaviors. Journal of Family Issues. 40, 1462–1485.

Grineski, S.E., Morales, D.X., Collins, T.W., & Rubio, R. (2018). Transitional dynamics of household food insecurity impact children's developmental outcomes. Journal of Developmental & Behavioral Pediatrics. 39, 715–725.

Heylen, E., Panicker, S.T., Chandy, S., Steward, W.T., & Ekstrand, M.L. (2015). Food insecurity and its relation to psychological well-being among south Indian people living with HIV. AIDS and Behavior. 19(8):1548–1558. https://doi.org/10.1007/s10461-014-0966-x.

Hossain, B., & Lamb, L. (2019). Economic security and psychological distress among Indigenous Canadians. Journal of Developing Areas. 53(1):109–125.

Hromi-Fiedler, A., Bermúdez-Millán, A., Segura-Pérez, S., & Pérez-Escamilla, R. (2011). Household food insecurity is associated with depressive symptoms among low-income pregnant Latinas. Maternal & Child Nutrition. 7, 421–430.

Huddleston-Casas, C., Charnigo, R., & Simmons, L.A. (2009). Food insecurity and maternal depression in rural, low-income families: A longitudinal investigation. Public Health Nutrition. 12, 1133–1140.

Jebena, M.G., Lindstrom, D., Belachew, T., Hadley, C., Lachat, C., & Verstraeten, R. (2016). Food insecurity and common mental disorders among Ethiopian Youth: Structural equation modeling. PLOS One. 11(11):e0165931. https://doi.org/10.1371/journal.pone.0165931.

Jones, A.D. (2017). Food insecurity and mental health status: A global analysis of 149 countries. American Journal of Preventive Medicine. 53(2):264–273. https://doi.org/10.1016/j.amepre.2017.04.008.

Kirkpatrick, S.I., McIntyre, L., & Potestio, M.L. (2010). Child hunger and long-term adverse consequences for health. Archives of Pediatrics & Adolescent Medicine. 164, 754–762.

Laraia, B., Vinikoor-Imler, L.C., & Siega-Riz, A.M. (2015). Food insecurity during pregnancy leads to stress, disordered eating, and greater postpartum weight among overweight women. Obesity. 23, 1303–1311.

Leroy, J.L., Ruel, M., Frongillo, E.A., Harris, J., & Ballard, T.J. (2015). Measuring the food access dimension of food security: A critical review and mapping of indicators. Food Nutrition. Bull. 36, 167–195.

Leung, C.W., Stewart, A.L., Portela-Parra, E.T., Adler, N.E., Laraia, B.A., & Epel, E.S. (2020). Understanding the psychological distress of food insecurity: A qualitative study of children's experiences and related coping strategies. Journal of the Academy of Nutrition and Dietetics. 120, 395–403.

Liu, Y., Njai, R.S., Greenlund, K.J., Chapman, D.P., & Croft, J.B. (2014). Relationships between housing and food insecurity, frequent mental distress, and insufficient sleep among adults in 12 US States, 2009. Preventing Chronic Disease. 11:E37. https://doi.org/10.5888/pcd11.130334.

Lund, T.B., Holm, L., Tetens, I., Smed, S., & Nielsen, A.L. (2017). Food insecurity in Denmark—Socio demographic determinants and associations with eating- and health-related variables. European Journal of Public Health. 28(2):283–288. https://doi.org/10.1093/eurpub/ckx121.

Martin, M.S., Maddocks, E., Chen, Y., Gilman, S.E., & Colman, I. (2016). Food insecurity and mental illness: Disproportionate impacts in the context of perceived stress and social isolation. Public Health. 132, 86–91.

Maynard, M.S., Perlman, C.M., & Kirkpatrick, S.I. (2019). Food insecurity and perceived anxiety among adolescents: An analysis of data from the 2009–2010 National Health and Nutrition Examination Survey (NHANES). Journal of Hunger & Environmental Nutrition. 14(3):339–351. https://doi.org/10.1080/19320248.2017.1393363.

McCurdy, K., Kisler, T., Gorman, K.S., & Metallinos-Katsaras, E. (2015). Food- and health-related correlates of self-reported body mass index among low-income mothers of young children. Journal of Nutrition Educcation and Behavior. 47, 225–233.

Melchior, M., Caspi, A., Howard, L.M., Ambler, A.P., Bolton, H., Mountain, N., & Moffitt, T.E. (2009). Mental health context of food insecurity: A representative cohort of families with young children. Pediatrics. 124, e564–e572.

Mittal, A., Krejci, C., & Craven, T. (2018). Logistics best practices for regional food systems: A review. Sustainability. 10, 168.

Moradi, S., Arghavani, H., & Issah, A. (2018). Food insecurity and anaemia risk: A systematic review and meta-analysis. Public Health Nutrition. 21, 3067–3079.

Moradi, S., Mirzababaei, A., & Dadfarma, A. (2019a). Food insecurity and adult weight abnormality risk: A systematic review and meta-analysis. European Journal of Nutrition. 58, 45–61.

Moradi, S., Mirzababaei, A., & Mohammadi, H. (2019b). Food insecurity and the risk of undernutrition complications among children and adolescents: A systematic review and meta-analysis. Nutrition. 62, 52–60.

Moraes, C.L., Marques, E.S., Reichenheim, M.E., Ferreira, M.F., & Salles-Costa, R. (2016). Intimate partner violence, common mental disorders and household food insecurity: An analysis using path analysis. Public Health Nutrition. 2016;19(16):2965–2974. https://doi.org/10.1017/s1368980016001178.

Morales, M.E., & Berkowitz, S.A. (2016). The relationship between food insecurity, dietary patterns, and obesity. Current Nutrition Reports. 5, 54–60.

Myers, C.A. (2020). Food insecurity and psychological distress: A review of the recent literature. Current Nutrition Reports. 9, 107–118.

Myers, A.M., & Painter, M.A. (2017). Food Insecurity in the United States of America: An examination of race/ethnicity and nativity. Food Security. 9, 1419–1432.

Nanama, S., & Frongillo, E.A. (2012). Altered social cohesion and adverse psychological experiences with chronic food insecurity in the non-market economy and complex households of Burkina Faso. Social Science & Medicine. 74, 444–451.

Pak, T.Y., & Kim, G. (2019). Food stamps, food insecurity, and health outcomes among elderly Americans. Preventive Medicine. 105871. https://doi.org/10.1016/j.ypmed.2019.105871.

Pawlak, K. (2016). Food security situation of selected highly developed countries against developing countries. Journal of Agribusiness and Rural Development. 40, 385–398.

Pérez-Escamilla, R., Villalpando, S., & Shamah-Levy, T. (2014) Household food insecurity, diabetes and hypertension among Mexican adults: Results from Ensanut 2012. Salud Publica Mexico 56, Suppl. 1, s62–s70.

Perkins, J.M., Nyakato, V.N., Kakuhikire, B., Tsai, A.C., Subramanian, S.V., & Bangsberg, D.R. (2018). Food insecurity, social networks and symptoms of depression among men and women in rural Uganda: A cross-sectional, population-based study. Public Health Nutrition. 21(5):838–848. https://doi.org/10.1017/S1368980017002154.

Ponda, M.P., Liang, Y., & Kim, J. (2017). A randomized clinical trial in vitamin D–deficient adults comparing replenishment with oral vitamin D3 with narrow-band UV type B light: Effects on cholesterol and the transcriptional profiles of skin and blood. The American Journal of Clinical Nutrition. 105, 1230–1238.

Rani, D., Singh, J.K., Acharya, D., Paudel, R., Lee, K., & Singh, S.P. (2018). Household food insecurity and mental health among teenage girls living in urban slums in Varanasi, India: A cross-sectional study. International Journal of Environmental Research and Public Health. 15(8). https://doi.org/10.3390/ijerph15081585.

Ritchie, H., Rosado, P., & Roser, M. (2023). Hunger and Undernourishment. Our World in Data. https://ourworldindata.org/hunger-and-undernourishment (accessed 24 June 2024).

Seligman, H.K., Bindman, A.B., & Vittinghoff, E. (2007). Food insecurity is associated with diabetes mellitus: Results from the National Health Examination and Nutrition Examination Survey (NHANES) 1999–2002. Journal of General Internal Medicine. 22, 1018–1023.

Sharkey, J.R., Johnson, C.M., & Dean, W.R. (2011). Relationship of household food insecurity to health-related quality of life in a large sample of rural and urban women. Women Health. 51, 442–460.

Sharpe, P.A., Whitaker, K., Alia, K.A., Wilcox, S., & Hutto, B. (2016). Dietary intake, behaviors and psychosocial factors among women from food-secure and food-insecure households in the United States. Ethnicity & Disease. 26, 139–146.

Shen, X., Gao, X., & Tang, W. (2015). Food insecurity and malnutrition in Chinese elementary school students. British Journal of Nutrition. 114, 952–958.

Sidebottom, A.C., Hellerstedt, W.L., Harrison, P.A., & Hennrikus, D. (2014). An examination of prenatal and postpartum depressive symptoms among women served by urban community health centers. Archives of Women's Mental Health. 17, 27–40.

Siefert, K., Finlayson, T.L., Williams, D.R., Delva, J., & Ismail, A.I. (2007). Modifiable risk and protective factors for depressive symptoms in low-income African American mothers. American Journal of Orthopsychiatry. 77, 113–123.

Synthesis of Technical Background Documents. (2018). https://www.fao.org/3/w2537e/w2537e02.htm (accessed 11 February 2024).

Tarasuk, V., Mitchell, A., McLaren, L., & McIntyre, L. (2013). Chronic physical and mental health conditions among adults may increase vulnerability to household food insecurity. Journal of Nutrition. 143, 1785–1793.

Trapp, C.M., Burke, G., Gorin, A.A., Wiley, J.F., Hernandez, D., Crowell, R.E., Grant, A., Beaulieu, A., & Cloutier, M.M. (2015). The relationship between dietary patterns, body mass index percentile, and household food security in young urban children. Childhood Obesity. 11, 148–155.

Tseng, K.K., Park, S.H., Shearston, J.A., Lee, L., & Weitzman, M.. (2017). Parental Psychological distress and family food insecurity: Sad dads in hungry homes. Journal of Developmental and Behavioral Pediatrics. 38(8):611–618. https://doi.org/10.1097/dbp.0000000000000481.

Weaver, L.J., & Hadley, C. (2009). Moving beyond Hunger and Nutrition: A systematic review of the evidence linking food insecurity and mental health in developing countries. Ecology of Food Nutrition. 48, 263–284.

Weaver, L.J., Owens, C., Tessema, F., Kebede, A., & Hadley, C. (2021). Unpacking the "black box" of global food insecurity and mental health. Social Science & Medicine. 282, 114042.
Weigel, M.M., Armijos, R.X., Racines, M., Cevallos, W., & Castro, N.P. (2016). Association of household food insecurity with the mental and physical health of low-income urban ecuadorian women with children. Journal of Environmental and Public Health. 2016:14. https://doi.org/10.1155/2016/5256084.
Whitaker, R.C., Phillips, S.M., & Orzol, S.M. (2006). Food insecurity and the risks of depression and anxiety in mothers and behavior problems in their preschool-aged children. Pediatrics. 118, e859–e868.
World Health Organization (WHO). (2017). Depression and Other Common Mental Disorders: Global Health Estimates. World Health Organization, Geneva. https://www.who.int/publications/i/item/depression-global-health-estimates

6 Microbial Innovations in the Fight Against Hunger

Shireen Adeeb Mujtaba Ali, Shaikh Shagufta Mujiburrehman, Amir Vazirizadeh, and Mukul Machhindra Barwant

6.1 INTRODUCTION

A sustainable future relies on food safety and security. To achieve this goal innovative solutions are required for food sustainability and security challenges upholding food standards and adhering to Sustainable Development Goals (SDGs).This goal focuses on zero hunger, promoting clean water, climate action in terrestrial and aquatic ecosystem (Boqvist et al., 2018). Microorganisms can play vital role for enhancing ecosystem and to combat climate change. By leveraging the beneficial effects of microorganisms, we can enhance agricultural output, as these organisms can stimulate plant growth, improve soil quality, and boost crop production. Microorganisms present a promising avenue for sustainable agriculture, minimizing environmental impact while enhancing food systems (Bernal, 2024; Dheeman et al., 2024). For sustainable food production, most important aspect is soil health. Microbial sources are being used to enhance organic matter decomposition and to improve soil fertility (Sáez-Sandino et al., 2023). Microorganism plays vital role in improvement of soil fertility by acting as plant growth-promoting microorganisms (PGPMs). PGPMs can enhance the plant growth by producing phytohormones such as cytokinin, ethylene, gibberellin, auxin, and salicylic acid. Soil nutrient level can be improved by microbes by stimulating plant growth which is primary producer of food chain (Nascimento et al., 2021; Bernal, 2024). PGPMs can form symbiotic relationship with plants and improves nutrient uptake from soil phosphate solubilization, mobilizing nitrogen and phosphorus (Zhao et al., 2023). Along with promoting plant growth nitrogen-fixing bacteria also increases the fertility of soil thus decreasing the use of synthetic biofertilizers (Jhu & Oldroyd, 2023).

This chapter aims to explore about how microbial innovation transforming food systems, from development of biofertilizers to single cell protein (SCP) and mycoprotein production. Harnessing bacterial utility for production of fermented food and pesticides.

6.2 ENHANCING AGRICULTURAL PRODUCTIVITY

6.2.1 Nitrogen-Fixing Bacteria

Nitrogen is an essential element for the growth and development of living organisms; however, the presence of a triple bond makes it challenging for plants to use it directly. Microorganisms must convert nitrogen into a form that plants can utilize. Two types of nitrogen fixation are: biological nitrogen fixation (BNF) and industrial nitrogen fixation (Zehr & Capone, 2020; Kebede, 2021; Ladha et al., 2022). BNF accounts for 80% nitrogen fixation on earth. BNF is an essential process that transforms atmospheric nitrogen into ammonia, a form that living organisms can directly use.

Two types of bacteria have been studied for nitrogen fixation (Kumar et al., 2025):

1. *Symbiotic Nitrogen-Fixing Bacteria*: These bacteria form mutualistic relationships with plants, such as legumes, and provide them with fixed nitrogen in exchange for carbohydrates.

DOI: 10.1201/9781003650881-6

2. *Nonsymbiotic Nitrogen-Fixing Bacteria*: These bacteria thrive in soil or water, fixing nitrogen without a symbiotic relationship with plants (Zehr & Capone, 2020; Soumare et al., 2020).

The mechanism of nitrogen fixation involves production of nitrogenase enzyme and its co-factors. The enzyme nitrogenase plays an important role in converting inert atmospheric nitrogen into ammonium (Kumar et al., 2025). This process involves an electron transport chain within the bacterial cell, which provides the necessary energy for nitrogen fixation. Through a series of electron transfers, nitrogenase catalyzes the reduction of nitrogen molecules, ultimately producing ammonia that can be utilized by living organisms. This enzymatic process is essential for making nitrogen available for biological use, supporting plant growth, and sustaining life on earth (Bellenger et al., 2020; Gu & Milton, 2020; Lawal, 2021).

Symbiotic nitrogen-fixing bacteria forms close relationship with host plant. The genera *Rhizobium* and *Bradyrhizobium* engage in symbiotic interactions with plants, leading to the formation of root nodules (Figure 6.1; Anelia & Galina, 2021) that facilitate nitrogen fixation. Additional genera, including *Sinorhizobium*, *Mesorhizobium*, and *Azorhizobium*, are also involved in this process (Rajkumari et al., 2022; Win et al., 2023). Twenty-seven families of plants from Papilionoidea including Fabaceae and other suborders have been reported for nodulation by nitrogen-fixing rhizobium. Nitrogen-fixing bacteria forms symbiotic relationship with grain legumes like soybean, peanut, chickpea, bean, lentil, and others form which enhances the soil fertility through BNF. This mutualistic beneficial relationship benefits both the plants and the soil ecosystem (Kaschuk et al., 2022; Huang, 2024). In agriculture system biofertilizer containing nitrogen fixers like *Rhizobia*, *Azospirillum*, *Azotobacter*, and *Bacillus*, in addition to phosphate-solubilizing bacteria can boost agricultural productivity (Soumare et al., 2020; Krauspe et al., 2021; Dheeman et al., 2024; Alomari et al., 2025).

Nitrogen-fixing bacteria used as biofertilizers offer several benefits like increased crop yield, improved soil fertility and they are often used as seed inoculants, especially for legumes, due to their symbiotic relationship with plants. Additionally, biofertilizers can be applied to non-legume crops like ginger and cassava, promoting sustainable agriculture practices (Ramasamy et al., 2020; Chaudhary et al., 2022; Dheeman et al., 2024; Alomari et al., 2025).

FIGURE 6.1 Nodulated roots of soybean plants from the field.

6.2.2 Mycorrhizal Fungi

Mycorrhizae can be defined as symbiotic relationship between fungi and plant root, which is described by Albert Bernhard Frank. Mycorrhizae are also known as fungus root in which plant root and fungi forms mutualistic relationship.

There are several types of mycorrhizal associations (Figure 6.2; Selosse & Le Tacon, 1998; Jha & Songachan, 2023), including:

1. Ectomycorrhiza
2. Ericoid mycorrhiza
3. Orchidaceous mycorrhiza
4. Arbuscular mycorrhiza

Each type has distinct characteristics and plays a unique role in plant-fungus interactions (Bagyaraj, 2014).

Glomus genus from arbuscular mycorrhizae, has widespread importance and diversity. Several Glomus species associated with plants cannabis varieties like Khardala and critical and forms symbiotic relationship (Msairi et al., 2023). Studies by scientist have suggested that Glomus species

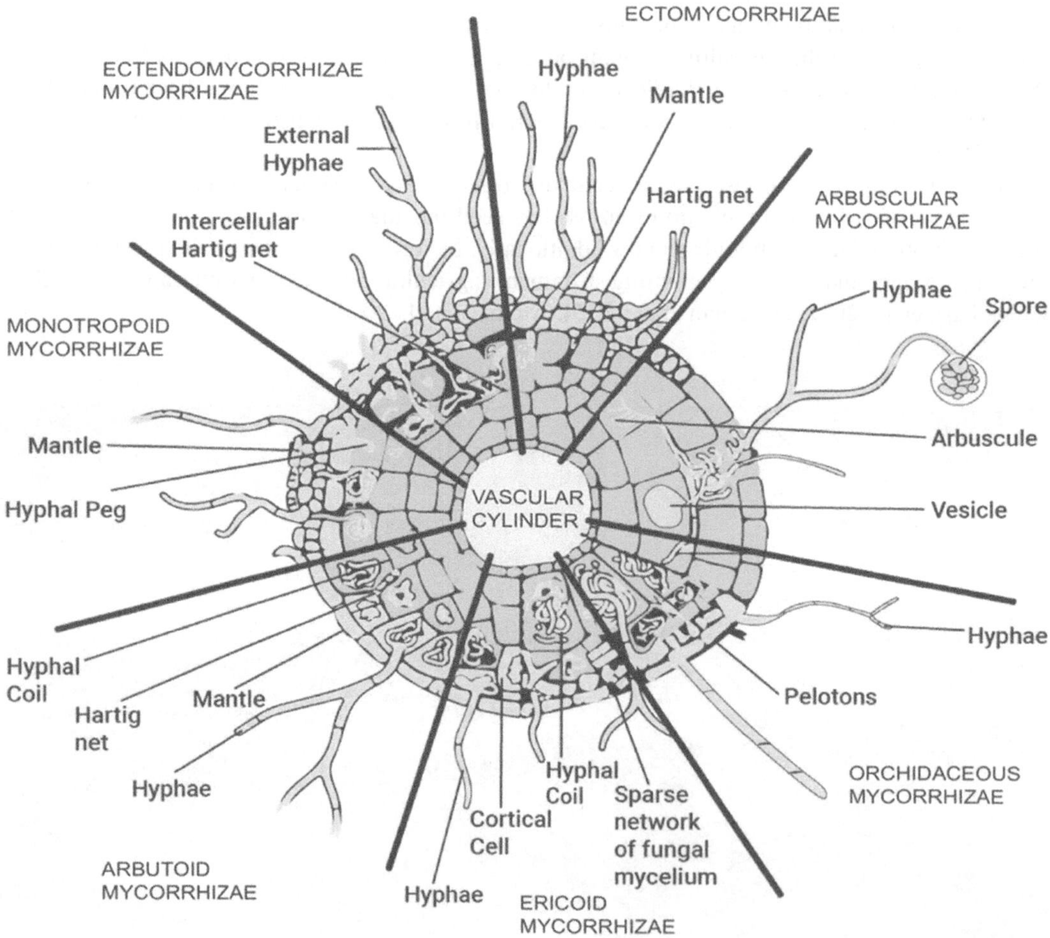

FIGURE 6.2 Types of mycorrhizae.

are more dominant in soil affected by draught and salinity (Hussain et al., 2025). This makes it an important tool for conserving agriculture and ecosystem land management (Chen et al., 2017). The impact of Glomus mycorrhizae enhances nutrition of hyphal transport of immobile mineral ions. It is essential especially for the phosphorus (Smith & Gianinazzi-Pearson, 1988).

6.2.3 Biofertilizers

Biofertilizers are ecofriendly alternative to chemicals fertilizers consist of beneficial microorganism obtained from rhizoplane and plants, it increase the growth of plants by 10–40% (Stewart & Roberts, 2012). Biofertilizers are being utilized to promote plant growth by improving soil fertility, protection of plant from phytopathogens (Maheshwari, 2015). They are beneficial to enhance seed survival, eliminates harmful chemical from soil, and shorten the flowering time (Table 6.1). Range of microorganisms such as molds, mushroom, cyanobacteria, phosphate solubilizers nitrogen-fixing bacteria, are routinely used as a biofertilizers (Odgerel, 2019) (Figure 6.3; Chaudhary et al., 2022).

TABLE 6.1
Examples of Biofertilizers and Mode of Action

Biofertilizers	Mechanism of Action	Groups	Example	References
Nitrogen-fixing organisms	Increases the rate of nitrogen in soil by biological nitrogen fixation. Fixed nitrogen used by plant for growth	Associative symbiotic, symbiotic, and free-living	*Acetobacter diazotrophicus Enterobacter*, *Aulosira bejerinkia*, *Azospirillum* spp., *Azotobacter*, *Herbaspirillum* spp., *Alcaligenes*, *Azoarcus* spp., *Trichodesmium*, *Frankia*, *and Desulfovibrio*	(Nosheen et al., 2021; Daniel et al., 2022)
Phosphorus solubilizers	Organic acid secreted by phosphorus solubilizers lowers the soil pH and make it usable for plants	Bacteria, fungi	*Agrobacterium*, *Microccocus*, *Pseudomonas striata*, *Trichoderma* spp., *Bacillus circulans*, *Bacillus*, *Mucilaginosus*, and *B. circulanscan*	(Nosheen et al., 2021; Daniel et al., 2022)
Potassium solubilizers	Make organic acids that break down silicates and help get rid of metals so that potassium (silicates) ions can be dissolved and used by plants	Bacteria	*Bacillus* spp., *Arthrobacter* spp.	(Etesami et al., 2017; Daniel et al., 2022)
Phosphorus mobilizing	Transfer of phosphorus takes place from soil to root. These groups of biosurfactant show variety of application	Mycorrhiza	*Gigaspora* spp. and *Sclerocystis* spp., *Arbuscular mycorrhiza*, *Acaulospora* spp., *Scutellospora* spp., *Glomus* spp.	(Nosheen et al., 2021; Daniel et al., 2022)
Potassium mobilizers	They transform potassium from the soil that is in accessible form for plant	Bacteria	*Bacillus* spp.	(Jha, 2017)
Plant growth-promoting	Production of hormones encourages the root development, increases nutrient availability and increases the crop yield	Plant growth-promoting rhizobacteria (PGPR)	*Enterobacter*, *Streptomyces*, *Agrobacterium*, *Arthrobacter*, *Erwinia*, *Pseudomonas fluorescens*, *Bacillus*, *Rhizobium*, *Pseudomonas* spp., and *Xanthomonas*	(Nosheen et al., 2021; Daniel et al., 2022)

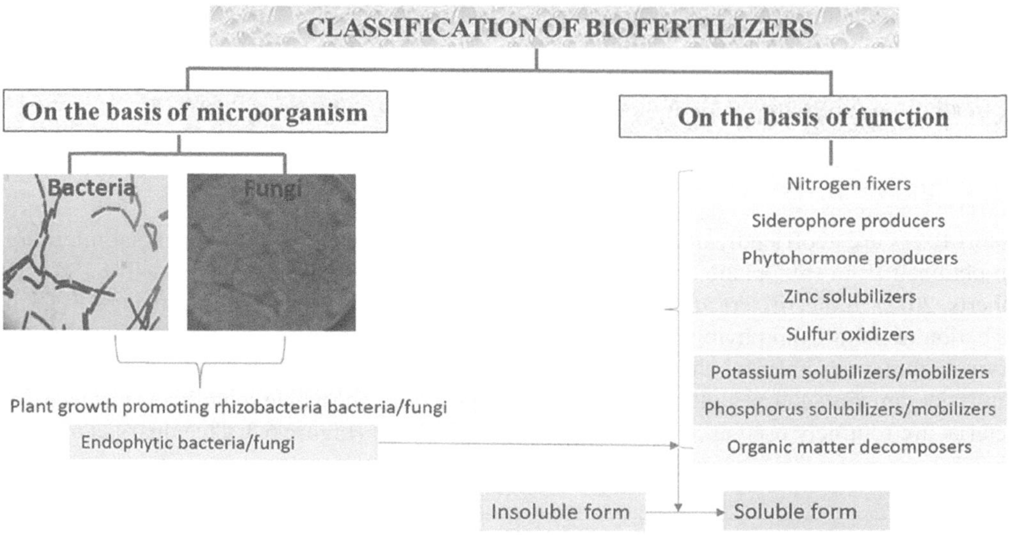

FIGURE 6.3 Classification of biofertilizers.

6.3 BIOCONTROL AGENTS FOR PEST MANAGEMENT

6.3.1 Biopesticides

The global population is rapidly increasing and is projected to reach 9.7 billion by 2050, with the majority of this growth occurring in Africa and Asia (United Nations Department of Economic and Social Affairs, 2015). This surge in population presents significant challenges for agriculture, such as meeting the rising demand for food and addressing environmental issues like soil degradation, water scarcity, and climate change (Pathak et al., 2018; Gonçalves, 2021). To overcome such challenges, sustainable agricultural practices and innovative solutions are needed such as:

1. *Biopesticides*: ecofriendly pest management agents
2. *Biostimulants*: substances that support the plant growth and stress resistance
3. *Organic-Based Treatments*: natural amendments that enhance soil fertility and plant health (Van Oosten et al., 2017)

Biopesticides are substances that occur naturally and are sourced from animals, plants, microalgae, cyanobacteria, and microorganisms to manage phytopathogens. According to the US Environmental Protection Agency, biopesticides are derived from natural materials like animals, plants, bacteria, and certain minerals. The genes or metabolic products of these biological substances can be utilized to protect crops from damage. Compared to traditional chemical pesticides, biopesticides are more environmentally friendly (Chang et al., 2003; Van Oosten et al., 2017).

Bacillus thuringiensis (Bt) is a gram-positive, sporulating bacterium known for producing crystal proteins (Cry), which are effective in killing various insects. This makes Bt a popular choice as a microbial insecticide for enhancing plant resistance through genetic engineering (Salehi Jouzani et al., 2008).

Mode of action of Bt proteinaceous includes gut paralysis of insect larvae, where gut enzymes activate toxin forming pores in cell membrane and results in sluggishness and ultimately leads to death (Vachon et al., 2012).

Bugs from the orders Coleoptera (beetles and weevils), Diptera (flies and mosquitoes), and Lepidoptera (butterflies and moths) can die from Cry proteins (Domínguez-Arrizabalaga et al., 2020). Besides Cry proteins, Bt also produces vegetative insecticidal proteins (Vip) and secreted insecticidal proteins (Sip), which are specific to certain targets, particularly species of Coleoptera and Lepidoptera, with the ability of biodegradability (Chakroun et al., 2016).

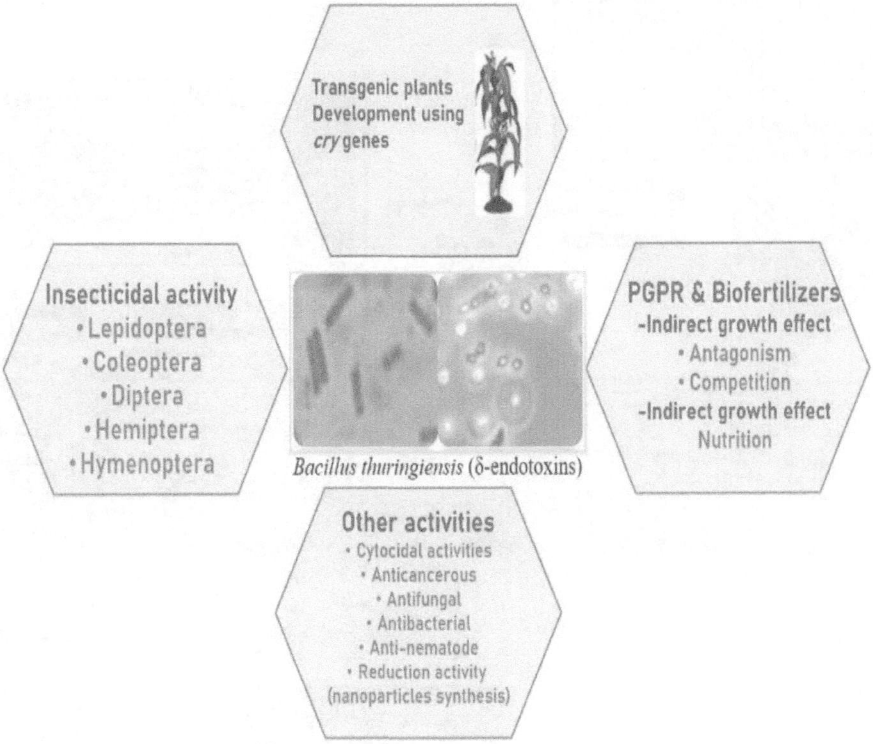

FIGURE 6.4 Use of *Bacillus thuringiensis* as a biocontrol agent.

Bt-based insecticides offer alternative use to chemical pesticides as it is harmful to environment and humans. Recent research is focused on developing effective genetically modified insect-resistant crops (Figure 6.4; Kumar et al., 2021). Bt proteins are specific to insect decreasing its side effects against other organisms (Bravo et al., 2011).

Along with Bt insecticides, Entomopathogenic fungi (EPF) also plays important role as biocontrol agent. EPF are eukaryotic, heterotrophic, unicellular or multicellular, filamentous microorganisms that can reproduce asexually, sexually, or both and generate range of spores (Bahadur, 2028). *Verticillium*, *Neozygites*, *Metarhizium*, *Nomuraea*, *Beauveria*, and *Entomophthora* are among the entomopathogenic taxa found in most taxonomic groupings (Deshpande, 1999).

Its history dates back in 1835, Agostino Bassi's discovered role of *Beauveria* in "white muscardine" diseases in Italy's silkworm. This fungus was initially named as Bassi by Vuillemin. Earliest use of *Beauveria* was in the United States for the control of blissus leucopterus chinch bugs during 1800s (Lord, 2005; Mascarin & Jaronski, 2016).

The mechanism of action or life cycle of *Beauveria* in insect host is briefly outlined in Figure 6.5 (Mascarin & Jaronski, 2016).

Beauveria and *Metarhizium* are widely used fungal insect pathogens used as biocontrol, accounting for approximately 70% of commercial mycoinsecticides (Table 6.2). The majority of these product formulations consist of 90% aerial conidia produced on solid substrates (Faria & Wraight, 2007).

6.4 CLIMATE-RESILIENT CROPS

6.4.1 Halotolerant Plant Growth-Promoting Rhizobacteria

Productivity of crop gets affected by environmental factors such as salinity (Dastageer et al., 2021), warmness (El-Saadony et al., 2022; AbuQamar et al., 2024), water deficiency (Roy et al., 2021), and metals (Khan et al., n.d.). Soil salinity is emerging as major threat to agricultural field and food

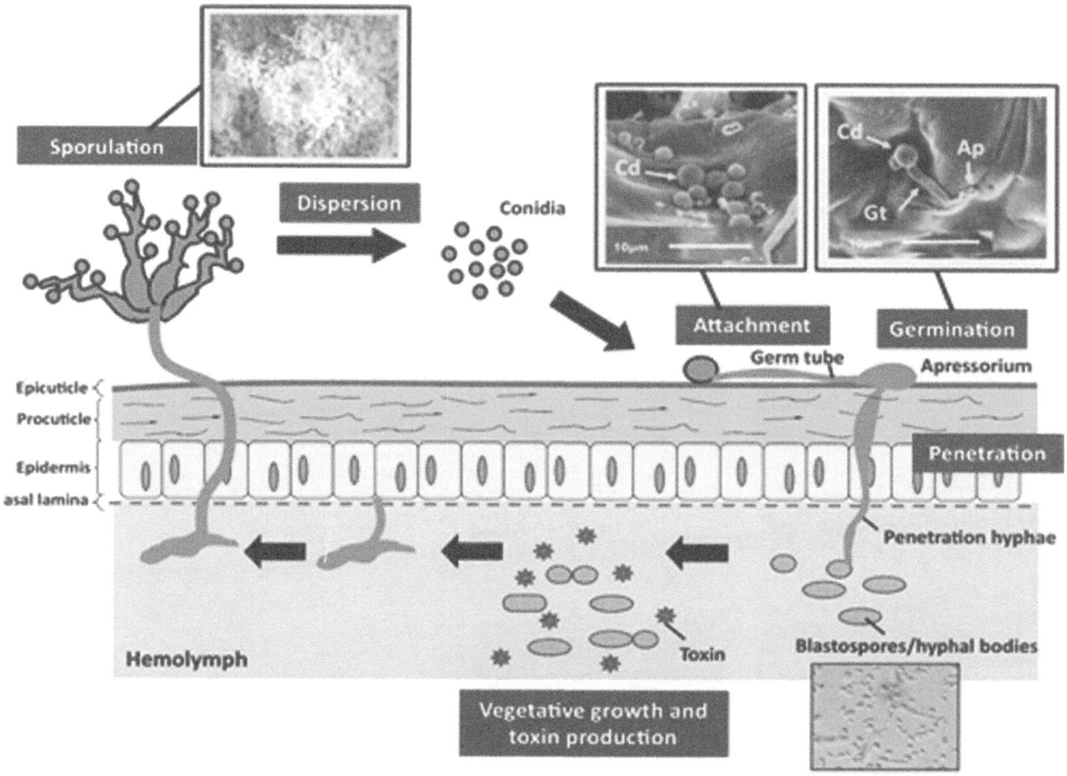

FIGURE 6.5 The mechanism of action or life cycle of *Beauveria* in insect host.

security, which affects 20% yield and estimated to reach 50% by 2050. Soil salt also affects diversity of soil and rhizosphere (Semida et al., 2021).

- Plant growth-promoting rhizobacteria (PGPR) enhance soil salt tolerance and support plant growth (Daliakopoulos et al., 2016; El-Tarabily et al., 2019). Halotolerant plant growth-promoting rhizobacteria (HT-PGPR) are crucial for boosting plant growth and yields, alleviating salt stress, and restoring degraded saline soils (Table 6.3; El-Tarabily et al., 2019; Kumar et al., 2023; AbuQamar et al., 2024).

TABLE 6.2
Sensitive Hosts of *B. bassiana*

Order	Susceptible Hosts
Lepidoptera	*Hypocrita jacobaea*, *Cydia nigricans*
Hymenoptera	*Ichneumonidae*, *Lasius fuliginosus*, *Vespula* spp., *Bombus pratorum*
Heteroptera	*Picromerus bidens*, *Anthocoris nemorum*
Diptera	*Leria serrata*
Homoptera	*Eulecanium* spp.
Coleoptera	*Calvia quattuordecimguttata*, *Anthonomus pomorum*, *Anthonomus pomorum*

Source: Keswani et al. (2013).

TABLE 6.3
Different Species of HT-PGPR and Their Effects on Improving Plant Salt Tolerance

Plant Species	Type of HT-PGPR	Mechanism	References
Solanum lycopersicum H 72	Pseudomonas fluorescens	Produces 1 aminocycloprpane-1-carboxylate deaminase and increased count of flowers and chlorophyll pigment	(Ali et al., 2014; AbuQamar et al., 2024)
Brassica campestris	*Enterobacter cloacae, Paenibacillus xylanexedens*	Modified ET and IAA levels; and enhanced nutrient uptake	(Yaish et al., 2016; AbuQamar et al., 2024)
Capsicum annuum	*Arthrobacter* sp., *Bacillus* sp.	Increased Pro levels and reduced osmotic stress	(Sziderics et al., 2007; AbuQamar et al., 2024)
Oryza sativa	*Pseudomonas pseudoalcaligenes*	Synthesis of EB; enhanced levels of antioxidant	(Jha et al., 2020; AbuQamar et al., 2024)
Brassica napus	*Enterobacter cloacae, Paenibacillus xylanexedens*	Improved photosynthesis, antioxidants; reduced proteins linked to membrane transport	(Cheng et al., 2012; AbuQamar et al., 2024)
Chlorophytum borivilianum	*Brachybacterium paraconglomeratum*	Tolerance to osmotic and oxidative damage, and enhanced level of IAA, chlorophyll, and nutrient uptake	(Barnawal et al., 2015; AbuQamar et al., 2024)
Glycine max	*Raoultella terrigena, P. putida*	Enhanced level of α-ketobutyrate, chlorophyll, germination and plant biomass	(Simarmata et al., 2018; AbuQamar et al., 2024)
Helianthus annuus	*Pseudomonas* sp.	Production of HCN, siderophore and IAA. Increased nitrogen fixation	(Tewari & Arora, 2016; AbuQamar et al., 2024)
Cicer arietinum	*P. rifietoensis, H. variabilis*	Production of biofilms and EPS	(Qurashi & Sabri, 2012; AbuQamar et al., 2024)

6.5 INNOVATIONS IN FOOD PRODUCTION

6.5.1 Enhancement of Nutritional Quality in Fermented Foods by Microorganisms

Fermentation: Fermentation is a metabolic process where microbes like yeasts, bacteria, and molds convert sugars and carbohydrates into substances, used as a product with well-characterized taste, smell, and texture of fermented food (Müller et al., 2012). Factors affecting food nutritional value are its digestibility and amount of nutrient present in food. Nutritional value and digestibility can be enhanced by the process of fermentation. Microbial enzymes play important role in fermentation process by breakdown of macronutrients into micronutrients (Hackmann & Zhang, 2023).

Lactobacillus play important role in food industry. Substrate utilized for lactic acid production is sugar and carbohydrates for the dairy product like yogurt and probiotics. In pharmaceutical industry lactic acid is used for pills, surgical sutures, and implants.

Fermented foods with live cultures, such as *Lactobacillus* offer numerous benefits for gut health, including:

1. Improved digestion
2. Enhanced immune system function
3. Reduced inflammation
4. Positive impact on mental health through the gut-brain connection (Vasudevan et al., 2025)

Yogurt, pickles, sauerkraut, kimchi are well-known fermented product produced by lactic acid fermentation. Yogurt has been classified into two types: standard culture yogurt made with *Lactobacillus bulgaricus*, and *S. thermophilus*. Yogurt products are diverse, available in various flavors, textures, and forms, making them suitable for different tastes and occasions, whether as a snack, dessert, or part of a meal (Banerjee et al., 2017). Kimhi another fermented product of Korea gaining global attention due to its unique flavor and health benefits. Kimchi production is performed by bacterial species such as *Leuconostoc*, *Weissella*, and *Lactobacillus* (Kang et al., 2022).

Along with bacterial fermented products, food from fungal origin also known worldwide. Among them worldwide famous is tempeh. Tempeh originated from Indonesia made from mass of fermented beans, cereals by using fungal mycelium (mostly *Rhizopus* species) (Nout & Kiers, 2005). *R. oryzae* or *R. oligosporus* has been widely studied for tempeh production for its relation with gut health, lung health, anemia, bone health, type II diabetes mellitus and skeletal muscle recovery (Teoh et al., 2024).

6.5.2 Single-Cell Proteins

The phrase SCP was introduced by Carol L. Wilson in 1966. Single-cell protein is derived from the dried cells of microorganisms. It serves as a protein supplement for both human and animal diets. SCP is composed of carbohydrates, nucleic acids, fats, minerals, and vitamins. It is particularly abundant in essential amino acids such as methionine and lysine, which are often absent in foods derived from plants and animals (Suman et al., 2015).

For the production of SCP' yeast are better organisms than bacteria and algae as it is easy to harvest, protein content (50% of them dry weight) and low amount of nucleic acid. *Saccharomyces cerevisiae* because of its ease of harvesting and greater size are widely used for production of SCP (Hamdy, 2013). SCP from *S. cerevisiae* considered one of the best solutions for world countries and developing countries that suffer from malnutrition problems food shortages (Bacha & Nasir, 2011; Malla Obaida, 2021).

Other sources of SCP production includes algae. Algae are autotrophic organism shows great genetic diversity. Their growth requirement includes inorganic nutrients like nitrogen and phosphorus, they also need water, carbon dioxide, and light (Radmer & Parker, 1994). Algae have been characterized into two categories micro-algae (unicellular) and macro-algae (multicellular) (Wood et al., 1991). The micro-algae have been reported to produce SCP (70%) by utilizing solar energy (Nasseri et al., 2011).

Spirulina is thread like structured *cyanobacterium* belonging to the oscillatoriaceae family of micro-algae. It's named spirulina due to spiral and helical structure of filaments. Spirulina is consumed as food source as it contains high amounts of protein (60–70%), vitamins (4%), antioxidants, and fatty acid (Kameshwari et al., 2020).

6.6 REDUCING FOOD WASTE

6.6.1 Lactic Acid: Nature's Preservative

Fermentation is oldest method of food preservation practiced for centuries. Fermentation involves breakdown of complex molecules into simpler one. Fermentation is catalyzed by enzymes produced by yeasts, bacteria, and yeasts in the absence of oxygen. Breakdown of this molecule not only imparts texture and flavors to food but also extends the shelf life of products.

Lactic acid fermentation, in particular, is widely used for:

1. *Vegetables*: sauerkraut, kimchi, pickles
2. *Fruits*: fermented fruit products
3. *Dairy*: yogurt, cheese, kefir (Archer, 2004; Brian, 2023)

In this process sugars are converted into lactic acid by lactic acid bacteria, which produces an acidic environment that prevents infections and spoiling. Because they are regarded as Qualified Presumption of Safety (QPS) in the EU and Generally Regarded as Safe (GRAS) in the United States, lactic acid bacteria (LAB) are emerging as a promising preservative substitute for chemical preservatives in the dairy industry. LAB has been researched for its antifungal properties in a variety of fermented foods, including bread, cheese, and yogurt (Schwenninger & Meile, 2004; Delavenne et al., 2013; Shi & Maktabdar, 2022).

Special bioreactors are designed for the optimal condition of fermentation for growth and activity of LAB. By providing optimal factors such as pH, temperature, and nutrient LAB actively convert carbohydrates into lactic acid. This process ensures high-level consistency for food producers to achieve lactic acid production which extends shelf life of product. Other than lactic acid other bio preservative compounds include acetic acid and propionic acid produced through fermentation (Luthra, 2024).

6.6.2 Mycoprotein Production

Mycoprotein are single-cell proteins naturally produced by fungus *Fusarium venenatum*. Mycoprotein is used as alternative to meat as its texture is similar to meat. Mycoprotein is emerging as efficient solution for global protein demands for meat, dairy, eggs, and other products due to their sustainability, nutritional value, and affordability (Finnigan et al., 2019; Hashempour-Baltork et al., 2020). Mycoprotein development has a rich history spanning over five decades (Figure 6.6; Akinsemolu & Onyeaka, 2025).

Quorn is well-known UK-based brand using mycoprotein as ingredient in various product. Quorn vegan option includes patties, chiqin fillets, mini sausage pastry puffs, diced chiqin pieces, and pineapple chipotle chiqin dippers (Akinsemolu & Onyeaka, 2025). All of these varieties offer best alternative to chicken like flavor and texture (Akinsemolu & Onyeaka, 2025).

Quorn's vegan products come in diverse forms, such as breaded patties, frozen foods, and ground foods. That's a significant difference! Quorn's production uses substantially less water compared to traditional ground beef, making it a more sustainable option. This reduced water footprint can contribute to a more environmentally friendly food choice. Quorn led the steps to followed for the commercial production of mycoprotein for other companies worldwide. One such company is US-based ***The Better Meat Co*** (Akinsemolu & Onyeaka, 2025).

6.6.3 Waste Valorization

Process of transforming waste material into utilizable product is called as waste valorization. Such process is being utilized from ancient times (Figure 6.7; Balu et al., 2021). Different valorization process is known for solving industrial demands. One of them is waste valorization (Serrano-Ruiz et al., 2012). Current major concern is related with disposal and management of food waste to landfill. Primary food waste treatment includes transformation of food waste into energy (anaerobic digestion), composting and animal feed (Arancon et al., 2016).

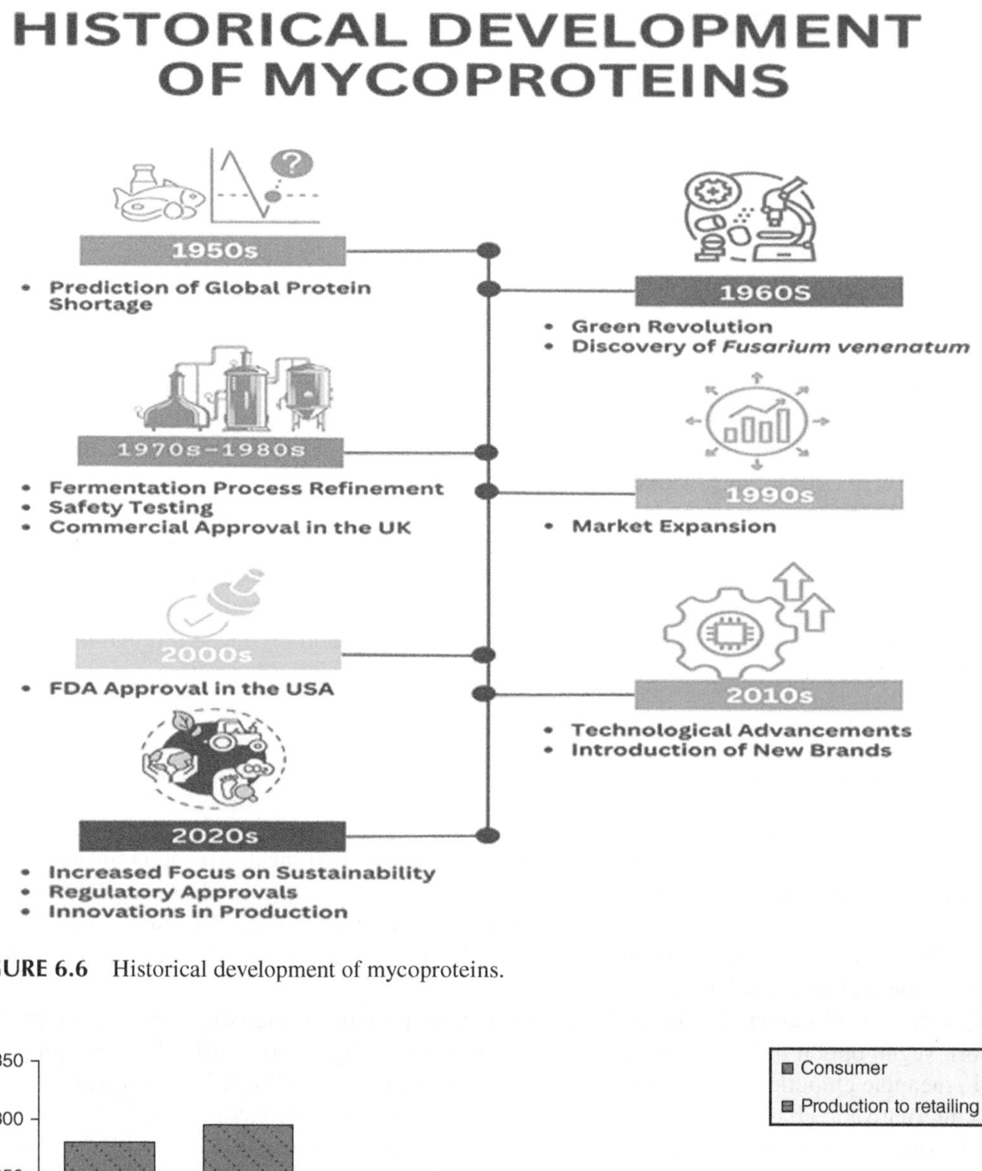

FIGURE 6.6 Historical development of mycoproteins.

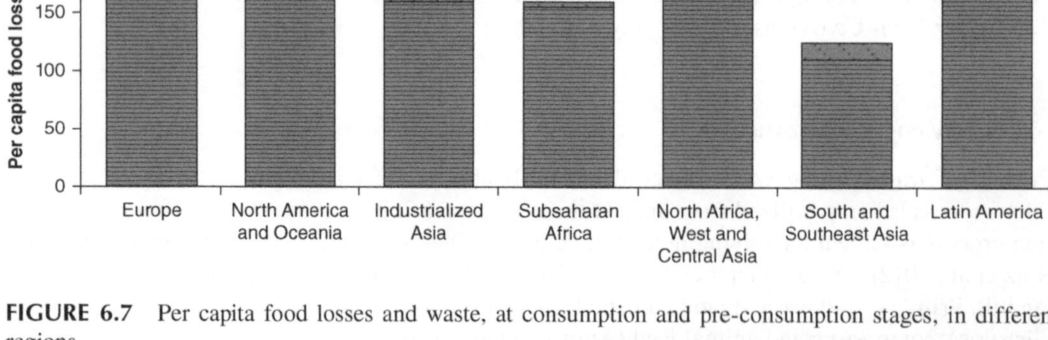

FIGURE 6.7 Per capita food losses and waste, at consumption and pre-consumption stages, in different regions.

TABLE 6.4
Agro-Industrial Waste Valorization Using Microorganisms

Substrate Utilized	Microorganism	Improved Area	Impact on Animal	References
Corn byproduct	*Bacillus subtilis*	Decreased amount of insoluble dietary fiber	Reduced abundance in bacterial community in the animal gut	(Liu et al., 2022; Yafetto et al., 2023)
Rapeseed meal	*Bacillus subtilis*, *Lactobacillus fermentum*	Increased amino acid Reduced isothiocyanates	Increased Immunoglobulin G and M, phosphorus, calcium; improved villus height, crypt depth; reduced urea nitrogen	(Xu et al., 2022; Yafetto et al., 2023)
Sesame meal	*Saccharomyces cerevisiae*, *Lactobacillus acidophilus*	Increased crude protein Decreased oxalate and phytate	Increased LAB and reduced coliform in ileum.	(Hajimohammadi et al., 2020; Yafetto et al., 2023)
Soybean meal	*Bacillus subtilis*	Decreased antigenic proteins Enhanced nutrient content	Improved weight gain and feed intake	(Zhang et al., 2016; Yafetto et al., 2023)
Sweet potato	*Saccharomyces boulardii*	Probiotic-enriched feed		(Jazi et al., 2018; Yafetto et al., 2023)
Wheat bran	*Trichoderma pseudokoningii*	Improved feed nutrient with increased cellulase and xylanase activity	Better feed nutrients with higher activity of cellulase and xylanase	(Chu et al., 2016; Yafetto et al., 2023)

Agro-industrial waste valorization offers a promising solution, leveraging microbial fermentation to convert waste into valuable products like succinic acid, bioplastics, biofuels (bioethanol, biodiesel), and animal feeds (Table 6.4; Balu et al., 2021).

6.7 BIOREMEDIATION

Bioremediation involves use of microorganisms for the removal of hazardous pollutant from contaminated soil. These microorganisms remediate soil pollutants: bacteria, yeast, and fungi are studied to recycle pollutant and xenobiotics from soil (Roccuzzo et al., 2020).

6.7.1 Bacterial Remediation for Heavy Metal

Heavy metals sources of includes anthropogenic activities and natural sources, heavy metals contamination can be found in water, soil, and atmosphere. Industrial sources include tanneries, fertilizers, varnishes, pharmaceutical, and pesticides producers (Chabukdhara & Nema, 2012). Metallurgical factories are significant pollution sources due to extraction and processing of metals, release of toxic substances, and environmental contamination (Wuana & Okieimen, 2014). Agro-based activities contribute to heavy metal contamination through: inorganic chemical use, phosphate fertilizers containing heavy metal impurities, and accumulation of toxic metals like As, Cd, Ni, Cr, and Zn (Nziguheba & Smolders, 2008).

TABLE 6.5
Heavy Metal Remediation by Bacteria

Microbial Species	Target Heavy Metals	References
Variovorax boronicumulans, Stenotrophomonas rhizophila	Cadmium, Lead	(Jalilvand et al., 2019; Tarfeen et al., 2020)
Microbacterium oxydans	Nickle, copper	(Heidari et al., 2020; Tarfeen et al., 2020)
Pseudomonas aeruginosa	Mercury	(Yin et al., 2016)
Bacillus sp., PZ-1	Chromium	(Ren et al., 2015; Sam Nirmala et al., 2025)
Pseudomonas sp., Cr13	Chromium	(Li et al., 2017)
Micrococcus luteus	Copper, Lead	(Puyen et al., 2012)
Bacillus cereus	Mercury	(Sinha et al., 2012; Sam Nirmala et al., 2025)
Rhodococcus opacus	Chromium, Copper, Lead	(Bueno et al., 2008)
Stenotrophomonas rhizophila	Zinc	(Ghosh & Saha, 2013)

Heavy metal toxicity affects cells by generating reactive oxygen species, destabilizing biomolecules, mimicking essential elements, and disrupting biological activities (Tarfeen et al., 2020). The severity of toxicity depends on factors like concentration, exposure route, and duration, potentially leading to cellular damage and harm to overall health (Prabhakaran et al., 2016).

Microorganisms like algae, bacteria, and fungi play important role in bioremediation, effectively cleaning up heavy metal-contaminated sites (Saraswat et al., 2020). Different microbial species exhibit unique abilities to uptake and remediate heavy metals, making them valuable tools for environmental restoration as shown in Table 6.5.

6.7.2 Bacterial Remediation of Pesticide

The word pesticides refer to different group of fungicides, defoliants, herbicides, fungicides, and nematicides (Tsaboula et al., 2019). Pesticide use is substantial, with herbicides (50%), insecticides (30%), and fungicides (15%) (Sharma et al., 2019; Tarfeen et al., 2020).

Industrial discharge of pesticide residues contaminates aquatic ecosystems, affecting biota, sediments, and ultimately, human health through water and food consumption. Insecticide remnants have been detected in drinking and groundwater globally, highlighting the issue's severity, as seen in China's Shanxi province with organochlorine pesticides like endosulfan-sulfate and aldrin (Li et al., 2015).Toxic pesticides include carbamates (CB), organophosphorus (OP), organochlorine (OC), and pose significant health risks by disrupting nerve function, primarily affecting the nervous system (Tarfeen et al., 2020). Exposure to pesticide can lead to severe health issues. Other types of pesticides, like rodenticides and nematicides, also contribute to the over 2 million tons used globally each year, impacting the environment and ecosystems (Ridolfi et al., n.d.). Exposure to pesticide has been linked to various health disorders, including Parkinson's disease, Hodgkin's and non-Hodgkin's disease, respiratory problems, endocrine disruption, and reproduction. These links draw attention to the possible long-term health hazards linked to pesticide exposure, highlighting the necessity of vigilance and cautious handling (Luo et al., 2016; Sabarwal et al., 2018).

Bacteria show amazing ability to break a range of toxicants, as they have catabolic genes to thrive on harsh condition. Presence of catabolic genes and their capacity to withstand extreme environmental conditions (Tarfeen et al., 2020). This strength of tolerance makes them highly suitable for the remediation of diverse pesticides as shown in Table 6.6, offering a promising solution for mitigating environmental pollution (Deng et al., 2007).

TABLE 6.6
Bacterial Remediation for Pesticide

Potential Bacteria Species	Target Pesticides R	References
Stenotrophomonas sp.	Tetrachlorvinphos, Chlorpyriphos	(Parekh et al., 1994; Yang et al., 2016)
Micrococcus sp.	Cypermethrin	(Tallur et al., 2007)
Serratia sp.	Tetrachlorvinphos	(Ortiz et al., 2010)
Sphingomonas sp.	Oxyfluorfen	(Keum et al., 2008)
Enterobacter sp.	Chlorpyrifos	(Singh et al., 2004)
Proteus sp.	Tetrachlorvinphos	(Ortiz et al., 2010)
Yersinia sp.	Tetrachlorvinphos	(Ortiz et al., 2010)
Rhodococcus sp.	Metamitron	(Parekh et al., 1994)
Streptomyces olivochromogenes, *Streptomyces chattanoogensis*	Chlorpyrifos	(Pailan et al., 2015)
Pseudomonas putida	Organophosphate	(Li et al., 2016)

6.8 CONCLUSION

By harnessing the power of microorganisms, we can improve agricultural productivity, develop sustainable pest control strategies, revolutionize food production, minimize food waste, and restore polluted environments through bioremediation. Integration of biotechnology in various sectors offers innovative solutions to fight hunger. By leveraging these advancements, we can improve food security, sustainability, and environmental health, ultimately creating a more resilient and productive future.

REFERENCES

AbuQamar, S. F., El-Saadony, M. T., Saad, A. M., Desoky, E. S. M., Elrys, A. S., El-Mageed, T. A., Semida, W. M., Abdelkhalik, A., Mosa, W. F., Al Kafaas, S. S., Naser, S., Ibrahim, E. H., Alshamsi, F. M., Mathew, B. T., & El-Tarabily, K. A. (2024). Halotolerant plant growth-promoting rhizobacteria improve soil fertility and plant salinity tolerance for sustainable agriculture—A review. *Plant Stress*, *12*. https://doi.org/10.1016/j.stress.2024.100482

Akinsemolu, A. A., & Onyeaka, H. N. (2025). *Discover Applied Sciences*, *7*(3). https://doi.org/10.1007/s42452-025-06614-0

Ali, S., Charles, T. C., & Glick, B. R. (2014). Amelioration of high salinity stress damage by plant growth-promoting bacterial endophytes that contain ACC deaminase. *Plant Physiology and Biochemistry*, *80*, 160–167. https://doi.org/10.1016/j.plaphy.2014.04.003

Alomari, S. M., Hasan, R. S., Al-Najim, A. N., & Saadi, A. M. (2025). Nitrogen-fixing bacteria and their applications in the environment: A review. *International Journal of Medical and All Body Health Research*, *5*(4), 178–187.

Anelia, I., & Galina, N. (2021). Biological nitrogen fixation in legumes: Understanding the process. https://www.legumehub.eu/is_article/biological-nitrogen-fixation-in-legumes/

Arancon, R., Lin, C., Chan, K., Kwan, T., & Luque, R. (2016). *Solid Waste Management*, 37–77. https://doi.org/10.1201/b19424-5

Archer, D. L. (2004). Freezing: An underutilized food safety technology? *International Journal of Food Microbiology*, *90*(2), 127–138. https://doi.org/10.1016/s0168-1605(03)00215-0

Bacha, U., Nasir, M., Khalique, A., Anjum, A. A., & Jabbar, M. A. (2011). Various Agro-Industrial Wastes for Saccharomyces Cerevisiae Biomass Production and its Quality Evaluation as Single Cell Protein. *Food Chemistry*, *5*(7).

Bagyaraj, D. J. (2014). Mycorrhizal fungi. *Proceedings of the Indian National Science Academy*, 415. https://doi.org/10.16943/ptinsa/2014/v80i2/55118

Bahadur, A. B. (2028). Entomopathogens: Role of insect pest management in crops. *Trends in Horticulture*, *1*(3). https://doi.org/10.24294/th.v1i4.833

Balu, A. M., Budarin, V., Shuttleworth, P. S., Pfaltzgraff, L. A., Waldron, K., Luque, R., & Clark, J. H. (2021). Valorisation of orange peel residues: Waste to biochemicals and nanoporous materials. *ChemSusChem*, *5*(9), 1694–1697. https://doi.org/10.1002/cssc.201200381

Banerjee, U., Malida, R., Panda, R., Halder, T., & Roymahapatra, G. (2017). Variety of yogurt and its health aspects – A brief review. *International Journal of Innovative Practice and Applied Research*, *7*, 56–66.

Barnawal, D., Bharti, N., Tripathi, A., Pandey, S. S., Chanotiya, C. S., & Kalra, A. (2015). ACC-deaminase-producing endophyte brachybacterium paraconglomeratum strain SMR20 ameliorates chlorophytum salinity stress via altering phytohormone generation. *Journal of Plant Growth Regulation*, *35*(2), 553–564. https://doi.org/10.1007/s00344-015-9560-3

Bellenger, J. P., Darnajoux, R., Zhang, X., & Kraepiel, A. M. (2020). Biological nitrogen fixation by alternative nitrogenases in terrestrial ecosystems: A review. *Biogeochemistry*, *149*(1), 53–73. https://doi.org/10.1007/s10533-020-00666-7

Bernal, P. (2024). How are microbes helping end hunger? *Microbial Biotechnology*, *17*(3). https://doi.org/10.1111/1751-7915.14432

Boqvist, S., Söderqvist, K., & Vågsholm, I. (2018). Food safety challenges and One Health within Europe. *Acta Veterinaria Scandinavica*, *60*, 1–13. https://doi.org/10.1186/s13028-017-0355-3

Bravo, A., Likitvivatanavong, S., Gill, S. S., & Soberón, M. (2011). Bacillus thuringiensis: A story of a successful bioinsecticide. *Insect Biochemistry and Molecular Biology*, *41*(7), 423–431. https://doi.org/10.1016/j.ibmb.2011.02.006

Brian, F. (2023). Fermentation techniques for the preservation and shelf-life extension of food. *Journal of Food Microbiology*, *7*(6). https://doi.org/10.35841/aafmy-7.6.176

Bueno, B., Torem, M., Molina, F., & de Mesquita, L. (2008). Biosorption of lead(ii), chromium(iii) and copper(ii) by R. Opacus: Equilibrium and kinetic studies. *Minerals Engineering*, *21*(1), 65–75. https://doi.org/10.1016/j.mineng.2007.08.013

Chabukdhara, M., & Nema, A. K. (2012). Assessment of heavy metal contamination in Hindon River sediments: A chemometric and geochemical approach. *Chemosphere*, *87*, 945–953. https://doi.org/10.1016/j.chemosphere.2012.01.055

Chakroun, M., Banyuls, N., Bel, Y., Escriche, B., & Ferré, J. (2016). Bacterial vegetative insecticidal proteins (VIP) from entomopathogenic bacteria. *Microbiology and Molecular Biology Reviews*, 329–350. https://doi.org/10.1128/mmbr.00060-15

Chang, J. H., Choi, J. Y., Jin, B. R., Roh, J. Y., Olszewski, J. A., Seo, S. J., O'Reilly, D. R., & Je, Y. H. (2003). An improved baculovirus insecticide producing occlusion bodies that contain *Bacillus thuringiensis* insect toxin. *Journal of Invertebrate Pathology*, *84*(1), 30–37. https://doi.org/10.1016/s0022-2011(03)00121-6

Chaudhary, P., Singh, S., Chaudhary, A., Sharma, A., & Kumar, G. (2022). Overview of biofertilizers in crop production and stress management for sustainable agriculture. *Frontiers in Plant Science*. https://doi.org/10.3389/fpls.2022.930340

Chen, M., Yang, G., Sheng, Y., Li, P., Qiu, H., Zhou, X., Huang, L., & Chao, Z. (2017). *Glomus mosseae* inoculation improves the root system architecture, photosynthetic efficiency and flavonoids accumulation of liquorice under nutrient stress. *Frontiers in Plant ScienceGlomus*, *8*. https://doi.org/10.3389/fpls.2017.00931

Cheng, Z., Woody, O. Z., McConkey, B. J., & Glick, B. R. (2012). Combined effects of the plant growth-promoting bacterium *Pseudomonas putida* UW4 and salinity stress on the brassica napus proteome. *Applied Soil Ecology*, *61*, 255–263. https://doi.org/10.1016/j.apsoil.2011.10.006

Chu, Y. T., Lo, C. T., Chang, S. C., & Tai, T. L. (2016). Effects of Trichoderma fermented wheat bran on growth performance, intestinal morphology and histological findings. *Italian Journal of Animal Science*, 82–92. https://doi.org/10.1080/1828051x.2016.1241133

Daliakopoulos, I., Tsanis, I., Koutroulis, A., Kourgialas, N., Varouchakis, A., Karatzas, G., & Ritsema, C. (2016). The threat of soil salinity: A European scale review. *Science of The Total Environment*, *573*, 727–739. https://doi.org/10.1016/j.scitotenv.2016.08.177

Daniel, A. I., Fadaka, A. O., Gokul, A., Bakare, O. O., Aina, O., Fisher, S., Burt, A. F., Mavumengwana, V., Keyster, M., & Klein, A. (2022). Biofertilizer: The future of food security and food safety. *Microorganisms*, *10*(6), 1220. https://doi.org/10.3390/microorganisms10061220

Dastageer, Z., Seleiman, M. F., Khan, I., Chattha, M. U., Ali, E. F., Alhammad, B. A., Jalal, R. S., Refay, Y., & Hassan, M. U. (2021). Glycine-betaine induced salinity tolerance in maize by regulating the physiological attributes, antioxidant defense system and Ionic homeostasis. *Notulae Botanicae Horti Agrobotanici Cluj-Napoca*, 12248. https://doi.org/10.15835/nbha49112248

Delavenne, E., Ismail, D., Pawtowski, A., Mounier, J., Barbier, G., & Le Blay, G. (2013). Assessment of lactobacilli strains as yogurt bioprotective cultures. *Food Control*, *30*(1), 206–213. https://doi.org/10.1016/j.foodcont.2012.06.043

Deng, L., Su, Y., Su, H., Wang, X., & Zhu, X. (2007). Sorption and desorption of lead (II) from wastewater by green algae Cladophora fascicularis. *Journal of Hazardous Materials*, *143*(1-2), 220–225. https://doi.org/10.1016/j.jhazmat.2006.09.009

Deshpande, M. V. (1999). Mycopesticide production by fermentation: Potential and challenges. *Critical Reviews in Microbiology*, *25*(3), 229–243. https://doi.org/10.1080/10408419991299220

Dheeman, S., Islam, M. T., Egamberdieva, D., & Siddiqui, N. (2024). *Soil Bacteria: Biofertilization and Soil Health*. Singapore: Springer.

Domínguez-Arrizabalaga, M., Villanueva, M., Escriche, B., Ancín-Azpilicueta, C., & Caballero, P. (2020). Insecticidal activity of bacillus thuringiensis proteins against Coleopteran pests. *Toxins*, *12*(7), 430. https://doi.org/10.3390/toxins12070430

El-Saadony, M. T., Saad, A. M., Soliman, S. M., Salem, H. M., Desoky, E. S. M., Babalghith, A. O., El-Tahan, A. M., Ibrahim, O. M., Ebrahim, A. A., Abd El-Mageed, T. A., Elrys, A. A., Elbadawi, A. A., El-Tarabily, K. A., & AbuQamar, S. F. (2022). Role of nanoparticles in enhancing crop tolerance to abiotic stress: A comprehensive review. *Frontiers in Plant Science*, *13*. https://doi.org/10.3389/fpls.2022.946717

El-Tarabily, K. A., Ayyash, M., AlKhajeh, A. S., Alnuaimi, L. H., Sham, A., ElBaghdady, K. Z., Tariq, S., & AbuQamar, S. F. (2019). Growth promotion of salicornia bigelovii by Micromonospora Chalcea UAE1, an endophytic 1-aminocyclopropane-1-carboxylic acid deaminase-producing actinobacterial isolate. *Frontiers in Microbiology*, *10*. https://doi.org/10.3389/fmicb.2019.01694

Etesami, H., Emami, S., & Alikhani, H. A. (2017). Potassium solubilizing bacteria (KSB): Mechanisms, promotion of plant growth, and future prospects – A review. *Journal of soil science and plant nutrition*, *17*(4), 897–911. https://doi.org/10.4067/s0718-95162017000400005

Faria, M. R., & Wraight, S. P. (2007). Mycoinsecticides and Mycoacaricides: A comprehensive list with worldwide coverage and International Classification of formulation types. *Biological Control*, *43*(3), 237–256. https://doi.org/10.1016/j.biocontrol.2007.08.001

Finnigan, T. J., Wall, B. T., Wilde, P. J., Stephens, F. B., Taylor, S. L., & Freedman, M. R. (2019). Mycoprotein: The future of nutritious nonmeat protein, a symposium review. *Current Developments in Nutrition*, *3*(6). https://doi.org/10.1093/cdn/nzz021

Ghosh, A., & Saha, P. D. (2013). Optimization of copper bioremediation by *Stenotrophomonas maltophilia* PD2. *Journal of Environmental Chemical Engineering*, *1*(3), 159–163. https://doi.org/10.1016/j.jece.2013.04.012

Gonçalves, A. L. (2021). The use of microalgae and cyanobacteria in the improvement of agricultural practices: A review on their biofertilising, biostimulating and biopesticide roles. *Applied Sciences*, *11*(2), 871. https://doi.org/10.3390/app11020871

Gu, W., & Milton, R. D. (2020). Natural and engineered electron transfer of nitrogenase. *Chemistry*, *2*(2), 322–346. https://doi.org/10.3390/chemistry2020021

Hackmann, T. J., & Zhang, B. (2023). The phenotype and genotype of fermentative prokaryotes. *Science Advances*, *9*(39). https://doi.org/10.1126/sciadv.adg8687

Hajimohammadi, A., Mottaghitalab, M., & Hashemi, M. (2020). Effects of microbial fermented sesame meal and enzyme supplementation on the intestinal morphology, microbiota, pH, tibia bone and blood parameters of broiler chicks. *Italian Journal of Animal Science*, *19*(1), 457–467. https://doi.org/10.1080/1828051x.2020.1755378

Hamdy, H. S. (2013). Production of mini-food by *Aspergillus niger*, *Rhizopus oryzae* and *Saccharomyces cerevisiae* using orange peels. *Romanian Biotechnological Letters*, *18*(1), 1–18.

Hashempour-Baltork, F., Khosravi-Darani, K., Hosseini, H., Farshi, P., & Reihani, S. F. (2020). Mycoproteins as safe meat substitutes. *Journal of Cleaner Production*, *253*, 119958. https://doi.org/10.1016/j.jclepro.2020.119958

Heidari, P., Mazloomi, F., & Sanaeizade, S. (2020). Optimization study of nickel and copper bioremediation *by Microbacterium oxydans* strain CM3 and CM7. *Soil and Sediment Contamination: An International Journal*, *29*(4), 438–451. https://doi.org/10.1080/15320383.2020.1738335

Huang, Q. (2024). *Enhancing Soil Health and Biodiversity through Nitrogen Fixation Symbiosis in Leguminous Plants*. https://doi.org/10.5376/mmr.2024.14.0006

Hussain, N., Hung, C.Y., & Wang, L. (2025). *Agricultural Nutrient Pollution and Climate Change: Challenges and Opportunities*. Cham: Springer.

Jalilvand, N., Akhgar, A., Alikhani, H., Rahmani, H. A., & Rejali, F. (2019). Removal of heavy metals zinc, lead, and cadmium by biomineralization of urease-producing bacteria isolated from Iranian mine calcareous soils. *Journal of Soil Science and Plant Nutrition*, *20*(1), 206–219. https://doi.org/10.1007/s42729-019-00121-z

Jazi, V., Ashayerizadeh, A., Toghyani, M., Shabani, A., Tellez, G., & Toghyani, M. (2018). Fermented soybean meal exhibits probiotic properties when included in Japanese quail diet in replacement of soybean meal. *Poultry Science*, *97*(6), 2113–2122. https://doi.org/10.3382/ps/pey071

Jha, Y. (2017). Potassium mobilizing bacteria: Enhance potassium intake in paddy to regulates membrane permeability and accumulate carbohydrates under salinity stress. *Brazilian Journal of Biological Sciences*. https://doi.org/10.21472/bjbs.040812

Jha, S. S., & Songachan, L. S. (2023). Exploring the potential of arbuscular mycorrhizal fungi as a biofertilizer. *Studies in Fungi*, *8*(1). https://doi.org/10.48130/sif-2023-0017

Jha, Y., Subramanian, R. B., & Patel, S. (2020). Combination of endophytic and rhizospheric plant growth promoting rhizobacteria in Oryza sativa shows higher accumulation of osmoprotectant against saline stress, *Acta Physiologiae Plantarum*, 797–802. https://doi.org/10.1007/s11738-010-0604-9

Jhu, M. Y., & Oldroyd, G. E. (2023). Dancing to a different tune, can we switch from chemical to biological nitrogen fixation for Sustainable Food Security? *PLOS Biology*, *21*(3). https://doi.org/10.1371/journal.pbio.3001982

Kameshwari, V., Selvaraj, S., & Sundaramoorthy, S. (2020). Single cell protein spirulina – A nutrient treasure. *Research Journal of Pharmacology and Pharmacodynamics*, *12*(2), 49. https://doi.org/10.5958/2321-5836.2020.00010.5

Kang, J. Y., Lee, M., Song, J. H., Choi, E. J., Kim, D. u., Lim, S. K., Kim, N., & Chang, J. Y. (2022. Lactic acid bacteria strains used as starters for kimchi fermentation protect the disruption of tight junctions in the Caco-2 cell monolayer model. *Journal of Microbiology and Biotechnology*, *32*(12), 1583–1585. https://doi.org/10.4014/jmb.2209.09026

Kaschuk, G., Auler, A. C., Vieira, C. E., Dakora, V. D., Jaiswal, S. K., & da Cruz, S. P. (2022). Coinoculation impact on plant growth promotion: A review and meta-analysis on coinoculation of rhizobia and plant growth-promoting bacilli in grain legumes. *Brazilian Journal of Microbiology*, *53*(4), 2027–2037. https://doi.org/10.1007/s42770-022-00800-7

Kebede, E. (2021). Contribution, utilization, and improvement of legumes-driven biological nitrogen fixation in agricultural systems. *Frontiers in Sustainable Food Systems*, *5*. https://doi.org/10.3389/fsufs.2021.767998

Keswani, C., Singh, S. P., & Singh, H. B. (2013). Beauveria bassiana: Status, mode of action, applications and safety issues. *Biotech Today*, *3*(1), 16. https://doi.org/10.5958/j.2322-0996.3.1.002

Keum, Y. S., Lee, Y. J., & Kim, J. H. (2008). *Journal of Agricultural and Food Chemistry*, *56*(16), 9146–9151.

Khan, I., Seleiman, M. F., Chattha, M. U., Jalal, R. S., Mahmood, F., Hassan, F. A., Izzet, W., A., B. A., Ali, E. F., Roy, R., Uslu, O. S., & Hassan, M. U. (n.d.). Enhancing antioxidant defense system of mung bean with a salicylic acid exogenous application to mitigate cadmium toxicity. *Notulae Botanicae Horti Agrobotanici Cluj-Napoca*, *49*(2), 12303. https://doi.org/10.15835/nbha49212303

Krauspe, V., Fahrner, M., Spät, P., Steglich, C., Frankenberg-Dinkel, N., Maček, B., Schilling, O., & Hess, W. R. (2021). Discovery of a small protein factor involved in the coordinated degradation of phycobilisomes in cyanobacteria. *Proceedings of the National Academy of Sciences*, *118*(5). https://doi.org/10.1073/pnas.2012277118

Kumar, D., Gupta, S., Hasan, W., Hussain, B., Ansari, M. J., & Singh, S. (2025). *Rhizosphere Revolution: Unveiling the Secrets of Insect Pheromones in Soil Health and Vermicompost Production*. Boca Raton, Delhi: CRC Press; Elite Publishing House.

Kumar, P., Kamle, M., Bohra, R., Kumar, D. M., & Sharma, B. (2021). Bacillus thuringiensis as microbial biopesticide: Uses and application for Sustainable Agriculture. *Egyptian Journal of Biological Pest Control*, *31*(1). https://doi.org/10.1186/s41938-021-00440-3

Kumar, V., Raghuvanshi, N., Pandey, A. K., Kumar, A., Thoday-Kennedy, E., & Kant, S. (2023). Role of halotolerant plant growth-promoting rhizobacteria in mitigating salinity stress: Recent advances and possibilities. *Agriculture*, *13*(1), 168. https://doi.org/10.3390/agriculture13010168

Ladha, J. K., Peoples, M. B., Reddy, P. M., Biswas, J. C., A. B., Jat, M. L., & Krupnik, T. J. (2022). Biological nitrogen fixation and prospects for ecological intensification in cereal-based cropping systems. *Field Crops Research*, *283*, 108541.

Lawal, I. (2021). Nitrogen fixing bacteria and their application for Heavy Metal Removal: A mini review. *Journal of Biochemistry, Microbiology and Biotechnology*, *9*(2), 43–47. https://doi.org/10.54987/jobimb.v9i2.617

Li, D., Xu, X., Yu, H., & Han, X. (2017). Characterization of $Pb2^{+}$ biosorption by psychrotrophic strain pseudomonas sp. I3 isolated from permafrost soil of Mohe Wetland in Northeast China. *Journal of Environmental Management*, *196*, 8–15. https://doi.org/10.1016/j.jenvman.2017.02.076

Li, X., Yin, X., & Lian, B. (2016). The degradation of dimethoate and the mineral immobilizing function for Cd^{2+} by Pseudomonas putida. *Geomicrobiology Journal*, *34*(3), 346–354. https://doi.org/10.1080/01490451.2016.1193571

Li, J., Zhang, C., Wang, Y., Liao, X., Yao, L., Liu, M., & Xu, L. (2015). Pollution characteristics and distribution of polycyclic aromatic hydrocarbons and organochlorine pesticides in groundwater at Xiaodian Sewage Irrigation Area, Taiyuan City. *Huan Jing Ke Xue*, *36*(1), 1–5.

Liu, P., Zhao, J., Guo, P., Lu, W., Geng, Z., Levesque, C. L., Johnston, L. J., Wang, C., Liu, L., Zhang, J., Ma, N., Qiao, S., & Ma, X. (2022). Dietary corn bran fermented by *Bacillus subtilis* MA139 decreased gut cellulolytic bacteria and microbiota diversity in finishing pigs. *Frontiers in Cellular and Infection Microbiology*, *7*. https://doi.org/10.3389/fcimb.2017.00526

Lord, J. C. (2005). From Metchnikoff to Monsanto and beyond: The path of microbial control. *Journal of Invertebrate Pathology*, *89*(1), 19–29. https://doi.org/10.1016/j.jip.2005.04.006

Luo, D., Zhou, T., Tao, Y., Feng, Y., Shen, X., & Mei, S. (2016). Exposure to organochlorine pesticides and non-Hodgkin lymphoma: A meta-analysis of observational studies. *Scientific Reports*, *6*(1). https://doi.org/10.1038/srep25768

Luthra, A. (2024). Role of fermenters in food preservation: Extending shelf life naturally. https://www.amergingtech.com/post/role-of-fermenters-in-food-preservation-extending-shelf-life-naturally

Maheshwari, D. K. (2015). *Bacterial Metabolites in Sustainable Agroecosystem*. Sustainable Development and Biodiversity. https://doi.org/10.1007/978-3-319-24654-3

Malla Obaida, B. A. (2021). Yeasts as a source of single cell protein production: A review. *PLANT ARCHIVES*, 324–328. https://doi.org/10.51470/plantarchives.2021.v21.s1.051

Mascarin, G. M., & Jaronski, S. T. (2016). The production and uses of *Beauveria bassiana* as a microbial insecticide. *World Journal of Microbiology and Biotechnology*, *32*(11). https://doi.org/10.1007/s11274-016-2131-3

Msairi, S., Rais, C., Maazouzi, S., Artib, M., El Gabardi, S., Mouden, N., Selmaoui, K., Benkirane, R., Ouazzani Touhami, A., & Douira, A. (2023). Arbuscular mycorrhizal symbiosis in two cannabis varieties (Khardala and critical) in Morocco. *Ecological Engineering & Environmental Technology*, *25*(5), 30–35. https://doi.org/10.12912/27197050/162701

Müller, M., Mentel, M., van Hellemond, J. J., Henze, K., Woehle, C., Gould, S. B., Yu, R. Y., van der Giezen, M., Tielens, A., & Martin, W. F. (2012). Biochemistry and evolution of anaerobic energy metabolism in eukaryotes. *Microbiology and Molecular Biology Reviews*, *76*(2), 444–495. https://doi.org/10.1128/mmbr.05024-11

Nascimento, F. X., Glick, B. R., & Rossi, M. J. (2021). Multiple plant hormone catabolism activities: An adaptation to a plant-associated lifestyle by *Achromobacter* spp. *Environmental Microbiology Reports*, *13*(4), 533–539. https://doi.org/10.1111/1758-2229.12987

Nasseri, A., Rasoul-Amini, S., Morowvat, M., & Ghasemi, Y. (2011). Single cell protein: Production and process. *American Journal of Food Technology*, *6*(2), 1–15.

Nosheen, S., Ajmal, I., & Song, Y. (2021). Microbes as biofertilizers, a potential approach for sustainable crop production. *Sustainability*, *13*(4), 1868. https://doi.org/10.3390/su13041868

Nout, M., & Kiers, J. (2005). Tempe fermentation, innovation and functionality: Update into. *Journal of Applied Microbiology*, *98*, 789–805. https://doi.org/10.1111/j.1365-2672.2004.02471.x

Nziguheba, G., & Smolders, E. (2008). Inputs of trace elements in agricultural soils via phosphate fertilizers in European countries. *Science of the Total Environment*, *390*(1), 53–57. https://doi.org/10.1016/j.scitotenv.2007.09.031

Odgerel, B. (2019). Effect of Chlorella vulgaris as a biofertilizer on germination of tomato and cucumber seeds. *International Journal of Aquatic Biology*, *7*, 95–99.

Ortiz, H., Laura, E., & Sanchez, S. (2010). Biodegradation of the organophosphate pesticide tetrachlorvinphos by bacteria isolated from agricultural soil in Mexico. *Revista Internacional de Contaminación Ambiental*, *26*, 1–12.

Pailan, S., Gupta, D., Apte, S., Krishnamurthi, S., & Saha, P. (2015). Degradation of organophosphate insecticide by a novel *Bacillus aryabhattai* strain SANPS1, isolated from soil of agricultural field in Burdwan, West Bengal, India. *International Biodeterioration & Biodegradation*, 191–195. https://doi.org/10.1016/j.ibiod.2015.05.006

Parekh, N., Walker, A., Roberts, S., & Welch, S. (1994). Rapid degradation of the triazinone herbicide metamitron by a *Rhodococcus* sp. isolated from treated soil. *Journal of Applied Bacteriology*, *77*(5), 467–475. https://doi.org/10.1111/j.1365-2672.1994.tb04389.x

Pathak, J., Rajneesh, Maurya, P. K., Singh, S. P., Häder, D. P., & Sinha, R. P. (2018). Cyanobacterial farming for environment friendly sustainable agriculture practices: Innovations and perspectives. *Frontiers in Environmental Science*, *6*. https://doi.org/10.3389/fenvs.2018.00007

Prabhakaran, P., Ashraf, M. A., & Aqma, W. S. (2016). Microbial stress response to heavy metals in the environment. *RSC Advances*, *6*(111), 109862–109877. https://doi.org/10.1039/c6ra10966g

Puyen, Z. M., Villagrasa, E., Maldonado, J., Diestra, E., Esteve, I., & Solé, A. (2012). Biosorption of lead and copper by heavy-metal tolerant *Micrococcus luteus* DE2008. *Bioresource Technology, 126*, 233–237. https://doi.org/10.1016/j.biortech.2012.09.036

Qurashi, A. W., & Sabri, A. N. (2012). Bacterial exopolysaccharide and biofilm formation stimulate chickpea growth and soil aggregation under salt stress. *Brazilian Journal of Microbiology, 43*(3), 1183–1191. https://doi.org/10.1590/s1517-83822012000300046

Radmer, R. J., & Parker, B. C. (1994). Commercial applications of algae: Opportunities and constraints. *Journal of Applied Phycology, 6*(2), 93–98. https://doi.org/10.1007/bf02186062

Rajkumari, J., Katiyar, P., Dheeman, S., Pandey, P., & Maheshwari, D. K. (2022). The changing paradigm of rhizobial taxonomy and its systematic growth upto postgenomic technologies. *World Journal of Microbiology and Biotechnology, 38*(11). https://doi.org/10.1007/s11274-022-03370-w

Ramasamy, M., Geetha, T., & Yuvaraj, M. (2020). Role of biofertilizers in plant growth and soil health. *Nitrogen Fixation*. https://doi.org/10.5772/intechopen.87429

Ren, G., Jin, Y., Zhang, C., Gu, H., & Qu, J. (2015). Characteristics of *Bacillus* sp. PZ-1 and its biosorption to Pb(II). *Ecotoxicology and Environmental Safety, 117*, 141–148. https://doi.org/10.1016/j.ecoenv.2015.03.033

Ridolfi, A. S., Álvarez, G. B., & Rodríguez Girault, M. E. (n.d.). Organochlorinated contaminants in general population of Argentina and other Latin American countries. *Bioremediation in Latin America*, 17–40. https://doi.org/10.1007/978-3-319-05738-5_2

Roccuzzo, S., Beckerman, A. P., & Trögl, J. (2020). New perspectives on the bioremediation of endocrine disrupting compounds from wastewater using algae-, bacteria- and fungi-based technologies. *International Journal of Environmental Science and Technology, 18*(1), 89–106. https://doi.org/10.1007/s13762-020-02691-3

Roy, R., Núñez-Delgado, A., Sultana, S., Wang, J., A. m., Battaglia, M. L., Sarker, T., Seleiman, M. F., Barmon, M., & Zhang, R. (2021). Additions of optimum water, spent mushroom compost and wood biochar to improve the growth performance of Althaea rosea in drought-prone coal-mined spoils. *Journal of Environmental Management, 295*, 113076. https://doi.org/10.1016/j.jenvman.2021.113076

Sabarwal, A., Kumar, K., & Singh, R. P. (2018). Hazardous effects of chemical pesticides on human health–cancer and other associated disorders. *Environmental Toxicology and Pharmacology*, 103–114. https://doi.org/10.1016/j.etap.2018.08.018

Sáez-Sandino, T., Delgado-Baquerizo, M., Egidi, E., & Singh, B. (2023). New microbial tools to boost restoration and soil organic matter. *Microbial Biotechnology, 16*(11). https://doi.org/10.1111/1751-7915.14325

Salehi Jouzani, G., Seifinejad, A., Saeedizadeh, A., Nazarian, A., Yousefloo, M., Soheilivand, S., Mousivand, M., Jahangiri, R., Yazdani, M., Amiri, R. M., & Akbari, S. (2008). Molecular detection of nematicidal crystalliferous bacillus thuringiensis strains of Iran and evaluation of their toxicity on free-living and plant-parasitic nematodes. *Canadian Journal of Microbiology, 54*(10), 812–822. https://doi.org/10.1139/w08-074

Sam Nirmala, N., Balakrishnan, N., Dhar, S., Bhar, S., & Lakkaboyana, S. K. (2025). Microbial inoculants for the management of pesticide toxicity in plants. *Microbial Inoculants*, 131–147. https://doi.org/10.1016/b978-0-443-23560-3.00009-7

Saraswat, R., Saraswat, D., & Yadav, M. (2020). A review on bioremediation of heavy metals by microbes. *International Journal of Advanced Research, 8*(7), 200–210. https://doi.org/10.21474/ijar01/11281

Schwenninger, S. M., & Meile, L. (2004). A mixed culture of *Propionibacterium jensenii* and *Lactobacillus paracasei* subsp. paracasei inhibits food spoilage yeasts. *Systematic and Applied Microbiology*, 229–237. https://doi.org/10.1078/072320204322881853

Selosse, M. A., & Le Tacon, F. (1998. The land flora: A phototroph-fungus partnership? *Trends in Ecology & Evolution, 13*(1), 15–20. https://doi.org/10.1016/s0169-5347(97)01230-5

Semida, W. M., Abd El-Mageed, T. A., Abdalla, R. M., Hemida, K. A., Howladar, S. M., Leilah, A. A., & Rady, M. O. (2021). Sequential antioxidants foliar application can alleviate negative consequences of salinity stress in *Vicia faba* L. *Plants*, 914. https://doi.org/10.3390/plants10050914

Serrano-Ruiz, J. C., Luque, R., Campelo, J. M., & Romero, A. A. (2012). Continuous-flow processes in heterogeneously catalyzed transformations of biomass derivatives into fuels and chemicals. *Challenges, 3*(2), 114–132. https://doi.org/10.3390/challe3020114

Sharma, A., Kumar, V., Shahzad, B., Tanveer, M., Sidhu, G. P., Handa, N., Kohli, S. K., Yadav, P., Bali, A. S., Parihar, R. D., Dar, O. I., Singh, K., Jasrotia, S., Bakshi, P., Ramakrishnan, M., Kumar, S., Bhardwaj, R., & Thukral, A. K. (2019). Worldwide pesticide usage and its impacts on ecosystem. *SN Applied Sciences, 1*(11). https://doi.org/10.1007/s42452-019-1485-1

Shi, C., & Maktabdar, M. (2022). Lactic acid bacteria as biopreservation against spoilage molds in dairy products – A review. *Frontiers in Microbiology, 12*. https://doi.org/10.3389/fmicb.2021.819684

Simarmata, R., Ngadiman, N., Rohman, S., & Simanjuntak, P. (2018). Amelioration of salt tolerance in soybean (*Glycine max* L.) by plant-growth promoting endophytic bacteria produce 1-aminocyclopropane-1-carboxylase deaminase. *Annales Bogorienses, 22*(2), 81. https://doi.org/10.14203/ann.bogor.2018.v22.n2.81-93

Singh, B. K., Walker, A., Morgan, J. A., & Wright, D. J. (2004). Biodegradation of chlorpyrifos by *Enterobacter* strain B-14 and its use in bioremediation of contaminated soils. *Applied and Environmental Microbiology*, 4855–4863. https://doi.org/10.1128/aem.70.8.4855-4863.2004

Sinha, A., Pant, K. K., & Khare, S. K. (2012). Studies on mercury bioremediation by alginate immobilized mercury tolerant *Bacillus cereus* cells. *International Biodeterioration & Biodegradation, 71*, 1–6. https://doi.org/10.1016/j.ibiod.2011.12.014

Smith, S. E., & Gianinazzi-Pearson, V. (1988). Physiological interactions between symbionts in vesicular-arbuscular mycorrhizal plants. *Annual Review of Plant Physiology and Plant Molecular Biology, 39*(1), 221–244. https://doi.org/10.1146/annurev.pp.39.060188.001253

Soumare, A., Diedhiou, A. G., Thuita, M., Hafidi, M., Ouhdouch, Y., Gopalakrishnan, S., & Kouisni, L. (2020). Exploiting biological nitrogen fixation: A route towards a sustainable agriculture. *Plants*, 1011. https://doi.org/10.3390/plants9081011

Stewart, W., & Roberts, T. (2012). Food security and the role of fertilizer in supporting it. *Procedia Engineering*, 76–82. https://doi.org/10.1016/j.proeng.2012.09.448

Suman, G., Nupur, M., Anuradha, S., & Pradeep, B. (2015). Single cell protein production: A review. *International Journal of Current Microbiology and Applied Sciences, 4*(9), 251–262.

Sziderics, A. H., Rasche, F., Trognitz, F., Sessitsch, A., & Wilhelm, A. (2007). Bacterial endophytes contribute to abiotic stress adaptation in pepper plants (*Capsicum annuum* L.). *Canadian Journal of Microbiology, 53*(11), 1195–1202. https://doi.org/10.1139/w07-082

Tallur, P. N., Megadi, V. B., & Ninnekar, H. Z. (2007). Biodegradation of cypermethrin by *Micrococcus* sp. strain CPN 1. *Biodegradation, 19*(1), 77–82. https://doi.org/10.1007/s10532-007-9116-8

Tarfeen, N., Nisa, K. U., Hamid, B., Bashir, Z., Yatoo, A. M., Dar, M. A., Mohiddin, F. A., Amin, Z., Ahmad, R. A., & Sayyed, R. Z. (2020). Microbial remediation: A promising tool for reclamation of contaminated sites with special emphasis on heavy metal and pesticide pollution: A review. *Processes, 10*(7). https://doi.org/10.3390/pr10071358

Teoh, S. Q., Chin, N. L., Chong, C. W., Ripen, A. M., How, S., & Lim, J. J. (2024). *Future Foods, 9*, 100330. https://doi.org/10.1016/j.fufo.2024.100330

Tewari, S., & Arora, N. K. (2016). *Fluorescent pseudomonas* sp. PF17 as an efficient plant growth regulator and biocontrol agent for sunflower crop under saline conditions. *Symbiosis, 68*(1-3), 99–108.

Tsaboula, A., Papadakis, E. N., Vryzas, Z., Kotopoulou, A., Kintzikoglou, K., & Papadopoulou-Mourkidou, E. (2019). Assessment and management of pesticide pollution at a river basin level part I: Aquatic ecotoxicological quality indices. *Science of The Total Environment, 653*, 1597–1611. https://doi.org/10.1016/j.scitotenv.2018.08.240

United Nations Department of Economic and Social Affairs. (2015). World population prospects: The 2015 revision. *United Nations Econ Soc Aff, 33*(2), 1–66.

Vachon, V., Laprade, R., & Schwartz, J. L. (2012). Current models of the mode of action of bacillus thuringiensis insecticidal crystal proteins: A critical review. *Journal of Invertebrate Pathology, 111*(1), 1–12. https://doi.org/10.1016/j.jip.2012.05.001

Van Oosten, M. J., Pepe, O., De Pascale, S., Silletti, S., & Maggio, A. (2017). The role of biostimulants and bioeffectors as alleviators of abiotic stress in crop plants. *Chemical and Biological Technologies in Agriculture, 4*(1). https://doi.org/10.1186/s40538-017-0089-5

Vasudevan, R., Sowmya, k., & D, H. (2025). Microbial fermentation: A novel attempt of enhancing food quality. *Journal of Microbiology & Experimentation, 13*(2), 34–40.

Win, K. T., Wasai-Hara, S., Tanaka, F., Oo, A. Z., Minamisawa, K., Shimoda, Y., & Imaizumi-Anraku, H. (2023). Synergistic N2-fixation and salt stress mitigation in soybean through dual inoculation of ACC deaminase-producing pseudomonas and bradyrhizobium. *Scientific Reports, 13*(1). https://doi.org/10.1038/s41598-023-43891-4

Wood, A., Toerien, D., & R. K, R. (1991). *Developments in Food Proteins* (Vol. 7). Elsevier Applied Science.

Wuana, R., & Okieimen, F. (2014). Heavy metals in contaminated soils. 1–50. https://doi.org/10.1201/b16566-3

Xu, F. Z., Zeng, X. Z., & Ding, X. L. (2022). Effects of replacing soybean meal with fermented rapeseed meal on performance, serum biochemical variables and intestinal morphology of broilers. *Asian-Australasian Journal of Animal Sciences, 25*(12), 1734–1741. https://doi.org/10.5713/ajas.2012.12249

Yafetto, L., Odamtten, G. T., & Wiafe-Kwagyan, M. (2023). Valorization of agro-industrial wastes into animal feed through microbial fermentation: A review of the Global and Ghanaian case. *Heliyon*, *9*(4). https://doi.org/10.1016/j.heliyon.2023.e14814

Yaish, M. W., Antony, I., & Glick, B. R. (2016). Isolation and characterization of endophytic plant growth-promoting bacteria from date palm tree (*Phoenix dactylifera* L.) and their potential role in salinity tolerance. *Antonie van Leeuwenhoek*, *107*, 1519–1532. https://doi.org/10.1007/s10482-015-0445-z

Yang, C., Liu, N., Guo, X., & Qiao, C. (2016). Cloning of MPD gene from a chlorpyrifos-degrading bacterium and use of this strain in bioremediation of contaminated soil. *FEMS Microbiology Letters*, *265*(1), 118–125. https://doi.org/10.1111/j.1574-6968.2006.00478.x

Yin, K., Lv, M., Wang, Q., Wu, Y., Liao, C., Zhang, W., & Chen, L. (2016). Simultaneous bioremediation and biodetection of mercury ion through surface display of carboxylesterase E2 from pseudomonas aeruginosa PA1. *Water Research*, *103*, 383–390. https://doi.org/10.1016/j.watres.2016.07.053

Zehr, J. P., & Capone, D. G. (2020). Changing perspectives in marine nitrogen fixation. *Science*, *368*(6492). https://doi.org/10.1126/science.aay9514

Zhang, Y., Shi, C., Wang, C., Lu, Z., Wang, F., Feng, J., & Wang, Y. (2016). Effect of soybean meal fermented with *Bacillus subtilis* BS12 on growth performance and small. *Food and Agricultural Immunology*, *29*(1), 133–146. https://doi.org/10.1080/09540105.2017.1360258

Zhao, B., Jia, X., Yu, N., Murray, J. D., Yi, K., & Wang, E. (2023). Microbe-dependent and independent nitrogen and phosphate acquisition and regulation in plants. *New Phytologist*, *242*(4), 1507–1522. https://doi.org/10.1111/nph.19263

7 Microbial Technology for Biofertilizers and Biopesticides in Sustainable Agriculture

Anburaj Raj, Roseline Jebapriya Gunaseelan, Santhanam Perumal, Vanmathi Vidhyashini Selvaraj, Raja Subramaniyan, Annamalai Neelamegam, and Elayaraja Dhandapani

7.1 INTRODUCTION

The degradation of the environment, the loss of biodiversity, loss of soil fertility, and the increase in food requirements with the growth of the population cannot be underestimated and demand sustainable agriculture practices (Srivastava et al., 2016; Rehman & Farooq, 2023). Over time, crop yields have also declined due to conventional farming, which includes the use of synthetic fertilizers and pesticides. While synthetic inputs have brought greater benefits to agriculture, they also cause very large-scale environmental and agricultural issues which stem from over-application and misuse (Tsion & Steven, 2019). The continuous use of chemical fertilizers will, in the long run, result in a loss of fertility and, ultimately, loss of production because of disruption to the diversity of soil microbial communities, organic matter, and structure influenced by the chemicals.

Overuse of fertilizers, particularly nitrogen and phosphorus, will discharge into water streams, resulting in algal blooms and eutrophication, subsequently decreasing aquatic ecosystems (Tal, 2018). Coupled with ineffective treatments due to the evolution of resistant insect communities, the persistently rising chemical pesticides, alongside pest management, negatively impact natural defenders, pollinators, and numerous other favorable organisms, disrupting vital roles and functions on local ecosystems (Whelan & Cunningham, 2020). The detection of pesticide residues in food and water has the potential to impact human health in numerous ways. Some of the ways human health can be impacted are cancer, neurological disorders, and endocrine disruption. Because of these harsh quantities, the focus is now on the need for more holistic methods of the environment that follow the core principles of sustainability (Sexton et al., 2007; Siddiqui et al., 2023).

Microbial agriculture has helped improve agricultural practices and turned the positives, as was the case with technology advances in microbial agricultural innovations (Barea, 2015). Mikiciuk et al. (2024) have also stated that most of the microorganisms including bacteria, fungi, actinomycetes, and cyanobacteria are important to maintain the sustenance of the soil, enriching it with nutrients, and enhancing the development of the plants. They are endowed with the inherent ability of nitrogen fixation, mobilization of potassium, fixation of phosphorus, and the generation of stimulating plant growth hormones. Furthermore, antagonistic or competitive mechanisms of plant pathogen reduction commonly described in Verma et al. (2017) bioantifertility mechanisms could also eliminate plant pathogens. Itelima et al. (2018) explain that the application of biofertilizer formulations that contain active beneficial microorganisms can aid in soil fertility, nutrient cycling, and reduce the need for chemical fertilizers.

Some examples of the biofertilizers that are practiced in agricultural management include phosphate-solubilizing bacteria in enhancing phosphorus availability to crops, *Azospirillum* and

DOI: 10.1201/9781003650881-7

Azotobacter in nitrogen fixation to non-leguminous crops, and *Rhizobium* in its symbiotic associations with legumes in atmospheric nitrogen fixation (Fukami et al., 2018; Nongthombam et al., 2021). Among the biopesticides and biofertilizers, the usefulness of bacteria as an aspect of growing food is the nutrification of the soil and the optimized yield of the crops, which is also done in a green, sustainable way. Additionally, microbial-based biopesticides create a safer alternative to chemical pesticide products (Ruiu, 2018). Microbial biopesticides such as *Bacillus thuringiensis*, *Trichoderma* spp., and *Beauveria bassiana* are great at controlling most insect pests, some plant diseases, and nematodes without affecting non-target species (Arthurs & Dara, 2019). In effect, microbial-based biopesticides are biodegradable, specific in their effect, and do not harm beneficial insects and soil microorganisms, allowing biopesticides to contribute to ecological balance. Biopesticides offer the potential of lowering chemical pesticide use and development of resistance as part of an integrated pest management (IPM) programs which support sustainable pest management (Glare & Nollet, 2023).

This review aims to understand the scope of microbial technology and the many applications that exist within the area of sustainable agriculture. This study explored the different microorganisms that play a role in developing soil fertility and plant growth, the specific type of advantages and the broader discussion of the modern need for pesticides. This study also explores the challenges, limitations, and future opportunities for potential microbial solutions in agriculture with climate change and resilient agricultural systems. This review considers the pathways of microbial treatments and may help to develop a more environmentally benign, productive, and sustainable agricultural paradigm.

7.2 MICROORGANISMS AND SOIL HEALTH

Microorganisms are critical for improving soil structure and achieving sustainable plant growth, as well as to support soil's functionality on longer time scales (Chen et al., 2024; Wei et al., 2024). Soil microorganisms, many of which are bacteria and fungi, secrete extracellular materials (polysaccharides, glycoproteins, and mucilage) that help to glue soil particles together to make aggregates (Hartmann & Six, 2023). Stable aggregates are necessary for improving soil conditions such as porosity, aeration, water infiltration, and resistance to erosion and compaction. Fungi, especially mycorrhizal fungi, also enhance soil structure by being physically entangled with the soil particles through their hyphae which assist in forming a soil matrix that is stable (Ritz & Young, 2004; Devi et al., 2020). Improving soil structure progresses root penetration, increases the water holding capacity of the soil, and enables better access to soil nutrients which each have the potential to allow for improved crop yield.

Soil bacteria break down organic matter including dead microbes, plant materials, animal waste into simple molecules of increased fertility, and enriched soil with additional micronutrients by releasing vital nutrients as forms that plants are able to uptake (Shrivastav et al., 2020). The natural activity of microbes increases availability of organic acids, enzymes, and bioactive compounds which further comobilize and enhance the chemical nature of the soil. In short, these microbes are turning dirt into a living, radius material that can support healthy plant growth with minimal need for artificial fertilizer.

7.2.1 MICROBES: ROLE IN NUTRIENT RECYCLING

Nutrient recycling is one of the key ecological services that microorganisms in the soil contribute (Saccá et al., 2017; Prasad et al., 2020). Biological nitrogen fixation is carried out by microbes in the nitrogen cycle and converts atmospheric nitrogen (N_2) that otherwise cannot be used directly by plants into bioavailable forms. Such free-living nitrogen fixers as *Azotobacter* and *Clostridium* perform equal tasks without any dependence on each other in the ground (Aasfar et al., 2021; Park & Jang, 2024). Phosphorus is also often present in the phosphorus cycle as the inorganic mineral form that cannot be easily consumed by plants (Duhamel, 2025). Phosphate-solubilizing microorganisms (PSMs), such as *Bacillus megaterium*, *Pseudomonas fluorescens*, and certain fungi, release organic acids, which degrade bound phosphorus compounds, making them soluble as phosphates, which again help plants to take up the phosphate (Sivasakthi et al., 2014; Ng et al., 2022).

Similarly, certain microorganisms release potassium required in the physiological functions of the plants, and they recycle the potassium present in mineral constituents as the functions of the soil. This also happens with sulfur when the microbes oxidize the organic and elementally bound sulfur to plant-available sulfate. Micronutrients (iron, zinc, manganese, etc.) can be recycled in microbial activity as well, to make sure that the plants have balanced food (Klotz et al., 2011; Anantharaman et al., 2018). The nutrients being continuously recycled through an array of various ways by the microbes are in bid to keep a balance of nutrients in the soil, avoid losses of the nutrients and increase the nutrient utilization in farming systems.

7.2.2 Microbes: Interaction with Plant Roots

The near environment adjacent to plant roots is rich in biological activities and diverse microbiological varieties, and this is the rhizosphere. In this microenvironment, microorganisms, soil particles, and plant roots come into close contact. Many organic compounds are exudates of the plant into the rhizosphere, which includes sugars, amino acids, organic acids, vitamins, and secondary metabolites. These exudates stimulate the development of microorganisms around the roots as they feed on those microorganisms (McNear, 2013; Prashar et al., 2014). There are beneficial microorganisms within the rhizosphere that react to their presence by stimulating plant growth and well-being in many ways. The production of some of these microorganisms ensures phytohormones, including gibberellins, cytokinins, and auxins, which promote branching and root lengthening to help the plant gain more access to nutrients and water (Spaepen, 2014; Egamberdieva et al., 2017).

Saha et al. (2013) assert that others produce siderophores, substances that bind iron in soil and make the metal more accessible to plants, at the same time inhibiting the development of potentially threatening, iron-requiring disease. In addition, microorganisms can precondition the plants to be more resistant to cope with biotic and abiotic stress by promoting systemic resistance. Ramamoorthy et al. (2001) indicate that the mycorrhizal fungi, particularly the arbuscular mycorrhizae (e.g., members of the genus *Glomus*) form symbiotic relationships with plant roots, hence extending the root system into the soil by means of their hyphal networks. They, combined, make the plants more resistant to drought and soil salinity and significantly raise the level of water, phosphorus, and micronutrients absorption. Thus, rhizosphere is an interactive interface in which the plant and the microbes are getting co-evolutionarily geared to mutual advantage of cooperation, forming an energetic association that is significant in sustainable agriculture (Van Loon et al., 1998; Pieterse et al., 2003).

There are a lot of microorganisms in the soil that must be present in order to enhance soil fertility, promote plant growth, and keep an ecosystem in balance. The beneficial microorganisms contribute to the prevention of plant diseases, the improvement of soil composition, and increase in agricultural productivity, besides the promotion of nutrient cycling (Sahu et al., 2017; Chauhan et al., 2023). In sustainable agriculture, biofertilizer properties of these microorganisms and the use of these organisms as biocontrol agents would reduce the use of chemicals and facilitate environmental well-being by intelligent utilization. Table 7.1 provides a summary of important soil microorganisms, their types, primary functions, and the specific crops they benefit.

TABLE 7.1
Examples of Helpful Soil Microbes and Their Functions

Microorganism	Primary Function	References
Rhizobium spp.	Fixes atmospheric nitrogen in legume root nodules	Yang et al. (2022)
Azospirillum spp.	Fixes nitrogen Produces growth-promoting substances	Fukami et al. (2018)
Azotobacter spp.	Nitrogen fixation Phytohormone production	Kizilkaya (2009); Aasfar et al. (2021)
Bacillus megaterium	Solubilizes insoluble phosphorus in soil	Saeid et al. (2018)
Pseudomonas fluorescens	Solubilizes phosphate Suppresses pathogens	Park et al. (2009)
Frateuria aurantia	Solubilizes potassium from minerals	Nowembabazi et al. (2021)
Glomus spp.	Enhances phosphorus and micronutrient uptake Drought resistance	Ruiz-Lozano et al. (1995)
Trichoderma spp.	Controls soil-borne pathogens Enhances nutrient uptake	de Medeiros et al. (2021)
Aspergillus niger	Solubilizes phosphate Promotes growth	Chuang et al. (2007)
Anabaena spp.	Fixes nitrogen in paddy fields Improves organic matter	Li et al. (2025)
Nostoc spp.	Nitrogen fixation Soil enrichment	Peng & Bruns (2019)

7.3 BIOFERTILIZERS: NATURAL BOOSTERS FOR PLANT GROWTH

Chaudhary et al. (2022) define biofertilizers as natural formulations comprising live microorganisms, which can enhance the availability of important nutrients of the plants thus enhancing their development and production. When colonizing the soil or the plant roots, these beneficial microorganisms ensure a nutrient cycle when using biological mechanisms such as potassium mobilization, phosphorus solubilization, and nitrogen fixation, among others (Malusà et al., 2016; Mahanty et al., 2017). Biofertilizers are much more natural and thus encourage soil fertility, sustainability, and ecological stability as compared to chemical fertilizers which are simply released on the ground in synthetic form. Their use can be a very important eco-friendly strategy in modern farming by not only reducing the detrimental effects of chemical fertilization but also maintaining or even raising agricultural yields (Shahwar et al., 2023).

7.3.1 Types and Examples of Biofertilizers

Broad grouping of biofertilizers is done depending on the specific nutrients that they contribute to release in addition to the individual biological roles that they have played in enhancing soil fertility and plant growth. These categories include potassium and micronutrient solubilizers, phosphorus solubilizers, and nitrogen fixers (Thomas & Singh, 2019). All types of biofertilizers include some specific type of microbial species that can operate by means of specific metabolic pathways, which convert some rare forms of nutrients into ones able to be used by the plants (Abbey et al., 2019). In order to exploit the concept behind the recent use of biofertilizer in sustainable farming, it is very necessary to understand the numerous types of biofertilizer and examples of biofertilizer. Nitrogen is also one of the key macronutrients that plants need when manufacturing proteins, nucleic acids, and other important substances of plant growth and development. Despite the fact that nitrogen is available freely in the environment in nitrogen gas (N_2) form, the plants cannot utilize nitrogen directly (Rashid et al.,

2014). The bacteria present in nitrogen-fixing biofertilizers can convert nitrogen that is present in the atmosphere into ammonia or other compounds that can be taken into the plant and utilized by the plant. These biofertilizers are highly critical in order to ensure that there is sufficient soil nitrogen in the absence of high quantity of synthetic nitrogen fertilizer (Shridhar, 2012). The roots of leguminous plants such as soybean, peanut, lentil, and chickpea specifically associate with *Rhizobium* species, which are nitrogen-fixing bacteria, on a symbiotic basis (Mejri et al., 2018; Lindström & Mousavi, 2020). Infection of roots by these bacteria results in nodule formation which then the bacteria contain N_2 in the atmosphere, converting it to ammonia with the help of an enzyme called nitrogenase.

Such symbiosis is not only beneficial to subsequent crops being grown in the same cycle by dint of the enriched surrounding soil but also serves to satiate the nitrogen requirement of the host plant. In this way, the application of *Rhizobium*-based biofertilizers significantly reduces synthetically produced nitrogen fertilizer in legume production (Lindström et al., 2010; Martínez-Hidalgo & Hirsch, 2017). Unlike *Rhizobium*, the *Azospirillum* species live and thrive in the vicinity of roots without forming nodules since they are involved in associative nitrogen. They are beneficial especially to non-leguminous crops such as rice, maize, wheat, and sorghum. Moreover, *Azospirillum* also produces phytohormones that enhance root elongation, root branching and overall root biomass, gibberellins, cytokinins, and indole-3-acetic acid (IAA) (Mehnaz, 2014). This increased root system enriches the ability to uptake water and nutrients; thus, good plant growth and increased yields are ready even in the nutrient-deficient soils.

Plants need phosphorus because it is involved with the transfer of energy, photosynthesis, and synthesis of nucleic acids and roots. Phosphorus, however, in the greater part of soils, is in the insoluble form and thus it is inaccessible to the plants. Organic acids, enzymes, other products of microorganism-mediated metabolism, and other activities break the insoluble phosphate compounds in aqueous media and convert them into their soluble forms that are easily absorbed by plants, such as orthophosphates (Kalayu, 2019). To ensure the solubilization of the bound phosphates, *B. megaterium* releases a range of organic acids, including citric acid and gluconic acid that lower the pH levels around soil particles. The microbial effect improves the supply of phosphorus to crops, such as wheat, rice, pulses, and oilseeds, leading to improved root growth, flowering, and fruit set (Soumare et al., 2020).

P. fluorescens can solubilize phosphorus, and it possesses the property of biocontrol against several soil-borne diseases (Noumavo et al., 2018). It increases the uptake of nutrients in addition to forming antibiotics, siderophores, and enzymes that do not allow dangerous bacteria and fungus to damage the plant. It has become an important biopesticide and biofertilizer in the integrated pest and nutrient management systems due to its twin role (Samanta et al., 2025).

Potassium is important in the physiological processes of plants including photosynthesis, osmoregulation, tolerance to stress, and the activation of enzymes. Similarly, micronutrients such as zinc, iron, manganese, and copper are required in the activities of enzymes, the process of using chlorophyll, and overall plant metabolism (Olaniyan et al., 2022). However, regrettably, owing to the interaction with the minerals in soil, these nutrients often are in forms that are not easy to uptake by plants. An example of a bacterium that is also known to be capable of potassium mobilization in the insoluble mineral forms, such as mica and feldspar, is the *Frateuria aurantia*. *F. aurantia* excretes potassium in the soil solution in such a way that plants can uptake it through the formation of organic acids and chelators. It is observed that the usage of such biofertilizer has resulted in significant enhancement in the crops such as sugarcane, banana, potato, and vegetables, especially in potassium deficient soils (Spoorthi et al., 2022; Teja & Saharan, 2025).

Along with potassium solubilization, many strains of *Bacillus* and *Pseudomonas* dissolve essential micronutrients and make them mobile. An example is that zinc, iron, and manganese are often found to be low in alkaline and calcareous soils, but can be enhanced by *Bacillus subtilis* and *Pseudomonas putida* (Verma et al., 2017; Maysoon & Hassan, 2024). These bacteria prevent nutrient constraints by secreting siderophores, organic acids, and enzymes, which ensure the plants well-balanced nutrition by either chelating or converting micronutrients into a form which could be easily assimilated by plants.

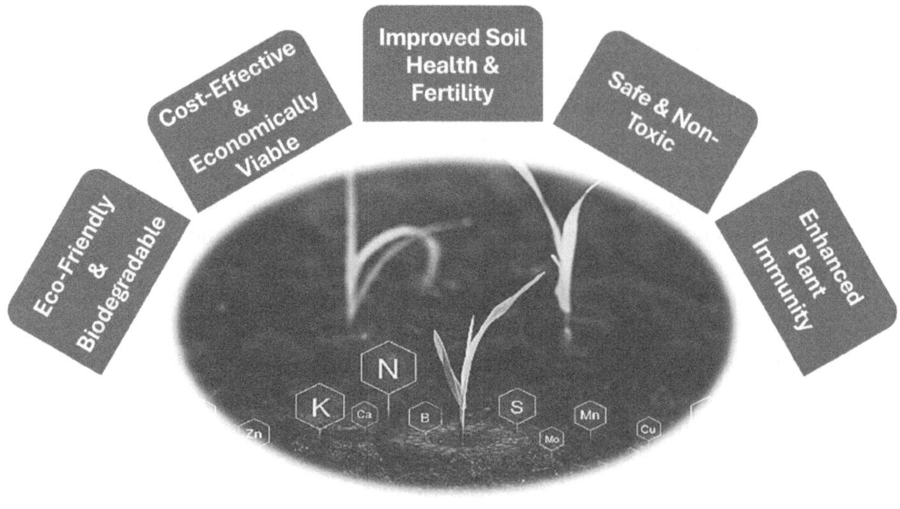

FIGURE 7.1 Advantages of biofertilizer.

7.3.2 Benefits of Biofertilizers over Chemical Fertilizers

Biofertilizers are an important part of sustainable agriculture because they have a number of important benefits over traditional chemical fertilizers (Figure 7.1). Their advantages go beyond only providing nutrients; they also help to preserve the environment, improve soil health, and save farmers' money (Suhag, 2016).

7.3.3 Impact of Biofertilizers on Crop Yield

This has been confirmed especially through numerous field studies and experimental trials in various crops that have raised significant evidence on the positive impacts of the biofertilizers on crop production. When *Rhizobium* is injected into the soybean seeds, e.g., yield gains in the range of 20%–30% have always been recorded, and the biological nitrogen fixation mechanism has been found to enhance nitrogen levels in the soil as well (Jaga & Sharma, 2015; Bertham et al., 2019; Htwe et al., 2019). As such, a combination of the phosphate-solubilizing bacteria and a nitrogen-fixing bacterium, *Azospirillum*, in rice cultivation has shown 15%–25% growth in crop yields in comparison with the untreated ones (Suhameena et al., 2020). Such a synergistic effect not only stimulated the growth of roots but also boosted the general vitality of the plant, which was able to absorb nutrients more effectively and cope with environmental conditions more effectively.

The application of potassium-solubilizing bacteria *F. aurantia* has amplified the yield of crops that have a high requirement of nutrients such as sugarcane, apples, and bananas (Nowembabazi et al., 2021). The use of this biofertilizer enhanced the availability of potassium in the soil, which raised the weight of the canes and sugar content, which is an important quality characteristic in the production of sugarcane (Haran & Thaher, 2019; Ng et al., 2022). Indeed, when compared to farms that had applied only chemical fertilizers, the phosphate-solubilizing strain *P. fluorescens* supported an increment of 12%–18% in grain yield during wheat cultivation when combined with *B. megaterium*. This advantage was attributed to the biocontrol potential of *Pseudomonas* and the higher phosphorus availability, which enabled it to act as a protective mechanism against soil-borne diseases on a crop (Khoshru et al., 2025). The considered synergy of potassium-solubilizing bacteria and reduced amount of the chemical fertilizer also contributed to banana cultivation (Verma et al., 2017; Pandey & Saharan, 2025). Economically, this strategy benefited farmers by making the cost of production go down but the results observed were equal to the results obtained when the full chemical fertilizer mix was used. Collectively, these case studies signify the ability of biofertilizers, to augment crop productivity, enhance the quality of the soil, and reduce the dependency on chemical fertilizers, all of which promote more sustainable and cost-effective farming practices (Priyadharsini & Muthukumar, 2016; Aloo et al., 2020; Mahmud et al., 2021).

7.4 BIOPESTICIDES FOR CROP PROTECTION

Biopesticides are made of living organisms, such as bacteria, fungi, and plant materials, that are used to manage pests in an environmentally responsible way. Unlike synthetic chemical pesticides, biopesticides are grounded in biological phenomena and are typically specific to the target pest or pathogen, thus causing little to no damage to humans, the environment and other helpful species (Gupta & Dikshit, 2010; Seiber et al., 2014). Sexton et al. (2007) state that there are diverse ways in which they are used, and among them are inhibiting the growth of diseases, producing toxins that act against specific pests, enhancing plant defense, or disrupting pest reproduction cycles. Biopesticides also contribute to IPM programs, which have a much more favorable effect on the environment because they provide a non-chemical solution to pesticide use and minimize exposure of food to pesticides and environmental contamination.

7.4.1 Common Microbial Biopesticides

Biopesticides can be microbial biological control agents which are particularly effective against microbially controlled diseases and insect pests of crops and weeds. They are prepared using naturally occurring microorganisms such as bacteria, fungi, and viruses. Microorganisms may swim free, compete with other microorganisms over space and nutrients, secrete toxins, parasitize plants, or activate plant defense systems, among others (Arthurs & Dara, 2019). Unlike the broad-spectrum conventional pesticides, microbial biopesticides are highly target-specific and ensure limited deleterious effects on helpful insects, non-target organisms, and the environment (Opender Koul, 2012). IPM techniques need to involve their use in order to sustainably protect crops. *B. thuringiensis* (Bt), *Trichoderma* sp., and *B. bassiana* are examples of microbial biopesticides considered to be well-known due to each mechanism operating on different pests and diseases (Kumar et al., 2019; Thakur et al., 2020; Vero et al., 2023).

B. thuringiensis (Bt) bacterial biopesticide is regarded as one of the most effective and commonly distributed biopesticides all over the world due to its capability of controlling a number of aggregations of insect pests (Kumar et al., 2021). Sporulating leaves the gram-positive and spore-forming bacteria Bt producing crystal (Cry) proteins. The larvae of some insect orders, such as Lepidoptera (butterflies and moths), Diptera (mosquitoes and flies), and Coleoptera (beetles), ingest these Cry proteins, and this leads to their poisoning. Upon their ingestion by unsuspecting insect larvae, the toxins bind to targets in the stomach of the insect, creating holes in the stomach wall of the insect and its digestive system is paralyzed and ultimately killed (Ruiu, 2018). Bt finds favor. It is regarded as an environmentally friendly pest control measure included in ecologically conscious pest management schemes, and organic agriculture, because it is highly specific, killing only inimical pests but not beneficial insects, pollinators, humans, and livestock. It has already been effective when applied to cotton, maize, vegetables, and fruit crops to control pests such as cabbage looper, corn borer, and cotton bollworm (Rosas-García, 2009). Moreover, biopesticides based on Bt do not contaminate the environment by leaving any toxic trace behind upon degrading in the environment.

Trichoderma species belong to beneficial fungi that are known to play a biologically controlling role to a wide range of soil-borne plant diseases. Other mechanisms by which these fungi control their effects include mycoparasitism (direct killing of the pathogens), competition with other organisms in nutrient acquisition and space occupation, and secretion of antifungal compounds, enzymes, and antibiotics which then prevent the growth of harmful fungi (de Medeiros et al., 2021).

The pathogens that are more easily controlled by *Trichoderma* are those that cause plant diseases that are destructive and pathological such as root rot, damping-off, and wilting such as *Rhizoctonia solani*, *Pythium* spp., and *Fusarium* spp. Along with illness suppression, *Trichoderma* enhances nutrient uptake, root growth, and the creation of systemic resistance, a fundamental characteristic of resistance to various environmental stressors and diseases (Lyubenova et al., 2023; Roseline Jebapriya et al., 2025). The dual purpose of *Trichoderma* makes it useful as an ingredient in improving crop vigor and crop yield, in addition to the protection of plants. It finds widespread use in crop production, either as a seed treatment, applied to the soil or as a root dip of seedling crops (cereals, legumes, vegetables, and plantation crops) (Kredics et al., 2024).

B. bassiana, an entomopathogenic fungus (killing bug fungus), is commonly used as a form of biological control, as a biopesticide, to target a wide range of insect pests. When fungal spores of the *B. bassiana* infect insects such as mealybugs, aphids, whiteflies, thrips, and beetles, they adhere to the cuticle and germinate and thus penetrate their exoskeleton once they contact their bodies (Mascarin & Jaronski, 2016). The fungus within the body of the insect then develops and then releases poisons that paralyze and kill. It is an effective fungus-based biopesticide which is especially effective on fruit crops, ornamental plants, field vegetables, and crops in the greenhouse. It fits perfectly in the IPM systems as it will perform effectively in diverse environmental conditions, and it coexists with other biological control agents. Notably, *B. bassiana* does not pose harmful effects on other helpful insects as a pollinator and natural predators, hence preserving ecological equilibrium (Kumar et al., 2025).

7.4.2 Advantages of Biopesticides over Chemical Pesticides

Natural sources including microbes, plant extracts, and certain minerals are used to make biopesticides, which have several benefits over traditional chemical pesticides (Figure 7.2). Their eco-friendliness, target specificity, and contribution to the long-term health of crops and ecosystems

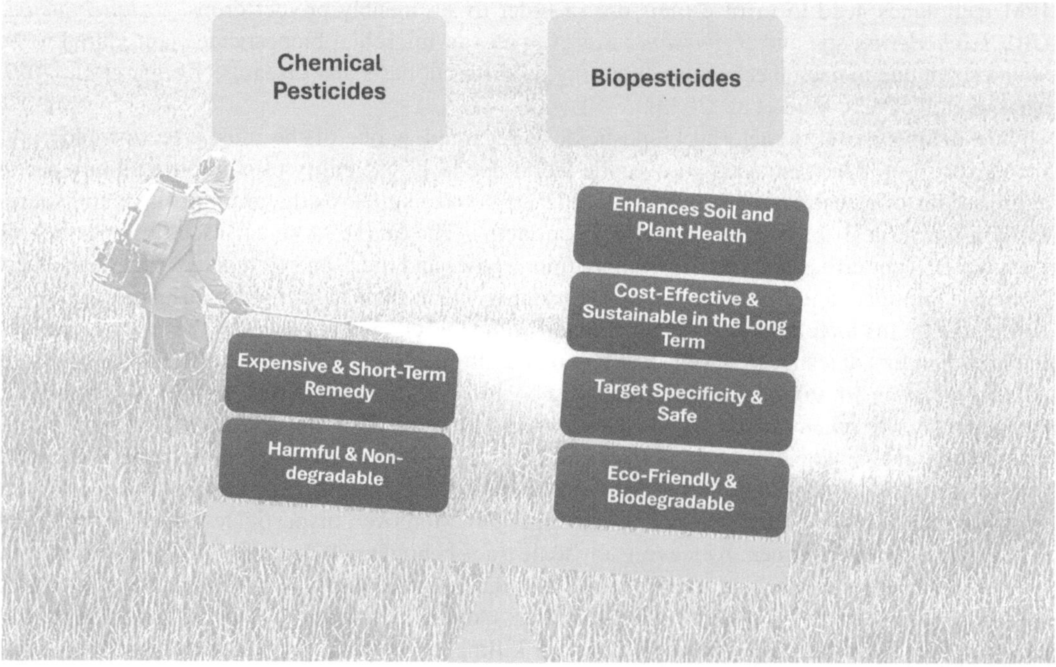

FIGURE 7.2 Benefits of biopesticides over traditional chemical pesticides.

have led to a growing recognition of their significance in sustainable agriculture (Ruiu, 2018; Wakefield, 2018; Thakur et al., 2020).

7.4.3 Impact of Biopesticide on Crop Protection

Various success stories involving the use of biopesticides in agriculture have been witnessed in many parts of the world and the product has proven to be very effective in controlling agricultural pests and diseases besides ensuring sustainability of the environment. Such real-life experiences highlight the fact that biopesticides can be incorporated to the mainstream farming to improve crop health as well as crop productivity and to curb chemical pesticide use. The application of *B. thuringiensis* (Bt) to control bollworm infestation in cotton crops falls under one of the most well-known success stories (Abro et al., 2004; Wang et al., 2009). The cotton bollworm (*Helicoverpa armigera*) has reduced significantly in such countries as China, India, and the United States due to the application of Bt-based biopesticides and Bt cotton; a genetically modified crop that produces Bt toxin. Since in India, the introduction of Bt cotton has led to a considerable rise in cotton production and a major decline in the use of chemical pesticides, which led to cost reduction in production and rise in the incomes of farmers (Gujar et al., 2007; Kathage & Qaim, 2012). Moreover, insecticide was not applied as much hence reduced the health risks to the farm workers and made the environment better.

The *Trichoderma* have been famous in India to manage pathogenic-related diseases in horticulture crops such as cucurbits, tomatoes, and chillies (Meher et al., 2021). Indicatively, Tamil Nadu farmers have documented that an overwhelming number of wilt, root rot, and damping-off diseases occur when they applied *Trichoderma*-based products (Kumar et al., 2017; Hegde & Vijaykumar, 2022). This made the plants more vigorous, soil healthier, and reduced the losses in the crops. *Trichoderma* is also becoming very common in banana farms as a preventive measure against Panama disease (*Fusarium* wilt) that would make the plant healthier and more high-yielding (Vijayasanthi et al., 2022). The *B. bassiana* has as well been effectively applied in controlling aphids, thrips, and whiteflies during the cultivation of vegetables (mainly in countries like the Netherlands and Spain) in a greenhouse environment (Prasad & Syed, 2010; Dannon et al., 2020). Greenhouse growers often prefer the use of this fungal biopesticide because it can be used along with the beneficial insects including the parasitoid wasps and the ladybugs besides the favorability of its owning environment to its humidity and temperatures. Use of *B. bassiana* in the greenhouse has not only resulted in making the intensive production of vegetables more sustainable, but also decreased the amount of chemical pesticides required to grow these fruits and vegetables, thereby making them more secure (Mascarin & Jaronski, 2016). *Metarhizium anisopliae* is a common entomopathogenic fungus in Brazil that is used to control spittlebugs on sugarcane (Bettiol, 2011). Spoilage bugs seriously damage plants and decrease their output through ingesting sap in the plants. The results of the adoption of *Metarhizium* formulations in these plantations included better yields of sugarcane, reduced applications of chemicals as pesticides, and effective insect control. Through this method of biocontrol, the quality of the soil and water were saved along with an increase in output (Mesquita et al., 2023).

P. fluorescens has effectively been utilized to control bacterial blight and sheath blight disease in rice farms in the rice-producing areas of South India. It has proven to be remarkably effective in enhancing plant immunity and mitigating pathogenic infections coupled with the use of foliar spray and seed treatment (David et al., 2018). Farmers who applied this biopesticide were able to increase their crop produce, produce healthier crops and reduced use of artificial fungicides and bactericides.

H. armigera and *Spodoptera* species have also been effectively suppressed using granulosis virus (GV) and nuclear polyhedrosis virus (NPV) in vegetables and pulses including pigeon pea (Nawaz et al., 2019). To give an example, NPV concoctions have been provided to the pulse growers in Karnataka, India and have been successful in managing pod perforators without any side effect of damage in the non-target insects (Srinivasa et al., 2008). The technique has assisted in ensuring that the production of pulses is sustainable and the food commodities have very low levels of residues of the pesticides.

7.5 COMBINING BIOFERTILIZERS AND BIOPESTICIDES

The combination of biofertilizers and biopesticides can help to achieve effective pest and disease control, high yields, and healthy crops (Abbey et al., 2019; Seenivasagan & Babalola, 2021). This mixture of biological products, rather than their reliance on only the use of chemicals, also helps to safeguard the plants against deadly infestations and insects as well as protect them against these insects and infestations without sacrificing the important nutrients they need. This twofold effect ensures long-term soil quality and safety of the environment and food security, preserving the principles of environmentally friendly agriculture (Balasubramanian & Karthickumar, 2017).

When applied together, biopesticides and biofertilizers create a synergic environment on the surface of a plant and in the soil which leads to enhanced defense and developmental processes. The availability of nutrients was optimized through the use of biofertilizers (*Rhizobium*, *Azospirillum*, *B. megaterium*, or *P. fluorescens*) through nitrogen fixation, phosphorus and other microelements solubilization, and potassium and other micronutrients mobilization. The implications of such practices are the resultant plants with powerful secure root formations and expanded uptake and absorption of nutrients (Lyubenova et al., 2023; Kredics et al., 2024; Kumar et al., 2025).

Meanwhile, biopesticides (*Trichoderma, B. bassiana, M. anisopliae*, and *B. thuringiensis*) help plants against pests and disease by parasitizing the target plants or secreting toxins that attack specific pests (Kumar et al., 2017; Dannon et al., 2020; Meher et al., 2021). This combined approach will reduce the use of chemical pesticides and fertilizer by keeping plants healthy by guarding them against biotic stressors and supplying them with plant food necessary to grow well. It has been found that biofertilizers and biopesticides can be used simultaneously to complement efficient crop production. Among them are enhanced germination levels, biomass, plant immunity, and yields as compared to the individual use of any input. With this combination, the environment is better preserved and better production developed, which contributes to sustainable agricultural practices.

7.6 CHALLENGES AND SOLUTIONS

Although a lot of advantages are associated with biopesticides and biofertilizers application in farming, there are several challenges being faced with respect to the use that have to be addressed so as to ensure increased application and efficient findings. As goods of microbial nature contain live organisms that are highly sensitive to environmental conditions such as temperature, humidity, and exposure to light, the shelf life and storage of these products represent major concerns. The ineffective application in the fields may be due to poor storage that reduces the viability of microbes. Against this, novel means of formulation to enhance stability and shelf life have been devised, including encapsulation, liquid formulations, and stable carrier materials like peat, lignite, or alginate beads (Ndolo et al., 2019; Chakraborty & Akhtar, 2021).

The quality of the product can also be safeguarded by establishing adequate storage centers, and even educating the farmers and distributors on the optimal methods to store them. Field performance instability caused by differences in soil type, pH, temperature, moisture available, and indigenous microbe population is also another major setback. Such environmental factors can play a role that determines whether the bacteria imported into the country will be able to survive and multiply creating an unbalanced situation in various geographical locations. The answers lie in choosing of site-specific microbial strains, designing of formulations that are relevant to a particular locality, and conducting wide field trials to identify the strains that best suit different agro-climatic conditions. Improving the effectiveness of biofertilizer and biopesticide can also be accomplished through proper choice of biopesticide and biofertilizer combination, and also proper dose, timing, and mode of application of biofertilizers and biopesticides. Moreover, non-availability of technical support, skills, and confidence of using biological products as compared to fast results and visible effects of chemical organzation is also a factor that still holds the farmers back in terms of awareness as well as acceptance. To address this, large scale information and education activities, pilot demonstrations and testimonies on the benefits, and practical applications of these bio-products must be brought forth. The non-governmental organizations (NGOs), research centers, and agricultural extension services are very important in providing and educating the farmers on the effective application of biological remedies. These challenges can be overcome through innovation, education, and regulatory measures to ensure that biofertilizers and biopesticides meet their potential, and offer sustainable agricultural production, protection of the environment, and improvement in farmer livelihoods.

7.7 FUTURE POSSIBILITIES

The fact is that technology advances, the necessity of climate-resistant agriculture, and the global determination of achieving the Sustainable Development Goals (SDGs) all testify to the luminous future of biofertilizers and biopesticides. Due to increased demands on climate change, soil degradation, and agriculture of sustainable food production, alternative practices are being developed to increase and enhance effectiveness and usage of microbial products. The key developments include new technologies such as liquid biofertilizers and nano-formulations. In comparing disadvantages and advantages of liquid biofertilizers with conventional solid formulations, it could be observed that there are several advantages such as high microbial viability due to the absence of hardened cytoplasmic membranes that accompanies drying and more easy and simplified administration and their long shelf lifespan.

When applied directly via foliar sprays, seed treatments, or irrigation systems, these formulations afford even spreading and favorable colonization on the African plant roots. Nano-formulations allow the incorporation of microbial agents into nano-sized carriers, improving their stability, their desired distribution, and their regulated release of the active substances. This approach enhances the performance of biofertilizers and biopesticides through the protection of the microorganisms against stress in the environment and ensuring that they remain active within the soil-plant ecosystem.

Technical advances are increasingly reliant on biopesticides and fertilizers as well as climate-resilient agriculture. Application of microbial products has the potential to assist crops adapt to stress conditions because of the fluctuating climate patterns which has been raising the incidence of drought, floods, and insect pandemics. Certain bacteria enhance soil infrastructure, augment water retention, and induce nutrient uptake under adverse conditions and thereby enhance plant resilience. Moreover, under the new pest dynamics due to climate change, the biopesticides can offer eco-friendly alternatives to chemical pesticides, reducing greenhouse emission associated with synthetic pesticides and sustainable control of pests.

The integration of biofertilizers and biopesticides is also in close correlation with the United Nations SDGs, namely, SDG 2 (Zero Hunger), SDG 12 (Responsible Consumption and Production), SDG 13 (Climate Action), and SDG 15 (Life on Land). Such biological materials assist to boost production of food stuffs without compromising quality of the environment by promoting environmentally friendly farming methods. They are found to be useful in not only enhancing food security and nutrition (SDG 2) but also enabling farmers to adopt to the impact of climate change (SDG 13), reducing pollution and chemical dependency (SDG 12), and soil health recovery (SDG 15).

7.8 CONCLUSION

Some of the benefits of employing microbial inputs in form of biopesticides and biofertilizers include the fact that they contribute immensely to sustainable production in agriculture in a greener way. These biological products are biologically safe because they enhance fertility in soil, stimulate plant growth, as well as provide effective ways of controlling pests and diseases, without compromising the environment. The inputs of microbes increase plant immunity and nutrient availability that in addition to increasing crop yields also leads to restoration of biodiversity and soil health.

To eliminate the use of chemical pesticides and fertilizers that often lead to the degradation of soils, pollution of waters, and even illnesses, there is a need to ensure that people turn to environmentally friendly agricultural practices using the microbial technology. Environmentally conscious farming encourages production of healthy and safe food, preservation of environment, and ensures sustainability of food production.

The right direction is to invest in their further research and work on developing complex formulations, microbe strain specific to a discrete geographical location, and new delivery formats such as liquid biofertilizers and nano-formulations. There is also the need to have extensive training of farmers, awareness creation, and demonstrations in the fields with the aim of increasing the uptake and appropriate use of these technology. Creating a high-quality control system, regulatory support, and public-private partnerships will help introduce microbial solutions into the traditional agricultural system more readily and clear the path to the rise of more resilient and sustainable agricultural practice in the future.

REFERENCES

Aasfar, A., Bargaz, A., Yaakoubi, K., Hilali, A., Bennis, I., Zeroual, Y., & Meftah Kadmiri, I. (2021). Nitrogen fixing *Azotobacter* species as potential soil biological enhancers for crop nutrition and yield stability. *Frontiers in Microbiology*, *12*, 628379.

Abbey, L., Abbey, J., Leke-Aladekoba, A., Iheshiulo, E. M. A., & Ijenyo, M. (2019). Biopesticides and biofertilizers: Types, production, benefits, and utilization. In *Byproducts from Agriculture and Fisheries: Adding Value for Food, Feed, Pharma, and Fuels* (pp. 479–500). Wiley.

Abro, G. H., Syed, T. S., Tunio, G. M., & Khuhro, M. A. (2004). Performance of transgenic Bt cotton against insect pest infestation. *Biotechnology*, *3*(1), 75–81.

Aloo, B. N., Tripathi, V., Makumba, B. A., & Mbega, E. R. (2022). Plant growth-promoting rhizobacterial biofertilizers for crop production: The past, present, and future. *Frontiers in Plant Science*, *13*, 1002448..

Anantharaman, K., Hausmann, B., Jungbluth, S. P., Kantor, R. S., Lavy, A., Warren, L. A., ... & Banfield, J. F. (2018). Expanded diversity of microbial groups that shape the dissimilatory sulfur cycle. *The ISME Journal*, *12*(7), 1715–1728.

Arthurs, S., & Dara, S. K. (2019). Microbial biopesticides for invertebrate pests and their markets in the United States. *Journal of Invertebrate Pathology*, *165*, 13–21.

Balasubramanian, P., & Karthickumar, P. (2017). Biofertilizers and biopesticides: A holistic approach for sustainable agriculture. In *Sustainable Utilization of Natural Resources* (pp. 255–284). CRC Press.

Barea, J. M. (2015). Future challenges and perspectives for applying microbial biotechnology in sustainable agriculture based on a better understanding of plant-microbiome interactions. *Journal of Soil Science and Plant Nutrition*, *15*(2), 261–282.

Bertham, R. R., Arifin, Z., & Nusantara, A. D. (2019). The improvement of yield and quality of soybeans in a coastal area using low input technology based on biofertilizers. *International Journal on Advanced Science, Engineering and Information Technology*, *9*(3), 787–791.

Bettiol, W. (2011). Biopesticide use and research in Brazil. *Outlooks on Pest Management*, *22*(6), 280–283.

Bonilla, I., & Bolanos, L. (2009). Mineral nutrition for legume-rhizobia symbiosis: B, Ca, N, P, S, K, Fe, Mo, Co, and Ni: A review. In *Organic Farming, Pest Control and Remediation of Soil Pollutants* (pp. 253–274). Springer.

Chakraborty, T., & Akhtar, N. (2021). Biofertilizers: Prospects and challenges for future. In *Biofertilizers: Study and Impact* (pp. 575–590). Wiley.

Chaudhary, P., Singh, S., Chaudhary, A., Sharma, A., & Kumar, G. (2022). Overview of biofertilizers in crop production and stress management for sustainable agriculture. *Frontiers in Plant Science*, *13*, 930340.

Chauhan, P., Sharma, N., Tapwal, A., Kumar, A., Verma, G. S., Meena, M., ... & Swapnil, P. (2023). Soil microbiome: Diversity, benefits and interactions with plants. *Sustainability*, *15*(19), 14643.

Chen, Q., Song, Y., An, Y., Lu, Y., & Zhong, G. (2024). Soil microorganisms: Their role in enhancing crop nutrition and health. *Diversity*, *16*(12), 734.

Chuang, C. C., Kuo, Y. L., Chao, C. C., & Chao, W. L. (2007). Solubilization of inorganic phosphates and plant growth promotion by *Aspergillus niger. Biology and Fertility of Soils, 43*(5), 575–584.

Dannon, H. F., Dannon, A. E., Douro-Kpindou, O. K., Zinsou, A. V., Houndete, A. T., Toffa-Mehinto, J., ... & Tamò, M. (2020). Toward the efficient use of *Beauveria bassiana* in integrated cotton insect pest management. *Journal of Cotton Research, 3*(1), 24.

David, B. V., Chandrasehar, G., & Selvam, P. N. (2018). *Pseudomonas fluorescens*: A plant-growth-promoting rhizobacterium (PGPR) with potential role in biocontrol of pests of crops. In *Crop Improvement through Microbial Biotechnology* (pp. 221–243). Elsevier.

de Medeiros, E. V., da Silva, L. F., da Silva, J. S. A., da Costa, D. P., de Souza, C. A. F., Berger, L. R. R., ... & Hammecker, C. (2021). Biochar and *Trichoderma* spp. in management of plant diseases caused by soilborne fungal pathogens: A review and perspective. *Research, Society and Development, 10*(15), e296101522465-e296101522465.

Devi, R., Kaur, T., Kour, D., Rana, K. L., Yadav, A., & Yadav, A. N. (2020). Beneficial fungal communities from different habitats and their roles in plant growth promotion and soil health. *Microbial Biosystems, 5*(1), 21–47.

Duhamel, S. (2025). The microbial phosphorus cycle in aquatic ecosystems. *Nature Reviews Microbiology, 23*(4), 239–255.

Egamberdieva, D., Wirth, S. J., Alqarawi, A. A., Abd_Allah, E. F., & Hashem, A. (2017). Phytohormones and beneficial microbes: Essential components for plants to balance stress and fitness. *Frontiers in Microbiology, 8*, 2104.

Fukami, J., Cerezini, P., & Hungria, M. (2018). *Azospirillum*: Benefits that go far beyond biological nitrogen fixation. *Amb Express, 8*(1), 73.

Glare, T. R., & Nollet, L. M. (2023). Types of biopesticides. In *Biopesticides Handbook* (pp. 7–24). CRC Press.

Gujar, G. T., Kalia, V., Kumari, A., Singh, B. P., Mittal, A., Nair, R., & Mohan, M. (2007). *Helicoverpa armigera* baseline susceptibility to *Bacillus thuringiensis* Cry toxins and resistance management for Bt cotton in India. *Journal of Invertebrate Pathology, 95*(3), 214–219.

Gupta, S., & Dikshit, A. K. (2010). Biopesticides: An ecofriendly approach for pest control. *Journal of Biopesticides, 3*(Special Issue), 186.

Haran, M. S., & Thaher, A. Z. T. (2019). Effect bio-fertilizer of *Bacillus, Azotobacter* and *Pseudomonas fluorescens* in the growth and production of corn plant (*Zea mays* L.). *Basrah Journal of Agricultural Sciences, 32*, 7–14.

Hartmann, M., & Six, J. (2023). Soil structure and microbiome functions in agroecosystems. *Nature Reviews Earth & Environment, 4*(1), 4–18.

Hegde, G. M., & Vijaykumar, K. N. (2022). Formulation, application and commercialization of biopesticides in India formulation, application and commercialization of biopesticides in India. *Asia Pacific Biofertilizers and Biopesticides Information Platform.* https://doi.org/10.56669/KSHW3457

Htwe, A. Z., Moh, S. M., Moe, K., & Yamakawa, T. (2019). Biofertilizer production for agronomic application and evaluation of its symbiotic effectiveness in soybeans. *Agronomy, 9*(4), 162.

Itelima, J. U., Bang, W. J., Onyimba, I. A., & Oj, E. (2018). A review: Biofertilizer; a key player in enhancing soil fertility and crop productivity. *Journal of Microbiology and Biotechnology Reports, 2*(1), 22–28.

Jaga, P. K., & Sharma, S. A. T. I. S. H. (2015). Effect of bio-fertilizer and fertilizers on productivity of soybean. *Annals of Plant and Soil Research, 17*(2), 171–174.

Kalayu, G. (2019). Phosphate solubilizing microorganisms: Promising approach as biofertilizers. *International Journal of Agronomy, 2019*(1), 4917256.

Kathage, J., & Qaim, M. (2012). Economic impacts and impact dynamics of Bt (*Bacillus thuringiensis*) cotton in India. *Proceedings of the National Academy of Sciences, 109*(29), 11652–11656.

Khoshru, B., Fallah Nosratabad, A., Mahjenabadi, V. A. J., Knežević, M., Hinojosa, A. C., Fadiji, A. E., ... & Mitra, D. (2025). Multidimensional role of *Pseudomonas*: From biofertilizers to bioremediation and soil ecology to sustainable agriculture. *Journal of Plant Nutrition, 48*(6), 1016–1042.

Kizilkaya, R. (2009). Nitrogen fixation capacity of *Azotobacter* spp. strains isolated from soils in different ecosystems and relationship between them and the microbiological properties of soils. *Journal of Environmental Biology, 30*(1), 73–82.

Klotz, M. G., Bryant, D. A., & Hanson, T. E. (2011). The microbial sulfur cycle. *Frontiers in Microbiology, 2*, 241.

Kredics, L., Büchner, R., Balázs, D., Allaga, H., Kedves, O., Racić, G., ... & Sipos, G. (2024). Recent advances in the use of *Trichoderma*-containing multicomponent microbial inoculants for pathogen control and plant growth promotion. *World Journal of Microbiology and Biotechnology, 40*(5), 162.

Kumar, K. K., Sridhar, J., Murali-Baskaran, R. K., Senthil-Nathan, S., Kaushal, P., Dara, S. K., & Arthurs, S. (2019). Microbial biopesticides for insect pest management in India: Current status and future prospects. *Journal of Invertebrate Pathology*, *165*, 74–81.

Kumar, M., Mehta, C. R., Agrawal, K. N., & Tripathi, M. K. (2025). Optimization of operating parameters for spraying microbial (*Bacillus thuringiensis* and *Beauveria bassiana*) based bio-pesticide solutions for foliar application. *International Journal of Pest Management*, *71*(3), 364–376.

Kumar, P., Kamle, M., Borah, R., Mahato, D. K., & Sharma, B. (2021). *Bacillus thuringiensis* as microbial biopesticide: Uses and application for sustainable agriculture. *Egyptian Journal of Biological Pest Control*, *31*(1), 95.

Kumar, T. V., Veena, S. S., Karthikeyan, S., & Sreekumar, J. (2017). Compatibility of *Trichoderma asperellum* with fungicides, insecticides, inorganic fertilizers and bio-pesticides. *Journal of Root Crops*, *43*(2), 68–75.

Li, S., Huang, W., Peng, C., Jing, X., Ding, J., Chen, T., … & Liang, Y. (2025). Enhancement of rice production and soil carbon sequestration utilizing nitrogen-fixing cyanobacteria. *Applied Soil Ecology*, *207*, 105940.

Lindström, K., & Mousavi, S. A. (2020). Effectiveness of nitrogen fixation in rhizobia. *Microbial Biotechnology*, *13*(5), 1314–1335.

Lindström, K., Murwira, M., Willems, A., & Altier, N. (2010). The biodiversity of beneficial microbe-host mutualism: The case of rhizobia. *Research in Microbiology*, *161*(6), 453–463.

Lyubenova, A., Rusanova, M., Nikolova, M., & Slavov, S. B. (2023). Plant extracts and *Trichoderma* spp: Possibilities for implementation in agriculture as biopesticides. *Biotechnology & Biotechnological Equipment*, *37*(1), 159–166.

Mahanty, T., Bhattacharjee, S., Goswami, M., Bhattacharyya, P., Das, B., Ghosh, A., & Tribedi, P. (2017). Biofertilizers: A potential approach for sustainable agriculture development. *Environmental Science and Pollution Research*, *24*(4), 3315–3335.

Mahmud, N. U., Surovy, M. Z., Gupta, D. R., & Islam, M. T. (2021). Advances in potassium nutrition in crop plants by potassium solubilizing bacteria. In *Agriculturally Important Microorganisms* (pp. 175–199). CRC Press.

Malusà, E., Pinzari, F., & Canfora, L. (2016). Efficacy of biofertilizers: Challenges to improve crop production. In *Microbial Inoculants in Sustainable Agricultural Productivity: Vol. 2: Functional Applications* (pp. 17–40). Springer India.

Martínez-Hidalgo, P., & Hirsch, A. M. (2017). The nodule microbiome: N2-fixing rhizobia do not live alone. *Phytobiomes Journal*, *1*(2), 70–82.

Mascarin, G. M., & Jaronski, S. T. (2016). The production and uses of *Beauveria bassiana* as a microbial insecticide. *World Journal of Microbiology and Biotechnology*, *32*(11), 177.

Masson-Boivin, C., & Sachs, J. L. (2018). Symbiotic nitrogen fixation by rhizobia—the roots of a success story. *Current Opinion in Plant Biology*, *44*, 7–15.

Maysoon, A. K., & Hassan, K. U. (2024). Synergistic effect of *Bacillus Mucilaginosus* and *Pseudomonas fluorescens* on the availability of soil potassium, growth and yield of *Cucurbita pepo* L. *Iraqi Journal of Agricultural Sciences*, *55*(5), 1650–1656.

McNear Jr, D. H. (2013). The rhizosphere-roots, soil and everything in between. *Nature education knowledge*, *4*(3), 1.

Meher, J., Rajput, R. S., Bajpai, R., Teli, B., & Sarma, B. K. (2021). *Trichoderma*: A globally dominant commercial biofungicide. In *Trichoderma: Agricultural Applications and Beyond* (pp. 195–208). Springer International Publishing.

Mehnaz, S. (2014). *Azospirillum*: A biofertilizer for every crop. In *Plant Microbes Symbiosis: Applied Facets* (pp. 297–314). Springer India.

Mejri, S., Saidi, M., & Belhadj, O. (2018). Potential of rhizobia in improving nitrogen fixation and yields of legumes. *Symbiosis*, *107*.

Mesquita, E., Hu, S., Lima, T. B., Golo, P. S., & Bidochka, M. J. (2023). Utilization of Metarhizium as an insect biocontrol agent and a plant bioinoculant with special reference to Brazil. *Frontiers in fungal biology*, *4*, 1276287.

Mikiciuk, G., Miller, T., Kisiel, A., Cembrowska-Lech, D., Mikiciuk, M., Łobodzińska, A., & Bokszczanin, K. (2024). Harnessing beneficial microbes for drought tolerance: A review of ecological and agricultural innovations. *Agriculture*, *14*(12), 2228.

Nawaz, A., Ali, H., Sufyan, M., Gogi, M. D., Arif, M. J., Ranjha, M. H., … & Ghramh, H. A. (2019). Comparative bio-efficacy of nuclear polyhedrosis virus (NPV) and Spinosad against American bollworm, *Helicoverpa armigera* (Hubner). *Revista Brasileira de Entomologia*, *63*(4), 277–282.

Ndolo, D., Njuguna, E., Adetunji, C. O., Harbor, C., Rowe, A., Den Breeyen, A., ... & Hospet, R. (2019). Research and development of biopesticides: Challenges and prospects. *Outlooks on Pest Management, 30*(6), 267–276.

Ng, C. W. W., Yan, W. H., Tsim, K. W. K., San So, P., Xia, Y. T., & To, C. T. (2022). Effects of *Bacillus subtilis* and *Pseudomonas fluorescens* as the soil amendment. *Heliyon, 8*(11).

Nongthombam, J., Kumar, A., Sharma, S., & Ahmed, S. (2021). *Azotobacter*: A complete review. *Bulletin of Environment, Pharmacology and Life Sciences, 10*(6), 72–79.

Noumavo, P. M. A., DagbÃ, G., Salami, H. A., Karimou, R., AlladÃ, A. M., Adedayo, O., ... & SaÃ, L. (2018). Biofertilising, plant-stimulating and biocontrol potentials of maize plant growth promoting rhizobacteria isolated in central and northern Benin. *African Journal of Microbiology Research, 12*(28), 664–672.

Nowembabazi, A., Taulya, G., & Tinzaara, W. (2021). Contribution of biofertiliser (*Frateuria auranta*) in an integrated potassium management package on growth of apple banana. *Open Journal of Soil Science, 11*(11), 521–537.

Olaniyan, F. T., Alori, E. T., Adekiya, A. O., Ayorinde, B. B., Daramola, F. Y., Osemwegie, O. O., & Babalola, O. O. (2022). The use of soil microbial potassium solubilizers in potassium nutrient availability in soil and its dynamics. *Annals of Microbiology, 72*(1), 45.

Opender Koul, O. K. (2012). Microbial biopesticides: opportunities and challenges. *Cabi Reviews (2011)*, 1–26.

Pandey, K., & Saharan, B. S. (2025). Potassium-solubilizing endophytes: Mechanisms and applications in enhancing sustainable agriculture and plant resilience. *Symbiosis, 95*(3), 1–15.

Park, K. H., Lee, C. Y., & Son, H. J. (2009). Mechanism of insoluble phosphate solubilization by *Pseudomonas fluorescens* RAF15 isolated from ginseng rhizosphere and its plant growth-promoting activities. *Letters in applied microbiology, 49*(2), 222–228.

Park, S., & Jang, J. (2024). Excessive extracellular ammonium production by a free-living nitrogen-fixing soil *Clostridium* sp. strain. *Microorganisms, 12*(12), 2634.

Peng, X., & Bruns, M. A. (2019). Development of a nitrogen-fixing cyanobacterial consortium for surface stabilization of agricultural soils. *Journal of Applied Phycology, 31*(2), 1047–1056.

Pieterse, C. M., van Pelt, J. A., Verhagen, B. W., Ton, J. U. R. R. I. A. A. N., van Wees, S. C., Léon-Kloosterziel, K. M., & Van Loon, L. C. (2003). Induced systemic resistance by plant growth-promoting rhizobacteria. *Symbiosis, 35*, 39–54.

Prasad, A., & Syed, N. (2010). Evaluating prospects of fungal biopesticide *Beauveria bassiana* (Balsamo) against *Helicoverpa armigera* (Hubner): An ecosafe strategy for pesticidal pollution. *Asian Journal of Experimental Biological Sciences, 1*(3), 596–601.

Prasad, S., Malav, L. C., Choudhary, J., Kannojiya, S., Kundu, M., Kumar, S., & Yadav, A. N. (2020). Soil microbiomes for healthy nutrient recycling. In *Current Trends in Microbial Biotechnology for Sustainable Agriculture* (pp. 1–21). Springer Singapore.

Prashar, P., Kapoor, N., & Sachdeva, S. (2014). Rhizosphere: its structure, bacterial diversity and significance. *Reviews in Environmental Science and Bio/Technology, 13*(1), 63–77.

Priyadharsini, P., & Muthukumar, T. (2016). Interactions between arbuscular mycorrhizal fungi and potassium-solubilizing microorganisms on agricultural productivity. In *Potassium Solubilizing Microorganisms for Sustainable Agriculture* (pp. 111–125). Springer India.

Ramamoorthy, V., Viswanathan, R., Raguchander, T., Prakasam, V., & Samiyappan, R. (2001). Induction of systemic resistance by plant growth promoting rhizobacteria in crop plants against pests and diseases. *Crop protection, 20*(1), 1–11.

Rashid, M. H. O., Krehenbrink, M., & Akhtar, M. S. (2014). Nitrogen-fixing plant-microbe symbioses. *Sustainable Agriculture Reviews, 15*, 193–234.

Rehman, A., & Farooq, M. (2023). Challenges, constraints, and opportunities in sustainable agriculture and environment. In *Sustainable Agriculture and the Environment* (pp. 487–501). Academic Press.

Ritz, K., & Young, I. M. (2004). Interactions between soil structure and fungi. *Mycologist, 18*(2), 52–59.

Rosas-García, N. M. (2009). Biopesticide production from *Bacillus thuringiensis*: An environmentally friendly alternative. *Recent Patents on biotechnology, 3*(1), 28–36.

Roseline Jebapriya, G., Anburaj, R., Prithiviraj, N., Santhanam, P., Jayanthi Kumar, & Sharangouda J. Patil. (2025). Biotechnological phytochemical synthesis: Innovations, challenges, advantages, implications. *Biotechnological and Phytochemical Prospects in Drug Discovery* (pp. 83–101). Springer International Publishing.

Ruiu, L. (2018). Microbial biopesticides in agroecosystems. *Agronomy, 8*(11), 235.

Ruiz-Lozano, J. M., Azcon, R., & Gomez, M. (1995). Effects of arbuscular-mycorrhizal *Glomus* species on drought tolerance: Physiological and nutritional plant responses. *Applied and environmental microbiology, 61*(2), 456–460.

Saccá, M. L., Barra Caracciolo, A., Di Lenola, M., & Grenni, P. (2017). Ecosystem services provided by soil microorganisms. In *Soil Biological Communities and Ecosystem Resilience* (pp. 9–24). Springer International Publishing.

Saeid, A., Prochownik, E., & Dobrowolska-Iwanek, J. (2018). Phosphorus solubilization by *Bacillus* species. *Molecules*, *23*(11), 2897.

Saha, R., Saha, N., Donofrio, R. S., & Bestervelt, L. L. (2013). Microbial siderophores: A mini review. *Journal of basic microbiology*, *53*(4), 303–317.

Sahu, N., Vasu, D., Sahu, A., Lal, N., & Singh, S. K. (2017). Strength of microbes in nutrient cycling: A key to soil health. In *Agriculturally Important Microbes for Sustainable Agriculture: Volume I: Plant-Soil-Microbe Nexus* (pp. 69–86). Springer Singapore.

Samanta, K., Islam, A., Banik, M., & Koley, S. (2025). Biofertilizers for sustainable agricultural practice in crop production: A review. *AGRIVITA Journal of Agricultural Science*, *47*(2), 410–424.

Seenivasagan, R., & Babalola, O. O. (2021). Utilization of microbial consortia as biofertilizers and biopesticides for the production of feasible agricultural product. *Biology*, *10*(11), 1111.

Seiber, J. N., Coats, J., Duke, S. O., & Gross, A. D. (2014). Biopesticides: State of the art and future opportunities. *Journal of Agricultural and Food Chemistry*, *62*(48), 11613–11619.

Sexton, S. E., Lei, Z., & Zilberman, D. (2007). The economics of pesticides and pest control. *International Review of Environmental and Resource Economics*, *1*(3), 271–326.

Shahwar, D., Mushtaq, Z., Mushtaq, H., Alqarawi, A. A., Park, Y., Alshahrani, T. S., & Faizan, S. (2023). Role of microbial inoculants as bio fertilizers for improving crop productivity: A review. *Heliyon*, *9*(6).

Shridhar, B. S. (2012). Nitrogen fixing microorganisms. *International Journal of Microbiology Research*, *3*(1), 46–52.

Shrivastav, P., Prasad, M., Singh, T. B., Yadav, A., Goyal, D., Ali, A., & Dantu, P. K. (2020). Role of nutrients in plant growth and development. In *Contaminants in Agriculture: Sources, Impacts and Management* (pp. 43–59). Springer International Publishing.

Siddiqui, Junaid Ali, Ruidong Fan, Hira Naz, Bamisope Steve Bamisile, Muhammad Hafeez, Muhammad Imran Ghani, Yiming Wei, Yijuan Xu, & Xiaoyulong Chen. (2023). Insights into insecticide-resistance mechanisms in invasive species: Challenges and control strategies. *Frontiers in Physiology*, *13*, 1112278.

Sivasakthi, S., Usharani, G., & Saranraj, P. (2014). Biocontrol potentiality of plant growth promoting bacteria (PGPR)-*Pseudomonas fluorescens* and *Bacillus subtilis*: A review. *African Journal of Agricultural Research*, *9*(16), 1265–1277.

Soumare, A., Boubekri, K., Lyamlouli, K., Hafidi, M., Ouhdouch, Y., & Kouisni, L. (2020). From isolation of phosphate solubilizing microbes to their formulation and use as biofertilizers: Status and needs. *Frontiers in bioengineering and biotechnology*, *7*, 425.

Spaepen, S. (2014). Plant hormones produced by microbes. In *Principles of Plant-Microbe Interactions: Microbes for Sustainable Agriculture* (pp. 247–256). Springer International Publishing.

Spoorthi, V. B., Ranadev, P., Ashwin, R., & Bagyaraj, J. D. (2022). Response of *Capsicum annuum* L. seedlings raised in pro trays to inoculation with AM fungus *Glomus bagyarajii* and K solubilizing bacterium *Frateuria aurantia*. *Seeds*, *1*(4), 315–323.

Srinivasa, M., Jagadeesh Babu, C. S., Anitha, C. N., & Girish, G. (2008). Laboratory evaluation of available commercial formulations of HaNPV against *Helicoverpa armigera* (Hub.). *Journal of Biopesticides*, *1*(2), 138–139.

Srivastava, P., Singh, R., Tripathi, S., & Raghubanshi, A. S. (2016). An urgent need for sustainable thinking in agriculture—an Indian scenario. *Ecological indicators*, *67*, 611–622.

Suhag, M. (2016). Potential of biofertilizers to replace chemical fertilizers. *International Advanced Research Journal in Science, Engineering and Technology*, *3*(5), 163–167.

Suhameena, B., Devi, S., Gowri, R., & Kumar, S. D. (2020). Utilization of *Azospirillum* as a biofertilizer—an overview. *International Journal of Pharmaceutical Sciences Review and Research*, *62*(2), 141–145.

Tal, A. (2018). Making conventional agriculture environmentally friendly: Moving beyond the glorification of organic agriculture and the demonization of conventional agriculture. *Sustainability*, *10*(4), 1078.

Tarafdar, J. C. (2022). Role of soil biology on soil health for sustainable agricultural production. In *Structure and Functions of Pedosphere* (pp. 67–81). Springer Nature Singapore.

Teja, A. P. S. N., & Saharan, B. S. (2025). Endophytic potassium-solubilizing microbes: A sustainable approach to enhancing crop nutrition and productivity. In *Metabolic Sustainability of Endophytes: Current Status, Challenges and Potential* (pp. 37–53). Springer Nature Singapore.

Thakur, N., Kaur, S., Tomar, P., Thakur, S., & Yadav, A. N. (2020). Microbial biopesticides: Current status and advancement for sustainable agriculture and environment. In *New and Future Developments in Microbial Biotechnology and Bioengineering* (pp. 243–282). Elsevier.

Thomas, L., & Singh, I. (2019). Microbial biofertilizers: Types and applications. In *Biofertilizers for Sustainable Agriculture and Environment* (pp. 1–19). Springer International Publishing.

Tsion, K., & Steven, W. (2019). An overview use and impact of organic and synthetic farm inputs in developed and developing countries. *African Journal of Food, Agriculture, Nutrition and Development*, *19*(3), 14517–14540.

Van Loon, L. C., Bakker, P. A. H. M., & Pieterse, C. M. J. (1998). Systemic resistance induced by rhizosphere bacteria. *Annual Review of Phytopathology*, *36*(1), 453–483.

Verma, P., Yadav, A. N., Khannam, K. S., Saxena, A. K., & Suman, A. (2017). Potassium-solubilizing microbes: Diversity, distribution, and role in plant growth promotion. In *Microorganisms for Green Revolution: Volume 1: Microbes for Sustainable Crop Production* (pp. 125–149). Springer Singapore.

Vero, S., Garmendia, G., Allori, E., Sanz, J. M., Gonda, M., Alconada, T., … & Wisniewski, M. (2023). Microbial biopesticides: Diversity, scope, and mechanisms involved in plant disease control. *Diversity*, *15*(3), 457.

Vijayasanthi, S., Akila, R., Ayyandurai, M., & Kannan, R. (2022). Survey, identification and management of *Fusarium* wilt of banana in Tamirabarani tract of Southern districts of Tamil Nadu. *Journal of Biological Control*, *36*(1).

Wakefield, M. E. (2018). Microbial biopesticides. In *Recent Advances in Stored Product Protection* (pp. 143–168). Springer Berlin Heidelberg.

Wang, Y., Ye, G., Luan, N., Xiao, J., Chen, Y., & Chen, D. (2009). Boll size affects the insecticidal protein content in *Bacillus thuringiensis* (Bt) cotton. *Field Crops Research*, *110*(2), 106–110.

Wei, X., Xie, B., Wan, C., Song, R., Zhong, W., Xin, S., & Song, K. (2024). Enhancing soil health and plant growth through microbial fertilizers: Mechanisms, benefits, and sustainable agricultural practices. *Agronomy*, *14*(3), 609.

Whelan, C. J., & Cunningham, J. J. (2020). Resistance is not the end: Lessons from pest management. *Cancer Control*, *27*(1), 1073274820922543.

Yang, J., Lan, L., Jin, Y., Yu, N., Wang, D., & Wang, E. (2022). Mechanisms underlying legume–rhizobium symbioses. *Journal of Integrative Plant Biology*, *64*(2), 244–267.

8 Increasing the Nutritional Value of Food with Microorganisms

A Focus on Food Fermentation and Probiotics

Alex Kamgain Tchuenchieu and Eridiong Onyenweaku

8.1 INTRODUCTION

Food insecurity, malnutrition, undernutrition, and diet-related diseases are key concepts often associated to the humanity living conditions and eating habits of populations. Their prevalence varies from one country to another. Public health policies are usually planned and implemented at national and international levels for their prevention and control. Food insecurity represents a lack of reliable access to sufficient, safe, and nutritious food by the population. Low- and middle-income countries are mostly concerned. In 2024, 8.2% of the global population faced hunger, not far from 8.5% in 2023, with improvements noticed in Asia and South America but continued increases in Africa and Western Asia (FAO, IFAD, UNICEF, WFP and WHO, 2025). Food insecurity in a region represents a risk factor to malnutrition among the population. This integrates undernutrition and overnutrition. Insufficient intake of energy, protein, or micronutrients is the characteristics of undernutrition leading to wasting, stunting, underweight, and micronutrient deficiencies. This is not the case with overnutrition which rather results from excessive consumption of calories, fats, or sugars. The consequences might then be overweight, obesity, and diet-related non-communicable diseases. While poverty, food insecurity, and infections are the commonly reported causes of undernutrition, overnutrition is rather associated with high-calorie diets, sedentary lifestyles, and urbanization. According to Word Health Organization, a total of 2.5 billion adults in the world were estimated to be overweight among which 890 million were obese in 2022, while 390 million were underweight. Their statistics also revealed that among children under 5 years old, 149, 45, and 37 million were stunted, wasted, and overweight or obese, respectively. Approximately, half of the deaths in this age group were linked to undernutrition, predominantly in developing countries (WHO, 2024). On the other hand, diet-related diseases represent the long-term health outcome of malnutrition. They can originate from poor diet quality, often due to overnutrition or imbalanced diets, and are recognized as world burden. Type 2 diabetes, hypertension, cardiovascular diseases, and cancer are the most prevalent diseases (Ma et al., 2025). Countries are more and more confronted with the triple burden of malnutrition. This consists of a simultaneous coexistence of undernutrition, micronutrient deficiencies, and overnutrition within the same population, household, or even individual. Low- and middle-income countries undergoing rapid nutritional and epidemiological transitions are increasingly concerned, which poses a significant public health challenge in those areas (The triple burden of malnutrition, 2023).

The availability of nutritious, healthy, and safe food is a target to help fight these burdens from one country to another. Fermentation is one of humanity's oldest food processing technique, practiced

 DOI: 10.1201/9781003650881-8

from generation to generation across nearly every culture as a natural method for preserving food and enhancing its sensory qualities (Bourdichon et al., 2021; Taveira et al., 2021). The direct positive impact is its use for reducing food waste from both animal and plant sources (Hadj Saadoun, 2021). Beyond preservation, fermentation has also been reported to significantly improve the nutritional value of foods. It relies on the metabolic activity of diverse microorganisms, such as bacteria, yeasts, and fungi, which transform raw substrates the target being to produce nutrient-rich and safe food products (Siddiqui et al., 2023). During the process, a breakdown of anti-nutritional factors and complex molecules is generally observed, thus releasing nutrients that will be more bioavailable to humans consuming the fermented products. Furthermore, many fermentative microbes synthesize essential vitamins that are either absent or present in limited quantities in raw materials (Sharma et al., 2020). Examples of fermented foods and beverages include beer, wine, cheese, yogurt, fermented porridges, cacao fermented for chocolate, and fermented cassava products, among others. Most traditional fermentation processes that have been identified have also been studied in detail, with the main microorganisms involved characterized. Some of these processes have even been adapted for industrial-scale production under controlled conditions using the specific microorganisms identified (Praveen & Brogi, 2025). Depending on the full production process, the final fermented product ready for consumption may still contain alive fermentative germs. When they are known as beneficial microorganisms that, when ingested in appropriate amounts, contribute to host health, they are qualified as probiotics. A common product containing probiotics is yogurt. The integration of these probiotics into both traditional and industrial fermented foods represents a sustainable pathway to deliver functional food to populations.

This chapter explores the microbial taxa involved in fermentation, the biochemical and molecular mechanisms by which they enhance food nutrition and safety, and the emerging evidence on the health-promoting properties of probiotics within fermented products.

8.2 FERMENTATIVE MICROORGANISMS AND FERMENTED FOODS CONSUMED IN THE WORLD

8.2.1 Microorganisms Associated to Food Fermentation

Lactic acid bacteria (LAB), yeasts, and filamentous fungi (molds) are the three main groups of microorganisms which are usually associated to food fermentation. When simultaneously involved in the process, germs are not independent actors but interact dynamically and differently from one food substrate to another, producing synergistic outcomes that cannot be achieved by any single taxon alone. The microbial interactions during fermentation are increasingly understood through metagenomics and metabolomics. The perspectives are precision fermentation strategies to maximize nutritional and sensory outcomes. The fermentation process in controlled conditions with specific microbes' species are nowadays common in industries for a wide range of fermented food products sold to population.

LAB are the main microorganisms responsible for fermenting a wide range of foods. Yogurt, cheese, pickles, and sauerkraut are examples of commonly enjoyed LAB-fermented foods. They are described as Gram-positive cells, non-motile, and non-spore-forming. Different genera dominate fermentation niches. They include *Lactobacillus*, *Bifidobacterium*, *Streptococcus*, and *Lacticaseibacillus*. Species belonging to *Leuconostoc*, *Lactococcus*, and *Enterococcus* are also quite often reported in the literature. LAB have the potential to convert food carbohydrates into lactic acid and other organic acids. This leads to an acidic environment through pH reduction and creates a selective environment that inhibit pathogenic and spoilage microbes (Aguirre-Garcia et al., 2024; Praveen & Brogi, 2025). Glucose catabolism proceeds here through either a homofermentative (8.1) or a heterofermentative (8.2 & 8.3) pathway, depending on the strain and surrounding environmental conditions. Homofermentative LAB convert carbohydrates exclusively into lactic acid. In contrast, heterofermentative strains might produce lactic acid together with ethanol, acetic acid, and carbon dioxide (Castellone et al., 2021).

$$C_6H_{12}O_6 \rightarrow 2C_3H_6O_3 + 2ATP \quad \text{Homofermentative (lactic acid)} \tag{8.1}$$

$$C_6H_{12}O_6 \rightarrow C_3H_6O_3 + C_2H_5OH + CO_2 + 1ATP \quad \text{Heterofermentative (ethanol pathway)} \tag{8.2}$$

$$C_6H_{12}O_6 \rightarrow C_3H_6O_3 + CH_3COOH + CO_2 + 2ATP \quad \text{Heterofermentative (acetic acid pathway)} \tag{8.3}$$

LAB ability to hydrolyze polysaccharides improves food quality, supports probiotic growth, and has applications in food, health, and fermentation industries. Contrary to earlier assumptions, LAB also possess starch metabolism pathways, with species such as *Lactiplantibacillus plantarum*, *Levilactobacillus brevis*, and *Weissella* contributing to starch hydrolysis in sourdough via amylases. LAB also act through protein degradation. This particularly the case in dairy fermentations, where proteolysis involves protein cleavage by cell envelope proteinases, peptide transport, and intracellular hydrolysis by peptidases. The resulting amino acids undergo diverse metabolic conversions (Wang et al., 2021).

Acetic fermentation is rather common in the production of vinegar, kombucha, water kefir, and cocoa. During this microbial process, bacteria mostly from the genus *Acetobacter*, transform organic substrates like sugars and alcohols into acetic acid (Sawant et al., 2025).

Yeasts are unicellular fungi which reproduce asexually by budding or division. *Saccharomyces cerevisiae* is the most common and commercialized yeast used for foods and beverages fermentation. However, other species have also been reported as having a food fermentative potential. *Saccharomyces boulardii*, *Zygosaccharomyces bailii*, *Candida* spp., and *Debaryomyces* spp. are examples. Yeasts principally ferment food through the alcoholic pathway (8.4) as it is the case in beer, wine, and bread production processes. Indeed, they produce ethanol and carbon dioxide, which contribute to leavening in bread and alcohol generation in beverages, while also creating anaerobic and anti-microbial conditions (Dashko et al., 2014; Sharma et al., 2020). A non-alcoholic pathway is also possible (8.5) as reported during cacao and coffee fermentation (Maicas, 2020). Beyond these primary roles, yeasts also have the ability to synthesize a diversity of volatile compounds (esters, aldehydes, and organic acids) that give aromas and flavors characteristic of fermented products (Praveen & Brogi, 2025). This is the case during coffee and cocoa fermentation where yeasts significantly influence their quality by metabolizing sugars and amino acids, producing flavor precursors and volatile compounds that define aroma and taste. Competition, quorum sensing, and the release of inhibitory substances are the different possible interactions between yeasts and other microorganisms in the fermentative medium. This affects population dynamics and metabolite production. Co-inoculation of *S. cerevisiae* with *Pichia kudriavzevii* improves chocolate quality by increasing acidity, fragrance, antioxidant capacity, and reducing undesirable compounds (de Melo et al., 2023).

$$C_6H_{12}O_6 \rightarrow 2C_2H_5OH + 2CO_2 \quad \text{Alcoholic pathway (ethanol)} \tag{8.4}$$

$$C_6H_{12}O_6 \rightarrow 2C_3H_8O_3 + CO_2 \quad \text{Non-alcoholic pathway (glycerol)} \tag{8.5}$$

Filamentous fungi (molds) are important contributors to some types of food fermentation. Asian fermentations like soy sauce, miso, sake, and rice vinegar are made using *Aspergillus oryzae* and *Aspergillus awamori*. Examples also extend to *Rhizopus oligosporus* and *Monascus purpureus* used to make Tempeh and red yeast rice, respectively. They act by secreting extracellular enzymes that could be amylases, proteases, or lipases. A release of amino acids, peptides, sugars, and fatty acids could therefore be observed resulting from the breakdown of macromolecules into bioavailable forms. Cheese ripening rather involves *Penicillium camemberti* and *Penicillium Roqueforti*, which confer specific textures, colors, and flavor notes. *Neurospora crassa* and *Geotrichum candidum* were reported to contribute to the ecology of surface-ripened cheeses. Importantly, these fungi often synergize with bacteria and yeasts, as in soybean fermentations, where mold hydrolysis products provide substrates for LAB and yeast growth (Yang et al., 2022).

8.2.2 Fermented Foods Consumed Worldwide

Table 8.1 provides an overview of diverse fermented foods and beverages that have been reported worldwide. Beer and wine are globally consumed beverages already produced on an industrial scale. They are obtained through fermentation of cereal grains or fruit sugars, respectively, using

TABLE 8.1
Fermented Food Around the World and the Associated Fermentative Microorganisms

Category	Fermented Food or Beverage (Region or Country Where They Are Found)	Raw Food or Substrate Used	Main Fermentative Microorganism(s) Identified or Used as Starter
1. Beverages	Beer (Global)	Cereal grains (barley, wheat, maize)	*Saccharomyces cerevisiae*, *Saccharomyces Pastorianus*
	Wine (Global)	Grapes or fruits	*S. cerevisiae*, *Saccharomyces bayanus*, *Oenococcus oeni*
	Pulque (Mexico)	Agave sap	*Zymomonas mobilis*, *S. cerevisiae*, LAB
	Kaffir beer (Southern Africa)	Sorghum, millet, maize	*S. cerevisiae*, LAB
	Palm wine (West and Central Africa; parts of Asia)	Palm sap	*S. cerevisiae*, LAB
2. Dairy	Cheese (Global)	Milk	*Lactococcus lactis*, *Streptococcus thermophilus*, *Lactobacillus helveticus*, *Propionibacterium freudenreichii*, *Penicillium* spp.
	Yogurt (Global)	Milk	*Lactobacillus delbrueckii* subsp. *bulgaricus*, *S. thermophilus*
	Dahi (South Asia: India, Pakistan, Bangladesh)	Milk	*L. lactis*, *Leuconostoc mesenteroides*, *Lactiplantibacillus plantarum*
	Kumiss (Central Asia: Mongolia, Kazakhstan)	Mare's milk	*L. delbrueckii*, *Lactobacillus acidophilus*, *S. cerevisiae*
	Acidophilus milk (Global; USA, Europe, Japan)	Milk	*L. acidophilus*
3. Legume-based	Tempeh (Southeast Asia: Indonesia)	Soybeans	*Rhizopus oligosporus*
	Natto (Japan)	Soybeans	*Bacillus subtilis*
	Kinema (Eastern Himalayas: Nepal, Bhutan, India)	Soybeans	*B. subtilis*
	Soy sauce (East Asia: Japan, China)	Soybeans, wheat	*Aspergillus oryzae*, LAB, yeasts
	Miso (Japan)	Soybeans, rice/barley	*A. oryzae*, LAB, yeasts
4. Cereal-based	*Ogi* (West Africa: Nigeria, Benin, Ghana)	Maize, sorghum, millet	*L. plantarum*, *Limosilactobacillus fermentum*, *S. cerevisiae*
	Idli (South India, Sri Lanka)	Rice and black gram	L. mesenteroides, L. plantarum
5. Vegetable-based	Kimchi (Korea)	Cabbage, radish	*L. plantarum*, *Leuconostoc* spp.
	Sauerkraut (Central and Eastern Europe; global)	Cabbage	*L. mesenteroides*, *L. plantarum*, *Pediococcus cerevisiae*
6. Root/Tuber-based	Garri (West Africa: Nigeria, Ghana, Cameroon)	Cassava	*L. plantarum*, *S. cerevisiae*

yeasts like *S. cerevisiae*. Traditional alcoholic drinks like Pulque, Kaffir beer, and palm wine are rather a result of a synergy of mixed fermentation involving a co-occurrence of yeasts and LAB. Dairy products made from milk fermentation are also diverse. Cheese and yogurt are widely known in contrast to region-specific products such as Dahi, Kumiss, and Acidophilus milk. These foods rely predominantly on LAB, including *Lactococcus*, *Lactobacillus*, *Streptococcus*, and also on mold species such as *Penicillium* especially in the case of cheese.

Another substrate that is also used in Asia to produce culturally fermented product is soybean. Tempeh, Natto, Kinema, Soy sauce, and miso are made from this legume. The filamentous fungi *A. oryzae*, LAB, and yeasts were reported to be involved in Soy sauce and miso (Sharma et al., 2020; Praveen & Brogi, 2025). Tempeh is obtained from white soybean seeds fermentation with *Rhizopus* mold (Dwiatmaka et al., 2021). *Bacillus subtilis* which is a non-LAB with probiotic properties is used as starter for soybean fermentation in order to produce Kinema and Natto (Sarkar et al., 1994; Wang et al., 2023). Cereal-based fermentations leading to Ogi rely on combinations of LAB and yeasts to transform staple grains like maize, sorghum, and millet into products with improved digestibility, organoleptic qualities, and shelf life. Similarly, Gari that is obtained from Cassava fermentation is a result of a combined action of *Lactobacillus plantarum* and *S. cerevisiae*. On the other side, Kimchi and sauerkraut which are derived from cabbage, and Idli from rice fermentation are the results of combined action of LAB, some with probiotic potential, namely *Leuconostoc mesenteroides* and *L. plantarum* (Sharma et al., 2020; Praveen & Brogi, 2025).

8.3 MECHANISMS OF NUTRITIONAL IMPROVEMENT THROUGH FERMENTATION

The nutritional enhancement conferred by fermentation is multi-dimensional. This occurs through the synergistic action of microbial metabolism and enzymatic modifications of food substrates. The positive impact is nowadays established. Globally, these transformations not only improve the availability of essential nutrients but also generate bioactive compounds with functional health benefits.

8.3.1 Enhancement of Minerals Bioavailability and Vitamins Content

One of the important nutritional improvement of foods observed after fermentation is the enzymatic degradation of anti-nutritional compounds (phytates, tannins, and oligosaccharides) which otherwise impair nutrient absorption. Phytates represent the main storage form of phosphorus, abundant in cereals, legumes, oil seeds, and nuts. They chelate minerals like iron, zinc, magnesium, and calcium, limiting their bioavailability (Hidevegi & Lastity, 2003; Gupta et al., 2015). During the process, bound minerals are released through phytases secretion by LAB and molds which hydrolyze phytic acid into inositol and phosphate (Sharma et al., 2020). As an illustration, significant reduction of phytate levels associated to an increase of iron and zinc absorption has been reported after lactic acid fermentation of different cereals (Atuna et al., 2021; Rodríguez-España et al., 2022). A degradation of tannins and flatulence-causing oligosaccharides in soybeans was also described during the fermentation step of Tempeh and miso production, improving digestibility and reducing gastrointestinal discomfort (Bensaada et al., 2024; Sawant et al., 2025). Similar effect was also reported with legumes. Fermentation of chickpea flour reduced phytic acid and tannin levels, improving mineral utilization with mung bean, an increase of calcium (21.5 → 49.30 mg/100 g) and iron (43.45 → 56.50 mg/100 g) was noticed. When the process was carried out with the mold *A. awamori*, the in vitro bioavailability of Fe and Zn was significantly enhanced, associated with reductions in phytates (from 0.40 to 0.11 mg/100 g) and saponins (from 0.18 to 0.03 mg/100 g). Fe and Zn assimilation were almost doubled after 96 h of fermentation with *A. oryzae* MTCC 3107. With black pea flour, fermentation for this duration increased Fe (32.892 to 38.421 μg/kg) and Zn (18.328 to 24.901 μg/kg) contents (Knez et al., 2023).

Many fermentative microorganisms are able to operate de novo synthesis of vitamins. Microbial production of B-complex vitamins is of high interest. Indeed, thiamine (B1), riboflavin (B2), folate (B9), and cobalamin (B12) act as coenzymes in several enzymatic processes in the human body (Hanna et al., 2022). Vitamin B2 and vitamin B9 were reported to be enhanced in yogurt and cheese by *Lactobacillus* and *Streptococcus* species during the fermentation. Similarly, *Propionibacterium freudenreichii* produces significant quantities of vitamin B12 in Swiss cheese (Praveen & Brogi, 2025). The content enrichment of traditional beverages like palm wine, Pulque, and Kaffir beer with thiamine, riboflavin, and niacin were attributed to fermentation (Sharma et al., 2020). The enrichment of vitamin B12 in particular (normally absent in plant-based foods) has profound implications for populations consuming predominantly vegetarian diets (Fernandes et al., 2024; Clemente-Suárez et al., 2025). Vitamin K (menaquinones) content, a nutrient important for bone health, cardiovascular protection, and blood coagulation, has also been described to have an enhanced content after fermentation. Unlike vitamin K_1 from leafy vegetables, K_2 is mainly synthesized by fermentative microbes. Natto made from soybeans fermented with *B. subtilis* is the richest source and provides high levels of menaquinone-7 (MK-7), a highly bioavailable form. *P. freudenreichii* also produces MK in cheeses. This is also the case for *Lactococcus lactis* and *Lactobacillus plantarum* in dairy and vegetable during fermentation (Sharma et al., 2020; Praveen & Brogi, 2025).

8.3.2 Generation of Bioactive Peptides and Functional Metabolites

Bioactive peptides (BAPs) are short amino acid sequences (2–20 residues) released from the hydrolysis of precursor proteins. During fermentation, smaller peptides and amino acids are released as the result of proteases secretion by microorganisms that break down proteins of the substrate. As outcome, digestibility of fermented foods is enhanced and amino acid balance improved. This is of high interest particularly in plant-based products where proteins may be tightly bound within fibrous matrices (Dallas et al., 2017; Dhiman et al., 2025). Antioxidant, immunomodulatory, antihypertensive, anti-microbial, and anti-thrombotic activities are associated to BAPs (Korhonen & Pihlanto, 2006; Sharma et al., 2020; Daroit & Brandelli, 2021). Those produced during the fermentation of milk exhibited angiotensin-converting enzyme (ACE)-inhibitory effects, therefore contributing to blood pressure regulation. BAPs with cholesterol-lowering and anti-inflammatory effects have been identified in fermented soybean products such as Natto and Tempeh. Others found in fermented meats (e.g., Spanish dry sausages, camel meat sausages, and venison) showed ACE-inhibitory, antioxidant, and anti-microbial activities. Proteolytic efficiency and bioactivity depend strongly on the bacterial strains and ripening conditions (Peres Fabbri et al., 2024).

Fermentation also generates a wide array of functional metabolites that extend health benefits beyond classical nutrition. Among these, γ-aminobutyric acid (GABA) that is produced by certain LAB such as *Lactobacillus brevis* (Cui et al., 2020). This compound is a neurotransmitter with neuroprotective effect. It is helpful in the reduction of stress, anxiety, and improved cardiovascular health. Fermented soybean foods showed enhanced isoflavone and saponin content during processing, which are linked to reduced risks of chronic diseases (Sharma et al., 2020; Praveen & Brogi, 2025). Indeed, fermentation can enhance antioxidant capacity of foods by releasing, promoting, and producing antioxidant compounds (Sarıtaş et al., 2024).

8.3.3 Fiber Modification and Prebiotic Potential

Fermentation modifies food fibers content by breaking down complex polysaccharides into prebiotic oligosaccharides. The latter will selectively stimulate beneficial gut microbes especially *Bifidobacterium* and *Lactobacillus* cells once in the human gut. During fermentation, insoluble fibers are transformed into soluble forms following enzymes (cellulases, xylanases, and β-glucosidases) secretion by microorganisms. This leads to galacto-oligosaccharides (GOS), fructo-oligosaccharides

(FOS), and xylo-oligosaccharides (XOS) formation in the substrate (Chelliah et al., 2024). During the sourdough bread process, LAB hydrolyze arabinoxylans into XOS (Michlmayr et al., 2013). This is also the case with Kimchi and sauerkraut where LAB-driven fermentation enriches soluble fibers fermentation and releases metabolites that promote *Lactobacillus* and *Bifidobacterium* bacteria (Xiang et al., 2019). These oligosaccharides resist digestion in the upper human gut and are later fermented in the colon to produce short-chain fatty acids (SCFAs). Indeed, acetate, propionate, and butyrate support gut health, improve mineral absorption, and suppress pathogens (Davani-Davari et al., 2019). The fermentation process therefore not only improves food digestibility but also strengthens the prebiotic potential of the final products. Diverse health benefits across multiple systems are attributed to prebiotics. In the gastrointestinal tract, they support beneficial microbes and improve barrier function. Their potential to alleviate conditions such as irritable bowel syndrome (IBS), Crohn's disease, and colorectal cancer risk have also been noted. Enhancement of immune responses by modulating cytokine production, antibody activity, and vaccine efficacy, while reducing infections and allergy risks, particularly in infants are other related properties. Through the gut–brain axis, prebiotics influence cognition, stress regulation, and mood, with emerging applications in autism and hepatic encephalopathy. Their effects extend to the skin where they improve hydration, barrier integrity, and protection against atopic dermatitis and UV damage (Davani-Davari et al., 2019).

8.4 MECHANISMS OF INTRINSIC FOOD SAFETY IMPROVEMENT

Food fermentation not only improves nutritional quality but also contributes to food safety through multiple mechanisms.

8.4.1 Inhibition of Pathogens

Inhibition of pathogenic microorganisms through microbial metabolites and physicochemical changes in the food matrix is a major mechanism of safety enhancement in fermented foods. LAB lower the pH of the environment by converting sugars into lactic acid and acetic acid. This acidification of the medium limits eventual growth of pathogenic of *Escherichia coli*, *Salmonella enterica*, *Listeria monocytogenes*, and *Clostridium botulinum* (Ibrahim et al., 2021; Fitsum et al., 2025). LAB also have the capacity to produce bacteriocins that disrupt the membranes of competing microbes. Nisin and pediocin, produced by *L. lactis* and *Pediococcus acidilactici*, respectively, are examples of these anti-microbial compounds (Simons et al., 2020; Darbandi et al., 2022). In fermented beverages, ethanol serves as preservative by impairing pathogen cell membranes and protein functions (Postigo et al., 2025). Sourdough bread spoilage is naturally controlled with the presence of antifungal organic acids, hydrogen peroxide, and cyclic dipeptides produced by certain LAB and yeasts which inhibit molds like *Aspergillus* and *Fusarium* (Nasrollahzadeh et al., 2022; Oleinikova et al., 2025). Fermentation process carried out in optimal conditions allow prolific growth of the desired microorganisms, which will then out-compete undesirable microorganisms. Properly prepared fermented foods are generally safe, but one cannot ignored eventual contamination from raw material or the environment especially in traditional uncontrolled conditions. Documented outbreaks include *Bacillus cereus* in douchi in China, *C. botulinum* in homemade fermented tofu in New York, *Salmonella* in fermented cashew nut cheese in Canada, and *E. coli* in Kimchi consumed in South Korean schools (Allwood et al., 2023).

8.4.2 Detoxification of Harmful Compounds

Fermentation also detoxifies naturally occurring or processing-derived toxicants. Dangerous mold growth can be controlled by other molds, yeasts, and bacteria through competitive exclusion

(Bianchini & Bullerman, 2009). While certain LAB are capable of binding or degrading mycotoxins such as Aflatoxins, Ochratoxin A, and Fumonisins. For instance, LAB in fermented maize products (e.g., Ogi and Kenkey) reduce mycotoxin bioavailability through adsorption onto cell walls or enzymatic transformation (Møller et al., 2021). Cassava roots (*Manihot esculenta*) processing into products like "Gari" or "Lafun" drastically reduces toxic cyanogenic glycosides (linamarin and lotaustralin) by enzymatic hydrolysis, preventing cyanide poisoning (Heuberger, 2005). Some fermentations carried out with selected starter cultures limit the accumulation of harmful biogenic amines (histamine, tyramine, and putrescine) that can cause hypertension and allergic reactions. Controlled cheese and fermented fish production process reduces amine levels by outcompeting amine-producing microbes (Kongkiattikajorn, 2015; Banicod et al., 2025).

8.5 PROBIOTICS IN FERMENTED FOODS

8.5.1 Probiotics: Definition and Health Importance

Some fermented foods and beverages ready for consumption are sources of probiotics. As mentioned earlier, they are beneficial microorganisms that, when ingested in appropriate amounts, contribute to host health (FAO/WHO, 2002). They have the ability to survive gastrointestinal passage by resisting acidic gastric conditions and bile salts. Beyond improving food preservation and sensory attributes, probiotics add functional value by modulating gut microbiota, enhancing immune function, and reducing risks of chronic diseases in humans (Hassan & Kovaleva, 2025). The intake of these positive microorganisms increases beneficial microbes inside gut while suppressing pathogens through competitive exclusion and the production of anti-microbial metabolites (Ma et al., 2023; Vinayamohan et al., 2024). A stimulation of immunoglobulin A (IgA) production, activation of dendritic cells, and enhancement of vaccine responses was associated to certain strains like *Lactobacillus rhamnosus* GG (Wang et al., 2017; Abavisani et al., 2024). They downregulate pro-inflammatory cytokines, while increasing anti-inflammatory mediators IL-10 known as beneficial in conditions like IBS and ulcerative colitis (Cristofori et al., 2021; Mazziotta et al., 2023). An improvement of lipid profiles, glucose regulation, and obesity management has been linked to a regular intake of probiotic-rich fermented foods. This is possible partly through modulation of SCFA production and bile salt metabolism (Wu et al., 2022; Nithya et al., 2023; Pyo et al., 2024).

8.5.2 Sources of Probiotics in Fermentation

Probiotics are naturally enriched in various traditional and industrial fermented foods. Some are commonly used as starter or adjunct cultures in dairy fermentations. *Lactobacillus* strains, *Streptococcus thermophilus*, and *Bifidobacterium* species are inoculated for yogurt, kefir, and cheese production (Jang et al., 2024). Sourdough, Ogi, and boza which are cereal-based products harbor LAB such as *Lactobacillus plantarum* and *Limosilactobacillus fermentum*, contributing to both preservation and probiotic benefits (Ogunremi et al., 2015; Setta et al., 2020). Vegetable fermentations like Kimchi, sauerkraut, and pickles are naturally rich in LAB of interest for gut health. *L. mesenteroides* and other LAB have been identified (Swain et al., 2014; Yuan et al., 2023). Fermented sausages and fish pastes also contain probiotics like *Pediococcus pentosaceus* and *Lactobacillus sakei* that also enhance their safety (Ray & Montet,2014; Munekata et al., 2020).

8.5.3 Stability and Viability in Fermented Products

The effectiveness of probiotics depends on their ability to survive food processing, storage, and gastrointestinal transit. Temperature as well as oxygen exposure and pH changes during food production can affect their viability (Tripathi & Giri, 2014). The nature of the food matrix is also an important factor. Dairy matrices such as milk and yogurt provide buffering capacity, whereas non-dairy matrices like vegetables and cereals may more require protective encapsulation (Sbehat et al., 2022; Gutiérrez-Álzate et al., 2024). Storage conditions are critical. Refrigeration improves survival, but prolonged storage can lead to a decline in viable counts (Tripathi & Giri, 2014; Al-Saadi, 2017). Gastrointestinal resistance is another essential aspect. Strains must withstand gastric acidity and bile salts to reach the intestine in effective doses, which means above 6 Log CFU/g (Naissinger da Silva et al., 2021). To address these challenges, microencapsulation, cryoprotection, and tailored starter culture selection are increasingly applied to ensure probiotic stability in both dairy and plant-based fermented foods.

8.6 CONCLUSION

Microorganisms are fundamental for food fermentation. This process is useful only by preserving food but also through the enhancement of nutritional quality by removing anti-nutrients, synthesizing vitamins, producing BAPs, and improving digestibility. The presence of microorganisms with probiotic properties in some fermented ready-to-eat products gives an added-value by supporting gut microbiota balance and modulating host immunity. Researches are nowadays turned toward the development of fermented foods that combine prebiotics with probiotics to create symbiotic effects that improve gut health. This synergistic system could play a significant role in combating food insecurity and malnutrition.

REFERENCES

Abavisani, M., Ebadpour, N., Khoshrou, A., Sahebkar, A. (2024). Boosting vaccine effectiveness: The groundbreaking role of probiotics. *Journal of Agriculture and Food Research, 16*, Article 101189. https://doi.org/10.1016/j.jafr.2024.101189

Aguirre-Garcia, Y. L., Nery-Flores, S. D., Campos-Muzquiz, L. G., Flores-Gallegos, A. C., Palomo-Ligas, L., Ascacio-Valdés, J. A., Sepúlveda-Torres, L., & Rodríguez-Herrera, R. (2024). Lactic acid fermentation in the food industry and bio-preservation of food. *Fermentation, 10*(3), 168. https://doi.org/10.3390/fermentation10030168

Allwood, J. G. Wakeling, L. T., Post, L. S., & Bean, D. C. (2023). Food safety considerations in the production of traditional fermented products: Japanese rice koji and miso. *Journal of Food Safety, 43*(4), e13048. https://doi.org/10.1111/jfs.13048

Al-Saadi, J. M. S. (2017). Encapsulation of probiotics (review). *Journal of Food Research, 6*(2), 75–81. https://doi.org/10.5539/jfr.v6n2p75

Atuna, R. A., Ametei, P. N., Bawa, A.-A., & Amagloh, F. K. (2021). Traditional processing methods reduced phytate in cereal flour, improved nutritional, functional and rheological properties. *Scientific African, 12*, e01063. https://doi.org/10.1016/j.sciaf.2021.e01063

Banicod, R. J. S., Ntege, W., Njiru, M. N., Abubakar, W. H., Kanthenga, H. T., Javaid, A., & Khan, F. (2025). Production and transformation of biogenic amines in different food products by the metabolic activity of the lactic acid bacteria. *International Journal of Food Microbiology, 428*, 110996. https://doi.org/10.1016/j.ijfoodmicro.2024.110996

Bensaada, S., Peruzzi, G., Cubizolles, L., Denayrolles, M., & Bennetau-Pelissero, C. (2024). Traditional and domestic cooking dramatically reduce estrogenic isoflavones in soy foods. *Foods (Basel, Switzerland), 13*(7), 999. https://doi.org/10.3390/foods13070999

Bianchini, A., & Bullerman, L. B. (2009). Biological control of molds and mycotoxins in foods. In M. Appell, D. F. Kendra, & M. W. Trucksess (Eds.), *Mycotoxin Prevention and Control in Agriculture* (ACS Symposium Series No. 1031). American Chemical Society. https://doi.org/10.1021/bk-2009-1031.ch001

Bourdichon, F., Arias, E., Babuchowski, A., Bückle, A., Bello, F. D., Dubois, A., Fontana, A., Fritz, D., Kemperman, R., Laulund, S., McAuliffe, O., Miks, M. H., Papademas, P., Patrone, V., Sharma, D. K., Sliwinski, E., Stanton, C., Von Ah, U., Yao, S., & Morelli, L. (2021). The forgotten role of food cultures. *FEMS Microbiology Letters, 368*(14), fnab085. https://doi.org/10.1093/femsle/fnab085

Castellone, V., Bancalari, E., Rubert, J., Gatti, M., Neviani, E., & Bottari, B. (2021). Eating fermented: Health benefits of LAB-fermented foods. *Foods (Basel, Switzerland), 10*(11), 2639. https://doi.org/10.3390/foods10112639

Chelliah, R., Kim, N. H., Park, S., Park, Y., Yeon, S.-J., Barathikannan, K., Vijayalakshmi, S., & Oh, D.-H. (2024). Revolutionizing renewable resources: Cutting-edge trends and future prospects in the valorization of oligosaccharides. *Fermentation*, *10*(4), 195. https://doi.org/10.3390/fermentation10040195

Clemente-Suárez, V. J., Redondo-Flórez, L., Martín-Rodríguez, A., Curiel-Regueros, A., Rubio-Zarapuz, A., & Tornero-Aguilera, J. F. (2025). Impact of vegan and vegetarian diets on neurological health: A critical review. *Nutrients*, *17*(5), 884. https://doi.org/10.3390/nu17050884

Cristofori, F., Dargenio, V. N., Dargenio, C., Miniello, V. L., Barone, M., & Francavilla, R. (2021). Anti-inflammatory and immunomodulatory effects of probiotics in gut inflammation: A door to the body. *Frontiers in Immunology*, *12*, 578386. https://doi.org/10.3389/fimmu.2021.578386

Cui, Y., Miao, K., Niyaphorn, S., & Qu, X. (2020). Production of gamma-aminobutyric acid from lactic acid bacteria: A systematic review. *International Journal of Molecular Sciences*, *21*(3), 995. https://doi.org/10.3390/ijms21030995

Dallas, D. C., Sanctuary, M. R., Qu, Y., Khajavi, S. H., Van Zandt, A. E., Dyandra, M., Frese, S. A., Barile, D., & German, J. B. (2017). Personalizing protein nourishment. *Critical Reviews in Food Science and Nutrition*, *57*(15), 3313–3331. https://doi.org/10.1080/10408398.2015.1117412

Darbandi, A., Asadi, A., Mahdizade Ari, M., Ohadi, E., Talebi, M., Halaj Zadeh, M., Darb Emamie, A., Ghanavati, R., & Kakanj, M. (2022). Bacteriocins: Properties and potential use as antimicrobials. *Journal of Clinical Laboratory Analysis*, *36*(1), e24093. https://doi.org/10.1002/jcla.24093

Daroit, D. J., & Brandelli, A. (2021). In vivo bioactivities of food protein-derived peptides – A current review. *Current Opinion in Food Science*, *39*, 1–7. https://doi.org/10.1016/j.cofs.2021.01.002

Dashko, S., Zhou, N., Compagno, C., & Piškur, J. (2014). Why, when, and how did yeast evolve alcoholic fermentation? *FEMS Yeast Research*, *14*(6), 826–832. https://doi.org/10.1111/1567-1364.12161

Davani-Davari, D., Negahdaripour, M., Karimzadeh, I., Seifan, M., Mohkam, M., Masoumi, S. J., Berenjian, A., & Ghasemi, Y. (2019). Prebiotics: Definition, types, sources, mechanisms, and clinical applications. *Foods*, *8*(3), 92. https://doi.org/10.3390/foods8030092

de Melo Pereira, G. V., de Lima, M. A., & de Oliveira, J. M. (2023). The essential role of spontaneous and starter yeasts in cocoa and coffee fermentation. *FEMS Yeast Research*, *23*(1), foad019. https://doi.org/10.1093/femsyr/foad019

Dhiman, S., Kaur, S., Thakur, B., Singh, P., & Tripathi, M. (2025). Nutritional enhancement of plant-based fermented foods: Microbial innovations for a sustainable future. *Fermentation*, *11*(6), 346. https://doi.org/10.3390/fermentation11060346

Dwiatmaka, Y., Lukitaningsih, E., Yuniarti, N., & Wahyuono, S. (2021). Fermentation of soybean seeds using *Rhizopus oligosporus* for Tempeh production and standardization based on isoflavones content [Preprint]. *Preprints.org*. https://doi.org/10.20944/preprints202104.0610.v1

FAO, IFAD, UNICEF, WFP and WHO. (2025). In Brief to the State of Food Security and Nutrition in the World 2025 – Addressing high food price inflation for food security and nutrition. Rome. https://doi.org/10.4060/cd6015en

Fernandes, S., Oliveira, L., Pereira, A., Costa, M. d. C., Raposo, A., Saraiva, A., & Magalhães, B. (2024). Exploring vitamin B12 supplementation in the vegan population: A scoping review of the evidence. *Nutrients*, *16*(10), 1442. https://doi.org/10.3390/nu16101442

Fitsum, S., Gebreyohannes, G., & Sbhatu, D. B. (2025). Bioactive compounds in fermented foods: Health benefits, safety, and future perspectives. *Applied Food Research*, *5*, Article 101097. https://doi.org/10.1016/j.afres.2025.101097

FAO-Food and Agricultural Organization of the United Nations and WHO-World Health Organization. (2002). Joint FAO/WHO working group report on drafting guidelines for the evaluation of probiotics in food. *Food and Agricultural Organization of the United Nations*. ftp://ftp.fao.org/es/esn/food/wgreport2.pdf

Gupta, R. K., Gangoliya, S. S., & Singh, N. K. (2015). Reduction of phytic acid and enhancement of bioavailable micronutrients in food grains. *Journal of Food Science and Technology*, *52*(2), 676–684. https://doi.org/10.1007/s13197-013-0978-y

Gutiérrez-Álzate, K., Beltrán-Cotta, L. A., Rekowsky, B. S. dos S., Cavalheiro, C. P., & Pereira da Costa, M. (2024). Micro- and nanoencapsulation of probiotics: Exploring their impact on animal-origin foods. *ACS Food Science & Technology*, *4*(12), 2799–2812. https://doi.org/10.1021/acsfoodscitech.4c00776

Hadj Saadoun, J., Bertani, G., Levante, A., Vezzosi, F., Ricci, A., Bernini, V., & Lazzi, C. (2021). Fermentation of agri-food waste: A promising route for the production of aroma compounds. *Foods (Basel, Switzerland)*, *10*(4), 707. https://doi.org/10.3390/foods10040707

Hanna, M., Jaqua, E., Nguyen, V., & Clay, J. (2022). B vitamins: Functions and uses in medicine. *The Permanente Journal*, *26*(2), 89–97. https://doi.org/10.7812/TPP/21.204

Hassan, M. Z., & Kovaleva, E. G. (2025). A comprehensive review of probiotics and human health: Current perspectives and applications. *Frontiers in Microbiology, 15*, Article 1487641. https://doi.org/10.3389/fmicb.2024.1487641

Heuberger, C. (2005). *Cyanide Content of Cassava and Fermented Products with Focus on Attiéké and Attiéké Garba* (Doctoral dissertation). Swiss Federal Institute of Technology Zurich. https://doi.org/10.3929/ethz-a-005182128

Hidevegi, M., & Lastity, R. (2003). Phytic acid content of cereals and legumes and interaction with proteins. *Periodica Polytechnica Chemical Engineering, 46*(1), 59–64.

Ibrahim, S. A., Ayivi, R. D., Zimmerman, T., Siddiqui, S. A., Altemimi, A. B., Fidan, H., Esatbeyoglu, T., & Bakhshayesh, R. V (2021). Lactic acid bacteria as antimicrobial agents: Food safety and microbial food spoilage prevention. *Foods (Basel, Switzerland), 10*(12), 3131. https://doi.org/10.3390/foods10123131

Jang, H. J., Lee, N. K., & Paik, H. D. (2024). Overview of dairy-based products with probiotics: Fermented or non-fermented milk drink. *Food Science of Animal Resources, 44*(2), 255–268. https://doi.org/10.5851/kosfa.2023.e83

Knez, E., Kadac-Czapska, K., & Grembecka, M. (2023). Effect of fermentation on the nutritional quality of the selected vegetables and legumes and their health effects. *Life (Basel, Switzerland), 13*(3), 655. https://doi.org/10.3390/life13030655

Kongkiattikajorn, J. (2015). Potential of starter culture to reduce biogenic amines accumulation in som-fug, a Thai traditional fermented fish sausage. *Journal of Ethnic Foods, 2*(4), 186–194. https://doi.org/10.1016/j.jef.2015.11.005

Korhonen, H., & Pihlanto, A. (2006). Bioactive peptides: Production and functionality. *International Dairy Journal, 16*(9), 945–960. https://doi.org/10.1016/j.idairyj.2005.10.012

Ma, H., Wang, M., Qin, C., Shi, Y., Mandizadza, O. O., Ni, H., & Ji, C. (2025). Trends in the burden of chronic diseases attributable to diet-related risk factors from 1990 to 2021 and the global projections through 2030: A population-based study. *Frontiers in Nutrition, 12*, 1570321. https://doi.org/10.3389/fnut.2025.1570321

Ma, T., Shen, X., Shi, X., Sakandar, H. A., Quan, K., Li, Y., Jin, H., Kwok, L.-Y., Zhang, H., & Sun, Z. (2023). Targeting gut microbiota and metabolism as the major probiotic mechanism: An evidence-based review. *Trends in Food Science & Technology, 129*, Article 106013. https://doi.org/10.1016/j.tifs.2023.06.013

Maicas S. (2020). The role of yeasts in fermentation processes. *Microorganisms, 8*(8), 1142. https://doi.org/10.3390/microorganisms8081142

Mazziotta, C., Tognon, M., Martini, F., Torreggiani, E., & Rotondo, J. C. (2023). Probiotics mechanism of action on immune cells and beneficial effects on human health. *Cells, 12*(1), 184. https://doi.org/10.3390/cells12010184

Michlmayr, H., Hell, J., Lorenz, C., Böhmdorfer, S., Rosenau, T., & Kneifel, W. (2013). Arabinoxylan oligosaccharide hydrolysis by family 43 and 51 glycosidases from *Lactobacillus brevis* DSM 20054. *Applied and Environmental Microbiology, 79*(21), 6747–6754. https://doi.org/10.1128/AEM.02130-13

Møller, C. O. de A., Freire, L., Rosim, R. E., Margalho, L. P., Balthazar, C. F., Franco, L. T., Sant'Ana, A. S. de S., Rattray, F. P., & Oliveira, C. A. F. de. (2021). Effect of lactic acid bacteria strains on the growth and aflatoxin production potential of *Aspergillus parasiticus*, and their ability to bind aflatoxin B_1, ochratoxin A, and zearalenone in vitro. *Frontiers in Microbiology, 12*, Article 655386. https://doi.org/10.3389/fmicb.2021.655386

Munekata, P. E. S., Pateiro, M., Zhang, W., Domínguez, R., Xing, L., Fierro, E. M., & Lorenzo, J. M. (2020). Autochthonous probiotics in meat products: Selection, identification, and their use as starter culture. *Microorganisms, 8*(11), 1833. https://doi.org/10.3390/microorganisms8111833

Naissinger da Silva, M., Tagliapietra, B. L., Flores, V. D. A., & Pereira Dos Santos Richards, N. S. (2021). *In vitro* test to evaluate survival in the gastrointestinal tract of commercial probiotics. *Current Research in Food Science, 4*, 320–325. https://doi.org/10.1016/j.crfs.2021.04.006

Nasrollahzadeh, A., Mokhtari, S., Khomeiri, M., & Saris, P. E. J. (2022). Antifungal preservation of food by lactic acid bacteria. *Foods (Basel, Switzerland), 11*(3), 395. https://doi.org/10.3390/foods11030395

Nithya, A., Misra, S., Panigrahi, C., Dalbhagat, C., Mishra, G., & Mishra, H. N. (2023). Probiotic potential of fermented foods and their role in non-communicable diseases management: An understanding through recent clinical evidences. *Food Chemistry Advances, 3*, Article 100381. https://doi.org/10.1016/j.focha.2023.100381

Ogunremi, O. R., Sanni, A. I., & Agrawal, R. (2015). Probiotic potentials of yeasts isolated from some cereal-based Nigerian traditional fermented food products. *Journal of Applied Microbiology, 119*(3), 797–808. https://doi.org/10.1111/jam.12875

Oleinikova, Y., Amangeldi, A., Zhaksylyk, A., Saubenova, M., & Sadanov, A (2025). Sourdough microbiota for improving bread preservation and safety: Main directions and new strategies. *Foods*, *14*(14), 2443. https://doi.org/10.3390/foods14142443

Peres Fabbri, L., Cavallero, A., Vidotto, F., & Gabriele, M. (2024). Bioactive peptides from fermented foods: Production approaches, sources, and potential health benefits. *Foods*, *13*(21), 3369. https://doi.org/10.3390/foods13213369

Postigo, V., García, M., Crespo, J., Canonico, L., Comitini, F., & Ciani, M. (2025). Bioactive properties of fermented beverages: Wine and beer. *Fermentation*, *11*(5), 234. https://doi.org/10.3390/fermentation11050234

Praveen, M., & Brogi, S. (2025). Microbial fermentation in food and beverage industries: Innovations, challenges, and opportunities. *Foods*, *14*(1), 114. https://doi.org/10.3390/foods14010114

Pyo, Y., Kwon, K. H., & Jung, Y. J. (2024). Probiotic functions in fermented foods: Anti-viral, immunomodulatory, and anti-cancer benefits. *Foods (Basel, Switzerland)*, *13*(15), 2386. https://doi.org/10.3390/foods13152386

Ray, R. C., & Montet, D. (Eds.). (2014). *Microorganisms and Fermentation of Traditional Foods*. CRC Press. https://doi.org/10.1201/b17307

Rodríguez-España, M., Figueroa-Hernández, C. Y., Figueroa-Cárdenas, J. D., Rayas-Duarte, P., & Hernández-Estrada, Z. J. (2022). Effects of germination and lactic acid fermentation on nutritional and rheological properties of sorghum: A graphical review. *Current Research in Food Science*, *5*, 807–812. https://doi.org/10.1016/j.crfs.2022.04.014

Sarıtaş, S., Portocarrero, A. C. M., Miranda López, J. M., Lombardo, M., Koch, W., Raposo, A., El-Seedi, H. R., de Brito Alves, J. L., Esatbeyoglu, T., Karav, S., & Witkowska, A. M. (2024). The impact of fermentation on the antioxidant activity of food products. *Molecules (Basel, Switzerland)*, *29*(16), 3941. https://doi.org/10.3390/molecules29163941

Sarkar, P. K., Tamang, J. P., Cook, P. E., & Owens, J. D. (1994). Kinema: A traditional soybean fermented food—Proximate composition and microflora. *Food Microbiology*, 11(1), 47–55. https://doi.org/10.1006/fmic.1994.1007

Sawant, S. S., Park, H.-Y., Sim, E.-Y., Kim, H.-S., & Choi, H.-S. (2025). Microbial fermentation in food: Impact on functional properties and nutritional enhancement—A review of recent developments. *Fermentation*, *11*(1), 15. https://doi.org/10.3390/fermentation11010015

Sbehat, M., Mauriello, G., & Altamimi, M. (2022). Microencapsulation of probiotics for food functionalization: An update on literature reviews. *Microorganisms*, *10*(10), 1948. https://doi.org/10.3390/microorganisms10101948

Setta, M. C., Matemu, A., & Mbega, E. R. (2020). Potential of probiotics from fermented cereal-based beverages in improving health of poor people in Africa. *Journal of Food Science and Technology*, *57*(11), 3935–3946. https://doi.org/10.1007/s13197-020-04432-3

Sharma, R., Garg, P., Kumar, P., Bhatia, S. K., & Kulshrestha, S. (2020). Microbial fermentation and its role in quality improvement of fermented foods. *Fermentation*, *6*(4), 106. https://doi.org/10.3390/fermentation6040106

Siddiqui, S. A., Erol, Z., Rugji, J., Taşçı, F., Kahraman, H. A., Toppi, V., Musa, L., Di Giacinto, G., Bahmid, N. A., Mehdizadeh, M., & Castro-Muñoz, R. (2023). An overview of fermentation in the food industry – Looking back from a new perspective. *Bioresources and Bioprocessing*, *10*(1), 85. https://doi.org/10.1186/s40643-023-00702-y

Simons, A., Alhanout, K., & Duval, R. E. (2020). Bacteriocins, antimicrobial peptides from bacterial origin: Overview of their biology and their impact against multidrug-resistant bacteria. *Microorganisms*, *8*(5), 639. https://doi.org/10.3390/microorganisms8050639

Swain, M. R., Anandharaj, M., Ray, R. C., & Parveen Rani, R. (2014). Fermented fruits and vegetables of Asia: A potential source of probiotics. *Biotechnology Research International*, *2014*, 250424. https://doi.org/10.1155/2014/250424

Taveira, I. C., Nogueira, K. M. V., Oliveira, D. L. G. D., & Silva, R. D. N. (2021). Fermentation: Humanity's oldest biotechnological tool. *Frontiers for Young Minds*, *9*,568656. https://doi.org/10.3389/frym.2021.568656

The triple burden of malnutrition. (2023). *Nature Food*, *4*(11), 925. https://doi.org/10.1038/s43016-023-00886-8

Tripathi, M. K., & Giri, S. K. (2014). Probiotic functional foods: Survival of probiotics during processing and storage. *Journal of Functional Foods*, *9*, 225–241. https://doi.org/10.1016/j.jff.2014.04.030

Vinayamohan, P., Joseph, D., Viju, L. S., Baskaran, S. A., & Venkitanarayanan, K. (2024). Efficacy of probiotics in reducing pathogenic potential of infectious agents. *Fermentation*, *10*(12), 599. https://doi.org/10.3390/fermentation10120599

Wang, C., Chen, J., Tian, W., Han, Y., Xu, X., Ren, T., Tian, C., & Chen, C. (2023). Natto: A medicinal and edible food with health function. *Chinese Herbal Medicines*, *15*(3), 349–359. https://doi.org/10.1016/j.chmed.2023.02.005

Wang, Y., Liu, L., Moore, D. J., Shen, X., Peek, R. M., Acra, S. A., Li, H., Ren, X., Polk, D. B., & Yan, F. (2017). An LGG-derived protein promotes IgA production through upregulation of APRIL expression in intestinal epithelial cells. *Mucosal Immunology*, *10*(2), 373–384. https://doi.org/10.1038/mi.2016.57

Wang, Y., Wu, J., Lv, M., Shao, Z., Hungwe, M., Wang, J., Bai, X., Xie, J., Wang, Y., & Geng, W. (2021). Metabolism characteristics of lactic acid bacteria and the expanding applications in food industry. *Frontiers in Bioengineering and Biotechnology*, *9*, 612285. https://doi.org/10.3389/fbioe.2021.612285

WHO. (2024). Malnutrition. https://www.who.int/news-room/fact-sheets/detail/malnutrition. (Last accessed 15th August 2025)

Wu, T., Wang, G., Xiong, Z., Xia, Y., Song, X., Zhang, H., Wu, Y., & Ai, L. (2022). Probiotics interact with lipids metabolism and affect gut health. *Frontiers in Nutrition*, *9*, 917043. https://doi.org/10.3389/fnut.2022.917043

Xiang, H., Sun-Waterhouse, D., Waterhouse, G. I. N., Cui, C., & Ruan, Z. (2019). Fermentation-enabled wellness foods: A fresh perspective. *Food Science and Human Wellness*, *8*(3), 1–10. https://doi.org/10.1016/j.fshw.2019.08.003

Yang, Q., Yao, H., Liu, S., & Mao, J. (2022). Interaction and application of molds and yeasts in Chinese fermented foods. *Frontiers in Microbiology*, *12*, 664850. https://doi.org/10.3389/fmicb.2021.664850

Yuan, Y., Yang, Y., Xiao, L., Qu, L., Zhang, X., & Wei, Y. (2023). Advancing insights into probiotics during vegetable fermentation. *Foods*, *12*(20), 3789. https://doi.org/10.3390/foods12203789

9 Biotechnology and Genetic Engineering in Food Security

Maria M. Adeyemi, Nonso Ugwuja, Victoria Bassey, Deborah Oni-Orisan, and Vanessa Williams

9.1 INTRODUCTION

The world faces troubling challenges that greatly impact food security. One of these challenges is an increase in the world's population and the rising number of countries that are food importers owing to their inability to produce the food they need to feed its growing population (Guiné, 2024). According to the United Nations World Population Prospects, the global population is projected to grow from approximately 8 billion in 2025 to 9 billion by 2050, and at least 10.5 billion by 2100. Such an increase in the world's population could pose a threat to ensure food security while managing the Earth's finite resources, which is only possible with more efficient food production, more sustainable use of natural resources, and minimization and re-utilization of food and agricultural waste (Guiné, 2024). Feeding a growing population requires not only producing more food but also doing so sustainably.

Other challenges include climate changes and food waste.

1. *Climate Changes*: These disrupts agricultural systems through altered precipitation patterns, increased frequency of extreme weather events, and shifting temperatures. Changes such as droughts, floods, and storms can exacerbate food insecurity, leading to reduced crop yields, heightened vulnerability of livestock to diseases, and displacement of agricultural zones (Godde et al., 2021).
2. *Food Waste*: In 2018, approximately 1 billion people (16% of the global population) suffered from chronic hunger in a time when there was more than enough food to feed everyone on the planet (McCarthy et al., 2018). Concurrently, about one-third of all food produced globally—equivalent to approximately 1.3 billion tons—is lost or wasted each year (Wagay, 2023). This paradox highlights the coexistence of food surplus and hunger: while enough food is produced to nourish the global population, logistical and systemic issues prevent its equitable distribution. The number of people impacted by hunger has been slowly rising since 2014, and thousands of tonnes of edible food are lost or wasted every day, therefore reducing food losses and waste is critical (Wagay, 2023). Food waste signifies lost nourishment and represents a squandered opportunity to combat hunger, as the fundamental rights to food is essential for ensuring food security and sovereignty (Onyeaka et al., 2024).

Traditional agricultural methods only may not be sufficient to meet the growing demand without causing severe environmental damage, such as deforestation, soil degradation, and excessive water consumption, which would undermine efforts to boost food production. Therefore, there is need to engage biotechnology and genetic engineering techniques in managing these challenges.

9.1.1 Combatting Food Insecurity

Food security is both a complex and challenging issue to resolve as it cannot be characterized or limited by geography nor defined by single grouping, such as, demography, education, geographic

DOI: 10.1201/9781003650881-9

location, or income (McCarthy et al., 2018). While carbohydrate-rich grains are vital for energy, a healthy human diet requires micronutrients—such as iodine, iron, zinc, and vitamins A, C, D, B1, and B3. These come either from plant uptake or from synthesis in plants and animals. Industrialized countries typically meet these needs through diverse diets including animal products. However, many developing countries lack access to such diversity, especially animal products, due to economic constraints. In 2000, 129 Heads of State endorsed the Millennium Development Goals (MDGs), including targets to reduce poverty, hunger, and child and maternal mortality by 2015. Despite broad support, the key nutrition-related goals—Goal 1 (eradicating extreme poverty and hunger) and Goals 4 and 5 (reducing child and maternal deaths)—were not achieved. In 2020, about 9.2% of the world population was facing chronic hunger and the United Nations set to achieve Zero Hunger by 2030 with the second Sustainable Development Goal (SDG) to transform our world, indicating urgent coordination action are necessary to transform the food systems, invest in sustainable agricultural practices, and mitigate its impact on global nutrition and food security.

To address food insecurity and environmental sustainability, a broad spectrum of agricultural approaches must be integrated. These include conventional breeding (even using mutagenesis), biotechnology, genetically modified organisms (GMOs), organic and inorganic fertilization, soil management, agro-ecology, and responsible pesticide use. No single method is sufficient on its own. Improving food production efficiency through biotechnology, precision farming, and sustainable agricultural practices is essential. Additionally, minimizing food and agricultural waste can help maximize the use of available resources, ensuring that more people are fed without increasing environmental strain. Biotechnology plays a crucial role in enhancing food security by improving crop yields, increasing nutritional value, and developing resistance to pests and diseases. Genetically modified (GM) crops have been engineered to exhibit desirable traits, such as increased nutrient content. A notable example is Golden Rice, engineered from normal rice by Ingo Potrykus and Peter Beyer in the 1990s to help improve human health by biosynthesizing beta-carotene, a pro-vitamin A when metabolized by humans (Baranski, 2013). This innovation aims to reduce the incidence of vitamin A deficiency (VAD), which is estimated to cause 670,000 deaths annually among children under five and lead to 500,000 cases of irreversible childhood blindness (WHO, 2023).

9.2 BIOTECHNOLOGY IN FOOD SECURITY

Biotechnology is broadly defined as a "technique that uses living organisms or substances from those organisms to make or modify a product, improve plants or animals, or develop microorganisms for specific uses" (Sairam & Prakash, 2005). It encompasses a wide range of procedures for modifying living organisms extending from traditional methods like fermentation to modern genetic engineering techniques. The field has evolved significantly, incorporating technologies such as recombinant DNA technology (combining DNA from different sources to create new genetic combinations, often used to produce therapeutic proteins, GMOs, and other applications), hybridoma technology (a method for producing large quantities of monoclonal antibodies by fusing antibody-producing B cells with immortal myeloma cells), enzyme engineering, and protein engineering. These advancements have profoundly impacted research in healthcare, diagnostics, and agriculture, and are making inroads into industries like petroleum, mining, food, and chemicals.

Although biotechnology applies across several fields, agricultural biotechnology is the most crucial for African countries particularly for resource-poor farmers whose sole livelihood depends on agriculture. Though the technique of biotechnology alone cannot solve all the problems associated with agricultural production, it has the potential to address specific issues such as increasing crop productivity, diversifying crops, enhancing the nutritional value of food, reducing the environmental impacts of agricultural production, and promoting market competitiveness (Abah et al., 2010).

Biotechnology has transformed agricultural practices worldwide, offering innovative solutions to enhance food security. It provides a revolutionary trend in life sciences and affects every facet of lives, from crop improvement to commerce, and drugs to sustainable development (Tonukari &

Omotor, 2010). The branch of biotechnology responsible for addressing food insecurity is agricultural biotechnology.

9.3 AGRICULTURAL BIOTECHNOLOGY

This aims to introduce sustainable agricultural practices with the best yield potential and minimal adverse effects on the environment (Gupta et al., 2016). It is the application of biotechnological techniques to modify organisms, such as crops and animals, improve their productivity, resistance to diseases, and nutritional value (James, 2017). It has applications in crop breeding, pest resistance, disease resistance, and environmental sustainability (FAO, 2020). This could be through genetic engineering, improved photosynthesis and optimized growth factors through genetic modified organisms, and improved nutrient utilization through biofortification.

9.3.1 Key Techniques in Agricultural Biotechnology

9.3.1.1 Genetic Engineering

This involves the modification of an organism's DNA to introduce new traits, such as pest resistance or enhanced nutritional content. GMOs are often developed using genetic engineering (Pellegrini & Fernandez, 2020). For example, combating pests was a major challenge. Thus, the gene from a bacterium, the Bt gene, that functions as an insect-resistant gene when inserted into crop plants like cotton, corn, and soybean helps prevent the invasion of pathogens. This management helps reduce the usage of potentially dangerous pesticides on the crop. Not only is the minimal or low usage of pesticides beneficial for the crop but also the load of the polluting pesticides on the environment is greatly reduced (Gupta et al., 2016).

1. *Molecular Markers*: Molecular markers are DNA sequences that can be linked to specific traits in plants and animals, such as drought resistance or higher yield. These markers are used in marker-assisted selection (MAS) to accelerate breeding programs (Abdul Aziz et al., 2022).
2. *Tissue Culture*: This technique involves growing plant cells or tissues in a controlled environment to propagate plants. It is commonly used to produce large numbers of uniform plants, especially for crops like bananas and potatoes (Bhardwaj et al., 2025).
3. *Recombinant DNA Technology*: Recombinant DNA involves combining genetic material from different sources. This technology is used to create transgenic crops with traits like herbicide resistance or improved nutritional content (FAO, 2020).

9.3.1.2 Genetic Modification Organisms

In the United States, the introduction of genetic modification organisms (GMOs) into agriculture began in 1993, when the FDA approved bovine somatotropin (BST), a growth hormone that boosts milk production in dairy cows. The next year, the FDA approved the first GM whole food, a tomato engineered for a longer shelf life. Since then, regulatory approval in the United States, Europe, and elsewhere has been won by dozens of agricultural GMOs, including crops that produce their own pesticides and crops that survive the application of specific herbicides used to kill weeds (Hussein, 2023). The unprecedented rapid adoption of transgenic crops during the initial five-year period (1996–2000), when GM crops were first adopted, reflects the significant multiple benefits realized by large and small farmers in industrial and developing countries that have grown transgenic crops commercially (ISAAA, 2017).

GM refers to the manipulation of an organism's genetic material using biotechnology to introduce, remove, or alter specific genes. Unlike traditional breeding, GM enables precise gene editing, often involving genes from different species. These organisms are produced through genetic engineering, where genes from one organism are inserted into another or a gene expression is reduced or prevented. CRISPR-Cas9, a gene-editing tool that allows scientists to cut and modify DNA at specific locations can also be used. A tool revolutionizing tomato breeding for improved quality, yield and stress ressistance (Larriba et al., 2024).

9.4 EXAMPLES OF GENETICALLY MODIFIED ORGANISMS

9.4.1 Golden Rice

Golden Rice was engineered from normal rice by Ingo Potrykus and Peter Beyer in the 1990s to help improve human health (Sowndarya, 2021). Golden Rice has an engineered multi-gene biochemical pathway in its genome. This pathway produces beta-carotene, a molecule that becomes vitamin A when metabolized by humans. Golden Rice is named for its golden color, which is caused by beta-carotene. Normal rice, *Oryza sativa*, does not express beta-carotene in its endosperm—the starchy and biggest part of the rice seed, which is usually an off-white color (Ye et al., 2000). Beta-carotene is part of a class of molecules called carotenoids, one of hundreds that plants naturally produce, and it has a yellow-orange hue. Carotenoids are essential nutrients for humans because they are precursors to molecules needed in metabolism.

9.4.2 *Bacillus Thuringiensis* (BT) Corn

Bt corn is maize (*Zea mays* L.) that has been genetically modified to express the cry1Ab gene from *Bacillus thuringiensis* (Bt) and produce a larvicidal toxin to kill lepidopteran pests, especially the European corn borer (ECB; *Ostrinia nubilalis*) (Deepak, 2001). The ECB is an economically important insect pest of corn (*Zea mays* L.) and other crops in North America and Europe, with yield losses up to 20% caused by larval feeding (Siegfried & Hellmich, 2012). Damage from larval infestations of corn was estimated to exceed US $1 billion annually in yield losses and control expenditures (USDA-NASS, 2011). Chemical insecticides often are not effective against *O. nubilalis* infestations because once the larvae tunnel into the corn stalk, they are protected from exposure. Transgenic corn plants that express insecticidal proteins from *B. thuringiensis* (Berliner) (Bt) have become an effective method of protecting corn plants from damage by *O. nubilalis* and have been widely adopted throughout the US corn belt (Hurley et al., 2002; USDA-NASS, 2011).

9.4.3 Rainbow Papaya

Papaya (*Carica papaya* L.) is produced commercially in many tropical and subtropical areas of the world for domestic consumption and for export by top producing countries like India, Brazil, Nigeria, Indonesia, and Mexico. Papaya is known as a "common man's fruit" because of its low cost and excellent nutritional value (Koul et al., 2022). The most limiting factor for papaya production worldwide is Papaya ringspot virus (PRSV), an aphid-transmitted potyvirus. Rainbow papaya (*C. papaya* L.) is a genetically engineered cultivar with resistance to PRSV. It is an F1 hybrid resulting from a cross between the transgenic Su and a yellow-fleshed non-transgenic cultivar, Kapoho (Kp) (Tripathi et al., 2011; Koul et al., 2022). These genetically engineered papaya has various nutrients including a higher amount of vitamin A when compared with non-genetically engineered papaya (Tripathi et al., 2011).

9.4.4 Aquabounty Salmon

The Aquabounty Salmon was developed by inserting into a fertilized Atlantic salmon egg, a transgene constructed from the growth hormone gene of a related species (Chinook salmon). The

Aquabounty genetic construct confers upon the Atlantic salmon a rapid growth phenotype and improved metabolic efficiencies associated with low but sustained levels of growth hormone. This allows the fast-growing Aquabounty Salmon to reach market size in half the time of a conventional Atlantic salmon. The Aquabounty Salmon will be available only as all-female, sterile, eyed eggs, and the resulting fish must be maintained in a freshwater, land-based, physically contained facility (Clifford, 2014).

9.4.5 Bt Cotton

Insect pests, particularly the bollworm (*Helicoverpa armigera*), have been a major threat to cotton production in China. Cotton growers have to struggle against these pests mainly by using pesticides. China's pest problems led the nation's scientists to seek new alternatives to manage the pests. When the possibility of incorporating Bt genes for resistance to insect pests was demonstrated, research and development of transgenic cotton was immediately initiated by the Chinese authorities. By 1997, under the financial support of the government, two Bt cotton varieties were developed using a modified Bt fusion gene (Cry lab and Cry (Ac) by the Chinese Academy of Agricultural Sciences (CAAS)), and commercialized. Since then, more than ten crop varieties incorporating the locally isolated Bt gene have been licensed for commercialization.

9.4.6 Flavr Savr Tomato

Tomato fruit quality such as appearance, flavor, nutritional profiles, and postharvest properties were improved by gene editing such as the utilization of zinc-finger nucleases (ZFN), transcription activator-like effector nucleases (TALEN) gene editing tool, and the CRISPR/Cas 9 (Clustered Regularly Interspaced Short Palindromic Repeats/CRISPR-associated protein 9) system. The FLAVR SAVR tomato was developed through the use of antisense RNA, where an additional copy of the polygalacturonase (PG) gene was inserted in the reverse orientation to regulate the expression of the PG enzyme in ripening tomato fruit (Larriba et al., 2024). This enzyme is one of the most abundant proteins in ripe tomato fruit and has long been thought to be responsible for softening in ripe tomatoes. The longer shelf life of the Flavr Savr tomato made it easier and cheaper to transport and handle, as it could be harvested at a riper stage and shipped long distances without spoilage.

9.5 BENEFITS OF GENETICALLY MODIFIED ORGANISMS

1. *Increased Agricultural Productivity*: GM crops often yield more and reduce crop loss. On-farm field trials carried out with *B. thuringiensis* (Bt) cotton in different states of India show that the technology substantially reduces pest damage and increases yields. The yield gains are much higher than what has been reported for other countries where GM crops were used mostly to replace and enhance chemical pest control (Qaim & Zilberman, 2003).
2. *Environmental Benefits*: Reduced pesticide uses and soil conservation through herbicide-tolerant crops. The adoption of GM insect resistant and herbicide tolerant technology has reduced pesticide spraying by 671.4 million kg (8.2%) and, as a result, decreased the environmental impact associated with herbicide and insecticide use on these crops (as measured by the indicator, the Environmental Impact Quotient (EIQ)) by 18.4% (Brookes & Barfoot, 2018).
3. *Enhanced Nutrition*: Golden Rice is engineered to combat VAD by producing beta-carotene, which the body converts to vitamin A, offering a potential solution for populations where rice is a staple food and VAD is prevalent (Shahbaz et al., 2020).

9.6 CONCERNS AND CONTROVERSIES TO GENETICALLY MODIFIED ORGANISMS

GMOs are required to undergo rigorous testing for safety, including allergenicity and toxicity, before approval. Regulatory bodies such as the FDA, EFSA, and WHO oversee GMO safety and labeling (FAO, 2020). Despite the benefits, GMOs raise ethical, ecological, and health-related concerns:

1. *Health Risks*: Potential allergens or long-term effects, though no conclusive evidence currently exists.
2. *Environmental Impact*: Deliberate or inadvertent releases of genetically engineered organisms or substances into the environment could have negative ecological effects under certain circumstances. Example: harm to non-target species, such as soil organisms, non-pest insects, birds, and other animals, disruption of biotic communities, including agro-ecosystems, irreparable loss or changes in species diversity or genetic diversity within species, etc.
3. *Socioeconomic Issues*: GM seed patenting can increase farmers' dependence on biotech companies. Similarly, the practices of science are no longer for the pursuit of knowledge but are firmly linked to commercial targets and objectives.

9.6.1 Biofortification and Food Security

Biofortification is the process of increasing the nutrient content of crops through conventional breeding, genetic engineering, or agronomic practices (Bouis & Welch, 2010). The focus is typically on key micronutrients such as iron, zinc, and provitamin A (vitamin A precursors), which are vital for human health but often deficient in staple-based diets. More than 2 billion people globally suffer from micronutrient deficiencies, leading to impaired cognitive development, weakened immunity, and increased morbidity (WHO/WFP/UNICEF, 2006). Biofortification offers a cost-effective, sustainable, and scalable solution, especially in rural areas where supplementation and industrial fortification are not always accessible (Bouis & Saltzman, 2017).

There are three primary methods employed in biofortification:

1. *Conventional Breeding*: This involves crossbreeding plant varieties that naturally contain higher levels of specific nutrients with high-yielding or locally adapted varieties. The aim is to develop crops that combine nutritional quality with desirable agronomic traits (Bouis & Welch, 2010). Example: Iron-rich beans in Rwanda and zinc-rich wheat in India have been developed through traditional breeding.
2. *Transgenic Approaches* (*Genetic Engineering*): This involves directly modifying the DNA of a plant by inserting genes that are responsible for producing or enhancing nutrients. It allows for the creation of traits not naturally found in a species. Example: Golden Rice, engineered to produce beta-carotene (a vitamin A precursor), is one of the most well-known transgenic biofortified crops (Ye et al., 2000; Sowndarya, 2021).
3. *Agronomic Biofortification*: This method enhances nutrient content by applying micronutrient-rich fertilizers (especially zinc, iron, selenium, etc.) to crops via soil or foliar sprays. It focuses on improving the plant's nutrient uptake from the environment, for example, zinc-enriched rice in Bangladesh was produced using zinc-containing fertilizers to increase zinc levels in the edible part of the grain (Mottaleb et al., 2025) (Table 9.1).

TABLE 9.1
Examples of Biofortified Crops

Crop	Nutrient Enhanced	Target Region	Impact
Golden Rice	Provitamin A	Southeast Asia	Reduces risk of blindness and mortality (Tang et al., 2009; Shahbaz et al., 2020; Sowndarya, 2021)
Iron Beans	Iron	Rwanda, Uganda	Improves iron intake among women and children (Nuwahereza et al., 2021)
Zinc Rice	Zinc	Bangladesh	Supports immune function and growth (Jongstra et al., 2022)
Orange-Fleshed Sweet Potato (OFSP)	Provitamin A	Sub-Saharan Africa	Enhances vitamin A status in children (Low et al., 2007)

9.6.1.1 Benefits of Biofortification

9.6.1.1.1 Sustainability

Biofortification offers a long-term, cost-effective solution to micronutrient deficiencies. Once a biofortified crop variety is developed and released, farmers can save and reuse seeds from season to season, just like traditional varieties. This makes biofortification self-sustaining and far less reliant on continuous external funding, unlike supplementation programs that require regular distribution of vitamins or fortified food products.

9.6.1.1.2 Accessibility

Biofortification is designed to reach those most in need—rural populations who often rely on a limited number of staple crops like rice, maize, or cassava. These populations may have limited access to diverse diets, vitamin supplements, or commercially fortified foods, especially in remote or economically disadvantaged areas (Saltzman et al., 2013). Because biofortified crops look and taste similar (or nearly similar) to local varieties, acceptance among communities can be high, especially when accompanied by nutrition education and outreach.

By adopting these opportunities, biofortification can maintain its status as a potent instrument for enhancing global food security. Incorporating biofortification into more comprehensive food security plans is essential to achieve sustainable development objectives for health and nutrition (Naik et al., 2024).

9.7 CONCLUSION

Modern agricultural biotechnology stands as one of the most significant advancements in contemporary science. When combined with conventional breeding techniques, it offers immense potential to boost crop yields, strengthen resistance to pests and diseases, and enhance tolerance to extreme weather conditions. Tropic's non-browning bananas represent a promising innovation in the fight against food waste, with the potential to reduce waste, improve food security, and contribute to sustainability efforts. The success of this product highlights the growing role of gene editing in agriculture, offering solutions to long-standing challenges related to perishability, food loss, and environmental impact. The future of food production lies in the continued development of more resilient, efficient, and sustainable crops. As gene editing technologies continue to evolve, they offer the possibility of transforming agricultural practices and addressing critical issues such as food waste, crop yields, and climate change. Biotechnology and genetic engineering presents innovative solutions essential for food security and sustainability. Its thoughtful application will help create a more equitable, nutritious, and resilient food system for future generation.

REFERENCES

Abah J., Ishaq M. N., & Wada J. C. (2010). The role of biotechnology in ensuring food security and sustainable agriculture. *African Journal of Biotechnology*, *9*(52), 8896–8900. http://www.academicjournals.org/AJB

Abdul Aziz, M., Brini, F., Rouached, H., & Masmoudi, K. (2022). Genetically engineered crops for sustainably enhanced food production systems. *Frontiers in Plant Science*, *13*, 1027828. https://doi.org/10.3389/fpls.2022.1027828

Baranski, M. (2013). Golden Rice. Embryo Project Encyclopedia ISSN: 1940-5030. https://hdl.handle.net/10776/6275

Bhardwaj, S., Kumari, M., T., C., Hoṣamani, M., Paul, M., Moirangthem, A., Dash, A. P., & Verma, S. (2025). Advances in micropropagation and tissue culture for horticultural crops: A Review. *Plant Cell Biotechnology and Molecular Biology*, *26*(11–12), 166–182. https://doi.org/10.56557/pcbmb/2025/v26i11-1210039

Bouis, H. E., & Saltzman, A. (2017). Improving nutrition through biofortification: A review of evidence from HarvestPlus, 2003 through 2016. *Global Food Security*, *12*, 49–58. https://doi.org/10.1016/j.gfs.2017.01.009

Bouis, H. E., & Welch, R. M. (2010). Biofortification—a sustainable agricultural strategy for reducing micronutrient malnutrition in the Global South. *Crop Science*, *50*(Supplement_1), S-20–S-32. https://doi.org/10.2135/cropsci2009.09.0531

Brookes, G., & Barfoot, P. (2018). Environmental impacts of genetically modified (GM) crop use 1996–2016: Impacts on pesticide use and carbon emissions. *GM Crops & Food*, *9*(3), 109–139. https://doi.org/10.1080/21645698.2018.1476792

Clifford, H. (2014). AquAdvantage® Salmon – a pioneering application of biotechnology in aquaculture. *BMC Proceedings*, *8*(Suppl 4), O31. https://doi.org/10.1186/1753-6561-8-S4-O31

Deepak S. (2001). Bt corn has a higher lignin content than non-Bt corn. *American Journal of Botany*. https://doi.org/10.2307/3558416

FAO (Food and Agriculture Organization of the United Nations). (2020). The State of Food and Agriculture: Moving Forward with Agricultural Biotechnology. https://www.fao.org

Godde, C. M., Mason-D'Croz, D., Mayberry, D. E., Thornton, P. K., & Herrero, M. (2021). Impacts of climate change on the livestock food supply chain: A review of the evidence. *Global Food Security*, *28*, Article 100488. https://doi.org/10.1016/j.gfs.2020.100488

Guiné, R. P. F. (2024). The challenges and strategies of food security under global change. *Foods*, *13*(13), 2083. https://doi.org/10.3390/foods13132083

Gupta, V., Sengupta, M., Prakash, J., & Tripathy, B. C. (2016). An introduction to biotechnology. *Basic and Applied Aspects of Biotechnology*, 1–21. https://doi.org/10.1007/978-981-10-0875-7_1

Hurley, T. M., Secchi, S., Babcock, B. A., & Hellmich, R. L. (2002). Managing the risk of European corn borer resistance to Bt corn. *Environmental and Resource Economics*, *22*, 537–58. http://dx.doi.org/10.1023/A:1019858732103

Hussein, M. R. (2023). The issue of genetically modified organism (GMO) and assess to food in Ethiopia. *Journal of Biotechnology and Immunology* 5(1).

ISAAA. (2017). *Global Status of Commercialized Biotech/GM Crops in 2017: Biotech Crop Adoption Surges as Economic Benefits Accumulate in 22 Years*. ISAAA Brief No. 53. ISAAA, Ithaca, NY.

James, C. (2017). Global Status of Commercialized Biotech/GM Crops: 2017. ISAAA Briefs No. 53. International Service for the Acquisition of Agri-biotech Applications (ISAAA). https://www.isaaa.org/

Jongstra, R., Hossain, M. M., Galetti, V., Hall, A. G., Holt, R. R., Cercamondi, C. I., Rashid, S. F., Zimmermann, M. B., Mridha, M. K., & Wegmueller, R. (2022). The effect of zinc-biofortified rice on zinc status of Bangladeshi preschool children: A randomized, double-masked, household-based, controlled trial. *The American Journal of Clinical Nutrition*, *115*(3), 724–737. https://doi.org/10.1093/ajcn/nqab379

Koul, B., Pudhuvai, B., Sharma, C., Kumar, A., Sharma, V., Yadav, D., & Jin, J.-O. (2022). *Carica papaya* L.: A tropical fruit with benefits beyond the tropics. *Diversity*, *14*(8), 683. https://doi.org/10.3390/d14080683

Larriba, E., Yaroshko, O., & Pérez-Pérez, J. M. (2024). Recent advances in tomato gene editing. *International Journal of Molecular Sciences*, *25*(5), 2606. https://doi.org/10.3390/ijms25052606

Low, J. W., Arimond, M., Osman, N., Cunguara, B., Zano, F., & Tschirley, D. (2007). A food-based approach introducing orange-fleshed sweet potatoes increased vitamin A intake and serum retinol concentrations in young children in rural Mozambique. *The Journal of Nutrition*, *137*(5), 1320–1327. https://doi.org/10.1093/jn/137.5.1320

McCarthy, U., Uysal, I., Badia-Melis, R., Mercier, S., O'Donnell, C. & Ktenioudaki, A. (2018). Global food security – Issues, challenges and technological solutions. *Trends in Food Science & Technology, 77.* ISSN 0924-2244. https://doi.org/10.1016/j.tifs.2018.05.002.

Mottaleb, K.A., Durand-Morat, A., Fatah, F.A., & Sarkar, M.A.R. (2025). Current Status of Zinc-Biofortified Rice Cultivation in Bangladesh. In Govindaraj, M., Govindan, V., & Palacios, N. (Eds.), *Breeding Zinc Crops for Better Human Health.* Springer. https://doi.org/10.1007/978-3-031-84342-6_6

Naik, B., Kumar, V., Rizwanuddin, S., Mishra, S., Kumar, V., Saris, P. E. J., Khanduri, N., Kumar, A., Pandey, P., Gupta, A. K., Khan, J. M., & Rustagi, S. (2024). Biofortification as a solution for addressing nutrient deficiencies and malnutrition. *Heliyon, 10*(9), e30595. https://doi.org/10.1016/j.heliyon.2024.e30595

Nuwahereza, C., Kamukama, S., Kayemba, C. N., Obai, G. (2021). Using iron bio-fortified beans to control iron deficiency anaemia among children 6–59 months in rural Uganda. *Research Square*, 1–15. https://doi.org/10.21203/rs.3.rs-446432/v1

Onyeaka, H., Adeboye, A. S., Bamidele, O. P., Onyeoziri, I., Adebo, O. A., Adeyemi, M. M., & Thera-Sekgweng, S. N. (2024). Beyond hunger: Unveiling the rights to food in sub-Saharan Africa. *Food and Energy Security, 13*(1), Article e530. https://doi.org/10.1002/fes3.530

Pellegrini, P., & Fernandez, M. (2020). Agricultural biotechnology and its role in sustainable food production. *Biotechnology Advances, 38*, 107764. https://doi.org/10.1016/j.biotechadv.2020.107764

Qaim M, & Zilberman, D. (2003). Yield effects of genetically modified crops in developing countries. https://doi.org/10.1126/science.1080609

Sairam, R. V., & Prakash, C. S. (2005). OBPC Symposium: Maize 2004 & beyond – Can agricultural biotechnology contribute to global food security? *In Vitro Cellular & Developmental Biology – Plant, 41*(4), 424–430. https://doi.org/10.1079/IVP2005663

Saltzman, A., Birol, E., Bouis, H. E., Boy, E., De Moura, F. F., Islam, Y., & Pfeiffer, W. H. (2013). Biofortification: Progress toward a more nourishing future. *Global Food Security, 2*(1), 9–17. https://doi.org/10.1016/j.gfs.2012.12.003

Shahbaz, U., Xiao-Bin, Y., Wasim, A., Regis, N., & Husnain, R. (2020). Golden Rice to Eradicate the Vitamin A deficiency in the Developing Countries. *European Journal of Nutrition & Food Safety, 12*(1): 53–63.

Siegfried, B. D., & Hellmich, R. L. (2012). Understanding successful resistance management: The European corn borer and Bt corn in the United States. *GM Crops & Food, 3*(3), 184–193. https://doi.org/10.4161/gmcr.20715

Sowndarya, K. (2021). Golden Rice: A genetically modified (GM) food. *Agriculture and Food: E-Newsletter, 3*(7), 87–89.

Tang, G., Qin, J., Dolnikowski, G. G., Russell, R. M., & Grusak, M. A. (2009). Golden Rice is an effective source of vitamin A. *The American Journal of Clinical Nutrition, 89*(6), 1776–1783. https://doi.org/10.3945/ajcn.2008.27119

Tonukari, J., & Omotor, G. (2010). Biotechnology and food security in developing countries. *Biotechnology and Molecular Biology Reviews, 5*(1). http://www.academicjournals.org/BMBR

Tripathi S., Suzuki J. Y., Carr J. B., McQuate G. T., Ferreira S. A., Manshardt R. M., Pitz, K. Y., Wall M. M., Gonsalves, M. (2011). Nutritional composition of Rainbow papaya, the first commercialized transgenic fruit crop. *Journal of Food Composition and Analysis, 24*(2), 140–147.

USDA-NASS (United States Department of Agriculture, National Agricultural Statistics Service). (2011). *Acreage.* Washington, DC, USA. http://usda.mannlib.cornell.edu/usda/current/Acre/Acre-06-30-2011.pdf

Wagay, A. H. (2023). Food waste: A critical element to end global hunger. *International Journal of Advanced Research in Biological Sciences.* https://doi.org/10.1007/978-3-031-86688-3_1

WHO Intervention. (2023). *Vitamin A Supplementation in Infants and Children 6–59 Months of Age.* World Health Organization, Geneva.

WHO/WFP/UNICEF. (2006). Preventing and controlling micronutrient deficiencies in populations affected by an emergency; Multiple Vitamin and Mineral Supplements for pregnant and lactating women, and for children aged 6 to 59 months. Joint statement by the World Health Organization, the World Food Programme and the United Nations Children's Fund.

Ye, Xudong, Al-Babili, Salim, Klöti, Andreas, Zhang, Jing, Lucca, Paola, Beyer, Peter, & Potrykus, Ingo. (2000). Engineering the provitamin A (β-carotene) biosynthetic pathway into (carotenoid-free) rice endosperm. *Science, 287*, 303–305. http://doi.org/10.1126/science.287.5451.303

10 The Role of Research and Innovation in the Future of Microbial Technologies

Odangowei Inetiminebi Ogidi, Balwant Singh, and Mukul Machhindra Barwant

10.1 INTRODUCTION

Food security is still a major concern with rising global problems including population increase, climate change, and declining arable land requiring creative and sustainable answers. Of these, microbial technologies are becoming a fascinating frontier. Food production, preservation, and safety have long depend on microorganisms (Barwant et al., 2025a,b,c). But developments in biotechnology and research are now releasing its wider potential in raising agricultural output, reducing post-harvest losses, and enhancing nutritional results. Research on microbial uses presents transforming opportunities in food systems both pre- and post-harvest. By acting as biocontrol agents, boosting nutrient absorption, and enhancing plant resilience, beneficial bacteria help to lower reliance on artificial fertilizers and pesticides (Bhattacharyya & Jha, 2012). These features are absolutely essential in building strong agricultural systems able to resist environmental pressures and keep high yields.

Microbial genomics and synthetic biology have made it possible for researchers to create targeted-functional microbial strains. From nitrogen fixation in non-legumes to the manufacture of important vitamins and amino acids, these modified bacteria can play functions ranging from such biotechnological solutions can solve nutritional shortages in underdeveloped areas, therefore complementing the objectives of world food security (Ogidi et al., 2024). Another area under increasing focus is the inclusion of microbial consortia into soil management techniques. Knowledge of microbial ecology helps scientists create synthetic communities that maximise soil fertility and plant health (Compant et al., 2010). Essential for sustainable agriculture, these microbial ecosystems are involved in carbon sequestration, nitrogen cycling, and insect resistance mechanisms (Iyiola et al., 2023).

Microbial discoveries have produced natural biopreservatives in post-harvest control that increase shelf life free from chemical additions. Lactic acid bacteria, for example, generate bacteriocins that stop food rotting organisms (Gálvez et al., 2007). This strategy not only cuts food waste but also satisfies consumer needs for items free of labels. Precision farming methods also help to revive an ancient microbial use: food fermentation. These developments enable the regulated manufacturing of high-value food components such as dairy proteins, meat analogues, and functional enzymes free from animal or plant inputs, therefore relieving some demand on conventional agricultural resources (Tamang et al., 2016).

Advancement of microbial technology depends on both public and commercial sector cooperation. Through finance, education, and policy frameworks, governments and research institutes have to make investments in basic science while creating surroundings fit for innovation (Godfray et al., 2010). Through the development of market-ready applications, the private sector may concurrently hasten commercialisation and scale. Not less significant is the part multidisciplinary research plays. Combining microbiology, data science, systems biology, and agronomy is generating fresh understanding of plant-microbe interactions and uses. Broadening the range of possible uses, technologies such as metagenomics and bioinformatics allow the investigation of unculturable bacteria (Delmont

 DOI: 10.1201/9781003650881-10

et al., 2015). Adoption of microbial technology will also follow the path influenced by society acceptance and legal structures. Public confidence has to be developed by open communication on the safety and effectiveness of microbial products. Harmonising worldwide standards can also help to enable trade and cross-border cooperation (Jansson & Hofmockel, 2020).

Developmentally, microbial technologies provide low-cost, locally produced substitutes for synthetic agrochemicals, therefore enabling smallholder farmers (Ogidi et al., 2023; Ogidi & Akpan, 2023). This democratisation of innovation promotes economic resilience in rural areas and helps to bridge yield gaps (Pretty et al., 2011). Another field where microbial innovation is absolutely essential is climate-smart agriculture. While protecting food production, microorganisms that increase drought resilience or lower methane emissions in rice fields can greatly help to slow down climate change (Singh et al., 2020). These line up with frameworks for food security and global climate targets.

Main public health issues still are foodborne disease and contamination. Development of phage-based detection systems and microbial biosensors can help to quickly identify infections in food products, hence enhancing safety monitoring throughout supply chains (Liu et al., 2020). These technologies raise customer confidence and food system resilience. Globally, microbial research is essential to reaching numerous United Nations Sustainable Development Goals (SDGs), including Zero Hunger, Good Health and Well-Being, and Climate Action. Nations can provide synergistic results across several objectives by including microbial innovation into agriculture policies and development projects.

Microbial technology has promise, but in low-resource environments, knowledge gaps, intellectual property (IP) restrictions, and inadequate infrastructure all challenge it. Equitable access and benefits depend on focused capacity-building and inclusive innovation methods (Altieri et al., 2015). Food security's future depends ultimately on our capacity to effectively and sustainably leverage biological processes. Understood by careful research and driven by ethical innovation, microbial technologies present a workable road ahead. These technologies could transform food systems and guarantee food for next generations by means of ongoing investment and teamwork (Barwant et al., 2025d,e).

10.2 CURRENT ADVANCES IN MICROBIAL RESEARCH

Driven by new technology and a better knowledge of microbial variety and functionality, microbial research has undergone a radical change recently. Long thought of as simple and mostly pathogenic, microorganisms have great ecological, industrial, and health-related value presently. Microbial research has been greatly expanded by the development of multidisciplinary techniques combining synthetic biology, bioinformatics, and genomics (Rinke et al., 2013). Table 10.1 and Figure 10.1 show the current advances in microbial research.

10.2.1 Role of Metagenomics and Microbiome Studies

The creation and use of metagenomics—the direct genetic study of genomes contained inside an environmental sample—has been a major advance in microbial research. Bypassing the need for cultivation, this approach helps researchers to find the hitherto undetectable microbial majority—which cannot be grown using conventional techniques (Handelsman, 2004). Accessing these unculturable bacteria has helped scientists better understand their complicated functions in ecosystems, health, and disease. Close relative to metagenomics, microbiome investigations concentrate on identifying the whole suite of microorganisms in a given environment, such as the human gut, soil, or plant rhizosphere (Ogidi & Felagha, 2024).

These investigations have shown the dynamic interactions between microorganisms and their hosts, therefore highlighting their vital contribution to immunity, digestion, and even neurological

TABLE 10.1
Current Advances in Microbial Research

Advance	Description	Key Benefits
Human Microbiome Research	Investigating the role of gut and skin microbiota in health and disease	Personalised medicine, probiotic therapies
Microbial Resistance Mechanisms	Understanding how microbes resist antibiotics and antimicrobials	Better antibiotic stewardship, novel treatment targets
Extremophile Microbiology	Studying microbes in extreme environments (heat, pressure, salinity, radiation)	Novel enzymes, industrial applications, astrobiology
Environmental Microbial Ecology	Exploring microbial roles in ecosystems and biogeochemical cycles	Climate change insights, soil health, conservation
Microbiome-Host Interactions	Analysing how microbes affect host physiology and immune response	Disease prevention, immune modulation strategies
Bioinformatics in Microbial Genomics	Using computational tools to interpret microbial genome data	Gene discovery, pathway mapping, drug development
Pathogen Surveillance Using Genomics	Real-time tracking of pathogen mutations and spread	Epidemic control, vaccine development
Metabolic Engineering of Microbes	Modifying microbial metabolism for targeted compound production	Biomanufacturing, synthetic food production
Microbial Communication (Quorum Sensing)	Studying how bacteria coordinate behaviour	Anti-virulence therapies, biofilm control
Industrial Microbiology Innovation	Enhancing microbial efficiency for production (enzymes, biofuels, etc)	Cost reduction, sustainability, green chemistry
Microbiota Transplantation	Transferring gut microbiota to restore microbial balance	Treating gut disorders, mental health implications
Antimicrobial Peptide (AMP) Research	Discovering natural microbial defence peptides	New antibiotics, immune enhancers
Space Microbiology	Examining microbial behaviour in microgravity and space conditions	Planetary protection, human health in space
Microbial Clustered Regularly Interspaced Short Palindromic Repeats (CRISPR) Systems	Exploring native CRISPR-Cas systems in prokaryotes	Novel gene-editing tools, antiviral defences
Synthetic Microbial Consortia	Engineering communities of microbes to perform complex tasks	Waste treatment, environmental remediation, agriculture

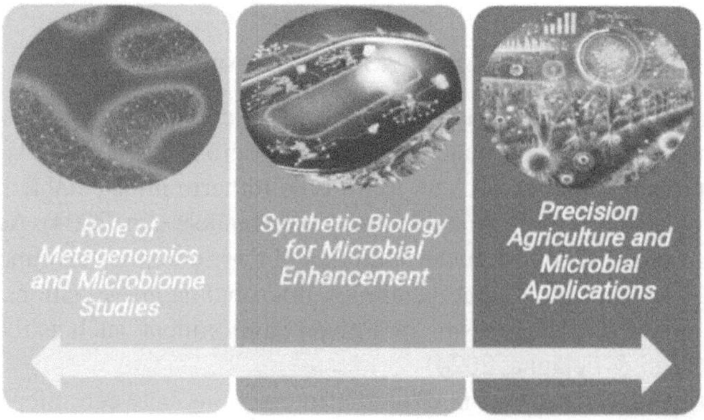

FIGURE 10.1 Some current advances in microbial research.

activity (Lloyd-Price et al., 2016). In agriculture, disease resistance and nutrient cycling are now understood to depend on the plant microbiome. Expanding microbiome research has been much aided by developments in next-generation sequencing (NGS). NGS technology have made high-throughput sequencing of microbial communities possible by reducing prices and increasing accuracy, hence enabling parallel analysis of hundreds of samples (Caporaso et al., 2011). This has spurred extensive research including the Human Microbiome Project, which mapped microbial diversity over several body regions.

10.2.2 Synthetic Biology for Microbial Enhancement

Within microbial study, synthetic biology has become a transforming field. Scientists may design bacteria to carry out particular functions, ranging from breaking down toxins to manufacturing drugs, by reprogramming their DNA. These modified strains are now being developed into synthetic routes for drug production, biosensors, and microbial factories producing sustainable biofuels (Cheng & Lu, 2012). Synthetic biology has been used in agriculture to improve microbial interactions supporting plant development. Microbes have been genetically engineered, for instance, to increase nitrogen fixation, therefore lowering reliance on artificial fertilizers. Others are designed to generate antimicrobial agents or plant hormones, therefore encouraging better crop growth (Oldroyd & Dixon, 2014).

Originating in microbial genomes, CRISPR-Cas systems have transformed genome editing. These instruments allow exact, programmable modifications to DNA, therefore allowing researchers to alter microbial genomes until unheard-of accuracy. In basic research as well as in applied disciplines as probiotics engineering and microbial therapies (Barrangou & Doudna, 2016), this has been very vital. Another horizon is microbial consortia engineering in which communities rather than single strains are designed. These days, scientists are building synthetic microbial ecosystems modelled by natural ones but tailored for stability and output. In bioremediation, wastewater treatment, and even space missions—where they can sustain closed-loop life support systems—such as consortia find application (Zhou et al., 2013).

10.2.3 Precision Agriculture and Microbial Applications

Within the field of precision farming, microbial research is transforming conventional methods of operation. More environmentally benign and sustainable biofertilizers and biopesticides are being developed from beneficial bacteria. Precision microbiome management seeks to improve yield by delivering the correct microorganisms to the correct crops under the appropriate conditions, hence minimising environmental effect (Singh et al., 2020). Using plant growth-promoting rhizobacteria (PGPR) is among the most exciting uses in this field. By either phytohormone production, solubilising nutrients, or pathogen suppression, these bacteria colonise plant roots and induce development. For seed coating and soil amendment, they are progressively finding place in commercial formulations (Bhattacharyya & Jha, 2012).

Furthermore, resulting from the junction of microbial studies and data science are prediction algorithms able to foresee microbial interactions and community dynamics. By use of machine learning algorithms, scientists can now forecast how variations in environmental circumstances or agricultural practices will influence microbial populations, therefore facilitating more focused interventions (Jansson & Hofmockel, 2020). Microbes are being used in environmental biotechnology to help with problems including pollution and climate change (Ogidi & Akpan, 2024; Ogidi & Richard-Nwachukwu, 2024). While some are developed to break down plastic trash or detoxify heavy metals from polluted soils, others are investigated for their capacity to oxidise methane, a strong greenhouse gas (Kumar et al., 2011). Another creative use is microbial fuel cells, which use microbial metabolism to turn organic waste into electricity. Low-energy devices and energy-positive treatment of wastewater could be powered by this technique. Such advances show the general value of microorganisms in solutions for clean energy (Logan et al., 2006).

Recent advances have also changed medical microbiology. The creation of microbiome-based diagnostics and treatments—such as faecal microbiota transplantation (FMT)—is changing how we treat diseases including *Clostridium difficile* infection, inflammatory bowel disease, and even mental health disorders (Smillie et al., 2018). Microbes are being used in fermentation techniques in industrial biotechnology to generate materials, enzymes, and high-value compounds. Now synthesising everything from insulin to biodegradable plastics, engineered yeast and bacteria help to lessen our dependence on petroleum-based products (Nielsen & Keasling, 2016).

Emerging field driven by international space exploration campaigns is space microbiology. Microbial survival in hostile conditions shapes not only space flight but also the hunt for life on distant worlds. These research also shed light on extremophile bacteria with biotechnology application (Horneck et al., 2010). Especially in microbial research, worldwide cooperation is hastening advancement. Projects aiming at cataloguing the microbial diversity of the earth include the Earth Microbiome Project and the Global Virome Project. These open-access materials enable collective scientific progress and reaction plans to newly developing diseases (Gilbert et al., 2014).

Though innovation is happening quickly, problems still exist. Still limiting the broad use of microbial technology are ethical questions surrounding genetically modified organisms, flaws in regulatory systems, and data standardising challenges. Policy and public participation help to address these issues; microbial research is experiencing a rebirth driven by modern technologies and multidisciplinary approaches. From metagenomics to precision agriculture, the uses are many and transforming. Constant research, infrastructure, and education will help to guarantee that microbiological science satisfies the urgent problems of the twenty-first century.

10.3 TECHNOLOGICAL INNOVATIONS IN MICROBIAL SOLUTIONS

Microbial solution technological advancements are transforming environmental sustainability, healthcare, and agriculture. The creation of metagenomics, which lets scientists examine genetic material from whole microbial communities without the necessity of cultivation, has been one of the most transforming developments. This method has revealed a large "microbial dark matter," shedding light on microbial interactions and purposes in intricate environments such as human gut or soil microbiomes (Rinke et al., 2013). Now driving microbiome-based medicines, precision agriculture, and bioremediation projects are metagenomic technologies. Another revolutionary idea that has surfaced is synthetic biology. Scientists can build or reprogramme microbial organisms to execute certain activities by means of genetic engineering. For example, designed bacteria may now generate plant growth hormones or fix atmospheric nitrogen more effectively, hence lessening dependency on chemical fertilizers (Oldroyd & Dixon, 2014). Synthetic bacteria are being produced in medicine to react to illness biomarkers inside the body and either operate as living diagnostics or administer focused medicinal treatments (Cheng & Lu, 2012). These features highlight the versatility of programmable bacteria in many different fields.

Precision of microbial innovation has been enhanced by CRISpen-Cas gene-editing technology. Originally developed from microbial immune systems, CRISPR technologies enable researchers to edit microbial genomes with hitherto unheard-of accuracy, hence accelerating the generation of novel strains for industrial and medical uses (Barrangou & Doudna, 2016). This has created fresh directions in vaccine manufacture, microbial fuel cells, and probiotic engineering. Personalised medicine and adaptable agriculture systems could benefit from the ability to make exacting genetic alterations in microorganisms.

Precision microbiome engineering is motivating sustainable approaches in agriculture. Targeted microbial consortia are helping to develop technological tools including microbial seed coverings, biofertilizers, and biopesticides. By restoring native microbial diversity, these developments improve soil health and lower the requirement for manufactured agrochemicals (Singh et al., 2020). By means of machine learning and sensor data, farmers may now precisely monitor microbial soil condition and implement microbial treatments, hence revolutionising conventional farming into a data-driven activity. Another field gaining from microbial creativity is environmental biotechnology. Oil spills are cleaned using engineered bacteria, plastic trash is broken down, and carbon dioxide from the atmosphere is sequestered (Kumar et al., 2011). Furthermore, providing low-cost energy solutions and trash treatment concurrently, microbial fuel cells turn organic waste into power using microorganisms (Logan et al., 2006). These discoveries not only show the adaptability of bacteria but also highlight their essential part in solving environmental and sustainability problems. Table 10.2 shows the technological innovations in microbial solutions.

TABLE 10.2
Technological Innovations in Microbial Solutions

Innovation	Description	Applications
CRISPR Gene Editing	Enables precise genetic modification in microbes	Biofuel production, enhanced probiotics
Synthetic Biology Platforms	Designs and constructs new microbial pathways	Custom enzyme production, biosynthesis of drugs
Metagenomics	Analyses genetic material from environmental samples	Soil health, microbial diversity profiling
Microbial Fuel Cells (MFCs)	Generates electricity using microbial metabolism	Renewable energy from wastewater
Bioreactor Engineering	Optimises environments for microbial growth and productivity	Large-scale fermentation, pharmaceuticals
AI-Powered Microbial Analytics	Uses AI to predict microbial interactions and behaviours	Drug discovery, food safety, industrial biotech
Next-Generation Sequencing (NGS)	Rapidly decodes microbial genomes	Pathogen tracking, microbiome research
Biosensors with Microbial Elements	Detects contaminants using engineered microbes	Water quality monitoring, environmental biosafety
Microbial Encapsulation	Protects microbes with nano or micro carriers for targeted delivery	Agriculture, probiotics, bioremediation
Biocontrol Agents	Utilises beneficial microbes to combat pests or pathogens naturally	Sustainable agriculture, integrated pest control
Microbial Desalination Cells (MDCs)	Uses microbial activity to treat saline water and generate electricity	Water purification, energy recovery
Microbiome Modulation Tools	Modifies microbial communities to promote health or productivity	Personalised medicine, animal nutrition
Phage Therapy Innovations	Employs bacteriophages to target harmful bacteria	Antibiotic alternatives, infection control
Omics Integration (Genomics, Proteomics, etc.)	Combines multi-omics data to map microbial functions	Precision microbiology, functional profiling
Microbial Nanotechnology	Uses microbes in the synthesis or delivery of nanoscale materials	Drug delivery, environmental sensing, biosynthesis
Microbial Carbon Capture	Harnesses microbes to absorb and store carbon from the atmosphere	Climate change mitigation, green manufacturing
Quorum Sensing Interference (QSI)	Disrupts microbial communication to prevent harmful behaviour	Infection control, anti-biofilm strategies
Fermentation Optimisation with Machine Learning (ML)	Uses ML algorithms to fine-tune microbial fermentation processes	Industrial-scale production of bio-based products
Waste-to-Resource Microbial Technologies	Converts waste products into valuable resources via microbial pathways	Bioplastics, biogas, fertilizers

10.3.1 AI and Big Data in Microbial Research for Microbial Solutions for Food Security

Big data analytics and artificial intelligence (AI) are fast changing the field of food security solutions by including into microbiological research. Concerning the complexity of microbial communities in food systems, soil, and plants, conventional analytical techniques usually struggle to interpret the great range of interactions. Particularly machine learning and deep learning algorithms, AI technologies are now allowing researchers to analyse vast microbiological datasets and expose hitherto unreachable trends and predictive models (Libbrecht & Noble, 2015). In microbial research, big data mostly comes from metagenomics, transcriptomics, and proteomics, and metabolomics—high-throughput methods. From food products to agricultural areas to environmental samples, these platforms create terabytes of raw data. Particularly in improving crop production and resilience, AI technologies are indispensable for managing, integrating, and analysing this data to find beneficial microbial strains and grasp their ecological activities (Zhao et al., 2020).

Within the framework of food security, AI-driven microbial analysis maps microbial populations crucial for nutrient cycling and plant health, therefore optimising soil condition. Customised biofertilizers that improve crop output without destroying natural ecosystems can be developed by means of machine learning models predicting how microbial assemblages affect soil fertility (Trivedi et al., 2020). By customising microbial treatments to particular crops and environments, this precision agricultural method increases sustainably boosting productivity. By means of microbiome analysis, AI technologies also help to detect plant diseases. Differentiating between pathogenic and benign bacteria allows AI systems to create biocontrol plans that reduce crop losses. In low-income areas where availability to synthetic agrochemicals is constrained and sustainable treatments are required for food security, this is especially important (Meena et al., 2020).

Big data and AI together have sped the identification of new microbial strains with agricultural promise. Applied to genomic and metabolomic information, deep learning algorithms can predict gene functions and metabolic pathways, so helping to identify bacteria able of nitrogen fixation, phosphorous solubilisation, or drought tolerance (Arango-Argoty et al., 2018). These functional bacteria can subsequently be used to boost plant development under environmental strain. Moreover, big data helps to forecast microbial interactions inside the rhizosphere. By simulating several scenarios involving microbial inoculants, AI helps researchers to predict how introduced strains would react in different soil kinds and environmental situations. This speeds the introduction of microbial-based products and lowers the possibility of failure in field tests (Jansson & Hofmockel, 2020).

Additionally, enabling real-time monitoring of microbial treatments in agricultural environments is AI. AI systems can monitor the performance of microbial inoculants on crops by combining sensor data with reports of microbial activity. Timely changes in farming methods made possible by this feedback loop guarantee that microbiological solutions constantly support food security goals (de Souza et al., 2020). AI is helping post-harvest food systems identify microbial contamination and rotting. A critical issue in world food security, predictive analytics based on microbiological fingerprints can estimate shelf life and spoiling risk, so enabling better inventory control and decrease in food waste (Drago et al., 2020). Furthermore, fast identification of pathogens made possible by microbial biosensors improved with AI interfaces guarantees food safety in supply networks (Ogidi et al., 2025b).

Furthermore, promoting democratisation of information is the inclusion of AI into microbiological study. AI-driven technologies and open-access platforms let researchers in low-resource environments examine challenging microbiological datasets without large computational capacity. This supports world cooperation and fair development of microbiological solutions for food security (Gilbert et al., 2014). Advancing microbiological solutions for food security depends critically on big data and AI. These technologies promote precision farming and food safety, fast identification and deployment of helpful bacteria, and improve our knowledge of microbial ecosystems. To fully realise the potential of microbial breakthroughs in guaranteeing a sustainable food future, ongoing investment in AI infrastructure and multidisciplinary cooperation will be crucial (Ogidi & Izah, 2024).

10.3.2 Smart Bioformulations and Microbial Delivery Systems for Food Security

Using cutting-edge microbial delivery technologies and smart bioformulations has become more and more important in tackling issues of world food security (Ogidi & Izah, 2024). These developments include creating exact, stable, and effective formulations of helpful bacteria that increase crop output, withstand environmental stress, and lower reliance on chemical inputs. These methods assist guarantee constant performance in many agricultural environments by delivering microbial inoculants successfully to the plant rhizosphere or phyllosphere (John et al., 2011). Usually involving mixtures of microbial strains—often PGPR, fungus-like mycorrhizae, or nitrogen-fixing bacteria—embedded in carriers that preserve the microbes and enable their release, smart bioformulations. These carriers—which might be polymers, nanoparticles, or organic compounds like alginate or peat—are chosen depending on their fit with bacteria and environmental conditions (Malusá et al., 2012). Such formulations seek to guarantee colonisation of the plant roots upon application, improve shelf life, and preserve viability during storage.

Modern microbiological delivery methods depend much on controlled-release technologies. Microencapsulation, nanoencapsulation, and emulsions are employed to shield microbial cells from abiotic stress like UV radiation, desiccation, and temperature swings (Vassilev et al., 2020). These targeted locations and gradual release of microorganisms by these encapsulated systems increase colonising efficiency and lower microbial loss. Microbial distribution has been improved even more by the combination of smart irrigation systems and precision agriculture equipment including drone-based spraying. These devices enable real-time bioformulation administration at ideal doses and exact timings, therefore matching microbial activity with stages of plant development or environmental cues. This reduces resource waste and raises the effectiveness of bioformulations (Bhattacharyya & Jha, 2012).

Furthermore, helping with sustainable pest and disease control are smart bioformulations. Applied to target pests, biopesticides developed using entomopathogenic bacteria or fungus cannot damage beneficial insects or soil condition. Unlike conventional pesticides, these biological control agents develop with pests and adapt to changing environmental conditions, therefore providing longer-term resilience (Glare et al., 2012). Multi-functional microbial consortia in which different bacteria are co-formulated to offer several advantages—such as nitrogen fixation, phosphorous solubilisation, and stress tolerance—are a new trend under development. These consortia cooperate cooperatively to enhance plant health and soil fertility and replicate natural microbial communities (Nautiyal et al., 2016). Modern delivery technologies guarantee that every consortium member stays functional and reaches their ecological habitat.

Another delivery method called biopriming treats seeds using microbial bioformulations prior to planting. Better crop establishment and yield follow from early root colonisation enhanced by this approach and seedling vigour promoted by which. Moreover, more resistant to abiotic stresses like salt or drought as well as to infections are bioprimed seedlings (Mahmood et al., 2016). This method fits very nicely in farming systems with limited resources where early crop performance is vital. Researchers are creating ready-to-use microbial compositions with little farmer involvement in order to solve scalability. These comprise dry spore-based powders that can be kept for long without refrigeration and liquid formulations including stabilising agents. Smallholder farmers in distant or resource-constrained environments especially depend on these developments (Bashan et al., 2014).

To guarantee the safety and efficacy of microbial bioformulations, regulatory systems and quality control procedures are under development. Offering incentives and expediting product registration procedures, governments and international bodies are helping bio-based agricultural inputs to grow. To guarantee consistent findings, quality criteria for microbiological count, formulation stability, and field efficacy are getting stricter concurrently (Glick, 2012). Smart bioformulations and microbial delivery systems are redefining food security and sustainable agriculture going forward. Improved precision, stability, and scalability of microbial applications let these technologies enable effective resource use and environmental preservation. Global mainstreaming of these solutions depends mostly on ongoing innovation and cooperation among scholars, business leaders, and legislators.

10.3.3 Next-Generation Biofertilizers and Biopesticides for Food Security

Emerging as main elements in sustainable agriculture systems meant to improve world food security are next-generation biofertilizers and biopesticides. Unlike their conventional equivalents, these cutting-edge products offer more focused, resilient, and environmentally friendly solutions by including developments in microbial science, formulation technologies, and delivery systems. Their creation directly responds to the increased demand for low-input, high-yield farming methods that cut reliance on synthetic agrochemicals (Mahanty et al., 2017). Using specialised strains of plant growth-promoting microorganisms (PGPMs), including nitrogen-fixing bacteria (*Rhizobium* and *Azotobacter*), phosphate-solubilising bacteria (*Pseudomonas* and *Bacillus*), and potassium-mobilising bacteria, next-generation biofertilizers. These bacteria increase the availability and absorption of nutrients in plants, therefore raising the crop output and quality. Recent developments include selectively developed or genetically altered microbial strains displaying more colonising efficiency, stress resistance, and multifarious properties (Vessey, 2003).

Apart from nutrient facilitation, these biofertilizers are under development using better carrier materials and techniques. For instance, encapsulation technologies permit regulated release in the rhizosphere and preserve microbial viability throughout storage. This guarantees constant performance under different surroundings. Long shelf life and simplicity of application make liquid biofertilizers enhanced with stabilisers increasingly appealing (Bhardwaj et al., 2014). Conversely, next-generation biopesticides target particular pests or diseases by means of microbial antagonists including *Trichoderma*, *Beauveria bassiana*, and *Bacillus thuringiensis*. Molecular characterisation and choice of high-potency strains with broad-spectrum or customised action are helping these products to be improved. By means of omic technologies and genetic markers, virulence determinants are identified, therefore enhancing strain selection for commercial application (Copping & Menn, 2000).

Enhancement of biopesticide efficacy depends critically on formulation science. Among the innovations are wettable powders, microencapsulations, and nanoemulsions that boost field product efficacy and stability. These formulations slow down degradation under UV or rain exposure and provide greater adhesiveness to plant surfaces. In tropical climes, where environmental unpredictability has previously hindered the application of biopesticides, such advances are especially advantageous (Chandler et al., 2011). Combining biopesticides with biofertilizers into integrated bioproducts is a trend with great significance. These multifarious inputs both improve plant development and provide defence against diseases. One formulation would call for nitrogen-fixing bacteria, for instance, together with chitinolytic fungus to control root diseases. For farmers, this whole strategy lowers input costs and streamlines application procedures (Glick, 2012).

Using these bioproducts fits rather nicely with precision farming. Nowadays, drone technology and Global Positioning System (GPS)-guided systems let biofertilizers and biopesticides be spatially targeted, hence lowering waste and increasing field-level efficacy. Furthermore, AI-based models can forecast nutrient shortages and disease outbreaks, therefore allowing the prompt use of microbial remedies to lower risk (Singh et al., 2020). From an environmental standpoint, next-generation bioproducts greatly lessen the impact of farming. Biofertilizers and biopesticides boost soil health and increase microbial variety, while chemical fertilizers and pesticides can lead to soil degradation, water pollution, and biodiversity loss. This conforms with ideas of climate-smart farming and regenerative agriculture (Altieri et al., 2015).

Adoption is still difficult, though. Product availability, variable performance under field circumstances, and lack of knowledge or training all provide challenges for farmers. Strong public-private collaborations, government incentives, and the inclusion of microbial inputs into agricultural extension programmes can help to break through these obstacles. Additionally changing to simplify the licensing and commercialisation of microbial bioproducts is regulatory systems (Malusá & Vassilev, 2014). In agricultural input technology, next-generation biofertilizers and biopesticides mark a radical change. They provide a sustainable route to improve food output while maintaining the integrity of ecosystems. Maintaining these microbial weapons for food security will depend critically on ongoing research, innovation, and favourable policy contexts as global stresses on agricultural systems becoming more intense.

10.4 CHALLENGES IN RESEARCH AND COMMERCIALISATION

By raising crop yields, improving soil condition, and lowering reliance on chemical inputs, microbial technologies show great potential to solve food security. Large-scale deployment of these technologies is hampered, nevertheless, by many difficulties related to research and commercialisation. From scientific and technical constraints to economic, legal, and societal impediments, these obstacles vary in nature (Malusá & Vassilev, 2014). Microbial ecosystem complexity presents one of the main scientific difficulties. Although the dynamic and context-specific character of microbial behaviour makes understanding the interplay between introduced bacteria and native soil communities difficult even if it is absolutely necessary. Even the most promising microbial strains could fail to grow or behave inconsistently across many environmental conditions (Finkel et al., 2017).

The low prediction of microbial efficacy under field circumstances presents still another important obstacle. Although many bacterial inoculants show strong effects in controlled laboratory or greenhouse environments, these findings can have little bearing on practical use. Variables such as soil composition, climate, plant genotype, and competition from indigenous bacteria greatly affect the outcome, hence making it challenging to generalise product performance (Bashan et al., 2014). Moreover, compromising the scalability of microbial products is the lack of strong formulation methods. Technically challenging is ensuring microbial viability throughout field application, transportation, and storage. Many formulas lose effectiveness or break down rapidly in response to temperature, humidity, or UV radiation. Emerging but still expensive and technologically demanding are innovations in encapsulation and controlled-release systems (John et al., 2011).

From a commercialisation standpoint, among the most important obstacles are legal ones. Many nations have complex, inconsistent, or poorly defined approval systems for microbial goods. Many times treating microbial inoculants as chemical pesticides or fertilizers, regulatory systems cause bottlenecks in registration and delays in product introduction (Chandler et al., 2011). The division of world norms and policies adds still another obstacle. While some areas—like the European Union—have sophisticated bio-based rules, others lack clear policies for microbial technologies. This legal difference inhibits the market spread of microbial products and hinders international trade, therefore deterring private company investment (Malusá & Vassilev, 2014).

Moreover, impeding microbial technology commercialisation are economic obstacles, particularly for small and medium-sized businesses (SMEs). Particularly in underdeveloped nations, the expenses related to R&D, field tests, registration, and distribution can be intolerable. Startups and innovative companies motivated from these financial hazards are discouraged from joining the market (Glick, 2012). Moreover, farmers' ignorance and lack of training still present a continuous obstacle. Many farmers, especially in low-income areas, are not familiar with microbial technologies or dubious of their effectiveness when compared to traditional agrochemicals. Effective products are available, but adoption rates remain low without strong extension services and demonstration trials (Mahanty et al., 2017).

Furthermore, restricting their spread are weak supply chains and distribution networks for microbial inputs. Unlike synthetic fertilizers and insecticides, microbial products need cautious handling and cold-chain storage—often lacking in rural regions. This logistical restriction influences consistency, availability of products, and finally, farmer confidence (Bhardwaj et al., 2014). Protection of IP raises still another issue. Once isolated and identified, microbials are challenging to patent because of their natural source. This makes it difficult for businesses to get competitive advantages, which discouragement of investment in strain research and development follows. Furthermore, open-access research can result in bio-piracy—that is, uncontrolled usage of proprietary strains—or other forms of exploitation (Chaudhary et al., 2018).

The course of microbiological technologies also depends on public opinion and society approval. Although often less controversial than genetically modified organisms, questions about safety, environmental impact, and long-term effects of microbial use remain existent. Building consumer and farmer confidence depends mostly on open communication, labelling, and regulatory control (Glare et al., 2012). Data integration presents difficulties for researchers trying to link ecological results with microbial activities. The great complexity and variety in field settings mean that, even with developments in metagenomics and bioinformatics, connecting particular microbial taxa or genes to agricultural advantages remains challenging (Jansson & Hofmockel, 2020).

The little cooperation among disciplines is another underdeveloped issue. Good development of microbial technologies calls for the cooperation of microbiologists, agronomists, engineers, economists, and legislators. But compartmentalised research and inadequate multidisciplinary funding restrict the speed and applicability of invention (Singh et al., 2020). Moreover, still exist public and private investment disparities. Although microbial research has expanded, translational research and commercialisation paths still get little support. To promote innovation pipelines from lab to field, bridging this gap calls for coordinated actions across governments, venture capitalists, and charitable organisations (Pretty et al., 2011). Although microbial technologies have good answers for food security, realising their full potential will need overcoming many obstacles. Essential are addressing scientific uncertainty, changing regulatory systems, improving farmer education, and boosting investment systems. Coordinated, multi-stakeholder efforts can propel microbial breakthroughs towards scalable, sustainable outcomes in agriculture forward.

10.5 FUTURE PERSPECTIVES AND RESEARCH DIRECTIONS

Future of sustainable agriculture and food security is likely shaped in great part by microbial technology. The creation of creative, microbe-based solutions provides a viable road forward as world populations increase and environmental stresses intensify. Future viewpoints in this subject centre not only on improving current technologies but also on investigating unrealised microbiological potential via multidisciplinary research and next-generation instruments (Finkel et al., 2017). Investigating new microbial species from harsh or understudied settings is one exciting path. Unique metabolic capacity of microorganisms from polar environments, deep-sea vents, and deserts allows one to use them for bioactive chemical synthesis, stress tolerance, and nutrition cycling. Advances in metagenomics and single-cell sequencing going forward will help to identify these "microbial dark matter" populations (Rinke et al., 2013).

Future studies will probably focus more on the functional characterising of microbial populations than on their mere variety. Developing successful microbial consortia requires an awareness of how molecules interact among plants, soil, and one another. Dealing with microbial functionality will depend critically on methods including transcriptomics, proteomics, and metabolomics (Trivedi et al., 2020). In agriculture, synthetic microbial communities should become really important. Unlike natural communities, which are sometimes erratic, synthetic consortia can be built for certain purposes as nitrogen fixation, disease control, or drought resilience. These created ecosystems provide a fresh degree of microbiome management precision since they can be tuned for specific crops and environmental situations. Computational modelling and AI-driven methods to forecast microbial interactions and maximise inoculant formulations will also grow in importance going forward. Large multi-omics datasets, anticipate microbial behaviour in different conditions, and custom microbial solutions for particular farming systems can all be found by machine learning algorithms (Zhao et al., 2020). Tools for genome editing and CRISPR will keep transforming the formation of microbial strains. Without changing plant genomes, these approaches let researchers improve microbial properties including stress tolerance, colonising efficiency, and metabolite synthesis. In areas where genetically altered crops run against social or legal opposition, this approach is especially crucial (Barrangou & Doudna, 2016).

Growing attention on creating climate-resilient microbial technology will help when climate change gets more severe. Microbes able to support crops under heat, drought, salinity, and flood stress will be the main emphasis of future study. By means of their identification and deployment, these microorganisms can assist to protect yields in sensitive areas and lower dependency on environmentally harmful activities (Singh et al., 2020). Regarding biopesticides, next-generation formulations with enhanced stability, shelf life, and specificity will be very important. Integration of nanotechnology and bioencapsulation will help to safeguard microbiological agents and improve their distribution (Ogidi et al., 2025a). These developments will increase the application of present biological control products and aid to overcome their constraints (Vassilev et al., 2020). One important area of study will be microbial biofortification—that is, utilising bacteria to improve crop nutritional value. Microbes able of generating or enhancing the absorption of micronutrients such as zinc, iron, and vitamin B12 help solve latent hunger in populations depending on staple crops. This fits health targets and worldwide nutrition policies (Goswami et al., 2016). Another crucial viewpoint is the interaction of regenerative agriculture and microbial technology. Using little intervention, this strategy stresses soil repair, carbon sequestration, and biodiversity (Okosodo & Ogidi, 2023). Microbial additions will be planned to restore soil health by working in concert with cover cropping, composting, and conservation tillage (Altieri et al., 2015).

Real-time monitoring and diagnostics will become standard in microbiological uses going forward. Field-based sequencing instruments and biosensors will let farmers quickly evaluate soil condition and microbial populations. These diagnostics will direct the site-specific, timely application of microbial products, hence enhancing efficiency and results (Drago et al., 2020). Phage treatment for plants—the use of bacteriophages to target certain pathogens—also attracts increasing interest. Unlike conventional pesticides, this precision biocontrol technique does not disturb helpful bacteria. Phage mixtures, delivery systems, and resistance management will all be investigated in future work (Buttimer et al., 2017). Regarding policies, the future calls for more solid regulatory systems acknowledging the special character of microbial technology. To encourage creativity and guarantee safe adoption of microbiological inputs, harmonised worldwide standards, accelerated approval paths, and evidence-based risk assessments are needed (Chandler et al., 2011). Still another area of study is the commercial scalability of microbial technologies. Low-cost, shelf-stable goods aimed at distribution in remote locations will be developed with effort. Scaling up these discoveries in the Global South will depend critically on public-private collaborations and international development projects (Pretty et al., 2011). Furthermore, influencing future paths will be capacity building and education. Critical is arming agripreneurs, extension agents, and farmers with knowledge and skills to apply microbial products successfully. Programmes for training, open-access sites, and mobile apps will assist to close the knowledge gap (Mahanty et al., 2017).

For further studies, ethical and environmental issues will take the stage. Scientists will have to assess how horizontal gene transfer, evolutionary processes, and microbial introductions affect native ecosystems over long terms. Protecting biodiversity will depend mostly on developing responsible innovation models (Glare et al., 2012; Ogidi & Akpan, 2022). Future studies will also investigate the function of endophytes—microbes living inside plant tissues—which provide special advantages in fostering resistance and growth. Since endophyte-based technologies are more closely related with plant metabolism than rhizosphere-based inoculants, they could outperform them in stability and efficacy (Santoyo et al., 2016).

Microbial technology meeting food supply chains will create fresh opportunities. Key elements of food security: food safety will be improved by cold-chain preservation, packaging materials including embedded probiotics, and microbial coatings to lower spoiling (Drago et al., 2020). Eventually, scaling advances will benefit from the democratisation of microbial science made possible by citizen science and participatory research. Open-source platforms, community-based trials, and inclusive innovation models will let several stakeholders participate to and gain from microbial discoveries (Gilbert et al., 2014). There are many chances for microbial technology to shape food security going forward. From cutting-edge genetic techniques to ecological stewardship and clever delivery systems, the path of this discipline will be shaped by the interaction of science, policy, and practice. Resilient food systems for the future depend on investments in inclusive, data-driven, ecologically aware research.

10.6 CONCLUSION

Research and invention in microbial technologies offer a ray of hope for sustainable, resilient agricultural systems in the framework of growing global food concerns. The use of microbial solutions is progressively seen as both a scientific need and a socio-economic possibility as natural resources get more rare and agricultural needs grow. By raising crop yields, boosting soil health, and lowering dependency on environmentally damaging inputs, these technologies could completely transform food output. Advances in science including metagenomics, synthetic biology, and bioinformatics have deepened our knowledge of microbial ecosystems and uses. Today, researchers have methods to separate beneficial strains, decipher difficult plant-microbe interactions, and create custom solutions for particular agroecological zones. Designing next-generation biofertilizers, biopesticides, and bioformulations that are more efficient, targeted, and flexible enough for changing environmental conditions has depended much on this understanding. Innovation goes beyond lab discoveries. Translational research and multidisciplinary collaboration has produced field-based solutions like precise delivery systems, seed biopriming, and microbial consortia engineering. These developments guarantee constant performance of microbial products under actual farming situations, therefore bridging the gap between theoretical possibility and practical use. Scaling impact and reaching world food security objectives depend on ongoing funding for such translational initiatives.

Realising the full possibilities of microbial technologies, though, calls on a strong ecosystem of policy support, funding, and education. Regulatory changes have to acknowledge the special character of microbial goods and enable quick approval and worldwide harmony. Likewise, crucial are farmer knowledge, training, and access to consistent microbial inputs. Achieving fair food systems depends on making sure these advances reach smallholder farmers—especially in the Global South. Multidisciplinary cooperation will be pillar of microbial innovation as we look ahead. Already changing our approach to agriculture are solutions combining biology, engineering, computer science, and agronomy. From genome-edited microbial strains to AI-driven predictive modelling, these partnerships are bringing in a new era of intelligent, biologically driven agriculture that matches output with environmental care. Furthermore, influencing future study paths will be ethical and environmental factors. Responsible innovation systems have to make sure that local farming methods are not unintentionally replaced or harmed by microbial technology. Building trust and encouraging long-term acceptance will depend on juggling scientific advancement with ecological purity and cultural sensitivity.

The dual pillars of research and invention will define microbial technologies for food security going forward. Microbial solutions can go from niche interventions to mainstream agricultural methods with ongoing scientific research, technical development, and cross-sector cooperation. They provide a road not just for sustainable world feeding but also for restoring natural balance and resistance against resource constraint and climate change. Accepting the full possibilities of microbial technologies is ultimately a dedication to rethink agriculture in accordance with nature rather than only a technological effort. To guarantee that none is left behind in the quest of world food security, the next decades will necessitate bold thinking, systematic reform, and constant investment in microbial science.

REFERENCES

Altieri, M. A., Nicholls, C. I., Henao, A., & Lana, M. A. (2015). Agroecology and the design of climate change-resilient farming systems. *Agronomy for Sustainable Development*, *35*(3), 869–890. https://doi.org/10.1007/s13593-015-0285-2

Arango-Argoty, G., Garner, E., Pruden, A., Heath, L. S., Vikesland, P., & Zhang, L. (2018). DeepARG: A deep learning approach for predicting antibiotic resistance genes from metagenomic data. *Microbiome*, *6*(1), 23. https://doi.org/10.1186/s40168-018-0401-z

Barrangou, R., & Doudna, J. A. (2016). Applications of CRISPR technologies in research and beyond. *Nature Biotechnology*, *34*(9), 933–941. https://doi.org/10.1038/nbt.3659

Barwant, M. M., Ogidi, O. I., Singh, B., Patil, V. S., & Sharma, S. A. B. (2025b). Climate change and pesticide dynamics. In B. R. Babaniyi & E. E. Babaniyi (Eds.), *The Interplay of Pesticides and Climate Change*. Springer. https://doi.org/10.1007/978-3-031-81669-7_7

Barwant, M. M., Ogidi, O. I., Yogita, C., & Munje, R. (2025c). Study of consumers choices on pesticides use and sustainability. In B. R. Babaniyi & E. E. Babaniyi (Eds.), *The Interplay of Pesticides and Climate Change*. Springer. https://doi.org/10.1007/978-3-031-81669-7_18

Barwant, M. M., Sharma, S. A. B., Lodh, D., Sil, D. C., & Ogidi, O. I. (2025e). Immedicable effects of organic pesticides on aquatic environment. In B. R. Babaniyi & E. E. Babaniyi (Eds.), *The Interplay of Pesticides and Climate Change*. Springer. https://doi.org/10.1007/978-3-031-81669-7_16

Barwant, M. M., Singh, S., Ogidi, O. I., Singh, S., & Naik, B. S. S. S. (2025d). Economic implications of climate change and pesticide use on agriculture. In B. R. Babaniyi & E. E. Babaniyi (Eds.), *The Interplay of Pesticides and Climate Change*. Springer. https://doi.org/10.1007/978-3-031-81669-7_19

Barwant, M. M., Singh, S., Singh, S., Naik, B. S. S. S., & Ogidi, O. I. (2025a). Resilience of ecosystems: Navigating the challenges of pesticide exposure and climate change. In B. R. Babaniyi & E. E. Babaniyi (Eds.), *The Interplay of Pesticides and Climate Change*. Springer. https://doi.org/10.1007/978-3-031-81669-7_3

Bashan, Y., de-Bashan, L. E., Prabhu, S. R., & Hernandez, J. P. (2014). Advances in plant growth-promoting bacterial inoculant technology: Formulations and practical perspectives (1998–2013). *Plant and Soil*, *378*(1), 1–33. https://doi.org/10.1007/s11104-013-1956-x

Bhardwaj, D., Ansari, M. W., Sahoo, R. K., & Tuteja, N. (2014). Biofertilizers function as key player in sustainable agriculture by improving soil fertility, plant tolerance and crop productivity. *Microbial Cell Factories*, *13*, 66. https://doi.org/10.1186/1475-2859-13-66

Bhattacharyya, P. N., & Jha, D. K. (2012). Plant growth-promoting rhizobacteria (PGPR): Emergence in agriculture. *World Journal of Microbiology and Biotechnology*, *28*, 1327–1350. https://doi.org/10.1007/s11274-011-0979-9

Buttimer, C., McAuliffe, O., Ross, R. P., Hill, C., O'Mahony, J., & Coffey, A. (2017). Bacteriophages and bacterial plant diseases. *Frontiers in Microbiology*, *8*, 34. https://doi.org/10.3389/fmicb.2017.00034

Caporaso, J. G., Lauber, C. L., Walters, W. A., Berg-Lyons, D., Lozupone, C. A., Turnbaugh, P. J., … & Knight, R. (2011). Global patterns of 16S rRNA diversity. *Nature*, *476*(7361), 457–463. https://doi.org/10.1038/nature10388

Chandler, D., Bailey, A. S., Tatchell, G. M., Davidson, G., Greaves, J., & Grant, W. P. (2011). The development, regulation and use of biopesticides for integrated pest management. *Philosophical Transactions of the Royal Society B: Biological Sciences*, *366*(1573), 1987–1998. https://doi.org/10.1098/rstb.2010.0390

Chaudhary, P., Yadav, D., & Nain, L. (2018). Challenges and opportunities in commercialization of biofertilizer products. In N. Meena et al. (Eds.), *Advances in PGPR Research* (pp. 243–259). Springer. https://doi.org/10.1007/978-981-10-6847-8_12

Cheng, A. A., & Lu, T. K. (2012). Synthetic biology: An emerging engineering discipline. *Annual Review of Biomedical Engineering*, *14*, 155–178. https://doi.org/10.1146/annurev-bioeng-071811-150118

Compant, S., Clément, C., & Sessitsch, A. (2010). Plant growth-promoting bacteria in the rhizo- and endosphere of plants: Their role, colonization, mechanisms involved and prospects for utilization. *Soil Biology and Biochemistry*, *42*(5), 669–678. https://doi.org/10.1016/j.soilbio.2009.11.024

Copping, L. G., & Menn, J. J. (2000). Biopesticides: A review of their action, applications and efficacy. *Pest Management Science*, *56*(8), 651–676. https://onlinelibrary.wiley.com/doi/full/10.1002/1526-4998(200008)56:8%3C651::AID-PS201%3E3.0.CO;2-U

Delmont, T. O., Quince, C., Shaiber, A., Esen, Ö. C., Lee, S. T. M., Rappé, M. S., … Eren, A. M. (2015). Nitrogen-fixing populations of Planctomycetes and Proteobacteria are abundant in surface ocean metagenomes. *Nature Microbiology*, *1*(1), 16041. https://doi.org/10.1038/s41564-018-0176-9

de Souza, R. S. C., Okura, V. K., Armanhi, J. S. L., Jorrín, B., Lozano, N., da Silva, M. J., & Arruda, P. (2020). Unlocking the bacterial and fungal communities assemblages of sugarcane microbiome. *Scientific Reports*, *6*, 28774. https://doi.org/10.1038/srep28774

Drago, E., Campardelli, R., Pettinato, M., & Perego, P. (2020). Innovations in Smart Packaging Concepts for Food: An Extensive Review. *Foods*, *9*(11), 1628. https://doi.org/10.3390/foods9111628

Finkel, O. M., Castrillo, G., Herrera Paredes, S., Salas González, I., & Dangl, J. L. (2017). Understanding and exploiting plant beneficial microbes. *Current Opinion in Plant Biology*, *38*, 155–163. https://doi.org/10.1016/j.pbi.2017.04.018

Gálvez, A., Abriouel, H., López, R. L., & Omar, N. B (2007). Bacteriocin-based strategies for food biopreservation. *International Journal of Food Microbiology*, *120*(1–2), 51–70. https://doi.org/10.1016/j.ijfoodmicro.2007.06.001

Gilbert, J. A., Jansson, J. K., & Knight, R. (2014). The Earth Microbiome Project: Successes and aspirations. *BMC Biology*, *12*, 69. https://doi.org/10.1186/s12915-014-0069-1

Glare, T. R., Caradus, J. R., Gelernter, W. D., Jackson, T. A., Keyhani, N. O., Köhl, J., ... & Stewart, A. (2012). Have biopesticides come of age? *Trends in Biotechnology*, *30*(5), 250–258. https://doi.org/10.1016/j.tibtech.2012.01.003

Glick, B. R. (2012). Plant growth-promoting bacteria: Mechanisms and applications. *Scientifica*, *2012*, 963401. https://doi.org/10.6064/2012/963401

Godfray, H. C. J., Beddington, J. R., Crute, I. R., Haddad, L., Lawrence, D., Muir, J. F., ... Toulmin, C. (2010). Food security: The challenge of feeding 9 billion people. *Science*, *327*(5967), 812–818. https://doi.org/10.1126/science.1185383

Goswami, D., Thakker, J. N., & Dhandhukia, P. C. (2016). Portraying mechanics of plant growth promoting rhizobacteria (PGPR): A review. *Cogent Food & Agriculture*, *2*(1), 1127500. https://doi.org/10.1080/23311932.2015.1127500

Handelsman, J. (2004). Metagenomics: Application of genomics to uncultured microorganisms. *Microbiology and Molecular Biology Reviews*, *68*(4), 669–685. https://doi.org/10.1128/MMBR.68.4.669-685.2004

Horneck, G., Klaus, D. M., & Mancinelli, R. L. (2010). Space microbiology. *Microbiology and Molecular Biology Reviews*, *74*(1), 121–156. https://doi.org/10.1128/MMBR.00016-09

Iyiola, A. O., Kolawole, A. S., Ajayi, F. O., Ogidi, O. I., & Ogwu, M. C. (2023). Chapter 22: Sustainable water use and management for agricultural transformation in Africa. In *Water Resources Management for Rural Development: Challenges and Mitigation* (Vol. 1, pp. 287–300). Elsevier. https://doi.org/10.1016/B978-0-443-18778-0.00001-5

Jansson, J. K., & Hofmockel, K. S. (2020). The soil microbiome—from metagenomics to metaphenomics. *Current Opinion in Microbiology*, *43*, 162–168. https://doi.org/10.1016/j.mib.2018.01.013

John, R. P., Tyagi, R. D., Brar, S. K., Surampalli, R. Y., & Prevost, D. (2011). Bio-encapsulation of microbial cells for targeted agricultural delivery. *Critical Reviews in Biotechnology*, *31*(3), 211–226.https://doi.org/10.3109/07388551.2010.513327

Kumar, M., Leon, V., Materano, A. D., & Ilzins, O. A. (2011). Enzymatic degradation of petroleum hydrocarbons. *Biotechnology Advances*, *29*(6), 944–960. https://doi.org/10.1016/j.biotechadv.2011.08.006

Libbrecht, M. W., & Noble, W. S. (2015). Machine learning applications in genetics and genomics. *Nature Reviews Genetics*, *16*(6), 321–332. https://doi.org/10.1038/nrg3920

Liu, Y., Wang, L., Liu, Y., Wang, Z., & Zhao, X. (2020). Phage-based biosensors for bacterial detection. *Trends in Biotechnology*, *38*(4), 337–339. https://doi.org/10.1016/j.tibtech.2019.11.006

Lloyd-Price, J., Abu-Ali, G., & Huttenhower, C. (2016). The healthy human microbiome. *Genome Medicine*, *8*(1), 51. https://doi.org/10.1186/s13073-016-0307-y

Logan, B. E., Hamelers, B., Rozendal, R., Schröder, U., Keller, J., Freguia, S., ... & Rabaey, K. (2006). Microbial fuel cells: Methodology and technology. *Environmental Science & Technology*, *40*(17), 5181–5192. https://doi.org/10.1021/es0605016

Mahanty, T., Bhattacharjee, S., Goswami, M., Bhattacharyya, P., Das, B., Ghosh, A., & Tribedi, P. (2017). Biofertilizers: Potential approach for sustainable agriculture development. *Environmental Science and Pollution Research*, *24*(4), 3315–3335. https://doi.org/10.1007/s11356-016-8104-0

Mahmood, A., Turgay, O. C., Farooq, M., & Hayat, R. (2016). Seed biopriming with plant growth promoting rhizobacteria: A review. *FEMS Microbiology Ecology*, *92*(8), fiw112. https://doi.org/10.1093/femsec/fiw112

Malusá, E., Sas-Paszt, L., & Czerny, M. (2012). The development of microbial inoculants for organic agriculture: A review. *Science for the Total Environment*, *439*, 360–370. https://doi.org/10.1016/j.scitotenv.2012.09.063

Malusá, E., & Vassilev, N. (2014). A contribution to set a legal framework for biofertilisers. *Applied Microbiology and Biotechnology*, *98*, 6599–6607. https://doi.org/10.1007/s00253-014-5828-y

Meena, K. K., Sorty, A. M., Bitla, U. M., Choudhary, K., Gupta, P., Pareek, A., & Singh, D. P. (2020). Abiotic stress responses and microbe-mediated mitigation in plants: The omics strategies. *Frontiers in Plant Science, 8*, 172. https://doi.org/10.3389/fpls.2017.00172

Nautiyal, C. S., Bhadauria, S., Bhavdish, N., & Dawar, S. (2016). Microbial consortia: Role in plant growth and sustainability. *Current Trends in Microbiology, 11*, 41–48.

Nielsen, J., & Keasling, J. D. (2016). Engineering cellular metabolism. *Cell, 164*(6), 1185–1197. https://doi.org/10.1016/j.cell.2016.02.004

Ogidi, O. I., & Akpan, U. M. (2022). Chapter 16: Aquatic biodiversity loss: Impacts of pollution and anthropogenic activities and strategies for conservation. In *Biodiversity in Africa: Potentials, Threats and Conservation*. Springer Nature Singapore Pte Ltd. https://doi.org/10.1007/978-981-19-3326-4_16

Ogidi, O. I., & Akpan, U. M. (2023). Impacts of chemical use in agricultural practices: Perspectives of soil microorganism and vegetation. In *One Health Implications of Agrochemicals and Their Sustainable Alternative*. Springer Nature Singapore Pte Ltd.

Ogidi, O. I., & Akpan, U. M. (2024). Microbial and phyto-remediation of crude oil contaminated soil. In *Eco-Restoration of the Environment*. CRC Press, Routledge Taylor and Francis Group.

Ogidi, O. I., & Felagha, I. (2024). Gut microbiome and diet interaction. In *Nutrition Controversies and Advances in Autoimmune Disease*. IGI Global.

Ogidi, O. I., & Izah, S. C. (2024). Emerging technologies in the Global South food industry: Prospects and challenges. In *Food Safety and Quality in Global South*. Springer Nature Singapore Pte Ltd. https://doi.org/10.1007/978-981-97-2428-4_7

Ogidi, O. I., Izah, S. C., & Akpan, U. M. (2023). Food quality and agrochemical use: Integrated monitoring, assessment, and management policies. In *One Health Implications of Agrochemicals and Their Sustainable Alternative*. Springer Nature Singapore Pte Ltd.

Ogidi, O. I., Izah, S. C., & Tobia, P. S. (2024). Biosurfactant production from cassava processing wastes. In *Sustainable Cassava: Strategies from Production through Waste Management*. Elsevier.

Ogidi, O. I., Okara, E., & Barwant, M. M. (2025a). Nanocarriers in plant genetic engineering. In J. Chen (Ed.), *Nanocarriers in Plant Science and Agriculture* (pp. 89–124). IGI Global Scientific Publishing. https://doi.org/10.4018/979-8-3693-9869-2.ch004

Ogidi, O. I., & Richard-Nwachukwu, N. (2024). Microbe assisted remediation of pesticide residues from soil and water. In *Eco-Restoration of the Environment*. CRC Press, Routledge Taylor and Francis Group.

Ogidi, O. I., Wenapere, C. M., & Chukwunonso, O. A. (2025b). Enhancing food safety and quality control with computer vision systems. In D. Bora & R. Bania (Eds.), *Computer Vision Techniques for Agricultural Advancements* (pp. 51–88). IGI Global Scientific Publishing. https://doi.org/10.4018/979-8-3693-8019-2.ch003

Okosodo, E. F., & Ogidi, O. I. (2023). Chapter 3: Biodiversity Conservation Strategies and Sustainability. In *Sustainable Utilization and Conservation of Africa's Biological Resources and Environment, Sustainable Development and Biodiversity*. Springer Nature Singapore Pte Ltd.

Oldroyd, G. E. D., & Dixon, R. (2014). Biotechnological solutions to the nitrogen problem. *Current Opinion in Biotechnology, 26*, 19–24. https://doi.org/10.1016/j.copbio.2013.08.006

Pretty, J., Toulmin, C., & Williams, S. (2011). Sustainable intensification in African agriculture. *International Journal of Agricultural Sustainability, 9*(1), 5–24. https://doi.org/10.3763/ijas.2010.0583

Rinke, C., Schwientek, P., Sczyrba, A., Ivanova, N. N., Anderson, I. J., Cheng, J. F., … & Woyke, T. (2013). Insights into the phylogeny and coding potential of microbial dark matter. *Nature, 499*(7459), 431–437. https://doi.org/10.1038/nature12352

Santoyo, G., Moreno-Hagelsieb, G., Orozco-Mosqueda, M. C., & Glick, B. R. (2016). Plant growth-promoting bacterial endophytes. *Microbiological Research, 183*, 92–99. https://doi.org/10.1016/j.micres.2015.11.008

Singh, B. K., Trivedi, P., Egidi, E., Macdonald, C. A., & Delgado-Baquerizo, M. (2020). Crop microbiome and sustainable agriculture. *Nature Reviews Microbiology, 18*(11), 601–602. https://doi.org/10.1038/s41579-020-00446-y

Smillie, C. S., Sauk, J., Gevers, D., Friedman, J., Sung, J., Youngster, I., … & Huttenhower, C. (2018). Strain tracking reveals the determinants of bacterial engraftment in the human gut following fecal microbiota transplantation. *Cell Host & Microbe, 23*(2), 229–240.e5. https://doi.org/10.1016/j.chom.2018.01.003

Tamang, J. P., Watanabe, K., & Holzapfel, W. H. (2016). Review: Diversity of microorganisms in global fermented foods and beverages. *Frontiers in Microbiology, 7*, 377. https://doi.org/10.3389/fmicb.2016.00377

Trivedi, P., Leach, J. E., Tringe, S. G., Sa, T., & Singh, B. K. (2020). Plant–microbiome interactions: From community assembly to plant health. *Nature Reviews Microbiology, 18*(11), 607–621. https://doi.org/10.1038/s41579-020-0412-1

Vassilev, N., Vassileva, M., Martos, V., García del Moral, L. F., Kowalska, J., Tylkowski, B., & Malusá, E. (2020). Formulation of microbial inoculants by encapsulation in natural polysaccharides: Focus on beneficial properties of carrier additives and derivatives. *Frontiers in Plant Science*, *11*, 270. https://doi.org/10.3389/fpls.2020.00270

Vessey, J. K. (2003). Plant growth promoting rhizobacteria as biofertilizers. *Plant and Soil*, *255*(2), 571–586. https://doi.org/10.1023/A:1026037216893

Zhao, X., Wu, D., Wu, Y., Wang, L., & Zhou, J. (2020). A review of AI applications in microbial genomics and metagenomics. *Frontiers in Microbiology*, *11*, 2066. https://doi.org/10.3389/fmicb.2020.02066

Zhou, J., Liu, W., Deng, Y., Jiang, Y.-H., Xue, K., He, Z., Van Nostrand, J. D., Wu, L., Yang, Y., & Wang, A. (2013). Stochastic assembly leads to alternative communities with distinct functions in a bioreactor microbial community. *mBio*, *4*(2), e00584–12. https://doi.org/10.1128/mBio.00584-12

11 Biological Control and Biopesticides
Microbial Protection Against Crop Pests

Urvashi Kashyap, Vivek Manyapu, Rupinder Kaur, Himanshi Gupta, and Mallikarjun Chanchi Prashanthkumar

11.1 INTRODUCTION

Global agricultural production faces major hurdles from a wide range of pests, including insects, weeds, nematodes, fungi, bacteria, and viruses. These pests significantly reduce yields, lower crop quality, and threaten food security and economic stability worldwide. Approximately 40% of global agricultural output is lost due to 67,000 different crop pest species, which include plant weeds, diseases, invertebrates, and a few vertebrate species (Oerke et al., 2012). Insect pests alone consume between 5% and 20% of major grain crops each year, causing economic losses exceeding $220 billion (Verma et al., 2020). Specific pests, such as the fall armyworm (*Spodoptera frugiperda*), have recently become invasive, posing serious threats in sub-Saharan Africa and parts of Asia. They severely affect staple crops like maize and raise widespread livelihood concerns (Fenibo et al., 2022). Beyond causing crop losses, pests hinder the sustainable increase of agriculture, which is vital as the global population is expected to reach 9.2 billion by 2050 (Beausang et al., 2017). Food production needs to increase on limited arable land and with restricted natural resources with effective pest management (Nazir et al., 2019). Moreover, in the 20th century, traditional methods like chemical pesticides transformed pest control and led to significant yield gains with various challenges, such as resistance, harm to beneficial insects and compromised food quality (Sharma et al., 2024; Kashyap & Reddy, 2025). However, Dependence on synthetic chemical pesticides raises various environmental, health, and agricultural issues that limit their long-term effectiveness (Bhatt et al., 2024). These include soil, air, and water contamination from pesticide residues, which can accumulate through food chains and threaten humans, wildlife, and beneficial species (Urvashi & Reddy, 2025). Persistent chemicals like organochlorines and organophosphates are toxic not only to pests but also to non-target organisms such as pollinators and natural pest enemies, thereby disturbing ecosystem stability (Yuan et al., 2019; Sheng et al., 2020; Brito et al., 2021).

Agricultural and horticultural crops are hosts to several different sucking and lepidopteran pests (Singh et al., 2025). Over the years, Farmers are utilizing chemical pesticides in order to control the crop pests which led the development of resistance in many pest populations, reduced effectiveness and requiring higher doses in future treatments (Jamtsho et al., 2021; Mateos Fernandez et al., 2022). The imprudent and excessive usage worsens pollution in the environment and endangers human health (Singh et al., 2025). Furthermore, exposure to pesticide residues to agricultural workers and consumers is associated with increased incidence of chronic diseases such cancer, liver and kidney disorders, and neurological disorders (Sabarwal et al., 2018). Therefore, there is an urgent demand for natural pest management strategies to combat insect damage (Fouda et al., 2016).

DOI: 10.1201/9781003650881-11

Biological control agents (BCAs) and biopesticides have become vital components of pest management, which are advantageous for the environment and also mitigate the issues associated with chemical pesticides. BCAs utilize living organisms, such as bacteria, fungi, viruses, insects, and nematodes, to reduce pest populations (Shahid et al., 2023). Microbial biopesticides, such as Bacillus, Trichoderma, Beauveria, Metarhizium, and baculoviruses, are widely recognized as safer and more environmentally friendly alternatives to chemical insecticides (Ruiu, 2018). They target pests with certain modes of action, which lessens off-target impacts (Kesho, 2020). This chapter integrates scientific research with technical and practical insights to elucidate the significant role of BCAs and microbial biopesticides in fostering ecologically sustainable, efficient, and economically viable pest control strategies in modern agriculture (Vero et al., 2023).

11.2 BIOLOGICAL CONTROL: PRINCIPLES AND STRATEGIES

11.2.1 Historical Background of Biological Control

Biological control is the exploitation of living agents to combat pests and invasive alien species organisms (including pathogens, pests and weeds) for diverse purposes to provide social, economic and environmental benefits (according to FAO). Biological control, or biocontrol, is the process of utilizing one organism to control the population of another organism. This technique has been employed for over 2,000 years, and it has become increasingly prevalent in pest control since the late 1800s The term "biological control" has been in use for over a century (Smith, 1919) and has been employed to manage nearly all types of pest species. These include invasive weeds, mosquitoes, insect pests, plant diseases and rodents (Labuschagne et al., 2016; Ingabire et al., 2017; Pertot et al., 2017; Jäkel et al., 2019).

In addition to agriculture, biocontrol principles are applied in a variety of fields, such as food safety (Jordan et al., 2014) and medicine (van Nood et al., 2013; Dedrick et al., 2019), although they are referred to by different terms. In recent years, there has been a significant increase in the demand for biocontrol solutions. The reason for this is that integrated pest management (IPM) is increasingly becoming the global standard for long-term crop protection (Stenberg, 2017) (Table 11.1).

11.2.2 Techniques in Biological Control

11.2.2.1 Classical Biological Control

Introduction of natural enemies from a pest's native habitat to control it in a new location where it has become invasive is called classical biological control. Once natural enemies can establish themselves, they generally keep the pest population under control. Example: In the 1880s, the vedalia beetle (*Rodolia cardinalis*) was brought from Australia to California to help control the cottony cushion scale (*Icerya purchasi*) in citrus plantations.

11.2.2.1.1 Augmentative Biological Control

Supplemental release of natural enemies to boost their populations when naturally occurring numbers are insufficient.

Two subtypes:

1. *Inoculative Release*: Small number of control agents are released at intervals to build up populations (e.g., in greenhouses).
2. *Inundative Release*: Large number released for immediate impact (like a biological pesticide).

Example: Releasing *Trichogramma* wasps to control caterpillar pests in crops.

TABLE 11.1
Historical Background of Biological Control

AD 900	First time red ant *Oecophylla smaragidna* was used to control the attack of leaf chewing insects on mandarin trees.
1602	Aldrovandi observed the hymenopteran parasite, *Apanteles glomeratus* depositing eggs in the pupae of the cabbage butterfly, *Pieris brassicae*.
1726	De Reaumur was the first person to find an insect pathogen. It was a Cordyceps fungus that grew on a noctuid.
1762	The Indian mynah bird, *Gracula religiosa*, was imported from India to Mauritius to help manage the red locust, *Nomadarcis septemfasciata*.
1776	The introduction of the predatory pentatomid *Picromerus bidens* in Europe successfully controlled the bed bug *Cimex lectularius*.
1888	The Vadalia beetle, *Rodolia cardinalis*, was introduced to California from Australia to control the cottony cushion scale, *Icerya purchasi*, on citrus trees. There was a well-planned and effective classical biological control attempt conducted.
1898	The coccinellid beetle *Cryptolaemus montrouzieri* was brought to India from Australia and released against the coffee green scale, *Coccus viridis*.
1920	A parasitoid known as *Aphelinus mali* was brought from England to India to get rid of the woolly aphid on apple, *Erisoma lanigerum*.
1929–31	*Rodolia cardinalis* was brought to India from the USA to help control the cotton cushiony scale *Icerya purchasi* on Wattle plants.
1957	The Regional station of commonwealth Institute of Biological Control (CIBC) was established at Bangalore.
1981	The Biocontrol Research Laboratory in Bangalore was the first private insectary in India.
1993	The National Bureau of Agriculturally Important Insects (formerly known as the Project Directorate of Biological Control, or PDBC) was set up in Bangalore, Karnataka, India.

11.2.2.1.2 Conservation Biological Control

Modification of the environment or existing practices to protect and enhance the effectiveness of natural enemies. Methods for conserving biological control include providing habitat (e.g., flowering plants for adult parasitoids), reducing use of broad-spectrum pesticides and managing crop diversity to support predator populations. Example: Planting hedgerows to support predatory insects like lady beetles or lacewings.

11.2.3 Mechanisms of Action of Biological Control

The mechanisms of action in biological control refer to the ways in which natural enemies suppress the population of pests. These mechanisms can vary depending on the type of natural enemy-predators, parasitoids, or pathogens, and how they interact with the pest organisms. Broadly, the mechanisms fall into four major categories: predation, parasitism, pathogenicity, and competition.

11.2.3.1 Predation

Predation involves a predator feeding on multiple prey individuals during its lifetime. Predators are usually free-living and consume their prey directly, causing immediate mortality. Common biological control predators include lady beetles, lacewings, spiders, and predatory mites. They are effective in reducing pest populations by consuming larvae, eggs, or adult stages of insects such as aphids, whiteflies, and thrips. Predation is often density-dependent meaning the more pests are present, the more food is available, increasing the predator population and impact.

11.2.3.2 Parasitism

Parasitism is the term used to describe the relationship between two species in which one organism lives in or on the other organism and derives benefits by causing some harm. Parasitism occurs when a parasitoid lays its eggs on or inside a host organism. Parasitism is a very specific mechanism, as many parasitoids target only certain pest species, making them ideal for targeted biological control.

11.2.3.2.1 Types of Parasitism

There are multiple types of parasitism, which are categorized according to their size, characteristics, interactions with their hosts, and life cycles.

11.2.3.2.2 Obligate Parasitism

This is a kind of parasitism in which the parasite is entirely reliant on the host to complete its life cycle. Obligate parasites are incapable of surviving in the absence of the host. Consequently, they do not cause significant damage to the host. Obligate parasitism is exhibited by bacteria, fungi, and viruses. For example, head lice die when they are removed from the human cranium.

11.2.3.2.3 Facultative Parasitism

This type of parasitism is characterized by the parasite's ability to persist in the absence of the host and its limited dependence on the host to complete its life cycle. *Strongyloides stercoralis*, a nematode species that is free-living, infects humans and causes the disease strongyloidiasis.

11.2.3.2.4 Ectoparasitism

Ectoparasitism is a phenomenon exhibited by parasites that reside outside the host's body. For instance, parasites and fleas.

11.2.3.2.5 Endoparasitism

Endoparasitism is a phenomenon exhibited by parasites that reside within the host's body. For instance, hookworms and nematodes.

11.2.3.2.6 Mesoparasitism

Mesoparasitism is exhibited by parasites that penetrate the host through the external apertures.

11.2.3.3 Pathogenicity (Microbial Control)

Pathogens such as bacteria, fungi, viruses, and nematodes infect and cause disease in pest organisms, leading to their death or reproductive failure. This is often referred to as microbial control. For example, *Bacillus thuringiensis* (Bt) is a bacterial pathogen widely used in biopesticides, especially against caterpillars. Entomopathogenic fungi like *Beauveria bassiana* invade and kill insects by penetrating their cuticle. Pathogens can be applied similarly to chemical pesticides and are especially useful in augmentative biocontrol approaches.

11.2.3.4 Competition

Although less direct, competition can also play a role in biological control. Natural enemies or introduced biocontrol agents may outcompete pests for resources such as food, space, or breeding sites. In microbial communities, beneficial microbes can outcompete plant pathogens in the rhizosphere (root zone), thereby protecting crops. For instance, *Pseudomonas fluorescens* competes with harmful fungi in the soil, suppressing root diseases (Table 11.2).

TABLE 11.2
Comparison of Biological Control Types

Type of Biological Control	Description	Mode of Action	Examples
Classical (Importation)	Introduction of natural enemies from the pest's native region to control invasive pests.	Long-term regulation by establishing natural enemy populations.	*Rodolia cardinalis*, (Vedalia beetle) for cottony cushion scale in citrus.
Augmentative	Mass rearing and periodic release of natural enemies. Can be inoculative or inundative.	Boosting natural enemy numbers for short or medium-term control.	*Trichogramma* spp. for controlling caterpillar pests in maize and vegetables.
Inoculative Release	A small number of agents released early in the season to establish over time.	Control develops gradually through reproduction of released enemies.	*Phytoseiulus persimilis* against spider mites in greenhouses.
Inundative Release	Large-scale release of natural enemies for immediate suppression.	Control achieved without relying on reproduction of the agents.	Bt spray for lepidopteran pests.
Conservation	Preservation and enhancement of existing natural enemies through habitat management.	Improving survival and efficiency of native beneficials.	Use of flowering plants or hedgerows to attract parasitoids and predators.
Microbial Control	Use of pathogenic microorganisms (bacteria, fungi, viruses, protozoa) against pests.	Causes disease or death in pests.	*Beauveria bassiana* for aphids and whiteflies; *Bt* for caterpillars.
Genetic Control	Use of genetic methods such as sterile males or engineered traits.	Suppresses pest reproduction or survival.	Sterile Insect Technique (SIT) in fruit flies; GM mosquitoes for malaria control.

11.3 BIOLOGICAL INTEGRATION WITH INTEGRATED PEST MANAGEMENT

Biological control is a fundamental component of IPM, which aims to manage pests using an ecologically balanced and economically viable approach. The integration of biological control into IPM strategies has become increasingly significant in recent years due to rising concerns over pesticide resistance, environmental degradation, and human health impacts caused by synthetic chemicals (Zhou et al., 2025).

In IPM, biological control refers to the use of natural enemies' predators, parasitoids, pathogens, and competitors to suppress pest populations (Khan et al., 2020). This method is often used in combination with other pest management tactics, such as cultural practices, mechanical controls, host plant resistance, and limited chemical use. The synergy between biological control and other IPM tools results in more sustainable, long-term pest suppression (Kashyap & Reddy, 2025) emphasized that the integration of habitat manipulation (such as floral strips and intercropping) supports the conservation of natural enemies and enhances biological control in open-field vegetable production systems (Figure 11.1). In protected cultivation (Ivezić et al., 2025) reported success in combining biological agents like *Encarsia formosa* with pheromone traps for managing whitefly infestations in greenhouse-grown tomatoes. Recent developments have enhanced the effectiveness of biological integration in IPM. For instance, the use of augmentative releases (e.g., *Trichogramma* spp., *Cryptolaemus montrouzieri*), conservation practices that protect native natural enemies, and biopesticides (such as Beauveria bassiana and Bt) have been widely adopted. Moreover, advances in molecular tools, artificial intelligence, and remote sensing are improving pest monitoring and enabling precise application of biological agents. Deguine et al. (2021) highlighted the need to assess the ecological risks of synthetic pesticides on non-target beneficial organisms to safeguard the success of IPM programs.

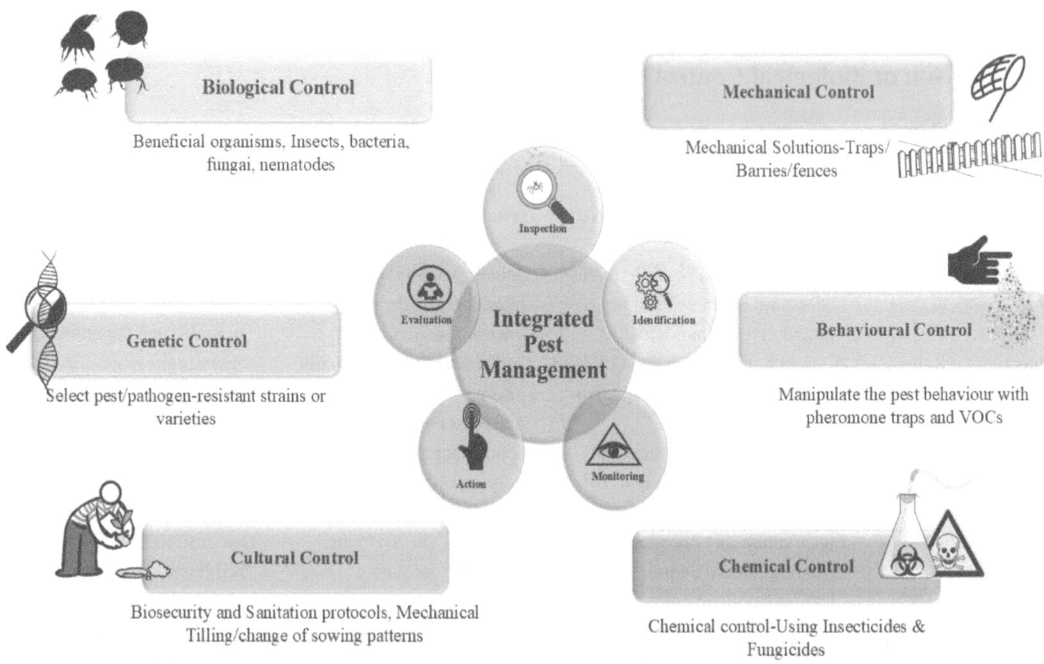

FIGURE 11.1 Different paradigms of integrated pest management with type of control.

11.4 CLASSIFICATION OF MICROBIAL BIOPESTICIDES

In the present world, agriculture plays a vital role in feeding the hungry people. There are mainly three challenges in the field of agriculture, namely the increasing population, stress on production and protection of the produced yield. To enhance yield production, farmers have adopted synthetic chemicals for yield protection, which, in turn, causes damage to the environment (Dhaliwal et al., 2015; Ranout et al., 2024). For decades, the farmer has relied on the use of artificial substances for insect pest management, offering rapid and broad-spectrum control of the insects. However, the widespread use of synthetic pesticides has raised critical concerns, including soil and water contamination, biomagnification, collateral damage to hosts and their surroundings, potential human health risks, and the development of resistance factors among insects/nematodes (Shah & Pell, 2003; 2014). Thus, these challenges provided a significant opportunity for the development of effective, safe biopesticides from microbes, including bacteria, fungi, viruses, protozoans, and nematodes.

Microbial biopesticides can be defined as biological pest control products that are obtained from microscopic living organisms or are naturally produced byproducts as their primary active compounds from bacteria, fungi, viruses, protozoa, and even entomopathogenic nematodes to control and suppress agricultural pests, insects, or plant diseases (Shahid et al., 2012). The natural compounds obtained from microbes leverage natural biological processes, causing disease in the pests and controlling them specifically (Sharma et al., 2023). Additionally, the compounds work as specific toxins, repellents or growth disruptors in the pest and help in controlling it specifically (Ali et al., 2017; Batista et al., 2018). Notably, these are known for their high specificity, environmental safety, and their efficient ability to integrate pest management. Based on the source of biopesticides from microorganisms, they are classified into four major categories and are mentioned below.

11.4.1 Bacterial Biopesticides

Bacterial biopesticides are chemical toxins or compounds possessing the ability to control pests/insects by various mechanisms of action, such as disruption of gut lining, neurotoxicity, etc. Bacterial biopesticides have seen a significant growth in sustainable agriculture, offering an environmentally friendly alternative to synthetic chemical pesticides (Baverstock et al., 2010). These microbial pesticides harness the natural pest control abilities of certain bacteria to protect crops.

11.4.2 Fungal Biopesticides

Fungal biopesticides are a group of toxins or fungi that attack the host, causing mortality of the host. These fungal-based biopesticides are naturally obtained from various fungal bodies, such as mycelium or spores. There are different kinds of toxins produced by the fungi which are listed below. The most effective fungal biopesticides are the Entomopathogenic fungi represent a vast group of over 750 known species that naturally parasitize different insects and pests. These fungi are classified across several phyla, namely Deuteromycota, Ascomycota, Zygomycota (Mantzoukas & Eliopoulos, 2020). Among these, the most used entomopathogenic fungus (EPF) in pest control are readily culturable fungi primarily from the Ascomycota phylum (Mascarin & Jaronski, 2016). Globally, various biopesticides derived from genera such as Metarhizium, Beauveria, Paecilomyces, Isaria, and Lecanicillium are widely employed. These fungi are highly versatile, capable of infecting a broad spectrum of arthropod species, regardless of the presence of natural predators (Khan et al., 2012; Castro et al., 2016). Key species currently incorporated into myco-insecticides includes *Beauveria bassiana*, *B. brongniartii*, *Metarhizium anisopliae*, *Lecanicillium* spp. (formerly *Verticillium lecanii*), *Isaria fumosorosea*, *Hirsutella thompsonii*, and *Cladosporium oxysporium*.

11.4.3 Viral Biopesticides

Viral biopesticides are a precise class of biopesticides that possess the natural ability to infect and cause mortality in insects and pests. These virus-based biopesticides are highly specific to particular insects. Among the viral biopesticides, there are four major families: Baculovirus, Entomopoxvirus, Cypovirus, and Iridovirus.

11.4.4 Protozoan and Nematode Agents

Protozoans are single-celled eukaryotic microorganisms which are known for their pathogenicity. Moreover, protozoans are commercialized as bacterial biopesticides and are known for being entomopathogenic. Similarly, Nematodes are also microscopic and are naturally found in the soil, dwelling in various life stages. Eventually, these are classified due to their pathogenicity caused to insects.

11.5 FORMULATION AND APPLICATION OF MICROBIAL BIOPESTICIDES

The efficacy of microbial biopesticides in the field apart from their pathogenicity, is contingent upon their formulation, application method, and adaptability to the field's environmental conditions (Bhattacharyya et al., 2024). In order to maximize their efficacy, extensive efforts have been made to develop protective and effective formulations, selecting optimal application methods, and timing interventions to ensure that microbes thrive and perform optimally across diverse agricultural ecosystems.

11.5.1 Types of Formulations

Developing formulations include the process of combining the active ingredient with inert carriers, additives, and protectants to enhance its stability in storage, facilitate its delivery, and perform better in the field (Vassilev et al., 2020). Formulations are crucial for safeguarding microbial propagules from dessication, ultraviolet (UV) radiation, and extremely high or low temperatures, as they are frequently vulnerable (Romano-Rodríguez et al., 2024). The formulations are generally categorized as Liquid formulations and Dry formulations. The liquid formulations of microbial pesticides include suspension concentrates (SC), emulsifiable concentrates (EC) while the dry formulations include wettable powders (WP), dusts, granules, etc. A few of the most significant and frequently used formulations are examined in this section.

11.5.1.1 Suspension Concentrates (SC)

These liquid formulations consist of wetting and dispersing agents along with insoluble active ingredient dispersed in water. The wetting agents ensure that the formulation spreads and penetrates the host surface, while the dispersing agents prevents settling and clumping.

11.5.1.2 Emulsifiable Concentrates (EC)

These are the widely used formulations of microbial pesticides due to their high potency and excellent stability. They are usually prepared by dissolving the AI in organic solvents. These are clear oily formulations that may additionally contain surfactants and other additives. Prior to use, these are diluted with water, which forms an oil-in-water emulsion.

11.5.1.3 Wettable Powders (WP)

WP consist of finely powdered AI, wetting agents, and dispersion agents, in addition to inert components such as clay or silica. WPs are solid dust-like dry compositions, forming suspensions when diluted in water. In contrast to certain liquid formulations, WPs are stable and do not settle during storage. This simplifies transportation and handling, and it eliminates the necessity for preservatives, which can be detrimental to microorganisms. The efficacy of WPs is further enhanced by their ability to retain a high concentration of the active microbial component.

11.5.1.4 Dusts

These are dry ready-to-use fine particles (dusts) possessing AI often in low concentrations. The AI is dried and blended along with inert carriers like talc, clay, kaolin, etc. They used to be the most widely used formulations of microbial pesticides however their use is now limited and banned in some countries due to the harm caused to the farmers or applicator if inhaled.

11.5.1.5 Granules

These are the coarse granular particles consisting of the AI consisting an inert carrier to enhance stability. Compared to some conventional pesticides, these are preferred for soil pests as they are made to release their AI gradually and frequently require soil moisture to activate. This leads to sustained pest suppression for prolonged time and a lower environmental impact.

11.5.2 Application Methods

The application methods of microbial pesticides are usually similar to chemical pesticides, however with microbial viability and formulation as the prime consideration. Common methods include foliar spraying, root dips, soil applications, and seed treatments. To shield seedlings against soil-borne pests in the early stages of development, microbial agents are coated to seeds prior to sowing. For long-term pest management of soil or root zone pests, granular formulations are applied to the soil (Sharma et al., 2015). The plant's foliage is directly sprayed with microbial pesticides for leaf defoliators or foliar infestations; however, the foliar spraying approach is typically preferred in the evening to prevent UV damage, particularly for fungus (Preininger et al., 2018). The root dip approach is used before transplanting a seedling or cutting because it allows microorganisms to colonize the root zone and shield the roots from pests and diseases (Srinivas & Lawande, 2007). Last but certainly not, incorporating microorganisms into drip irrigation systems enables consistent delivery through irrigation, making it ideal for precision agriculture and greenhouse environments (Wang et al., 2009).

11.5.3 Factors Affecting Efficacy

Numerous biological, environmental, and application-related factors influence how well microbial insecticides work in the field. Temperature is among the most significant factors. A certain temperature range is ideal for the development and pathogenicity of each microbial species. For instance, entomopathogenic fungi such as *Beauveria bassiana* and *Metarhizium anisopliae* show greater infectivity in temperatures between 20°C and 30°C (Kikankie et al., 2010). However, if the temperature rises over 35°C or falls below 10°C, their efficacy drastically decreases (Kikankie et al., 2010). While low temperatures tend to slow down metabolic processes and postpone germination, excessive heat can denature proteins and dry up spores. Likewise, the viability and effectiveness of microbial formulations may be jeopardized if they are administered or kept in environments that exceed their tolerance limitations.

The second most important factor, particularly for fungal biopesticides is the humidity. For their spores to develop and effectively penetrate the host cuticle, these microorganisms need enough moisture. The extended periods of low humidity can significantly reduce their efficacy, therefore humid locations or seasons are better suited for these biopesticides. Another important limiting factor is exposure to UV radiation. This is especially true for products that are applied to exposed foliage (Kavusi et al., 2023). The DNA, proteins, and pigments of microbial propagules can be harmed by UV radiation, which lowers their viability and shortens their permanence on plant surfaces. Due to their rapid degradation in bright sunlight, fungal spores and viral occlusion bodies are particularly vulnerable. Applications are frequently scheduled for early morning or late evening to reduce such losses, and several formulations use encapsulation technology or UV-protective components to safeguard the microorganisms. Microbial activity and survival are also impacted by the pH of the surrounding environment, whether it be in the soil or on the surfaces of leaves. Many helpful bacteria thrive in environments that are neutral to slightly acidic, but they may find it difficult to survive in environments that are extremely acidic or alkaline. Excessive pH levels might hinder the germination of fungus spores or disrupt the viability of bacteria (Liu et al., 2018). In order to assist maintain a favourable microenvironment, buffering agents are frequently included in formulations.

The target pest's behaviour and life stage also play a significant role. Due to their weaker immune systems and thinner cuticles, younger insect instars are typically more prone to illness. Nonetheless, certain pests exhibit behaviours that reduce their exposure to biopesticides, such as hiding beneath leaves, digging into plant tissues, or only being active at night (Senthil-Nathan, 2014). Beyond these biological interactions, the formulation and application technique selection are still crucial because they allow the microorganisms to reach and endure at the target location, protect them during storage, and ensure their survival throughout handling and delivery.

11.6 ADVANTAGES AND LIMITATIONS OF MICROBIAL BIOPESTICIDES

Microbial pesticides are environmentally safe as they decompose naturally without leaving harmful residues (Perumal et al., 2024). In contrast to conventional chemical insecticides, their mode of action involves the production of toxin and physical invasion, which reduces the possibility of pests developing resistance (Ansari & Butt, 2011). In addition, these pesticides are highly specific, indicating that they only affect specific insect pests and have minimal or no impact on non-target organisms. Additionally, they are generally regarded secure for livestock and pets and pose minimal risks to human health, rendering them suitable for use in organic farming (Santos et al., 2017). It has been well documented that the application of these pesticides can promote the development of beneficial microbial communities in the soil, thereby improving the health and fertility of the soil (Dara, 2019). Additionally, these can be synergistically incorporated with other pest control strategies, including cultural and mechanical controls, to improve overall pest management strategies (Skinner et al., 2014).

Microbial biopesticides present numerous advantages and stand as an environmentally preferable alternative to chemical pesticides. However, in the real world, microbial biopesticides present challenges that complicate their practical applications. Microbes are biologically active living organisms or molecules that can be influenced by a number of environmental factors like temperature, humidity, UV radiation, etc. (Shah & Pell, 2014). For example, fungal spores may be incapable of germination in low humidity, and UV radiation can rapidly exterminate propagules applied to the surface (Quesada-Moraga et al., 2024). Therefore, apart from their pathogenicity, the efficacy of these microbial biopesticides in the field is contingent upon their formulation, application method, and adaptability to the field's environmental conditions. Their mode of action is typically slower, which may not be ideal for the rapid control of pest populations that are spreading rapidly. Their products frequently have a shorter shelf-life when compared to chemical pesticides, necessitating meticulous storage and handling to preserve their effectiveness and viability. Additionally, the introduction of non-native strains can result in the disruption of existing microbial communities or the impact on non-target organisms, which can pose a threat to local ecosystems. The potential for microbial pesticides to be restricted in use, particularly in large-scale commercial agriculture, is due to the fact that the production and formulation are more costly than synthetic insecticides. The use of microbial pesticides is subject to strict guidelines in different regions, result in a complex and time-consuming registration and approval process.

11.7 RECENT ADVANCES AND FUTURE PROSPECTUS

Since 1973 (the year when first publication was published on microbial pesticides by (Upholt et al., 1973), the microbial pesticides have gained the research interest at a rate of 5.16% annually (Sawangproh et al., 2025). The trend is increasing at a very faster rate from the last 10 years that brought a revolution in crop pest management, specially by focusing efficiency, sustainability and technological integrity. The upgradation of bacteria and fungi as pesticides through genetic engineering and suitable consortia has been achieved for the maximum impact.

11.7.1 Genetic Engineering of Microbial Agents

Recombinant DNA technology is being recommended as part of the ongoing effort to create novel biomolecules and increase the effectiveness of the biopesticides that have been found. Next generation biopesticides are now being developed using novel fusion proteins. This approach only permits the addition of a toxin (which is non-toxic to higher species) to the carrier protein, rendering it lethal to pests only when ingested orally. Additionally, it was toxic when the predator injected it into the intended prey's body (Tomar et al., 2024).

11.7.1.1 CRISPR/Cas9

An efficient and precise genome editing tool is CRISPR/Cas9 that can allow DNA modification at the target site of any living entity, even microorganisms. Currently, this technology is gaining significant interest in microbial DNA modifications (Komal et al., 2023). Genes related to toxin secretion, resilience to environmental factors, specificity of host, etc., can be edited using this powerful technology (Zhang, 2024). CRISPR/Cas9 has a wide application in agriculture for producing crops with pest resistance, thus nullifying the dependency on chemical pesticides that in turn leads to sustainable agriculture system (Borrelli et al., 2018).

In one of the instances Bt was engineered using CRISPR as microbial biopesticide (Zhang, 2024). Overall, CRISPR/Cas9 accelerates the development of safer, more effective biopesticides and contributes to environmentally friendly pest management strategies. An endogenous type I-Bt in CRISPR-Cas system which was actively identified and characterized by Wang et al. (2025) from *Saccharopolyspora spinosa* using bioinformatics, transcriptomics, and plasmid interference assays. The system's protospacer adjacent motifs were determined, and the CRISPR-Cas operon was shown to be actively expressed. By introducing editing plasmids containing a mini-CRISPR array and repair templates, the researchers achieved 100% gene deletion efficiency in *Saccharopolyspora spinosa*. Three restriction modification (RM) system genes (one type I and two type II) systematically deleted. Simultaneous deletion of these three RM systems increased plasmid transformation efficiency by four folds for both replicative and integrative plasmids.

11.7.1.2 RNA Interference (RNAi)

RNAi also gained equal importance and interest as CRISPR for the upgradation of new and powerful generation of microbes that are highly specific and resistant to harsh environment. Generally, RNAi involves two main approaches (Beernink et al., 2024): (1) host induced gene silencing where transgenic crops are developed to produce double stranded (ds)RNA that can target pest genes directly, and (2) spray induced gene silencing that involves foliar application of dsRNA. There are many examples like *Escherichia coli*, *Saccharomyces cerevisiae*, *Bacillus subtilis*, Bt, and symbiotic bacteria like *Rhodococcus rhodnii* and *Snodgrassella alvi*. Recently, the advancements in RNAi strategies are achieved by combining recombinant DNA technology where dsRNA is synthesized using engineered bacterial, fungal and viral vectors that are introduced into host plants for specifically targeting the pests. To make the RNAi based biopesticide economically feasible, there is a need of ability to produce highly pure dsRNAs at high quantity but at low cost. Because of their high adaptability, growth rate and low maintenance cost, microbial biopesticides are the direct solution. The specialty of RNAi is that just small quantity of dsRNA is enough to commence the RNAi machinery in both larval and adult stages (Nitnavare et al., 2021).

11.7.1.2.1 Bacterial Vectors

The dsRNA expressing bacterial species can be produced in mass and sprayed in the affected crop fields. Bacteria that are RNAse III deficient were capable of producing dsRNA at high stability and are able to efficiently pass into the gut of *Rhodnius prolixus* and *Frankliniella occidentalis* leading to the death of the insects (Whitten et al., 2016).

11.7.1.2.2 Yeast Vectors

the most common yeast vector is *Saccharomyces cerevisiae* that can express dsRNA to target a specific gene (e.g., tubulin gene in case of *Drosophila suzukii*) (Murphy et al., 2016). There are some instances where engineered yeasts proven detrimental for Colorado potato beetle (*Leptinotarsa decemlineata*) (Parmar & Gilchrist, 2023).

11.7.1.2.3 Viral Vectors

There are many plant diseases that are affected by viruses transferred by different pests. Therefore, scientist have come up with an idea of engineering RNA plant virus as a powerful target for RNAi, that results in virus specific small interfering RNAs (Nitnavare et al., 2021). For example, recombinant tobacco mosaic virus was used against potato psyllid (*Bactericera cockerelli*) (Wuriyanghan & Falk, 2013) and citrus mealybug (*Planococcus citri*) (Khan et al., 2013).

11.7.2 Use of AI, IoT, and Remote Sensing for Biocontrol

The computerized technology is also much helpful and crucial in the role of biopecides and biocontrolling. In the day today life artificial intelligence (AI) is now trending in each and every field. It is known to bring a revolution by its significant role in identifying, monitoring, and augmenting biocontrol agents. Large number of datasets related to crop fields, genomics, environmental factors can be analysed by AI-based platforms for the pest outbreak, thus optimizing the time and dosage of the biopesticide to be applied for instant effect (Cui et al., 2024; Gurr et al., 2024) (Figure 11.2a). The real time pest identification from the pictures and data analytical tools are advancing rapidly. This in turn reduces lots of manpower, cost, and time with instant output. It's even easier for the scientist to formulate new biopesticide if the results are not satisfactory.

The Internet of Things (IoT) applies for the automatic system of pest management both in greenhouse and open farmlands. It can connect the networks of sensor and smart devices for continuous intelligently managing and monitoring pest infestation. Furthermore, IoT can nullify human effort and can solve many problems even if in the absence of human (Wang et al., 2024) (Figure 11.2b).

One of the highly valuable technology is remote sensing that use three approaches: (1) monitoring and mapping of the fields, (2) forecasting and prediction models, and (3) management and decision making based on the data collected (Abd El-Ghany et al., 2020). Satellites and multi-imaging devices are used to obtain synoptic view, rapid survey, precise data, cover inaccessible areas. Airborne remote sensing is highly accurate for the pest infestations, which can be monitored in no time. Generally, remote sensing can detect the behaviour of the insects like sooty mould deposition on the honeydew produced by insects, defoliation of plants, change in leaf colour, geometric shape distortion of the plants, etc. (Abd El-Ghany et al., 2020).

11.7.3 Nanotechnology in Biopesticide Formulations

Nanoparticles (NPs) have peculiar characteristics that make it beneficial for producing formulations that are convenient and easy yet effective for pest control. Therefore, the concept of nano-pesticide was initiated. The benefits of the nano-pesticides over conventional pesticides are (1) reduced quantity to be applied on the crops, (2) less storage space, (3) reduced contamination, (4) efficient targeting and killing, (5) lower maintenance, (6) reduced or zero residual toxicity, and (7) sustainability. There are mainly four categories of NPs that are applicable for disease and pest control.

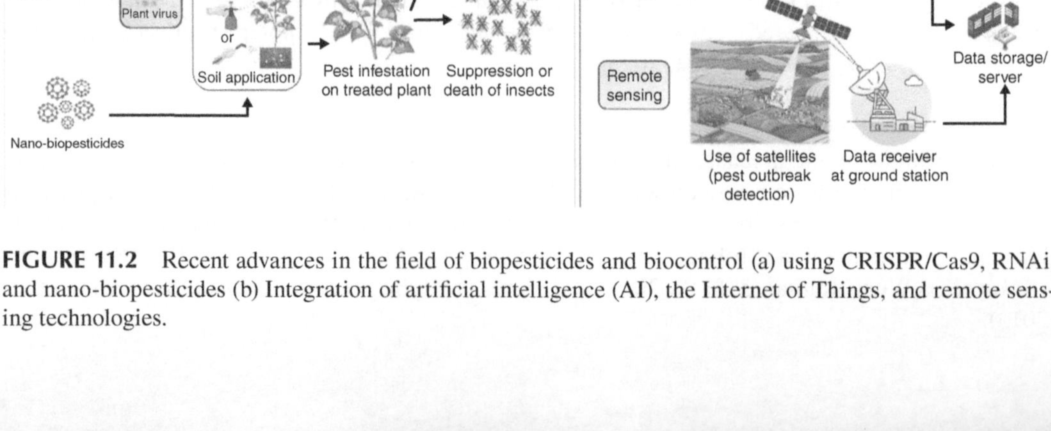

FIGURE 11.2 Recent advances in the field of biopesticides and biocontrol (a) using CRISPR/Cas9, RNAi, and nano-biopesticides (b) Integration of artificial intelligence (AI), the Internet of Things, and remote sensing technologies.

11.7.3.1 Metal Based Nano-Pesticides

1. Silver NPs (Bharani & Namasivayam, 2017; Kalimuthu et al., 2017; Benelli, 2018)
2. Zinc based NPs (Malaikozhundan et al., 2017)
3. Copper NPs (Muthamil Selvan et al., 2018; Chakrabarti & Patra, 2020),
4. Iron/magnetic NPs (Zhao et al., 2017)
5. Gold NPs (Gandhi et al., 2019)
6. Magnesium based (Chakrabarti & Patra, 2020; Fouda et al., 2021)
7. Titanium-based NPs (Ghareeb et al., 2024)

11.7.3.2 Non-Metal-Based Nano-Pesticides

1. Silica-based NPs (Qin et al., 2016; Shoaib et al., 2018)
2. Calcium-based NPs (Hua et al., 2015)
3. Abamectin formulations (Cherif et al., 2022)

11.7.3.3 Polymeric and Biopolymeric Nano-Pesticides

1. Zein (Pascoli et al., 2020)
2. Chitosan (Campos et al., 2018)

11.7.3.4 Carbon-Based Nano-Pesticides

1. Carbon nanotubes (Jangid et al., 2021)
2. Carbon quantum dot (Mohapatra et al., 2018; Chandra et al. 2022)
3. Graphen (Nana et al., 2021; Lampiri et al., 2024)

11.7.4 Sustainable Development Goals Achieved

Indeed, the biocontrol and biopesticides contribute to different sustainable development goals (SDGs) as their approaches are promisingly sustainable and eco-friendly.

11.7.4.1 SDG-1

No poverty: In many places, agriculture is associated with rural areas. Efficiently controlling the pests using eco-friendly approaches like biocontrol and biopesticides, there will be enhanced crop production. When there is development in agriculture and yield, the rural livelihood gains financial progress, and thus eliminates poverty (Cai and Dimopoulos, 2024).

11.7.4.2 SDG-2

Zero hunger: When there will be ample of agricultural food produce, there will be no hunger thus it can achieve food security (Cai and Dimopoulos, 2024).

11.7.4.3 SDG-3

Good health and well-being: the biopesticides and biocontrol are known to reduce the residual toxicity from the conventional harmful chemicals (Cai and Dimopoulos, 2024). When the residual toxicity is reduced then the bioaccumulation and biomagnification will no longer be crucial leading to healthy life.

11.7.4.4 SDG-6

Clean water and sanitation: In the conventional methods, the pesticides were sprayed over the fields in large amounts that were used to contaminate ground water by surface runoff and leaching, thus making the ground water unfit for drinking. Fortunately, the biopesticides and biocontrol agents do not have any residual toxicity; rather some of the agents (NPs) can help in degradation of toxic pollutants (Fenibo et al., 2021).

11.7.4.5 SDG-12

Responsible consumption and production: The biopesticides and biocontrol agents prevent reliance on the synthetic pesticides that are detrimental to the environment, harmful for consumption and unsuitable for crop production on a long run (Fenibo et al., 2021).

11.7.4.6 SDG-13

Climate action: Synthetic chemicals require lot of chemical reactions and as a byproduct they release toxic fumes and wastes (Zhou et al., 2025). Preventing the usage and in turn reducing the production of chemical pesticides will definitely reduce the production of toxic and greenhouse gases leading benefit the climate as a whole (Fenibo et al., 2021).

11.7.4.7 SDG-14

Life under water: The residual toxicity that used to flow from the agricultural farms into nearby water bodies via surface runoff and results in impactful health effects on aquatic ecosystem. This impact can be nullified with the use of biopesticides and biocontrol agents to a great extent and further, they can be helpful in degradation of the existing contaminants (Fenibo et al., 2021).

11.7.4.8 SDG-15

Life on land: Since the crops are grown on soil, soil health is of utmost importance. Biopesticides and biocontrol agents can enhance the biodiversity by regenerating beneficial soil microorganisms. They also help in maintaining ecological balance and diversify the agroecosystems (Hulot and Hiller, 2021).

11.7.5 Future Prospects and Research Needs

Future studies and the use of biopesticides should give priority to the development of novel microbial strains with enhanced characteristics, such as increased effectiveness and wider target spectrum. Understanding the mechanisms of action of biopesticides is essential to maximizing their effectiveness and lowering the likelihood of resistance. Biopesticides can enhance overall pest control when used in conjunction with other strategies in comprehensive pest management programs.

11.8 CONCLUSIONS

The availability of biopesticides that combat a variety of crop pests is crucial for managing agroecosystems in a way that protects both human health and the environment. Consequently, various strategies for crop protection are consistently being explored and evaluated. Specifically, approaches that promote sustainable agricultural practices are garnering increased focus. In this context, the interest in biopesticides has increased significantly. In recent years, there has been a notable rise in the number and variety of formulations utilizing these biopesticides, accompanied by increased safeguarding from major chemical corporations. Moreover, farmers' increased demand is also supported by a growing market offering newly launched and improved products that can be used alone, in rotation, or alongside traditional chemicals Abdelfattah et al. (2018), Malacrino et al. (2018), and Bennett et al. (2018). The biopesticide industry is experiencing significant growth in both academic and commercial investments, with several discoveries being developed into new biopesticide products that expand the global market supply (Abdelfattah et al., 2018). This includes creating innovative solutions for new targets or using new active ingredients that enhance the effectiveness of existing ones (Abdelfattah et al., 2018; Malacrino et al., 2018). Conversely, recent research has deepened our understanding of biopesticides and biological control methods for managing agricultural and horticultural pests through proper and practical application (Bennett et al., 2018).

REFERENCES

Abd El-Ghany, N. M., Abd El-Aziz, S. E., & Marei, S. S. (2020). A review: Application of remote sensing as a promising strategy for insect pests and diseases management. *Environmental Science and Pollution Research*, *27*(27), 33503–33515.

Abdelfattah, A., Malacrinò, A., Wisniewski, M., Cacciola, S.O., & Schena, L. (2018). Metabarcoding: A powerful tool to investigate microbial communities and shape future plant protection strategies. *Biological Control*, 120, 1–10.

Ali, S., Zhang, C., Wang, Z., Wang, X. M., Wu, J. H., et al. (2017). Toxicological and biochemical basis of synergism between the entomopathogenic fungus *Lecanicillium muscarium* and the insecticide matrine against *Bemisia tabaci* (Gennadius). *Scientific Reports*, *7*, (1) 46558.

Ansari M. A., & Butt T. M. (2011). Effects of successive subculturing on stability, virulence, conidial yield, germination and shelf-life of entomopathogenic fungi. *Journal of Applied Microbiology*, *110*(6), 1460–1469.

Batista, A. C. D. L., Souza Neto, F. E. D., & Paiva, W. D. S. (2018). Review of fungal chitosan: Past, present and perspectives in Brazil. *Polímeros*, 28, 275–283.

Baverstock, J., Roy, H. E., & Pell, J. K. (2010). Entomopathogenic fungi and insect behaviour: From unsuspecting hosts to targeted vectors. *Biological Control*, *55*, 89–102.

Beausang, C., Hall, C., & Toma, L. (2017). Food waste and losses in primary production: Qualitative insights from horticulture. *Resources, Conservation and Recycling*, *126*, 177–185.

Beernink, B. M., Amanat, N., Li, V. H., Manchur, C. L., Whyard, S., & Belmonte, M. F. (2024). SIGS vs. HIGS: Opportunities and challenges of RNAi pest and pathogen control strategies. *Canadian Journal of Plant Pathology*, *46*(6), 675–689.

Benelli, G. (2018). Mode of action of nanoparticles against insects. *Environmental Science and Pollution Research*, *25*(13), 12329–12341.

Bennett, A.E., Orrell, P., Malacrino, A., & Pozo, M.J. (2018). Fungal-mediated above–belowground interactions: The community approach, stability, evolution, mechanisms, and applications. In Aboveground–belowground community ecology, (pp. 85–116). Cham: Springer International Publishing.

Bharani, R. A., & Namasivayam, S. K. R. (2017). Biogenic silver nanoparticles mediated stress on developmental period and gut physiology of major lepidopteran pest *Spodoptera litura*. *Journal of Environmental Chemical Engineering*, *5*(1), 453–467.

Bhatt, P., Chauhan, N., Sendri, N., Bhandari, P., & Reddy, S. G. E. (2024). *Thalictrum foliolosum* DC against aphid, *Aphis craccivora* Koch and mealybug, *Planococcus lilacinus* Cockerell: A potential bioinsecticide. *Biocatalysis and Agricultural Biotechnology*, *60*, 103326.

Bhattacharyya, P. N., Baishya, B., Borgohain, D. J., Sarma, B., Bhattacharyya, L. H., Garganese, F., & Barman, T. S. (2024). Unveiling the virulence and stress tolerance of Beauveria bassiana against major insect pests in tea with an emphasis on bibliometrics using data mining tools. In *Entomopathogenic fungi: Prospects and challenges* (pp. 247–298). Springer. https://doi.org/10.1007/978-981-97-5991-0_11

Borrelli, V. M., Brambilla, V., Rogowsky, P., Marocco, A., & Lanubile, A. (2018). The enhancement of plant disease resistance using CRISPR/Cas9 technology. *Frontiers in Plant Science*, *9*, 1245.

Brito, V. D., Achimón, F., Pizzolitto, R. P., Ramírez Sánchez, A., Gómez Torres, E. A., Zygadlo, J. A., & Zunino, M. P. (2021). An alternative to reduce the use of the synthetic insecticide against the maize weevil *Sitophilus zeamais* through the synergistic action of *Pimenta racemosa* and *Citrus sinensis* essential oils with chlorpyrifos. *Journal of Pest Science*, *94*(2), 409–421.

Cai, P., & Dimopoulos, G. (2024). Microbial biopesticides: A one health perspective on benefits and risks. *One Health*, *100962*. https://doi.org/10.1016/j.onehlt.2024.100962

Campos, E. V., Proença, P. L., Oliveira, J. L., Melville, C. C., Della Vechia, J. F., De Andrade, D. J., & Fraceto, L. F. (2018). Chitosan nanoparticles functionalized with β-cyclodextrin: A promising carrier for botanical pesticides. *Scientific Reports*, *8*(1), 2067.

Castro, T., Mayerhofer, J., Enkerli, J., Eilenberg, J., Meyling, N. V., de Andrade Moral, R., Demétrio, C. G. B., & Delalibera Jr, I., (2016). Persistence of Brazilian isolates of the entomopathogenic fungi *Metarhizium anisopliae* and *M. Robertsii* in strawberry crop soil after soil drench application. *Agriculture, Ecosystems & Environment*, *233*, 361–369.

Chakrabarti, A., & Patra, P. (2020). Relative larvicidal property of common oxide nanostructures against *Culex quinquefasciatus*. *IET Nanobiotechnology*, *14*(5), 389–395.

Chandra, S., Bano, D., Sahoo, K., Kumar, D., Kumar, V., Yadav, P. K., & Hasan, S. H. (2022). Synthesis of fluorescent carbon quantum dots from Jatropha fruits and their application in fluorometric sensor for the detection of chlorpyrifos. *Microchemical Journal*, *172*, 106953.

Cherif, A., Mansour, R., Sun, C., & Grissa-Lebdi, K. (2022). Lethal effects of nano and commercial formulations of abamectin on *Tuta absoluta* (Meyrick) and its mirid predators *Macrolophus pygmaeus* and *Nesidiocoris tenuis*. *International Journal of Tropical Insect Science*, *42*(3), 2183–2193.

Cui, J. L., Li, H., He, Q., Jin, B. Y., Liu, Z., Zhang, X. M., & Zhang, L. (2024). Integrating classic AI and agriculture: A novel model for predicting insecticide-likeness to enhance efficiency in insecticide development. *Computational Biology and Chemistry*, *112*, 108113.

Dara, S. K. (2019). Non-entomopathogenic roles of entomopathogenic fungi in promoting plant health and growth. *Insects*, *10*(9), 277.

Dedrick, R. M., Guerrero-Bustamante, C. A., Garlena, R. A., Russell, D. A., Ford, K., Harris, K., Gilmour, K. C., Soothill, J., Jacobs-Sera, D., Schooley, R. T., & Hatfull, G. F. (2019). Engineered bacteriophages for treatment of a patient with a disseminated drug-resistant *Mycobacterium abscessus*. *Nature Medicine*, *25*(5), 730–733.

Deguine, J. P., Aubertot, J. N., Flor, R. J., Lescourret, F., Wyckhuys, K. A., & Ratnadass, A. (2021). Integrated pest management: Good intentions, hard realities. A review. *Agronomy for Sustainable Development*, *41*(3), 38.

Dhaliwal, G. S., Jindal, V., & Mohindru, B. (2015). Crop losses due to insect pests: Global and Indian scenario. *Indian Journal of Entomology*, *77*, 165–168.

Fenibo, E. O., Ijoma, G. N., & Matambo, T. (2021). Biopesticides in sustainable agriculture: A critical sustainable development driver governed by green chemistry principles. *Frontiers in Sustainable Food Systems*, *5*, 619058.

Fenibo, E. O., Ijoma, G. N., Nurmahomed, W., & Matambo, T. (2022). The potential and green chemistry attributes of biopesticides for sustainable agriculture. *Sustainability*, *14*(21), 14417.

Fouda, E. A., Abou-Yousef, H. M., Abdallah, I. S., & Kandil, M. A. (2016). Resistance monitoring and enzyme activity in three field populations of cowpea aphid (*Aphis craccivora*) from Egypt. *Crop Protection*, *81*, 163–167.

Fouda, A., Awad, M. A., Eid, A. M., Saied, E., Barghoth, M. G., Hamza, M. F., … Hassan, S. E. D (2021). An eco-friendly approach to the control of pathogenic microbes and *Anopheles stephensi* malarial vector using magnesium oxide nanoparticles. *International Journal of Molecular Sciences*, *22*(10), 5096.

Gandhi, A. D., Murugan, K., Umamahesh, K., Babujanarthanam, R., Kavitha, P., & Selvi, A. (2019). Lichen *Parmelia sulcata* mediated synthesis of gold nanoparticles: An eco-friendly tool against *Anopheles stephensi* and *Aedes aegypti*. *Environmental Science and Pollution Research*, *26*(23), 23886–23898.

Ghareeb, A., Fouda, A., Kishk, R. M., & El Kazzaz, W. M. (2024). Unlocking the potential of titanium dioxide nanoparticles: An insight into green synthesis, optimizations, characterizations, and multifunctional applications. *Microbial Cell Factories*, *23*(1), 341.

Gurr, G. M., Liu, J., & Pogrebna, G. (2024). Harnessing artificial intelligence for analysing the impacts of nectar and pollen feeding in conservation biological control. *Current Opinion in Insect Science*, *62*, 101176.

Hua, K. H., Wang, H. C., Chung, R. S., & Hsu, J. C. (2015). Calcium carbonate nanoparticles can enhance plant nutrition and insect pest tolerance. *Journal of Pesticide Science*, *40*(4), 208–213.

Hulot, J. F., & Hiller, N. (2021). Exploring the benefits of biocontrol for sustainable agriculture–A literature review on biocontrol in light of the European Green Deal. *Institute for European Environmental Policy: Brussels, Belgium*, pp. 1–42.

Ingabire, C. M., Hakizimana, E., Rulisa, A., Kateera, F., Van Den Borne, B., Muvunyi, C. M., Mutesa, L., Van Vugt, M., Koenraadt, C. J., Takken, W., & Alaii, J. (2017). Community-based biological control of malaria mosquitoes using *Bacillus thuringiensis* var. israelensis (Bti) in Rwanda: Community awareness, acceptance and participation. *Malaria Journal*, *16*(1), 399.

Ivezic, A., Popovic, T., Trudic, B., Krndija, J., Barosevic, T., Sarajlic, A., Stojacic, I., & Kuzmanovic, B. (2025). Biological control agents in greenhouse tomato production (*Solanum lycopersicum* L.): Possibilities, challenges and policy insights for Western Balkan region. *Horticulturae*, *11*(2), 155.

Jäkel, T., Promkerd, P., Sitthirath, R., Guedant, P., & Khoprasert, Y. (2019). Biocontrol of rats in an urban environment in Southeast Asia using Sarcocystis singaporensis. *Pest Management Science*, *75*(8), 2148–2157.

Jamtsho, T., Banu, N., & Kinley, C. (2021). Critical review on past, present and future scope of diamondback moth management. *Plant Archives*, *21*, 1199–1210.

Jangid, K., Sahu, R. P., Pandey, R., Chen, R., Zhitomirsky, I., & Puri, I. K. (2021). Multiwalled carbon nanotubes coated with nitrogen–sulfur co-doped activated carbon for detecting fenitrothion. *ACS Applied Nano Materials*, *4*(5), 4781–4789.

Jordan, K., Dalmasso, M., Zentek, J., Mader, A., Bruggeman, G., Wallace, J., De Medici, D., Fiore, A., Prukner-Radovcic, E., Lukac, M., & Axelsson, L. (2014). Microbes versus microbes: control of pathogens in the food chain. *Journal of the Science of Food and Agriculture*, *94*(15), 3079–3089.

Kalimuthu, K., Panneerselvam, C., Chou, C., Tseng, L. C., Murugan, K., Tsai, K. H., & Benelli, G. (2017). Control of dengue and zika virus vector *Aedes aegypti* using the predatory copepod *Megacyclops formosanus*: Synergy with *Hedychium coronarium*-synthesized silver nanoparticles and related histological changes in targeted mosquitoes. *Process Safety and Environmental Protection, 109*, 82–96.

Kashyap, U., & Reddy, S. G. E. (2025). Metabolic profiling and insecticidal activities of *Rosmarinus officinalis* L. for the management of *Aphis craccivora* Koch and *Planococcus lilacinus* Cockerell. *Chemical and Biological Technologies in Agriculture, 12*(1), 46.

Kavusi, E., Shahi Khalaf Ansar, B., Dehghanian, Z., Asgari Lajayer, B., Nobaharan, K., Ma, Y., & Glick, B. R. (2023). Delivery of beneficial microbes via seed coating for medicinal and aromatic plant production: A critical review. *Journal of Plant Growth Regulation, 42*(2), 575–597.

Kesho, A. (2020). Microbial bio-pesticides and their use in integrated pest management. *Chemical and Biomolecular Engineering, 5*(1), 26–34.

Khan, A. M., Ashfaq, M., Kiss, Z., Khan, A. A., Mansoor, S., & Falk, B. W. (2013). Use of recombinant tobacco mosaic virus to achieve RNA interference in plants against the citrus mealybug, *Planococcus citri* (Hemiptera: Pseudococcidae). *PLOS One, 8*(9), e73657.

Khan, S., Guo, L., Maimaiti, Y., Mijit, M., & Qiu, D. (2012). Entomopathogenic fungi as microbial biocontrol agent. *Molecular Plant Breeding, 3*, 63–79.

Khan, S., Uddin, M. N., Rizwan, M., Khan, W., Farooq, M., Shah, A. S., Subhan, F., Aziz, F., Rahman, K. U., Khan, A., & Ali, S. (2020). Mechanism of insecticide resistance in insects/pests. *Polish Journal of Environmental Studies, 29*, 2023–2030.

Kikankie, C. K., Brooke, B. D., Knols, B. G., Koekemoer, L. L., Farenhorst, M., Hunt, R. H., & Coetzee, M. (2010). The infectivity of the entomopathogenic fungus *Beauveria bassiana* to insecticide-resistant and susceptible *Anopheles arabiensis* mosquitoes at two different temperatures. *Malaria Journal, 9*(1), 71.

Komal, J., Desai, H. R., Samal, I., Mastinu, A., Patel, R. D., Kumar, P. D., & Bhoi, T. K. (2023). Unveiling the genetic symphony: Harnessing CRISPR-Cas genome editing for effective insect pest management. *Plants, 12*(23), 3961.

Labuschagne, L., Swanepoel, L. H., Taylor, P. J., Belmain, S. R., & Keith, M. (2016). Are avian predators effective biological control agents for rodent pest management in agricultural systems? *Biological Control, 101*, 94–102.

Lampiri, E., Yap, P. L., Berillis, P., Athanassiou, C. G., & Losic, D. (2024). Graphene powders as new contact nanopesticides: Revealing key parameters on their insecticidal activity for stored product insects. *Chemosphere, 364*, 143200.

Liu, J., Sui, Y., Wisniewski, M., Xie, Z., Liu, Y., You, Y., & Wang, Q. (2018). The impact of the postharvest environment on the viability and virulence of decay fungi. *Critical Reviews in Food Science and Nutrition, 58*(10), 1681–1687.

Malacrinò, A., Campolo, O., Medina, R.F., & Palmeri, V., 2018. Instar-and host-associated differentiation of bacterial communities in the Mediterranean fruit fly *Ceratitis capitata*. PloS one, *13*(3), e0194131.

Malaikozhundan, B., Vaseeharan, B., Vijayakumar, S., & Thangaraj, M. P. (2017). *Bacillus thuringiensis* coated zinc oxide nanoparticle and its biopesticidal effects on the pulse beetle, *Callosobruchus maculatus*. *Journal of Photochemistry and Photobiology B: Biology, 1*(174), 306–314.

Mantzoukas, S., & Eliopoulos, P. A. (2020). Endophytic entomopathogenic fungi: A valuable biological control tool against plant pests. *Applied Sciences, 10*, 360.

Mascarin, G. M., & Jaronski, S. T. (2016). The production and uses of *Beauveria bassiana* as a microbial insecticide. *World Journal of Microbiology and Biotechnology, 32*, 177.

Mateos Fernandez, R., Petek, M., Gerasymenko, I., Jutersek, M., Baebler, S., Kallam, K., & Patron, N. J. (2022). Insect pest management in the age of synthetic biology. *Plant Biotechnology Journal, 20*(1), 25–36.

Mohapatra, S., Bera, M. K., & Das, R. K. (2018). Rapid "turn-on" detection of atrazine using highly luminescent N-doped carbon quantum dot. *Sensors and Actuators B: Chemical, 263*, 459–468.

Murphy, K. A., Tabuloc, C. A., Cervantes, K. R., & Chiu, J. C. (2016). Ingestion of genetically modified yeast symbiont reduces fitness of an insect pest via RNA interference. *Scientific Reports, 6*(1), 22587.

Muthamil Selvan, S., Vijai Anand, K., Govindaraju, K., Tamilselvan, S., Kumar, V. G., Subramanian, K. S., & Raja, K. (2018). Green synthesis of copper oxide nanoparticles and mosquito larvicidal activity against dengue, zika and chikungunya causing vector *Aedes aegypti*. *IET Nanobiotechnology, 12*(8), 1042–1046.

Nana, L., Ruiyi, L., Qinsheng, W., Yongqiang, Y., Xiulan, S., Guangli, W., & Zaijun, L. (2021). Colorimetric detection of chlorpyrifos in peach based on cobalt-graphene nanohybrid with excellent oxidase-like activity and reusability. *Journal of Hazardous Materials, 415*, 125752.

Nazir, T., Khan, S., & Qiu, D. (2019). Biological control of insect pest. *Pests Control and Acarology, 21*, 78–87.

Nitnavare, R. B., Bhattacharya, J., Singh, S., Kour, A., Hawkesford, M. J., & Arora, N. (2021). Next generation dsRNA-based insect control: Success so far and challenges. *Frontiers in Plant Science*, *12*, 673576.

Oerke, E. C., Dehne, H. W., Schönbeck, F., & Weber, A. (2012). *Crop production and crop protection: Estimated losses in major food and cash crops*. Elsevier.

Parmar, R., & Gilchrist, E. (2023). Yeast-based biopesticide could be future control method for PMPs. *Pest Management*. Prof. https://www.mypmp.net/yeast-based-biopesticide-could-be-future-control-method-for-pmps/ (accessed 7.4.25).

Pascoli, M., de Albuquerque, F. P., Calzavara, A. K., Tinoco-Nunes, B., Oliveira, W. H. C., Gonçalves, K. C., & Fraceto, L. F. (2020). The potential of nanobiopesticide based on zein nanoparticles and neem oil for enhanced control of agricultural pests. *Journal of Pest Science*, *93*(2), 793–806.

Pertot, I., Giovannini, O., Benanchi, M., Caffi, T., Rossi, V., & Mugnai, L. (2017). Combining biocontrol agents with different mechanisms of action in a strategy to control Botrytis cinerea on grapevine. *Crop Protection*, *97*, 85–93.

Perumal, V., Kannan, S., Pittarate, S., & Krutmuang, P. (2024). A review of entomopathogenic fungi as a potential tool for mosquito vector control: A cost-effective and environmentally friendly approach. *Entomological Research*, *54*(3), e12717.

Preininger, C., Sauer, U., Bejarano, A., & Berninger, T. (2018). Concepts and applications of foliar spray for microbial inoculants. *Applied Microbiology and Biotechnology*, *102*(17), 7265–7282.

Qin, X., Xiang, X., Sun, X., Ni, H., & Li, L. (2016). Preparation of nanoscale *Bacillus thuringiensis* chitinases using silica nanoparticles for nematicide delivery. *International Journal of Biological Macromolecules*, *82*, 13–21.

Quesada-Moraga, E., González-Mas, N., Yousef-Yousef, M., Garrido-Jurado, I., & Fernández-Bravo, M. (2024). Key role of environmental competence in successful use of entomopathogenic fungi in microbial pest control. *Journal of Pest Science*, *97*(1), 1–15.

Ranout, A. S., Kaur, R., Kumar, R., & Nadda, G. (2024). Pathogenicity and compatibility studies of native *Tolypocladium inflatum* and *Clonostachys krabiensis* against *Tetranychus urticae*. *Journal of Applied Entomology*, *148*(10), 1210–1222.

Romano-Rodríguez, E., Navarro-Torre, S., Mateos-Naranjo, E., Flores-Duarte, N. J., Rodríguez-Llorente, I. D., Pajuelo, E., & Redondo-Gómez, S. (2024). From salty soils to scientific breakthroughts: *Halocnemum strobilaceum* and its novel plant-growth promoting bacterial species. Sociedad Española de Fijación del Nitrógeno (SEFIN) In *Book of abstracts* (p. 22).

Ruiu, L. (2018). Microbial biopesticides in agroecosystems. *Agronomy*, *8*(11), 235.

Sabarwal, A., Kumar, K., & Singh, R. P. (2018). Hazardous effects of chemical pesticides on human health–Cancer and other associated disorders. *Environmental Toxicology and Pharmacology*, *63*, 103–114.

Santos, L., Alves, A., & Alves, R. (2017). Evaluating multi-locus phylogenies for species boundaries determination in the genus Diaporthe. *PeerJ*, *5*, e3120.

Sawangproh, W., Paejaroen, P., Afifah, L., & Phaenark, C. (2025). Microbial pesticides: A bibliometric analysis of global research trends (1973–2024). *Egyptian Journal of Biological Pest Control*, *35*(1), 2.

Senthil-Nathan, S. (2014). A review of biopesticides and their mode of action against insect pests. Environmental sustainability: Role of green technologies, 49–63.

Shah, P. A., & Pell, J. K. (2003). Entomopathogenic fungi as biological control agents. *Applied Microbiology and Biotechnology*, *61*, 413–423.

Shah, P. A., & Pell, J. K. (2014). Entomopathogenic fungi as biological control agents. *Applied Microbiology and Biotechnology*, *98*, 3515–3523.

Shahid, A. A., Rao, Q. A., Bakhsh, A., & Husnain, T. (2012). Entomopathogenic fungi as biological controllers: New insights into their virulence and pathogenicity. *Archives of Biological Sciences*, *64*, 21–42.

Shahid, M., Shaukat, F., Shahid, A., Sohail, A., & Nadeem, M. (2023). Biopesticides: A potential solution for the management of insect pests. *Agrobiological Records*, *13*, 7–15. https://doi.org/10.47278/journal.abr/2023.022

Sharma, A., Sharma, S., & Yadav, P. K. (2023). Entomopathogenic fungi and their relevance in sustainable agriculture: A review. *Cogent Food & Agriculture*, *9*(1), 2180857.

Sharma, K. K., Singh, U. S., Sharma, P., Kumar, A., & Sharma, L. (2015). Seed treatments for sustainable agriculture-A review. *Journal of Applied and Natural Science*, *7*(1), 521.

Sharma, M., Urvashi, Gupta, H., Anmol, Sharma, U., & Reddy, S. (2024). Chemical Composition and insecticidal potential of essential oil from *Murraya koenigii* (L.) obtained by natural deep eutectic solvents. *Neotropical Entomology*, *53*(6), 1318–1331.

Sheng, Z., Jian, R., Xie, F., Chen, B., Zhang, K., Li, D., & Hong, W. D. (2020). Screening of larvicidal activity of 53 essential oils and their synergistic effect for the improvement of deltamethrin efficacy against *Aedes albopictus*. *Industrial Crops and Products*, *145*, 112131.

Shoaib, A., Elabasy, A., Waqas, M., Lin, L., Cheng, X., Zhang, Q., & Shi, Z. H. (2018). Entomotoxic effect of silicon dioxide nanoparticles on *Plutella xylostella* (L.) (Lepidoptera: Plutellidae) under laboratory conditions. *Toxicological & Environmental Chemistry*, *100*(1), 80–91.

Singh, P. P., Urvashi, Sangeetha, K. S., Sharma, U., & Reddy, S. G. E. (2025). Insecticidal activities and mechanism of action of steroidal saponins from *Trillium govanianum* wall. ex D. Don against *Plutella xylostella* (L.) and *Aphis craccivora* Koch. *Neotropical Entomology*, *54*(1), 67.

Skinner, M., Parker, B. L., & Kim, J. S. (2014). Role of entomopathogenic fungi in integrated pest management. *Integrated Pest Management*, 169–191.

Srinivas, P. S., & Lawande, K. E. (2007). Seedling root dip method for protecting onion plants from thrips. *Indian Journal of Plant Protection*, *35*(2), 206–209.

Stenberg, J. A. (2017). A conceptual framework for integrated pest management. *Trends in Plant Science*, 22:759–769. https://doi.org/10.1016/j.tplants.2017.06.010

Tomar, P., Thakur, N., Jhamta, S., Chowdhury, S., Kapoor, M., Singh, S., & Yadav, A. N. (2024). Bacterial biopesticides: Biodiversity, role in pest management and beneficial impact on agricultural and environmental sustainability. *Heliyon*, *10*(11), e31550.

Upholt, W. M., Engler, R. A., & Terbush, L. E. (1973). Regulation of microbial pesticides. *Annals of the New York Academy of Sciences*, *217*(1), 234–234. https://doi.org/10.1111/j.1749-6632.1973.tb32764.x

Urvashi, & Reddy, S. G. E. (2025). Chemical profiling, insecticidal and enzyme inhibition activities of *Rosmarinus officinalis* and *Ocimum sanctum* against aphid, mealy bug and diamondback moth. *Chemistry & Biodiversity*, *22*(5), e202402647.

van Nood, E., Vrieze, A., Nieuwdorp, M., Fuentes, S., Zoetendal, E. G., de Vos, W. M., Visser, C. E., Kuijper, E. J., Bartelsman, J. F., Tijssen, J. G., & Speelman, P. (2013). Duodenal infusion of donor feces for recurrent *Clostridium difficile*. *New England Journal of Medicine*, *368*(5), 407–415.

Vassilev, N., Vassileva, M., Martos, V., Garcia del Moral, L. F., Kowalska, J., Tylkowski, B., & Malusá, E. (2020). Formulation of microbial inoculants by encapsulation in natural polysaccharides: Focus on beneficial properties of carrier additives and derivatives. *Frontiers in Plant Science*, *11*, 270.

Verma, D., Banjo, T., Chawan, M., Teli, N., & Gavankar, R. (2020). Microbial control of pests and weeds. In *Natural remedies for pest, disease and weed control* (pp. 119–126). Academic Press.

Vero, S., Garmendia, G., Allori, E., Sanz, J. M., Gonda, M., Alconada, T., Wisniewski, M. (2023). Microbial biopesticides: Diversity, scope, and mechanisms involved in plant disease control. *Diversity*, *15*(3), 457.

Wang, W., He, H., Liu, H., Gao, Y., Dang, F., Zhao, X., & Lu, Y. (2025). Developing a robust genome editing tool based on an endogenous type IB CRISPR-Cas system in *Saccharopolyspora spinosa*. *Science China Life Sciences*, *68*(5), 1324–1336.

Wang, X., Zhu, H., Reding, M. E., Locke, J. C., Leland, J. E., Derksen, R. C., Krause, C. R. (2009). Delivery of chemical and microbial pesticides through drip irrigation systems. *Applied Engineering in Agriculture*, *25*(6), 883–893.

Wang, Z., Qiao, X., Wang, Y., Yu, H., & Mu, C. (2024). IoT-based system of prevention and control for crop diseases and insect pests. *Frontiers in Plant Science*, *15*, 1323074.

Whitten, M. M., Facey, P. D., Del Sol, R., Fernández-Martínez, L. T., Evans, M. C., Mitchell, J. J., & Dyson, P. J. (2016). Symbiont-mediated RNA interference in insects. *Proceedings of the Royal Society B: Biological Sciences*, *283*(1825), 20160042.

Wuriyanghan, H., & Falk, B. W. (2013). RNA interference towards the potato psyllid, *Bactericera cockerelli*, is induced in plants infected with recombinant tobacco mosaic virus (TMV). *PLoS One*, *8*(6), e66050.

Yuan, L., Yang, X., Yu, X., Wu, Y., & Jiang, D. (2019). Resistance to insecticides and synergistic and antagonistic effects of essential oils on dimefluthrin toxicity in a field population of *Culex quinquefasciatus* Say. *Ecotoxicology and Environmental Safety*, *169*, 928–936.

Zhang, W. (2024). CRISPR-based gene editing in Bt for improved insecticidal properties. *Bioscience Methods*, *15*(5), 216–215.

Zhao, X., Meng, Z., Wang, Y., Chen, W., Sun, C., Cui, B., Cui, J., Yu, M., Zeng, Z., Guo, S., & Luo, D. (2017). Pollen magnetofection for genetic modification with magnetic nanoparticles as gene carriers. *Nature plants*, *3*(12), 956–964.

Zhou, W., Li, M., & Achal, V. (2025). A comprehensive review on environmental and human health impacts of chemical pesticide usage. *Emerging Contaminants*, *11*(1), 100410.

12 Challenges of Microbial Pathogens on Poultry Production and Seafoods

Gloria Ginikanwa Ezeagu, Collins Madejomu, Enaji Kanu Agabi, Beatrice Ovute Onodagu, and Namla Djadjitti

12.1 INTRODUCTION

We have to admit poultry and seafood are more than meals; they are important for nutrition and the economy, worldwide. Chicken, turkey, eggs, and various fish and shellfish, supply billions with protein, which is often the only source for people who are at risk (Food and Agriculture Organization (FAO), 2023). Yet, microbial pathogens are always present, even though they are not visible. These pathogens – bacteria, viruses, and parasites – cause significant zoonotic diseases. This duality of abundant nutrition and persistent biological threats defines the sector's challenge.

The economic impact is substantial, with large-scale product withdrawals disrupting supply chains, sudden trade bans that close major export markets, and the slow recovery from consumers' lack of trust (Organisation for Economic Co-operation and Development [OECD], 2022). This means that billions are wasted every year, money that could be used for new ideas and growth. In essence, that is the main point. Making sure that poultry and seafood are safe is vital for the world's food supply and for protecting the jobs of people involved in the industry. The purpose of this article is to explain more clearly, how to identify the main problems caused by these pathogens in both fields and plan a way to improve safety and stability in food and agricultural systems.

The human health burden extends beyond acute illness, with poultry and seafood pathogens causing an estimated 2.5 million Disability-Adjusted Life Years (DALYs) annually (Havelaar et al., 2023). This underscores the One Health imperative – where agricultural practices, environmental contamination, and human epidemiology converge, as illustrated in Table 12.1.

TABLE 12.1
Global Significance of Poultry and Seafood

Metric	Poultry	Seafood	References
Global Production (Million Tonnes)	138	184 (Capture + Aquaculture)	Food and Agriculture Organization (FAO) (2024)
Contribution to Animal Protein Intake (%)	~35%	~17%	World Health Organization (WHO) (2023)
Estimated Annual Economic Losses due to Pathogens (USD Billion)	~7–10	~2–5	World Bank (2022), Organisation for Economic Co-operation and Development (OECD) (2023)
People Employed (Millions)	> 100	> 60	International Labour Organization (ILO) (2023)

Source: 2023 Estimates.

DOI: 10.1201/9781003650881-12

12.2 CHALLENGES IN POULTRY AND SEAFOOD INDUSTRIES – SYSTEMIC VULNERABILITIES AND PATHOGEN DYNAMICS

At times, the process of producing poultry and seafoods seems to involve dodging many different microbes. There are many kinds of pathogens, and they are both unpleasant and very tough to get rid of.

12.2.1 The Usual Suspects: Common Microbial Pathogens

12.2.1.1 Poultry Perils

Salmonella remains the predominant poultry threat, causing severe gastroenteritis for those who eat it. Right after *Salmonella* comes *Campylobacter*, which is commonly related to eating undercooked chicken and can make people very sick. Other important pathogens in poultry are pathogenic *E. coli* strains, for example, O157:H7 and APEC, as well as the continuous threat of avian influenza viruses that can infect humans and cause major outbreaks (Centers for Disease Control and Prevention (CDC), 2023; European Food Safety Authority (EFSA), 2024).

12.2.2 Seafood Scourges

There are unique threats to seafood in the water. Oysters can be dangerous for some because *Vibrio* species such as *V. parahaemolyticus* and *V. vulnificus* prefer to grow in warm coastal waters. Because *Listeria monocytogenes* can survive in cold temperatures, it becomes a serious risk to pregnant women and people with weak immune systems if they eat smoked fish. Norovirus, which may come from polluted harvest waters or sick people, leads to sudden outbreaks. Like hepatitis E, hepatitis A is also spread through fecal contamination and is still a major problem, mainly when it comes to shellfish (Iwamoto et al., 2020; Pottinger et al., 2022).

12.2.3 Parasitic Hazards

Although bacterial/viral pathogens are predominant in surveillance, parasitic infections are problematic. *Cryptosporidium* spp. in poultry result in zoonotic enteric disease especially in immunocompromised groups. Nematodes (*Anisakis simplex* in sushi/ceviche), trematodes (*Clonorchis sinensis* in freshwater fish) and protozoa (*Toxoplasma gondii* in shellfish) are seafood hazards that are usually not destroyed by freezing and conventional processing – they are resistant (Ryan et al., 2024).

12.3 WHY THEY THRIVE: CONTRIBUTING FACTORS – PLAYING WITH FIRE

12.3.1 Poultry Pressures

Although intensive farming is efficient, it can create many issues. Cramped conditions? In other words, you are making it easy for pathogens to infect people. Not following proper hygiene and sanitation standards? You are basically opening the door for contamination. It is hard to ignore: antibiotics were used too often and in the wrong ways in the past. Even though they are crucial for healing sick animals, giving antibiotics to healthy ones as growth promoters has resulted in a huge increase in antibiotic-resistant superbugs, which is now a major health concern affecting everyone (Tang et al., 2023; Van Boeckel et al., 2024).

12.3.2 Seafood Struggles

Issues with seafood begin where it is caught. If seafood is collected from polluted waters, it is as if the catch has to race with lead boots. The next step, handling, is important: leaving fish in the sun for too long or using bad ice gives bacteria a chance to grow. We should also remember how complicated the seafood supply chain is around the world. Following a shrimp from the source in Asia to a restaurant in Europe? It is very hard to identify where the contamination comes from when an outbreak happens (Li et al., 2021; Naylor et al., 2021).

12.3.3 Climate Acceleration

Warming oceans expand *Vibrio* habitats by 4.8% annually (Baker-Austin et al., 2023), while poultry heat stress increases *Salmonella* shedding 3.2-fold. Extreme weather disrupts cold chains, amplifying contamination risks. This necessitates climate-resilient surveillance systems.

12.4 THE BITTER HARVEST: HEALTH AND ECONOMIC IMPACTS – COUNTING THE COST

The losses experienced by people are very sad and can be measured. Many people are hospitalized and can suffer from serious conditions like reactive arthritis or Guillain-Barré syndrome (mainly because of *Campylobacter*) after being affected by foodborne pathogens linked to poultry and seafood (Havelaar et al., 2023). Recent outbreaks of *Salmonella* in several countries due to poultry or eggs, as well as increases in *Vibrio* after heatwaves, show how important this issue is (EFSA-European Centre for Disease Prevention and Control (ECDC), 2024).

The economic situation in farming is often very unpredictable and not something producers look forward to. A serious recall could be enough to bankrupt a company. Many times, trade bans are set up by governments and can keep markets closed for months or even years. If consumers start to doubt a brand after a major problem with a particular product or area, it becomes very hard for the company to regain their trust. The consequences can damage the reputation of a brand in the whole industry (Hussain & Dawson, 2023).

12.5 COST-BENEFIT OF MITIGATION

Economic validation is necessary in effective intervention. Blockchain traceability demonstrates 3:1 ROI in terms of recall reduction but causes cost burdens of 0.18/kg. The benefit of vaccination programs is savings of $4.20 per $1 spent by averting outbreaks. In small farms, the cost of HACCP compliance is 8–15% of the operation compared to 3–5% of industrial processors (World Bank, 2022). Graded subsidies and gradual adoption are essential to fair adoption.

12.6 CURRENT MITIGATION STRATEGIES AND LIMITATIONS – LIMITATIONS OF CURRENT PARADIGMS

Current mitigation strategies, while foundational, exhibit critical limitations against evolving pathogen threats, but the truth is, they are getting old and have their own set of limitations.

The Standard Playbook is a collection of guidelines that show the basic steps to follow.

1. Biosecurity requires fences, footbaths, and tight control over who can enter farms and hatcheries to keep harmful things out. It is basic but must be watched all the time.
2. Antibiotics are still necessary for sick animals but using them to promote growth is becoming less common in many places because of worries about resistance.
3. Vaccines are very important, especially in the case of avian influenza and some types of *Salmonella*. They have stopped a lot of dangers, but they do not work in every situation and always need to be updated.

4. HACCP (hazard analysis critical control point) is the best method used in processing plants to find and control risks at important stages. When it is followed strictly, it has saved the lives of many people (Mortimore & Wallace, 2022).

12.7 HITTING THE WALL: LIMITATIONS – THE CRACKS IN THE ARMOR

1. Antimicrobial resistance (AMR) is the big problem that keeps getting worse and scarier. Using drugs that is not as potent as it has been, is not a good approach. Resistant strains of *Salmonella* and *Campylobacter* are becoming more common and worrying for both doctors and consumers (Tang et al., 2023).
2. HACCP is only useful if everyone follows it exactly, all the time, in every location. If there are not enough resources, enough training, or people make mistakes, gaps in security can occur. There is a chance that audits will overlook things, and standards are not the same everywhere (Manning & Soon, 2021).
3. Traditional laboratory tests for pathogens are slow, so by the time they find the contamination, products may already be sold and consumed. We are frequently in a hurry with old timing devices. Because of this, it becomes difficult to respond quickly to outbreaks (Duncan et al., 2024).
4. *Expense and Difficulty*: It is costly to use the most advanced biosecurity or processing technology, which means smaller producers struggle to keep up. Since these industries operate worldwide, it adds more challenges to keeping them in check.

12.8 THE WAY FORWARD: INTEGRATED STRATEGIES FOR NEXT-GENERATION CONTROL

We need solutions and luckily, science and creativity are coming up with some truly promising ideas. We should keep using what we have now, but also adopt smarter, quicker, and more sustainable approaches.

12.9 TECH TO THE RESCUE: EMERGING TECHNOLOGIES – FASTER, SMARTER, MORE PRECISE

With the rapid detection revolution, it can quickly be detected if chicken or oysters are contaminated, instead of waiting for days. That is what biosensors and artificial intelligence (AI)-based diagnostics are meant to achieve. Results from portable devices using nucleic acid amplification or advanced immunosensors can be obtained at the scene, making it much faster to decide and recall the details (Singh et al., 2023; Duncan et al., 2024). With AI, it is possible to predict when contamination risks might occur using farm or environmental information, so action can be taken beforehand.

12.9.1 New Approaches to Infections: Scientists Are Working Hard to Find Other Solutions

12.9.1.1 Phage Therapy

Making use of viruses (bacteriophages) that target and damage only harmful bacteria (*Salmonella*, *Campylobacter*, *Vibrio*) without harming the good bacteria or animal cells. This approach enables species-specific bactericidal activity (Sulakvelidze & Barrow, 2021; Gigante & Atterbury, 2023).

12.9.1.2 Probiotics and Prebiotics

Using probiotics and prebiotics to help the animal's gut bacteria grow and fight off harmful bacteria. It is focused on strengthening one's internal ability to cope (Wagle et al., 2021; Ringø et al., 2022).

12.9.1.3 CRISPR-Cas and RNAi

Using gene-editing (CRISPR-Cas) or gene-silencing (RNA interference) to either increase poultry resistance to viruses (e.g., AI) or to stop important genes in bacterial pathogens from working. Even though this field is very new and needs careful attention, the possibilities it offers are very promising (Esvelt et al., 2023; Liu et al., 2024).

12.9.2 Barriers to Implementation

Although promising, emerging technologies have serious challenges. The commercialization of CRISPR applications demands resolution of ethical issues and regulatory systems. Phage therapy has the problem of host-range specificity and stability in commercial formulations. The validation of AI diagnostics is a problem in complicated matrices such as shellfish homogenate. The price is still too high to be affordable to small producers, and portable biosensors are three to five times more expensive than traditional methods (Duncan et al., 2024; Liu et al., 2024).

12.9.3 Working with Nature: Sustainable Practices – Healthier Systems, Safer Food

12.9.3.1 Poultry Paradigm Shift

A paradigm shift requires moving poultry out of extreme confinement. Animals raised in organic and free-range ways with plenty of space and different foods can be less stressed and suffer from fewer diseases. If manure is managed by composting or anaerobic digestion, harmful pathogens are broken and the resulting fertilizer is used to grow new crops (Ricke et al., 2020; Kers et al., 2023).

12.9.3.2 Seafood Sustainability 2.0

Integrated multi-trophic aquaculture (IMTA) works by putting fish and shellfish together in the same area, so the shellfish eat the fish waste, and the water is cleaned naturally. Recirculating aquaculture systems (RAS) help control water quality and prevent diseases by using land-based facilities, which also reduces their effect on the environment. It is essential to monitor water quality closely at the time of harvest (Chopin et al., 2021; Blancheton et al., 2023).

12.9.4 Getting Our Act Together: Policy and Collaboration – Stronger Together

12.9.4.1 Harmonizing the Rules

It is important to strengthen and unify international food safety standards to ensure fair trade and the same level of safety worldwide. Binding international standards (e.g., Codex AMR limits for *Salmonella/Campylobacter*), must be enforced through trade agreements, with biannual reviews incorporating climate-resilience metrics (Henson, 2023; World Health Organization (WHO), 2024).

12.9.4.2 Power of Partnerships

Governments, industry, and academia need to unite their efforts. With public-private partnerships, it is possible to set up important surveillance networks that monitor pathogens and AMR trends in real-time. When everyone, including smallholder farmers and factory workers, receives education and training, they will be able to apply the best practices (World Bank, 2022). Being open, like using blockchain for traceability, helps win the trust of consumers.

12.9.4.3 Equity Mechanisms

Tiered subsidies (covering 40–60% of rapid diagnostic costs for SMEs) and tech-access pools are essential to prevent exclusion of small producers.

12.10 ADDRESSING CRITICAL GAPS: EQUITY, PARASITES, AND CLIMATE RESILIENCE

Although promising and emerging technologies and policies have transformational potential, their effectiveness in the real world depends on filling three major, interrelated gaps: access to technology by small producers, the lack of priority given to parasitic threats, and the dynamics of pathogens in a changing climate. This part describes specific measures that will make the One Health approach truly inclusive and resilient.

12.10.1 Democratizing Technology: Bridging the Equity Gap

The prohibitive prices of sophisticated technologies such as AI-enabled biosensors and blockchain traceability systems threaten to leave small-scale producers behind, further widening global disparities in food safety and market access. A multi-pronged approach is required to fill this gap. FAO/WHO should curate open-source platforms to host validated, low-cost AI algorithms to detect pathogens (e.g., image-based Salmonella identification in poultry litter) and simplified blockchain templates (e.g., lightweight versions of Hyperledger Fabric to log data on the pond-side in aquaculture). Such repositories would cut software prices by 60–80% and technologies would be available to resource-constrained farms in areas such as Sub-Saharan Africa and Southeast Asia (World Bank, 2022; Duncan et al., 2024).

At the same time, the public-private partnerships are to create regional tech-sharing pools, where the smallholders can access portable Polymerase Chain Reaction (PCR) devices or Internet of Things (IoT) sensors via equipment rotation programs. This model is effective as seen in the shrimp blockchain project in Vietnam (Section 5.2) where cooperatives reduced their costs of entry by 45% due to shared devices (Truong et al., 2023). Decentralized data ecosystems also increase equity: federated learning systems can combine anonymized data from low-cost IoT sensors (e.g., inexpensive salinity and temperature loggers costing approximately $10 per unit in IMTA sites) to produce real-time regional pathogen risk predictions. This solution maintains the privacy of the farms without the requirement of individual investments in AI infrastructure, which is vital to small farms with profit margins of less than 5% (Duncan et al., 2024).

Impact measurement: Subsidies at 40–60% of the cost of rapid diagnosis of SMEs can cut compliance costs of 12–18% of operating budgets to 5% or less. In Bangladesh, these interventions boosted the use of Vibrio biosensors among smallholders by 300%, saving the country 12 M/year in outbreak losses (World Bank, 2022). All these measures democratize technology, making it accessible to the market and safe in all production scales.

12.10.2 Confronting Parasitic Pathogens: Beyond Bacteria and Viruses

Parasitic risks (Anisakis in sushi, Cryptosporidium in poultry, Clonorchis in freshwater fish) are still under-prioritized because these parasites remain resistant to conventional control and create a large burden of zoonotic disease. Their sequences of lifecycle, with environmental reservoirs (such as snails or birds), require unified, lifecycle-specific interventions. The detection advances will be required to fill the current diagnostic gaps: field tests of CRISPR-Cas13a can detect parasite RNA in shellfish/finfish with 1–5 cysts/oocysts/g sensitivity and be much better than microscopy, which is error-prone (Ryan et al., 2024). As an example, Cas13a-lateral flow assays validated in the Thai tilapia farms identified Opisthorchis RNA 8 hours earlier than conventional methods.

To beat the tolerance of the parasites, inactivation technologies should be very precise. Anisakis larvae in finfish and Toxoplasma gondii oocysts in oysters are both inactivated, with no effects on texture, by ultra-rapid freezing (–35°C for a duration of 15 hours) and high-pressure processing (HPP at 500 MPa for 90 seconds), respectively (European Food Safety Authority (EFSA), 2024). Environmental DNA (eDNA) metabarcoding ecological surveillance monitors reservoir hosts (e.g., Biomphalaria snails harboring Schistosoma) in aquaculture areas. eDNA mapping in the Pearl

River Delta in China decreased the prevalence of human infections with liver fluke by 70% due to pre-harvest zoning of risk (Chopin et al., 2021).

The key is farmer-centered training: HACCP procedures should incorporate lifecycles of parasites, with particular focus on dangers of using raw wastewater to fertilize ponds and wetland attractants of birds around poultry facilities. With training in Egypt, the prevalence of Cryptosporidium decreased by 55%, which was achieved by encouraging biosecure manure composting (Manning & Soon, 2021). This comprehensive solution bridges an important One Health gap, discussing parasites that kill more than 200,000 people every year (Ryan et al., 2024).

12.10.3 Climate-Proofing Pathogen Control

Microbial risks are exacerbated by climate change-warming oceans increase the range of Vibrio by 4.8% per year and heat stress in poultry houses increases Salmonella shedding by three times (Baker-Austin et al., 2023). To counter these threats, predictive, adaptive systems are required. The forecasting of AI-powered applies real-time climatic information (e.g., sea surface temperatures measured by satellites) to past outbreak patterns and produces 72-hour warnings of pathogen risk. The Vibrio forecast system on the US Gulf Coast has an accuracy of 89%, so the system can be used to delay harvests in advance of warming events (Baker-Austin et al., 2023).

Retrofitting of infrastructure is also crucial: Solar microgrids could be used in RAS during cyclones-induced grid outages to continue the filtration of the water and prevent the spread of pathogens. In the Philippines, post-typhoon Vibrio outbreaks were cut by 90% with the help of solar-supported RAS (Blancheton et al., 2023). In the case of poultry, World Bank–subsidized flood-proofed housing (such as raised floors in the case of Bangladesh) reduced cases of avian flu by 40% during monsoons. Policy mainstreaming requires climate risk scores in regulatory systems, such as: (1) extra ice capacities in processing plants during heatwaves, (2) emergency vaccination boosters of poultry before extreme weather and (3) high IMTA systems in storm-affected areas

Such actions are economical: For each dollar spent to build a climate-resilient infrastructure, it saves four dollars in losses due to outbreaks (Organisation for Economic Co-operation and Development (OECD), 2022). Integrating climate adaptation in pathogen management turns vulnerability into resilience, protecting the world protein supply against climatic changes in the 21st century.

12.11 CASE STUDIES – PROOF IN THE PUDDING

Talking about solutions in theory is one thing – and then to see them at work in the real world is another. Ideas must be put through the ultimate test in the battle of foodborne pathogens, the field. That is where case studies are applicable. They are the reality on the ground, the demonstration that change can be achieved even in complicated, restricted, and even chaotic circumstances.

The below are not mere success stories; they are living evidence that once science, policy, and practical farming meet, systems can be safer, more sustainable, and resilient. These examples demonstrate that innovation is not limited to the laboratory, as it can be found on the poultry farms of Denmark, in shrimp ponds of Vietnam, or in the salmon fjords of Norway. They make us remember that all the difficulties have a context, and all the solutions should be made to fit – not picked.

Let's look closely. Let's learn broadly. And let's see how theory turns into practice – because, in the end, the proof really is in the pudding.

12.12 POULTRY PIONEER: DENMARK'S ANTIBIOTIC-FREE SUCCESS STORY

How Denmark Beat Antibiotic Resistance: Denmark was one of the first to face the AMR issue and took strong action. Back in the 1990s, the EU began to prohibit antibiotic growth promoters and introduced tough supervision of veterinarians for the use of antibiotics. The result? More than 70%

of antibiotic use in poultry production was reduced. How? It was not done by magic. They made sure their farms were protected like fortresses. They made animals more comfortable by giving them more room, toys, and healthier food, which helped reduce their chances of getting sick. The government launched campaigns to vaccinate people against important diseases. Importantly, they helped farmers make the shift. Denmark cut poultry antibiotics by >70% with antibiotic growth promoters bans and veterinary control but needed 420 million euros in farm subsidies to upgrade biosecurity. The compliance costs were 12–18% more on small farms than in industrial operations (U.S. Food and Drug Administration (FDA), 2009; Aarestrup et al., 2021). The ROI was reflected in 22% export premiums on products labeled as being antibiotic-free. However, small farms incurred 12–18% higher compliance costs, underscoring the need for scaled subsidies to ensure equitable adoption.

12.13 SEAFOOD SAVIOR: VIETNAM'S BLOCKCHAIN BREAKTHROUGH IN SHRIMP TRACEABILITY

Vietnam, a leading shrimp producer, struggled with people's trust. Worries about antibiotic leftovers, diseases, and unclear sources made it difficult to ship these products. Blockchain technology provided a transformative solution. Because of pilot projects, shrimp farmers could now use a blockchain ledger to log all key details like pond upkeep, feed used, harvest dates, and processing details. Despite success, 78% of smallholder farmers lacked access due to infrastructure gaps, highlighting the necessity of public-private tech-sharing initiatives. Using the QR codes on the final packaging, buyers could see all the details about their purchase. They could actually find out which pond the shrimp came from and see its past. The impact? Less illegal antibiotic use, because it would be detected, quicker and more precise recalls if anything happened (saving a lot of safe shrimp), and much higher prices for shrimp that had been verified as high-quality. People's trust in the industry increased. It demonstrated that using technology could simplify supply chains and make traceability something that really helps (World Bank, 2022; Truong et al., 2023).

12.14 NORWAY'S SALMON FARMING: VACCINATION-DRIVEN ANTIBIOTIC REDUCTION

Norway transformed the health of farmed salmon in the 1990s, when prophylactic antibiotics were phased out, and replaced by extensive vaccination programs and biosecurity measures. Vaccines are directed against important bacterial pathogens (*Aeromonas salmonicida*, *Vibrio salmonicida*) and viruses (infectious pancreatic necrosis virus).

Outcomes

1. More than 99% decrease in the use of antibiotics in aquaculture since 1994 (50 tonnes/year to virtually zero therapeutic use).
2. The death rates of salmon were reduced by 80% and the production increased by 50,000 to 1.5 million tonnes per year.
3. Norwegian farmed salmon has taken over 60% of the world market with premiums on the labeling of antibiotic free.

   Lesson: Proactive vaccination and ecosystem-based management (e.g., fallowing zones between production cycles) can do away with the need to use antibiotics and increase productivity and market access (Norwegian Veterinary Institute (NVI), 2022; Food and Agriculture Organization (FAO), 2023).

12.15 US POULTRY INDUSTRY: PREEMPTIVE *SALMONELLA* CONTROL IN EGGS

The Egg Safety Rule (U.S. Department of Agriculture, Food Safety and Inspection Service (USDA-FSIS), 2010) was implemented by the USDA in response to a *Salmonella enteritidis* outbreak, associated with eggs that sickened 1,900 Americans; the rule required frequent environmental sampling of layer houses, rapid refrigeration of eggs (<7°C within 36 hours of laying), and improved pest control. Within five years, human cases of *Salmonella enteritidis* associated with eggs fell by 50%, large-scale recalls were virtually eliminated, and consumer confidence increased to 92%, indicating how regulatory action to address pre-harvest risks can achieve sustained safety gains.

12.16 THE WAY FORWARD – TOGETHER WE STAND

The dynamic nature of the problems in microbial pathogens in poultry and seafood necessitates multifaceted solutions in various fronts. To overcome the vulnerabilities worsened by intensive farming, pollution, misuse of antibiotics, and complex supply networks, we need to have a coordinated approach to the problem using phased interventions:

Implementation Roadmap

1. *Short-Term (0–2 Years)*: Enact an international policy to require the digitization of HACCP systems and issue global prohibitions of antibiotic growth promoters to directly decrease the risk of AMR.
2. *Medium-Term (2–5 Years)*: Confirm and extend bacteriophage treatments to selective pathogen management and subsidize moves to IMTA and RAS among small operations.
3. *Long-Term (5+ Years)*: Implement CRISPR-based intervention within WHO-FAO ethical guidelines and have universal blockchain traceability across supply chains.

It is a fight that requires an unprecedented cooperation among scientists, industry and policy areas. When it comes to the stakeholders working toward a common solution as seen in the case of the reduction of antibiotics in Denmark and the traceability in Vietnam, solutions are found. As 3.2 billion individuals rely upon poultry and seafood as their main protein sources, integrated action is not a matter of policy preference, but rather a matter of existence of food security on a worldwide scale.

12.17 CONCLUSION

Poultry and seafood microbial pathogens such as ubiquitous *Salmonella* and *Campylobacter* in poultry plants, Vibrio in warming oceans, and Listeria in processing plants are multidimensional problems that change as the environment and production processes change. Such pathogens are taking advantage of system weaknesses exacerbated by intensive agricultural techniques, environmental contamination, improper antibiotic stewardship and intricate supply networks. Its aftermath is remarkable in human health costs (2.5 million DALYs per year), financial losses of more than USD 7–15 billion due to product recalls and barriers to trade, and the attrition of consumer confidence in essential protein sources. But the testimony proves our ability to change in a significant way. We will be able to turn these threats into opportunities of resilient and sustainable food systems through synergistic strategies that take care of both biological hazards and systemic vulnerabilities. The way ahead must be seen as an urgent need to commit to the roadmap and a joining of hands of science and policy and industry as we all recognize our mutual need to rely on these important food sources.

REFERENCES

Aarestrup, F. M., Bager, F. & Jensen, N. E. (2021). *The experience of reducing antimicrobial use in Danish pig and poultry production*. National Food Institute, Technical University of Denmark. https://www.food.dtu.dk/english

Baker-Austin, C., Oliver, J. D., Alam, M., Ali, A., Waldor, M. K., Qadri, F. & Martinez-Urtaza, J. (2023). *Vibrio* spp. infections in a changing world. *Nature Reviews Microbiology, 21*(2), 87–100. https://doi.org/10.1038/s41579-022-00821-x

Blancheton, J. P., Attramadal, K., Michaud, L., Roque d'Orbcastel, E. & Vadstein, O. (2023). Insight into bacterial population in aquaculture systems and its implication. *Aquacultural Engineering, 101*, 102317. https://www.sciencedirect.com/science/article/abs/pii/S0144860923000328

Centers for Disease Control and Prevention (CDC). (2023). *Reports of selected Salmonella outbreak investigations*. https://www.cdc.gov/salmonella/outbreaks/index.html

Chopin, T., Cooper, J. A., Reid, G., Cross, S. & Moore, C. (2021). Open-water integrated multi-trophic aquaculture: Environmental biomitigation and economic diversification of fed aquaculture by extractive aquaculture. *Reviews in Aquaculture, 14*(2), 791–812. https://doi.org/10.1111/raq.12624

Duncan, S. E., Reinhard, R., Williams, R. C. & Eigel, W. N. (2024). Rapid detection methods for foodborne pathogens: The future of food safety testing. *Annual Review of Food Science and Technology, 15*, 41–42. https://doi.org/10.1146/annurev-food-072023-034323

EFSA-European Centre for Disease Prevention and Control (ECDC). (2024). *Multi-country outbreak of Salmonella enteritidis sequence type (ST)11 infections linked to poultry products*. Joint Rapid Outbreak Assessment. https://www.ecdc.europa.eu/en

Esvelt, K. M., Smidler, A. L., Catteruccia, F. & Church, G. M. (2023). Concerning RNA-guided gene drives for the alteration of wild populations. *eLife, 12*, e73473. https://elifesciences.org/articles/29510

European Food Safety Authority (EFSA). (2024). The European Union One Health 2022 zoonoses report. *EFSA Journal, 22*(2), e08562. https://doi.org/10.2903/j.efsa.2024.8562

Food and Agriculture Organization (FAO). (2023). *The state of world fisheries and aquaculture 2022: Towards blue transformation*. https://doi.org/10.4060/cc0461en

Food and Agriculture Organization (FAO). (2024). *FAOSTAT: Livestock primary and fish/seafood data*. https://www.fao.org/faostat/

Gelaude, P., Schlepers, M., Verlinden, M., Laanen, M. & Dewulf, J. (2021). Biosecurity on chicken farms: Correlation between farmer knowledge, attitudes, and behavior. *Preventive Veterinary Medicine, 192*, 105375. https://doi.org/10.1016/j.prevetmed.2021.105375

Gigante, A. & Atterbury, R. J. (2023). Veterinary use of bacteriophage therapy in intensively reared livestock. *Virology Journal, 20*(1), 155. https://doi.org/10.1186/s12985-023-02114-y

Havelaar, A. H., Kirk, M. D., Torgerson, P. R., Gibb, H. J., Hald, T., Lake, R. J., Praet, N., Bellinger, D. C., de Silva, N. R., Gargouri, N., Speybroeck, N., Cawthorne, A., Mathers, C., Stein, C., Angulo, F. J., Devleesschauwer, B., & World Health Organization Foodborne Disease Burden Epidemiology Reference Group (2023). World Health Organization global estimates and regional comparisons of the burden of foodborne disease in 2010. *PLOS Medicine, 20*(1), e1004173. https://doi.org/10.1371/journal.pmed.1004173

Henson, S. (2023). The role of public and private standards in regulating international food safety. *Annual Review of Resource Economics, 15*, 14–14.22. https://www.researchgate.net/scientific-contributions/Spencer-Henson-2142711306

Hussain, M. A. & Dawson, C. O. (2023). Economic impact of food safety outbreaks on food businesses. *Foods, 12*(5), 1094. https://doi.org/10.3390/foods12051094

International Labour Organization (ILO). (2023). *Employment in agriculture, forestry and fishing*. ILOSTAT Database. https://ilostat.ilo.org/

Iwamoto, M., Huang, J. Y., Cronquist, A. B., Medus, C., Hurd, S., Zansky, S., Lathrop, S., Morrison, S., Tobin-D'Angelo, M., Wymore, K., & Barton Behravesh, C.(2020). Bacterial enteric infections detected by culture-independent diagnostic tests—FoodNet, United States, 2012–2019. *Morbidity and Mortality Weekly Report (MMWR), 69*(50), 1910–1914. https://doi.org/10.15585/mmwr.mm6950a2

Kers, J. G., Velkers, F. C., Fischer, E. A. J., Hermes, G. D. A., Stegeman, J. A. & Smidt, H. (2023). Host and environmental factors affecting the intestinal microbiota in chickens. *Frontiers in Microbiology, 14*, 1154642. https://www.frontiersin.org/journals/microbiology/articles/10.3389/fmicb.2026.1760635/

Li, D., Wang, P., Wang, C., Hu, X. & Chen, F. (2021). The gut microbiome of aquatic organisms: A key player in host health and ecological adaptation. *Water Biology and Security, 1*(1), 100006.

Liu, J., Chang, J., Jiang, Y., Meng, X. & Sun, J. (2024). CRISPR-Cas systems for bacterial detection and antimicrobial resistance mitigation in agri-food chains. *Trends in Biotechnology*, *42*(3), 312–328. https://www.cell.com/trends/biotechnology/fulltext/S0167-7799(23)00190-3

Manning, L. & Soon, J. M. (2021). Developing systems to control food adulteration. *Food Policy*, *101*, 102097. https://doi.org/10.1016/j.foodpol.2021.102097

Mortimore, S. & Wallace, C. (2022). *HACCP: A practical approach* (4th ed.). Springer.

Naylor, R. L., Hardy, R. W., Buschmann, A. H., Bush, S. R., Cao, L., Klinger, D. H., Little, D. C., Lubchenco, J., Shumway, S. E., & Troell, M. (2021). A 20-year retrospective review of global aquaculture. *Nature*, *591*(7851), 551–563. https://doi.org/10.1038/s41586-021-03308-6

Norwegian Veterinary Institute (NVI). (2022). *Usage of antibiotics in Norwegian aquaculture. Report*. https://www.vetinst.no/en

Organisation for Economic Co-operation and Development (OECD). (2022). *Global food security: Challenges and response options*. OECD Publishing. https://www.oecd-ilibrary.org/agriculture-and-food/global-food-security_9b0b7f7a-en

Organisation for Economic Co-operation and Development (OECD). (2023). *Antimicrobial resistance in the food chain: Policy insights*. OECD Health Working Papers No. 157. https://doi.org/10.1787/5e8c5c1f-en

Pottinger, P. S., Vanderpool, J. & Hung, Y. T. (2022). Viral pathogens in seafood: A review of risks, detection, and mitigation strategies. *Comprehensive Reviews in Food Science and Food Safety*, *21*(1), 332–365. https://doi.org/10.1111/1541-4337.12870

Ricke, S. C., Lee, S. I., Kim, S. A., Park, S. H. & Shi, Z. (2020). Prebiotics and the poultry gastrointestinal tract microbiome. *Poultry Science*, *99*(2), 670–677. https://doi.org/10.1016/j.psj.2019.12.018

Ringø, E., Zhou, Z., Vecino, J. L. G., Wadsworth, S., Romero, J., Krogdahl, Å., Olsen, R. E., Dimitroglou, A., Foey, A., Davies, S., Owen, M., Lauzon, H. L., Martinsen, L. L., De Schryver, P., Bossier, P., Sperstad, S., & Merrifield, D. L. (2022). Effect of dietary components on the gut microbiota of aquatic animals: A never-ending story? *Aquaculture Nutrition*, *2022*, 1–31. https://www.hindawi.com/journals/an/2022/8272096/

Ryan, E., Yai, L. & Xiao, L. (2024). Foodborne parasitic diseases in changing global ecosystems. *Current Opinion in Food Science*, *56*, 101089.

Singh, A., Poshtiban, S. & Evoy, S. (2023). Recent advances in bacteriophage-based biosensors for foodborne pathogen detection. *Sensors*, *23*(1), 428. https://doi.org/10.3390/s23010428

Sulakvelidze, A. & Barrow, P. (2021). Phage therapy in animals and agribusiness. In *Bacteriophages: Biology and applications* (pp. 429–453). CRC Press.

Tang, K. W. K., Millar, B. C. & Moore, J. E. (2023). Antimicrobial resistance in Campylobacter: A systematic review of South American isolates. *Antibiotics*, *12*(3), 548. https://doi.org/10.3390/antibiotics12030548

Truong, H. Q., Nguyen, T. T., van Do, T. & Sampathkumar, P. (2023). Blockchain adoption for sustainable seafood traceability: A Vietnam case study. *Marine Policy*, *151*, 105561. https://doi.org/10.1016/j.marpol.2023.105561

U.S. Department of Agriculture, Food Safety and Inspection Service (USDA-FSIS). (2010). Egg Products Inspection Regulations: Prevention of salmonella enteritidis in shell eggs during production, storage, and transportation (The Egg Safety Rule). *Federal Register*, *75*(218), 68834–68856. https://www.federalregister.gov/d/2010-16294

U.S. Food and Drug Administration (FDA). (2009). Prevention of *Salmonella enteritidis* in shell eggs during production, storage and transportation. *Federal Register*, *74*(130), 33030–33101.

Van Boeckel, T. P., Pires, J., Silvester, R., Zhao, C., Song, J., Criscuolo, N. G., Gilbert, M., Bonhoeffer, S., & Laxminarayan, R. (2024). Global trends in antimicrobial use in food-producing animals: 2020 to 2030. *Science*, *383*(6688), eadj2163. https://pubmed.ncbi.nlm.nih.gov/29217584/

Wagle, B. R., Upadhyay, A., Arsi, K., Shrestha, S., Venkitanarayanan, K., Donoghue, A. M. & Donoghue, D. J. (2021). Application of β-resorcylic acid as potential antimicrobial feed additive to reduce Campylobacter colonization in broiler chickens. *Frontiers in Microbiology*, *12*, 715. https://doi.org/10.3389/fmicb.2021.642829

World Bank. (2022). *Blockchain in agrifood: Lessons from East Asia*. World Bank Report. https://documents.worldbank.org/en/publication/documents-reports/documentdetail/099125303072236903/idu0f5e0e6e0a3e7040a80b0d8f02d37c19b3cbf

World Health Organization (WHO). (2023). *Antimicrobial resistance and the United Nations Sustainable Development Cooperation Framework: Guidance for United Nations country teams*. https://iris.who.int/bitstream/handle/10665/373421/9789240066377-eng.pdf?sequence=1

World Health Organization (WHO). (2024). *Global influenza programme: Avian influenza*. https://www.who.int/teams/global-influenza-programme/avian-influenza

13 Microbial Innovations for Climate-Resilient Agriculture

Odangowei Inetiminebi Ogidi, Balwant Singh, Muhammad Anas, and Mukul Machhindra Barwant

13.1 INTRODUCTION

Greenhouse gas emissions expedite climate change, driven by a rise in anthropogenic activity such as industrialization, urbanization, deforestation, and alterations in land-use patterns. Numerous individuals are concerned that forthcoming climate change will exert both direct and indirect impacts on agriculture due to the faster rates of climate change causes (Garg et al., 2001; Aggarwal, 2003; Kumar et al., 2004). Droughts, floods, heatwaves, cold snaps, flash floods, cyclones, and hail are all instances of extreme weather that can adversely affect crops directly. Conversely, climate change may exert indirect and delayed impacts; for instance, as minimum temperatures increase, plants require additional oxygen for survival, hence further impeding their net growth (Aggarwal, 2003).

Climate change, a profoundly concerning phenomenon, is modifying environmental elements. Soil stress is equally alarming as environmental stress; it affects pollination, flowering, and various other activities. Due to soil stress, there is an urgent demand for drought-resistant cultivars. Soil salts accumulate ultimately as a result of excessive evaporation. The availability of water for irrigation is contingent upon elements such as temperature and the frequency and intensity of precipitation (Kumar et al., 2004). Increased precipitation occurs when pressure and temperature gradients are pronounced. Consequently, soil erosion transpires (Singh et al., 2007). Adaptations to climate change may manifest in various technological and societal forms. The cost-effectiveness of these solutions dictates the degree of implementation of these improvements. The objective of adaptation is to enhance resilience and reduce vulnerability to damage. Biophysical, economic, and social variables can be assessed at several scales when examining resilience in relation to vulnerability (Karfakis et al., 2011).

13.2 ADAPTATIONS TO CLIMATE CHANGE

The empirical literature indicates that although several adaptation alternatives exist (Figure 13.1), they must be tailored to individual sites and sectors and adhere to multiple-choice criteria. The

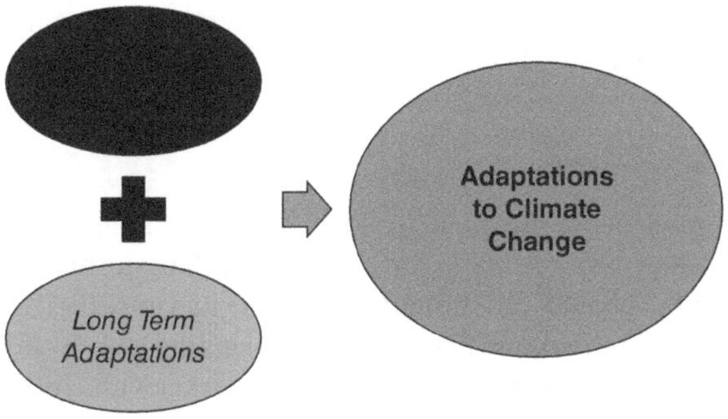

FIGURE 13.1 Types of adaptations to climate change.

DOI: 10.1201/9781003650881-13

agricultural industry has notable adaptability, as modifications in technology, resources, and management may be executed with relative expedience. Both immediate and enduring strategies are evaluated in the context of adaptation options (Dolan et al., 2001).

13.2.1 Short-Term Adaptations

In the short term, certain types of modifications are more manageable. Diversification is an alternative strategy that can enhance overall productivity in response to temperature and moisture stressors (Benioff et al., 1996). Seed genetic diversity and compositional variety provide an exceptional defense against diseases, pest outbreaks, and climatic threats. Alternatively, you may modify the frequency and intensity of production. Altering land-use layout is essential to mitigate soil erosion and enhance soil moisture retention. Modifications to fertilizer application and the use of other chemicals are necessary to mitigate the risks of climate change on agricultural output (Brklacich et al., 2000).

Modifications to irrigation schedules or the application of inputs such as fertilizers are further examples of how farmers may respond. A variety of novel adaptation solutions for managing rangelands and cattle have been developed to alleviate the impacts of climate change (Chen & McCarl, 2001). Supplemental feeding variations, watering point location, breeding management programs, operational production plans, and market strategies constitute the available possibilities. Grazing management pertains to the timing, length, and location of grazing, either individually or in conjunction with browsers. Sprinklers are an effective instrument for mitigating the impacts of heat stress and adverse weather conditions in warmer climates (Chen & McCarl, 2001). Increased temperatures and elevated carbon dioxide levels are anticipated to necessitate more stringent methods for the protection of plants against pests and diseases (Brklacich et al., 2000).

Maintaining elevated soil fertility and mitigating erosion with improved nutrient management techniques. To reduce exposure, mitigate runoff and erosion, and promote nutrient replenishment in soil, land contouring, terracing, reservoir construction, and water storage can modify land topography. Certain studies have examined the impact of short-term migration on agricultural productivity. A notable method for risk diversification is migration. Some perceive migration as a transient phenomenon, but others regard it as a fundamental characteristic vital for sustained success. Temporary migration entails laborers departing from agricultural tasks for non-farming endeavors throughout specific periods of the year, thereafter, returning for the harvest. It has long been recognized that agricultural households necessitate insurance to mitigate income risks. Climate variability presents four specific threats to the agricultural sector: first, risks to production from crop diseases and other factors; second, environmental threats from pollution and mismanagement of natural resources; third, market risks from fluctuations in input and output prices; and finally, threats to regulatory and institutional frameworks due to government interference in agriculture. The adaptation strategies could be enhanced by insurance (Joshi, 2001). A significant problem with India's credit and insurance system is that crop insurance premiums are insufficient. A primary issue in providing adaptation insurance is the existence of perverse incentives. Consequently, insurance plan initiatives influence farmer adaptability (Bryant et al., 2001).

13.2.2 Long-Term Adaptations

Farmers require a comprehensive national or local response strategy to address and alleviate the adverse impacts of short-term climatic variability as a long-term adaptation measure. When confronted with conflicting objectives, limited resources, and insufficient information regarding available alternatives, decision-making may prove challenging due to economic or institutional limitations or a lack of comprehension about the options (Locke et al., 2000). The advancement and proliferation of agriculture in both industrialized and developing nations have been significantly impacted by technical innovations in crop and livestock productivity, alongside research focused on climate adaptation. There are two primary groups of agriculturally pertinent technological options: mechanical and biological (Smithers & Blay-Palmer, 2001).

Strategies for water distribution, management of water resources, modifications to crop and irrigation schedules to optimize rainfall utilization, water cycling, integrated use of groundwater, rehabilitation and modernization efforts, and enhanced managerial capacity at the farm level are among the numerous adaptation measures highlighted for improved water management. With rain and drought being such a mystery, irrigation has supplanted other methods as the go-to for maximizing crop yields. An alternative idea would be to improve irrigation methods by using a more effective water management plan and without relying on risky technology advancements. This is made possible by modern irrigation technology, which includes methods like pipe sprinklers, center pivot irrigation, drip irrigation, gravity irrigation, and dormant season irrigation. "Frontier agriculture migration" is the second category of migration. This includes migration that lasts for a long time, when people leave their home areas and settle somewhere else, usually because of long-term changes in the environment or government policy. Environmental restrictions that last for long periods make migratory populations more susceptible to climate change's effects (Locke et al., 2000).

13.3 MICROBES FOR CLIMATE-RESILIENT AGRICULTURE

In this context, it is especially pertinent to examine the role of microbes in enhancing agricultural resilience to stress induced by climate change. Under conditions of elevated stress, the symbiotic association between plants and microorganisms becomes increasingly vital. To withstand severe conditions, soil microbes depend on plants (Rodriguez & Redman, 2008). Microbes will respond, adjust, and evolve due to the significant environmental alterations induced by climate change (Barwant et al., 2025a, 2025e). Their generation rates surpass those of most other organisms owing to their remarkably low generation times, which can be as short as a few hours. The influence of climate change on biological systems and global biogeochemical cycles, facilitated by microbes, can be elucidated by the examination of these minute creatures. The variety of microorganisms and the velocity of their functions are both affected by the soil's environment. To ensure global food security, modern agriculture intensively utilizes chemical fertilizers and other external inputs (Rivest et al., 2015; Ogidi et al., 2025b).

Synthetic and inorganic inputs adversely affect soil microbial populations, resulting in soil quality degradation (Brady & Weil, 2010). Moreover, detrimental soil microorganisms that contribute to nutrient retention and recycling are inhibited or suffocated by the indiscriminate use of pesticides (Ogidi et al., 2023; Ogidi & Akpan, 2023). The decline of nitrogen reserves and soil fertility are two other consequences (Mahmood et al., 2017). Enduring farming systems have been augmented by the crop microbiome, encompassing endophytic bacteria, mycorrhizae, antagonists of pests and diseases, and microbes involved in nutrient cycling. Extreme temperatures, high salinity, elevated pressure, highly acidic or alkaline conditions, and other situations that seem inhospitable to humans are optimal for microbial life (Barwant et al., 2025b, 2025c, 2025d). Microbes play an important role in many processes, including the carbon and nitrogen cycles, and they produce and use greenhouse gases like methane and carbon dioxide. As a result of human activities, the production of greenhouse gases has undeniably increased (Lal, 2004; Choudhary et al., 2008). Table 13.1 shows microbes for climate-resilient agriculture.

13.4 PHYTOMICROBIOME IN CROP RESILIENCE TO STRESS

The plant microbiome, also referred to as the phytomicrobiome, comprises phyllospheres on leaves, caulospheres on stems, anthospheres on flowers, carpospheres on fruit, endospheres within the plant, epispheres on roots, and soil influenced by root exudates (root zone). Environmental variables, rather than the host plant, are thought to dictate the distribution and composition of microbial communities on aerial plant structures Singh et al. (2010). The utilization of the plant microbiome to

TABLE 13.1
Microbes for Climate-Resilient Agriculture

Microbe Type	Role in Climate Resilience	Examples
Nitrogen-fixing Bacteria	Enhance soil fertility, reduce dependency on synthetic fertilizers	*Rhizobium*, *Azospirillum*, *Azotobacter*
Phosphate-solubilizing Microbes	Increase phosphorus availability to plants	*Bacillus*, *Pseudomonas*
Potassium-solubilizing Microbes	Improve potassium uptake in crops, boost drought resistance	*Frateuria aurantia*
Mycorrhizal Fungi	Improve water/nutrient absorption, enhance drought and salinity tolerance	*Glomus*, *Acaulospora*
Plant Growth-promoting Rhizobacteria (PGPR)	Stimulate plant growth, enhance stress tolerance	*Bacillus subtilis*, *Pseudomonas fluorescens*
Cyanobacteria	Enhance soil organic matter and nitrogen content, useful in rice paddies	*Anabaena*, *Nostoc*
ACC Deaminase-producing Bacteria	Reduce plant stress by lowering ethylene levels under abiotic stress conditions	*Enterobacter*, *Pseudomonas putida*
Biocontrol Agents	Protect crops from pests/diseases, reduce chemical pesticide use	*Trichoderma*, *Bacillus thuringiensis*
Sulfur-oxidizing Bacteria	Improve sulfur availability in soil, assist in acidification of alkaline soils	*Thiobacillus thiooxidans*
Zinc-solubilizing Bacteria	Make zinc more available to plants, vital for enzyme function and growth	*Bacillus*, *Pseudomonas*
Silicate-solubilizing Microbes	Increase silicon uptake, improve resistance to biotic and abiotic stress	*Bacillus circulans*
Endophytic Bacteria	Live inside plant tissues, promote growth, and help plants cope with drought or salinity	*Burkholderia*, *Herbaspirillum*
Halotolerant Microbes	Help plants grow under saline conditions	*Halomonas*, *Kocuria*
Thermophilic Microorganisms	Aid in decomposition and nutrient cycling under high-temperature environments	*Thermus aquaticus*, *Thermomyces lanuginosus*
Cellulolytic Microbes	Decompose crop residues efficiently, aiding carbon cycling and improving soil organic matter	*Cellulomonas*, *Trichoderma reesei*

enhance crop yield is particularly appealing in the context of climate change, as it offers a low-input technique that can be implemented significantly more swiftly than traditional plant breeding or genetic engineering methods. Addressing a wide array of issues sometimes necessitates collaborative interactions between host plants and the bacteria constituting the Phyto microbiome (Lebeis, 2014). The beneficial effects of symbiotic bacteria become more evident when plants experience stress (Coleman-Derr & Tringe, 2014). Symbiotic microorganisms enhance plant growth through mechanisms such as antagonism against phytopathogens, improved food availability, hormone synthesis, and biological nitrogen fixation (Wang et al., 2012; Ogidi & Izah, 2024).

Microbial symbioses are essential, if not imperative, for plants to endure severely harsh environments (Mabood et al., 2014). Individuals residing in phytomicrobiomes that are especially susceptible to biotic and abiotic stresses have evolved to coexist together. Microorganisms in the microbiomes of plants native to severe environments may aid in their survival. These encompass agave, alpine mosses, lichens, primroses, and flora that thrive in geothermal soil. Enhancing agricultural productivity in adverse conditions can be accomplished by leveraging the phytomicrobiome's intrinsic adaptations to environmental challenges, much to utilizing genetic adaptations in organisms for traditional breeding or genetic modification (Lebeis, 2014). This is attributable to the microbiome's pivotal function in plant health and resilience. Stringent regulations can render genetic engineering a costly method of crop production, despite its seeming efficiency and wider

TABLE 13.2
Phytomicrobiome in Crop Resilience to Stress

Component of Phytomicrobiome	Type of Stress Resisted	Mechanisms Involved	Examples
Rhizobacteria (PGPR)	Drought, salinity, heavy metals	Produce phytohormones, improves nutrient uptake, induces systemic tolerance	*B. subtilis, P. fluorescens*
Mycorrhizal Fungi	Drought, nutrient deficiency	Enhance root water uptake, phosphorus solubilization, improves soil structure	*Glomus* spp., *Acaulospora* spp.
Endophytes (Bacteria and Fungi)	Heat, cold, drought, salinity	Live within plant tissues, modulate stress-responsive genes, promote growth	*Burkholderia* spp., *Piriformospora indica*
Actinobacteria	Pathogen stress, drought	Antibiotic production, siderophore release, promotes root development	*Streptomyces* spp.
Cyanobacteria	Salinity, drought, nutrient-poor soils	Fix atmospheric nitrogen, improve soil organic matter	*Anabaena, Nostoc*
Halotolerant Bacteria	Salinity	Osmoprotectant production, reduces ion toxicity	*Halomonas, Marinobacter*
Phosphate-solubilizing Microbes	Nutrient deficiency	Solubilize insoluble phosphorus compounds, enhance root development	*Pseudomonas* spp., *Bacillus* spp.
Silicate-solubilizing Microbes	Abiotic stress (mechanical, thermal)	Release silicon, strengthen cell walls, enhance stress resistance	*B. circulans*
Biocontrol Microbes	Pathogen attack, insect stress	Produce antimicrobial compounds, induce systemic resistance	*Trichoderma* spp., *B. thuringiensis*
ACC Deaminase-producing Bacteria	Drought, salinity, flooding	Decrease stress ethylene levels in plants, promote root elongation	*P. putida, Enterobacter* spp.
Volatile Organic Compound (VOC) Producers	Heat, pathogens, drought	Emission of stress signaling volatiles, priming plant defenses	*Bacillus* spp., *Paenibacillus* spp.

array of genetic options. The vast genetic reservoir of plant-associated microorganisms is the sole impediment to the speed, adaptability, and utility of plant-microbiome interventions (Lebeis, 2014; Ogidi & Felagha, 2024). Table 13.2 shows the phytomicrobiome in crop resilience to stress.

13.4.1 Microbes as Biofertilizers

Biofertilizers are substances composed of microorganisms. No hazardous ingredients are present in the microbiological composition (Vessey, 2003). Biofertilizers, which facilitate plant growth enhancement, are typically produced using indigenous microorganisms (Ogidi & Richard-Nwachukwu, 2024). The concept of biofertilizers was founded with the discovery of phosphate solubilizers and nitrogen-fixing *Azospirillum*. Sustainable agricultural development necessitates the scientific integration of microbial fertilizers and their application (Ogidi & Akpan, 2024).

13.4.1.1 Plant Growth Promoting Activity by Biofertilizers

A variety of soil microorganisms, including nitrogen fixers and phosphate solubilizers, are recognized for their ability to produce plant growth-promoting substances (PGPS). A few of the many benefits of bioinoculants include a reduction in the occurrence of plant diseases and nematode infestations, improved nitrogen acquisition through biological nitrogen fixation, increased phosphate solubilization, and the production of hormones that promote plant growth. Phytohormones enhance both plant growth and dry matter production. A primary reason for the enhanced growth and increased dry matter production in maize following *Azospirillum* inoculation is the bacteria's ability to fix nitrogen and synthesize plant growth regulators. Indole-3-acetic acid (IAA) was the primary hormone produced. Numerous gibberellins, indole-3-lactic acid, indole-3-methanol (Xie et al., 2005), abscisic acid (ABA), cytokinins, indole-3-butyric acid (IBA), and unidentified indole compounds were detected at markedly reduced quantities.

However, the above finding remained physiologically relevant Studies have demonstrated the inoculation of *Pseudomonas aeruginosa*, *Pseudomonas cepacia*, and *Pseudomonas hydrophila*. After being inoculated with *P. fluorescens* and *P. putida*, winter wheat plants in the growth chamber showed an increase in height, root and shoot biomass, and tiller count (Vessey, 2003; Xie et al., 2005). Researchers discovered that *Azospirillum* isolates synthesized increased levels of IAA in the presence of tryptophan in the medium. Govindarajan and Kavitha (2004) discovered that *Azospirillum* brasiliense Cd and A exhibited enhanced growth and increased production of indole compounds. Lipoferum Br17 resulting from mild aeration. The search for early root colonizers that facilitate plant growth, either directly or indirectly, has garnered heightened attention in recent years. Rhizobacteria that aid in phytoremediation belong to genera such *Azospirillum*, *Azotobacter*, *Bacillus*, *Bradyrhizobium*, *Erwinia*, *Pseudomonas*, *Rhizobium*, and *Serratia*.

13.4.1.2 Nitrogen Fixers

Three of the most prominent nitrogen-fixing organisms include *Azolla*, *Azospirillum*, and *Rhizobium*. The spiral-shaped, microaerophilic, gram-negative bacterium *Azospirillum* engages in a symbiotic interaction to fix atmospheric nitrogen. The genus *Dodereinera* consists of multiple species. Historically, acidic soil was the predominant environment for *Azospirillum*. Laterite soil had the highest population, whereas extremely acid sulfate saline kari soil demonstrated the lowest. Piao et al. (1992) examined the impact of two new bacterial strains, *Beijerinckia mobilis* and *Clostridium* sp., on plant growth enhancement. They found that these strains significantly influenced the growth of several crops when applied to the plants. Isolate from the rhizosphere of pea plants. *Clostridium* species and *Movovius* species. Crop output increased by 1.5–2.5 times when cultures were utilized in conjunction with mineral fertilizers.

13.4.1.2.1 Mechanism of Nitrogen Fixation

Worldwide, non-leguminous crops make extensive use of biological nitrogen fixation. *Azotobacter* and *Azospirillum* are examples of non-symbiotic nitrogen-fixing bacteria that carry out the process, which involves a complicated enzyme system. By feeding on root exudates and fixing a large amount of atmospheric nitrogen, the rhizospheric bacterium *Azospirillum* multiplies in crop plants. Metabolic reactions generate the requisite protons and electrons, subsequently catalyzed by the enzyme nitrogenase. Nitrogen fixation can solely be accomplished by eukaryotic organisms in association with prokaryotic microorganisms; hence, eukaryotic species cannot independently benefit from nitrogen fixing. The acetylene reduction test was primarily employed to assess nitrogen fixation by *Azospirillum* and to evaluate their capacity for nitrogen fixation.

13.4.1.3 Phosphate Solubilizers

Fungus species like *Aspergillus* and *Pseudomonas* help plants get insoluble phosphorus, while bacteria like *Bacillus* and *Pseudomonas* are great at solubilizing phosphate. Soil microbes can break down insoluble phosphate mostly through their capacity to produce organic acids. Soil pH can be decreased by *Pseudomonas* strains by the formation of organic acids, as found by Burgess and Lowe (1996). The effects of phosphate solubilizers on plants can be influenced by a number of factors. These include the release of phytohormones, ease of nitrogen fixation, mineralization, and mobilization of extra nutrients, resistance to plant diseases, and improvement of microorganisms in the rhizosphere that promote plant growth. Due to their ability to mobilize phosphate and solubilize phosphorus, mycorrhizae have sparked interest among agricultural experts in incorporating them into nutrient management systems (Ogidi et al., 2024). Using an enrichment strategy, he went on to create a medium with ammonium sulfate as the nitrogen and glucose as the carbon. Soil and root-zone microbes that produce acid and bacteria that can dissolve phosphate were isolated using specialized media created by Hedge and Dwivedi (1994) and Rashid et al. (2004).

The insoluble inorganic phosphate can be successfully dissolved by actinomycetes, bacteria, and fungi. The capacity of PGPR to display a variety of phenotypes is influenced by soil and ambient factors, interactions between microbes and plants, and interactions among microorganisms themselves. Among the phosphate-solubilizing bacteria found in *Bacillus licheniformis* and the wheat rhizosphere, researchers identified *P. aeruginosa*, *P. cepacia*, *P. fluorescens*, and *P. putida*. Those living in the root zone of rice plants (Narsian & Patel, 1995). The capacity of phosphate-solubilizing bacteria to generate PGPS has brought them widespread recognition. The production of auxin-type PGPS was achieved by growing P-solubilizing *Bacillus polymyxa* in a liquid medium that was rich in tryptophan (Dey et al., 2004). While P-solubilizing *B. polymyxa* produced a considerable amount of IAA and gibberellic acid (GA).

13.4.1.3.1 Mechanism of Phosphate Solubilization

Microorganisms that solubilize phosphate in a liquid medium were found to generate various acids from simple carbohydrates. There were a variety of acids included, including lactic and gluconic monocarboxylic acids, succinic and oxalic dicarboxylic acids, maleic and maleic dicarboxylic acids, and citric tricarboxylic hydroxyl acids. Citric and fumaric acids exhibited the highest capacity for P-solubilization, but aliphatic acids shown greater efficacy than phenolic acids in this aspect. Inorganic phosphorus can be transformed into a bioavailable form for crop nutrition through the organic acid generated by phosphorus-solubilizing bacteria (PSB). Studies identified this as the principal mechanism for establishing acidic soil conditions that retain higher phosphorus levels (Dey et al., 2004; Ali & Vidhale, 2013).

13.4.2 Microbes as Biocontrol Agents

A biological control agent (BCA) denotes microbial antagonists utilized in plant pathology; the term has also been applied in other biological disciplines. Beneficial bacteria may diminish the nutrient availability for phytopathogens by competing for space and resources, hence inhibiting their growth. *Aspergillus*, *Penicillium*, *Trichoderma*, and antagonistic actinomycetes are recognized as potent sources of several antibiotics with distinct mechanisms of action (Nakouti et al., 2012). Antagonistic bacteria generally predominate in the microflora of disease-suppressing soils (Zivkovic et al., 2010). Various *Trichoderma* species are proficient at capitalizing on emerging chances (Rahul et al., 2014). Studies indicate that the biological control strain WCS417r of R. colonizes the rhizosphere has been recorded. *Pseudomonas fluorescens* WCS417r–treated plants exhibited significant suppression of *Rhizobium radiobacter*. The proliferation of syringae in diseased leaves, along with the resistance response elicited by the plant, significantly reduced the symptoms induced by the challenging pathogens *Fusarium oxysporum* and *Pseudomonas syringae* (Nakouti et al., 2012; Rahul et al., 2014).

13.4.3 Microbial Role on Plants Abiotic Stresses

Numerous compounds synthesized by bacteria that assist plants in managing abiotic stresses encompass signaling molecules, thuricin 17, and bacteriocin. Microbially generated plant hormones can enhance plant tolerance to abiotic stresses. Improved plant health under stress is a direct outcome of increased nutrient absorption, which is made possible by the synthesis of IAA and GA, which in turn increases root length, surface area, and root tip quantity (Egamberdieva & Kucharova, 2009). Rhizosphere microbes produce indole acetic acid and volatile fatty acids, which change the root's shape and architecture, making them more resilient and enhancing their ability to absorb water and nutrients. Furthermore, cytokinins produced by microorganisms can stimulate plants to synthesize ABA, which induces the closure of stomata and decreases water loss by transpiration (Egamberdieva et al., 2016). Recent studies have identified bacteria from multiple genera, including *Rhizobium* sp., *Bacillus* sp., *Pseudomonas* sp., *Pantoea*, *Paenibacillus*, and *Burkholderia*, that enhance host plant tolerance to diverse forms of abiotic stress (Choudhary et al., 2011, 2007).

13.4.4 Microbes for Biotic Stress on Plants

In response to both natural and artificial threats, plants actively display their defensive mechanisms. In both induced systemic resistance (ISR) and systemic acquired resistance (SAR), plants are prepared to face future challenges by being exposed to specific chemicals or bacteria before an infection or treatment begins (Marasco et al., 2012). By vying for resources and space, beneficial bacteria might restrict the availability of accessible nutrients, hence inhibiting the proliferation of plant illnesses (Mohseni et al., 2013). Antagonistic bacteria capable of generating compounds generally dominate soil microflora (Mercado-Blanco et al., 2004). A variety of antibiotics can be produced by antagonistic bacteria, such as *Aspergillus*, *Penicillium*, *Trichoderma*, and *Actinomycetes*.

Pathogens that infect soil-dwelling plants can be eliminated biostatically and biocidally by antagonistic microorganisms. Beneficial soil-borne microorganisms not only diminish local diseases but also enhance plant resistance to herbivores and a wide array of pathogens in systemic plant tissues (Rybakova et al., 2016). Plants can enhance their disease resistance when subjected to biotic stimuli from diverse plant growth-promoting microorganisms (PGPMs). These induce an involuntary state of defense response (ISSR) in plants upon contact with their roots. This is typically referred to as ISSR facilitated by rhizobacteria. Cucumbers infected with *Colletotrichum orbiculare*, halo blight, and Fusarium wilt were the initial subjects to demonstrate PGPR-elicited ISR, succeeded by common beans and carnations. Root-colonizing plant pathogens, such as *Staphylococcus aureus*, *Pseudomonas* sp., and *Bacillus pumilus* help prevent foliar diseases. (Beneduzi et al., 2012; Berg et al., 2013b).

13.4.5 Microbes for Food Security and Ecosystem Resilience

Microbial communities, commonly referred to as microbiomes, provide numerous critical services, including nutrient cycling, disease control, and the removal of hazardous substances, which are essential for live organisms and ecosystems. Climate change-induced environmental alterations jeopardize the microbiome, resulting in the reduction of beneficial bacteria and the proliferation of diseases (Berg & Cernava, 2022). The increasing human population, which demands a steady food supply, complicates this work significantly. Climate change, according to Zhuang et al. (2021) and Rodell and Li (2023), has a major influence on agricultural output due to the following: increased temperatures, decreased water availability, increased pest and disease pressures, nutritional deficits, soil salinity, and less stable yields (Iyiola et al., 2023). Using microbes is a sustainable way to meet the nutritional needs of a growing world population. Through the intentional management of microbiomes using therapies like microbial probiotics or environmental alterations, microbiome stewardship seeks to preserve biodiversity and improve ecosystem functions (Ogidi & Akpan, 2022; Okosodo & Ogidi, 2023).

According to Peixoto et al. (2022), microbiomes impact the host organism's and its environment's nutrition cycle, disease resistance, and stress tolerance. Ecosystem health and resilience are fundamental to this methodology's premise that microbiomes are paramount. Targeted microbiome enhancement can augment or rehabilitate ecosystem functions, which are crucial to human health and well-being. Soil health, crop yields, dependency on chemical fertilizers, and carbon sequestration are all improved when beneficial bacteria are introduced into agricultural practices (Berg et al., 2013a, 2021; Cavicchioli et al., 2019; Allard et al., 2020). As a result, eco-friendly procedures are encouraged. According to Menegat et al. (2022), synthetic nitrogen fertilizers are responsible for approximately 2.1% of the world's emissions and 10% of greenhouse gas emissions in agriculture. Microorganisms used as biofertilizers can boost plant development and decrease need for artificial fertilizers, leading to higher crop yields (Shah et al., 2021).

According to Olmo et al. (2022), using biofertilizers in soybean farming lowered Brazil's yearly carbon emissions by about 430 million tons of CO_2 equivalent. Soil bacteria have the potential to cut down on N_2O emissions by as much as 50%, according to studies (Wu et al., 2018; Kavadia et al., 2020; Hiis et al. 2024), which could lead to a drop in greenhouse gas emissions from agricultural soils. Biochar is made by pyrolyzing organic biomass without oxygen, which results in a carbon-dense substance. This method's resistance to decomposition makes it an excellent long-term carbon sink and carbon storage option. The use of biochar in conjunction with microbial inoculants is a crucial part of adapting to climate change. This practice improves soil fertility, increases plant development, decreases emissions of greenhouse gases, and lessens the impact of soil-borne diseases (Bolan et al., 2023; Zhu et al., 2024). Profitability, environmental sustainability, and improved health outcomes have been demonstrated through the intentional cultivation and use of beneficial microbes in various ecosystems and hosts, including agriculture, aquaculture, and human health (Hai, 2015; Fenster et al., 2019; Wang et al., 2022; Garcías-Bonet et al., 2024).

The worldwide biofertilizer market is expected to increase from an initial worth of 6.3 billion USD at a CAGR of 12.2% by the end of 2032, according to Fortune Business Insights (2024). Successful business strategies for creating and implementing microbial communities in various domains can provide insights for environmental and wildlife applications. Bioremediation, which involves breaking down harmful compounds and cleaning up polluted areas, can improve ecosystems and bring economic benefits if microbial agents are used more extensively and selectively (Parnian et al., 2022; Madison et al., 2023; Helmy & Kardena, 2024). Using microbial probiotics can be beneficial for a wide range of crops and agroecological zones. Low- and middle-income countries can significantly gain from their adaptability. By developing strains tailored to local conditions, we may involve local communities in conservation initiatives and mitigate losses impacting their livelihoods. Smallholder farmers can acquire microbial inoculants, which serve as economical substitutes for costly chemical inputs (Sarkar et al., 2021). It is crucial to mitigate the combinatorial effects of environmental degradation by enhancing microbial responses to both global and local influences, which typically result in a transition toward pathogenic assemblages. A crucial mechanism is a positive feedback loop that accomplishes this objective.

13.5 CHALLENGES AND FUTURE DIRECTIONS

The difficulties of incorporating microbes into farming methods must be recognized while considering their potential. The delicate protein capsid, tail, and nucleic acid core of bacteria make them vulnerable to physical stresses in agricultural settings, which can reduce their stability and effectiveness. In most cases, virions have a core made of RNA or DNA enclosed in a protein shell called a capsid. To aid infection, a protein tail extends from the capsid (Burrell et al., 2017). The viability and usefulness of microorganisms in agricultural settings are threatened by a number of environmental factors. According to Gomez et al. (2022), Gayder et al. (2023), and Jo et al. (2023), the stressors include ultraviolet light, temperature changes, desiccation, moisture availability, acidic conditions, and other variables. Reasons such as poor nutrition, unfavorable conditions, and the start of slow growth phases reduce the production of phage infections (Fister et al., 2016).

Most bacteria are susceptible to acidity at pH values below 4, even if this varies among bacteria depending on characteristics like shape (Jurczak-Kurek et al., 2016). Phage specificity is generally a good thing, but it means that phages only work against certain types of bacteria, so there have to be different phages for different strains of the same species (Zhan et al., 2015; Jeon et al., 2016). Engineering approaches are being explored to overcome the phage treatment's limited specificity (Hibstu et al., 2022). Expanding the host range of phages is one approach. Because bacteria can develop phage resistance—similar to antibiotic resistance—through spontaneous mutations and restriction–modification mechanisms, it is essential to continuously monitor and adjust phage treatments (Ling et al., 2022). To improve their accessibility and agricultural utility, phage compositions can need some tweaking. According to Moye et al. (2018), temperature control is crucial since phage preparations usually need refrigeration (2–8°C) to keep them effective. The use of microbes in farming often faces difficulties in obtaining regulatory clearance. Numerous nations necessitate clear standards and principles to comprehend the advantages and disadvantages of regulatory procedures and to navigate them effectively.

According to Liu et al. (2023) and Yang et al. (2023), microbes are required to follow good manufacturing practices (GMPs) to guarantee that their products are consistently made under control and meet quality standards. In order to include phages as biocontrol agents into these systems and practices, it may be necessary to educate farmers and make adjustments to current agricultural operations. Instruction on the use of phages in biocontrol should be provided to farmers since research shows that they are more effective than conventional chemical antibacterials in reducing bacterial crop diseases (Kuek et al., 2022; Vázquez et al., 2022). Agricultural extension organizations must be included in bacteriophage-based biocontrol application in order to help farmers by disseminating information (Kuek et al., 2022). Incorporating microorganisms as biocontrol agents should be a smooth process that doesn't interfere with current production methods, doesn't affect farmers much, and is financially feasible (Huss & Raman, 2020). Lin et al. (2017), Jia et al. (2023), and Ranveer et al. (2024) discuss how researchers are studying phage diversity to find phages with high selectivity against pathogenic bacteria and how to optimize delivery methods to make sure phages reach their targets on plants (Jończyk et al., 2011; Ly-Chatain, 2014; Selcuk et al., 2024).

Scientists are investigating what makes phages stable and how long they may survive in different environments. Scientists are looking into creating new phage strains to fight against the effects of climate change on farming. Potential adaptations to climate change that these strains may have include resistance to heat and drought (Svircev et al., 2018), increased stress tolerance, flexibility to different growing seasons, the ability to produce adaptable phage products suitable for different environments (Villalobos-López et al., 2022), and ecological consciousness (Meaden & Koskella, 2013). Together, these factors strengthen agriculture's ability to withstand and adapt to climate change. Last but not least, adding microbes to food as a preservative can make it last longer. There might be a wide variety of these preservatives at various stores. A phage combination that targets a degrading strain of *P. fluorescens* can increase the salad's shelf life by 50%, which is the equivalent of an extra day, according to research by NexaBiome. This impact is strong when the phages are linked to the packaging. Improvements to this phage combination in the future may allow for longer storage times (NexaBiome 2022). The constraints, including their restricted host range and the necessity to address regulatory and public acceptance issues, must be carefully considered in the development of phage therapy. Utilizing bacteriophage for climate-resilient agriculture necessitates research, the establishment of regulatory frameworks, and public engagement (Ogidi et al., 2025a).

13.6 CONCLUSION

Recent research have enhanced our comprehension of how high-quality soil can support vital ecological processes such as organic matter synthesis and decomposition, carbon storage and sequestration, and the mitigation of atmospheric carbon dioxide levels. The carbon content in soil is influenced by various factors, including microbial populations, climatic conditions, material characteristics, soil classification, vegetation age and type, and landform. Changes in the frequency and severity of agricultural pests and diseases are an indirect result of climate change. Limited information exists regarding the mechanisms by which climate change induces soil erosion. Limited research has investigated the effects of climate change variability or fluctuations in the intensity of extreme events such as droughts and floods.

REFERENCES

Aggarwal, P. K. (2003). Impact of climate change on Indian agriculture. *Journal of Plant Biology, 30*(2), 189–198.

Ali, S. S., & Vidhale, N. N. (2013). Antagonistic activity of rhizosphere fluorescent pseudomonads against soil-borne fungal pathogens of chickpea and their plant growth-promoting activities. *International Journal of Current Microbiology and Applied Sciences, 2*(12), 303–311.

Allard, S. M., Costa, M. T., Bulseco, A. N., Helfer, V., Wilkins, L. G. E., Hassenrück, C., … Bowman, J. (2020). Introducing the mangrove microbiome initiative: Identifying microbial research priorities and approaches to better understand, protect, and rehabilitate mangrove ecosystems. *mSystems, 5*(6), e00658–20. Cham, Switzerland: Springer. https://doi.org/10.1128/mSystems.00658-20

Barwant, M. M., Ogidi, O. I., Singh, B., Patil, V. S., & Sharma, S. A. B. (2025a). Climate change and pesticide dynamics. In B. R. Babaniyi & E. E. Babaniyi (Eds.), *The Interplay of Pesticides and Climate Change.* Cham: Springer. https://doi.org/10.1007/978-3-031-81669-7_7

Barwant, M. M., Ogidi, O. I., Yogita, C., & Munje, R. (2025b). Study of consumers choices on pesticides use and sustainability. In B. R. Babaniyi & E. E. Babaniyi (Eds.), *The Interplay of Pesticides and Climate Change.* Cham: Springer. https://doi.org/10.1007/978-3-031-81669-7_18

Barwant, M. M., Sharma, S. A. B., Lodh, D., Sil, D. C., & Ogidi, O. I. (2025c). Immedicable effects of organic pesticides on aquatic environment. In B. R. Babaniyi & E. E. Babaniyi (Eds.), *The Interplay of Pesticides and Climate Change.* Cham: Springer. https://doi.org/10.1007/978-3-031-81669-7_16

Barwant, M. M., Singh, S., Ogidi, O. I., Singh, S., & Naik, B. S. S. S. (2025d). Economic implications of climate change and pesticide use on agriculture. In B. R. Babaniyi & E. E. Babaniyi (Eds.), *The Interplay of Pesticides and Climate Change.* Cham: Springer. https://doi.org/10.1007/978-3-031-81669-7_19

Barwant, M. M., Singh, S., Singh, S., Naik, B. S. S. S., & Ogidi, O. I. (2025e). Resilience of ecosystems: Navigating the challenges of pesticide exposure and climate change. In B. R. Babaniyi & E. E. Babaniyi (Eds.), *The Interplay of Pesticides and Climate Change.* Cham: Springer. https://doi.org/10.1007/978-3-031-81669-7_3

Beneduzi, A., Ambrosini, A., & Passaglia, L. M. P. (2012). Plant growth-promoting rhizobacteria (PGPR): Their potential as antagonists and biocontrol agents. *Genetics and Molecular Biology, 35*(4, Special issue), 1044–1051.

Benioff, R., Guill, S., & Lee, J. (Eds.). (1996). *Vulnerability and Adaptation Assessments: An International Handbook.* Dordrecht, the Netherlands: Kluwer Academic Publishers.

Berg, G., Alavi, M., Schmidt, C. S., Zachow, C., Egamberdieva, D., & Kamilova, F. (2013a). Mining the plant microbiome: Diversity, function and its potential for plant biotechnology. In F. J. de Bruijn (Ed.), *Plant Microbiome: By de Bruijn F. J.* (pp. 587–592). New York, NY: John Wiley & Sons, Inc.

Berg, G., & Cernava, T. (2022). The plant microbiota signature of the Anthropocene as a challenge for microbiome research. *Microbiome, 10*, 54. https://doi.org/10.1186/s40168-021-01224-5

Berg, G., Kusstatscher, P., Abdelfattah, A., Cernava, T., & Smalla, K. (2021). Microbiome modulation—Toward a better understanding of plant microbiome response to microbial inoculants. *Frontiers in Microbiology, 12*, 650610. https://doi.org/10.3389/fmicb.2021.650610

Berg, G., Zachow, C., Müller, H., Philipps, J., & Tilcher, R. (2013b). Next-generation bio-products sowing the seeds of success for sustainable agriculture. *Agronomy, 3*(4), 648–656. https://doi.org/10.3390/agronomy3040648

Bolan, S., Hou, D., Wang, L., Hale, L., Egamberdieva, D., Tammeorg, P., … Bolan, N. (2023). The potential of biochar as a microbial carrier for agricultural and environmental applications. *Science of the Total Environment, 886*, 163968. https://doi.org/10.1016/j.scitotenv.2023.163968

Brady, N. C., & Weil, R. R. (2010). *The Nature and Properties of Soils* (14th ed.). Upper Saddle River, NJ: Pearson Education International.

Brklacich, M., Bryant, C. R., Veenhof, B., & Beauchesne, A. (2000). Agricultural adaptation to climatic change: A comparative assessment of two types of farming in central Canada. In H. Millward, K. Beesley, B. Ilbery, & L. Harrington (Eds.), *Agricultural and Environmental Sustainability in the New Countryside* (pp. 40–51). Winnipeg, Canada: Hignell Printing Ltd.

Bryant, C. R., Smit, B., Brklacich, M., Johnston, T. R., Smithers, J., Chiotti, Q., & Singh, B. (2001). Adaptation in Canadian agriculture to climatic variability and change. *Climatic Change, 45*, 181–201.

Burgess, B. K., & Lowe, D. J. (1996). Mechanism of molybdenum nitrogenase. *Chemical Reviews, 96*(7), 2983–3012.

Burrell, C. J., Howard, C. R., & Murphy, F. A. (2017). *Fenner and White's Medical Virology* (5th ed.). Amsterdam: Elsevier/Academic Press.

Cavicchioli, R., Ripple, W. J., Timmis, K. N., Azam, F., Bakken, L. R., Baylis, M., … Webster, N. S. (2019). Scientists' warning to humanity: Microorganisms and climate change. *Nature Reviews Microbiology*, *17*, 569–586. https://doi.org/10.1038/s41579-019-0222-5

Chen, C. C., & McCarl, B. A. (2001). An investigation of the relationship between pesticide usage and climate change. *Climatic Change*, *50*, 475–478.

Choudhary, D. K., Prakash, A., & Johri, B. N. (2007). Plant growth-promoting abilities of rhizobacterial strains under salt stress. *Indian Journal of Microbiology*, *47*, 289–297.

Choudhary, D. K., Sharma, K. P., & Gaur, R. K. (2011). Enhanced seed germination and plant growth by thermophilic *Bacillus* strains. *Biotechnology Letters*, *33*, 1905–1910.

Choudhary, A. K., Thakur, R. C., & Kumar, N. (2008). Effect of integrated nutrient management on soil physical and hydraulic properties in a rice–wheat cropping sequence in the north-western Himalayas. *Indian Journal of Soil Conservation*, *36*(2), 97–104.

Coleman-Derr, D., & Tringe, S. G. (2014). Building a better phyllosphere: Leveraging the plant microbiome to reduce drought stress in plants. *Frontiers in Microbiology*, *5*, 283.

Dey, R., Pal, K. K., Bhatt, D. M., & Chauhan, S. M. (2004). Growth promotion and yield enhancement of chickpea (*Cicer arietinum* L.) by phosphate-solubilizing fluorescent pseudomonads under field conditions. *Microbiological Research*, *159*, 371–394.

Dolan, A. H., Smit, B., Skinner, M. W., Bradshaw, B., & Bryant, C. R. (2001). *Adaptation to Climate Change in Agriculture: Evaluation of Options* (Occasional Paper No. 26). Department of Geography, University of Guelph, Guelph, Ontario, Canada.

Egamberdieva, D., Jabborova, D., & Berg, G. (2016). Role of root-associated bacteria and fungi in improving plant salt stress response. *Plant and Soil*, *405*, 35–45.

Egamberdieva, D., & Kucharova, Z. (2009). Contribution of beneficial microbes to nutrient uptake in arid soils. *Biology and Fertility of Soils*, *45*, 561–573.

Fenster, K., Freeburg, B., Hollard, C., Wong, C., Rønhave Laursen, R., & Ouwehand, A. C. (2019). The production and delivery of probiotics: A review of a practical approach. *Microorganisms*, *7*(3), 83. https://doi.org/10.3390/microorganisms7030083

Fister, S., Robben, C., Witte, A. K., Schoder, D., Wagner, M., … Rossmanith, P. (2016). Influence of environmental factors on phage–bacteria interaction and the efficacy and infectivity of phage P100. *Frontiers in Microbiology*, *7*, 1152. https://doi.org/10.3389/fmicb.2016.01152

Fortune Business Insights. (2024). *Biofertilizers Market Size, Share & Industry Analysis, By Type (Nitrogen Fixing, Phosphate Solubilizers, and Others), By Microorganism (Rhizobium, Azotobacter, Azospirillum, Pseudomonas, Bacillus, VAM, and Others), By Form (Liquid, Powder/Dry, Carrier Based and Encapsulated/Gel based), By Application (Seed Treatment, Soil Treatment, and Others), By Crop Type (Cereals, Pulses & Oilseeds, Fruits & Vegetables, Fiber Crops, Turfs and Ornamentals and Others), and Regional Forecast, 2025-2032*. https://www.fortunebusinessinsights.com/industry-reports/biofertilizers-market-100413 (Accessed 2024 November 15).

Garcías--Bonet, N., Roik, A., Tierney, B., García, F. C., Villela, H. D. M., Dungan, A. M., … Peixoto, R. S. (2024). Horizon scanning the application of probiotics for wildlife. *Trends in Microbiology*, *32*(3), 252–269. https://doi.org/10.1016/j.tim.2023.11.002

Garg, A., Shukla, P. R., Bhattacharya, S., & Dadhwal, V. K. (2001). Regional and sectoral assessment of greenhouse gas emissions in India. *Atmospheric Environment*, *35*, 703–713.

Gayder, S., Kammerecker, S., & Fieseler, L. (2023). Biological control of the fire blight pathogen *Erwinia amylovora* using bacteriophages. *Journal of Plant Pathology*, *106*, 1–17. https://doi.org/10.1007/s42161-023-01478-y

Gomez, M., Szewczyk, A., Szamosi, J., Leung, V., Filipe, C., & Hosseinidoust, Z. (2023). Stress exposure of evolved bacteriophages under laboratory versus food processing conditions highlights challenges in translatability. *Viruses*, *15*(1), 113. https://doi.org/10.3390/v15010113

Govindarajan, K., & Kavitha, K. (2004). Studies on *Azospirillum* associated with rice varieties. In S. Kannaiyan, K. Kumar, & K. Govindarajan (Eds.), *Biofertilizers Technology for Rice-Based Cropping System* (pp. 221–224). Coimbatore, India: Tamil Nadu Agricultural University (TNAU).

Hai, N. V. (2015). The use of probiotics in aquaculture. *Journal of Applied Microbiology*, *119*(4), 917–935. https://doi.org/10.1111/jam.12893

Hedge, D. M., & Dwivedi, B. S. (1994). Crop response to biofertilizers in irrigated areas. *Fertilizer News*, *39*(4), 19–26.

Helmy, Q., & Kardena, E. (2024). Enhancing field-scale bioremediation of weathered petroleum oil-contaminated soil with biocompost as a bulking agent. *Case Studies in Chemical and Environmental Engineering*, *9*, 100735. https://doi.org/10.1016/j.cscee.2024.100735

Hibstu, Z., Belew, H., Akelew, Y. et al. (2022). Phage therapy: A different approach to fight bacterial infections. *Biologics: Targets and Therapy*, *16*, 173–186. https://doi.org/10.2147/BTT.S381237

Hiis, E. G., Vick, S. H. W., Molstad, L., Røsdal, K., Jonassen, K. R., Winiwarter, W., & Bakken, L. R. (2024). Unlocking bacterial potential to reduce farmland N_2O emissions. *Nature*, *630*, 421–428. https://doi.org/10.1038/s41586-024-07464-3

Huss, P., & Raman, S. (2020). Engineered bacteriophages as programmable biocontrol agents. *Current Opinion in Biotechnology*, *61*, 116–121. https://doi.org/10.1016/j.copbio.2019.11.013

Iyiola, A. O., Kolawole, A. S., Ajayi, F. O., Ogidi, O. I., & Ogwu, M. C. (2023). Chapter 22: Sustainable water use and management for agricultural transformation in Africa. In S. Madhav, A. L. Srivastav, S. C. Izah & E. D. van Hullebusch (Eds.), *Water Resources Management for Rural Development: Challenges and Mitigation* (pp. 287–300). Elsevier: Amsterdam, Netherlands. https://doi.org/10.1016/B978-0-443-18778-0.00001-5

Jeon, J., Ryu, C. M., Lee, J. Y. et al. (2016). In vivo application of bacteriophage as a potential therapeutic agent to control OXA-66-like carbapenemase-producing *Acinetobacter baumannii* strains belonging to sequence type 357. *Applied and Environmental Microbiology*, *82*(14), 4200–4208. https://doi.org/10.1128/AEM.00526-16

Jia, H. J., Jia, P. P., Yin, S. et al. (2023). Engineering bacteriophages for enhanced host range and efficacy: Insights from bacteriophage–bacteria interactions. *Frontiers in Microbiology*, *14*, 1172635. https://doi.org/10.3389/fmicb.2023.1172635

Jo, S. J., Kwon, J., Kim, S. G. et al. (2023). The biotechnological application of bacteriophages: What to do and where to go in the middle of the post-antibiotic era. *Microorganisms*, *11*(9), 2311. https://doi.org/10.3390/microorganisms11092311

Jończyk, E., Kłak, M., Międzybrodzki, R. et al. (2011). The influence of external factors on bacteriophages—Review. *Folia Microbiologica (Praha)*, *56*, 191–200. https://doi.org/10.1007/s12223-011-0039-8

Joshi, S. (2001). Towards a pro-poor housing finance system. *The Hindu Financial Daily.*

Jurczak-Kurek, A., Gąsior, T., Nejman-Faleńczyk, B. et al. (2016). Biodiversity of bacteriophages: Morphological and biological properties of a large group of phages isolated from urban sewage. *Scientific Reports*, *6*, 34338. https://doi.org/10.1038/srep34338

Karfakis, P., Knowles, M., Smulders, M., & Capaldo, J. (2011). *Effects of Global Warming on Vulnerability to Food Insecurity in Rural Nicaragua* (ESA–FAO Working Paper No. 11-18). Food and Agriculture Organization; European Space Agency.

Kavadia, A., Omirou, M., Fasoula, D., & Ioannides, I. M. (2020). The importance of microbial inoculants in a climate-changing agriculture in Eastern Mediterranean region. *Atmosphere*, *11*(12), 1340. https://doi.org/10.3390/atmos11121340

Kuek, M., McLean, S. K., & Palombo, E. A. (2022). Application of phages in food production and their potential as biocontrol agents in the organic farming industry. *Biological Control*, *165*, 104817. https://doi.org/10.1016/j.biocontrol.2021.104817

Kumar, K. K., Kumar, K. R., Ashrit, R. G., Deshpande, N. R., & Hansen, J. W. (2004). Climate impacts on Indian agriculture: A review. *International Journal of Climatology*, *24*(11), 1375–1393.

Lal, R. (2004). Soil carbon sequestration impacts on global climate change and food security. *Science*, *304*(5677), 1623–1627. https://doi.org/10.1126/science.1097396 (scirp.org)

Lebeis, S. L. (2014). The potential for microbiome-mediated drought resistance in plants. *Frontiers in Plant Science*, *5*, 287.

Lin, D. M., Koskella, B., & Lin, H. C. (2017). Phage therapy: An alternative to antibiotics in the age of multi-drug resistance. *World Journal of Gastrointestinal Pharmacology and Therapeutics*, *8*(3), 162–173. https://doi.org/10.4292/wjgpt.v8.i3.162

Ling, H., Lou, X., Luo, Q. et al. (2022). Recent advances in bacteriophage-based therapeutics: Insight into the post-antibiotic era. *Acta Pharmaceutica Sinica B*, *12*(10), 4348–4364. https://doi.org/10.1016/j.apsb.2022.05.007

Liu, S., Quek, S., & Huang, K. (2023). Advanced strategies to overcome the challenges of bacteriophage-based antimicrobial treatments in food and agricultural systems. *Critical Reviews in Food Science and Nutrition*, 1–25. https://doi.org/10.1080/10408398.2023.2254837

Locke, C., Adger, W. N., & Kelly, P. M. (2000). Changing places: Migration's social and environmental consequences. *Environment*, *42*(7), 24–35.

Ly-Chatain, M. H. (2014). The factors affecting the effectiveness of treatment in phage therapy. *Frontiers in Microbiology*, *5*, 51. https://doi.org/10.3389/fmicb.2014.00051

Mabood, F., Zhou, X., & Smith, D. L. (2014). Microbial signaling and plant growth promotion. *Canadian Journal of Plant Science*, *94*(6), 1051–1063. https://doi.org/10.4141/CJPS2013-148

Madison, A. S., Sorsby, S. J., Wang, Y., & Key, T. A. (2023). Increasing in situ bioremediation effectiveness through field-scale application of molecular biological tools. *Frontiers in Microbiology*, *13*, 1084380. https://doi.org/10.3389/fmicb.2022.1005871

Mahmood, F., Khan, I., Ashraf, U., Shahzad, T., Hussain, S., Shahid, M., … Ullah, S. (2017). Effects of organic and inorganic manures on maize and their residual impact on soil physico-chemical properties. *Journal of Soil Science and Plant Nutrition*, *17*(1), 22–32. https://doi.org/10.4067/S0718-95162017005000002

Marasco, R., Rolli, E., & Ettoumi, B. (2012). A drought resistance-promoting microbiome is selected by root system under desert farming. *PLoS ONE*, *7*(10), e48479. https://doi.org/10.1371/journal.pone.0048479 (academia.edu)

Meaden, S., & Koskella, B. (2013). Exploring the risks of phage application in the environment. *Frontiers in Microbiology*, *4*, 358. https://doi.org/10.3389/fmicb.2013.00358

Menegat, S., Ledo, A., & Tirado, R. (2022). Greenhouse gas emissions from global production and use of nitrogen synthetic fertilisers in agriculture. *Scientific Reports*, *12*, 14490. https://doi.org/10.1038/s41598-022-18773-w

Mercado-Blanco, J., Rodríguez-Jurado, D., Hervás, A., & Jiménez-Díaz, R. M. (2004). Suppression of Verticillium wilt in olive planting stocks by root-associated fluorescent *Pseudomonas* spp. *Biological Control*, *30*(3), 474–486.

Mohseni, M., Norouzi, H., Hamedi, J., & Roohi, A. (2013). Isolation and characterization of phosphate-solubilizing and plant growth-promoting bacteria from saline–sodic soils of Iran. *International Journal of Molecular and Cellular Medicine*, *2*(2), 64–71.

Moye, Z., Woolston, J., & Sulakvelidze, A. (2018). Bacteriophage applications for food production and processing. *Viruses*, *10*(4), 205. https://doi.org/10.3390/v10040205

Nakouti, I., Sihanonth, P., & Hobbs, G. (2012). A new approach to isolating siderophore-producing actinobacteria. *Letters in Applied Microbiology*, *55*(1), 68–72.

Narsian, V., & Patel, H. H. (1995). Screening of Actinomycetes for antagonism against plant pathogenic fungi. *Indian Journal of Microbiology*, *35*, 127–132.

Nexabiome. (2022). Bacteriophage benefit—from food to therapeutics. *Nexabiome Blog*. https://nexabiome.com/blog/bacteriophage-benefit-from-food-to-therapeutics/ (Accessed March 13, 2024).

Ogidi, O. I., & Akpan, U. M. (2022). Chapter 16: Aquatic biodiversity loss: Impacts of pollution and anthropogenic activities and strategies for conservation. In S. C. Izah (Ed.), *Biodiversity in Africa: Potentials, Threats and Conservation*. Springer Nature Singapore Pte Ltd. https://doi.org/10.1007/978-981-19-3326-4_16

Ogidi, O. I., & Akpan, U. M. (2023). Impacts of chemical use in agricultural practices: Perspectives of soil microorganism and vegetation. In M. C. Ogwu & S. C. Izah (Eds.), *One Health Implications of Agrochemicals and Their Sustainable Alternative*. Springer Nature Singapore Pte Ltd.

Ogidi, O. I., & Akpan, U. M. (2024). Microbial and phyto-remediation of crude oil contaminated soil. In S. V. Rathod (Ed.), *Eco-Restoration of the Environment*. CRC Press, Routledge Taylor and Francis Group.

Ogidi, O. I., & Felagha, I. (2024). Gut microbiome and diet interaction. In S. Patnaik, A. M. Hamad, D. Paul, P. K. Dutta & M. Shafiq (Eds.), *Nutrition Controversies and Advances in Autoimmune Disease*. IGI Global.

Ogidi, O. I., & Izah, S. C. (2024). Emerging technologies in the global South food industry: Prospects and challenges. In M. C. Ogwu, S. C. Izah & N. R. Ntuli (Eds.), *Food Safety and Quality in Global South*. Springer Nature Singapore Pte Ltd. https://doi.org/10.1007/978-981-97-2428-4_7

Ogidi, O. I., Izah, S. C., & Akpan, U. M. (2023). Food quality and agrochemical use: Integrated monitoring, assessment, and management policies. In M. C. Ogwu & S. C. Izah (Eds.), *One Health Implications of Agrochemicals and Their Sustainable Alternative*. Springer Nature Singapore Pte Ltd.

Ogidi, O. I., Izah, S. C., & Tobia, P. S. (2024). Biosurfactant production from cassava processing wastes. In M. C. Ogwu, S. C. Izah, A. A. C. Alves & S. C. Babu (Eds.), *Sustainable Cassava: Strategies from Production Through Waste Management*. Elsevier.

Ogidi, O. I., Okara, E., & Barwant, M. M. (2025a). Nanocarriers in plant genetic engineering. In J. Chen (Ed.), *Nanocarriers in Plant Science and Agriculture* (pp. 89–124). IGI Global Scientific Publishing. https://doi.org/10.4018/979-8-3693-9869-2.ch004

Ogidi, O. I., & Richard-Nwachukwu, N. (2024). Microbe assisted remediation of pesticide residues from soil and water. In S. V. Rathod (Ed.), *Eco-Restoration of the Environment*. CRC Press, Routledge Taylor and Francis Group.

Ogidi, O. I., Wenapere, C. M., & Chukwunonso, O. A. (2025b). Enhancing food safety and quality control with computer vision systems. In D. Bora & R. Bania (Eds.), *Computer Vision Techniques for Agricultural Advancements* (pp. 51–88). IGI Global Scientific Publishing. https://doi.org/10.4018/979-8-3693-8019-2.ch003

Okosodo, E. F., & Ogidi O. I. (2023). Chapter 3: Biodiversity conservation strategies and sustainability. In S. C. Izah & M. C. Ogwu (Eds.), *Sustainable Utilization and Conservation of Africa's Biological Resources and Environment, Sustainable Development and Biodiversity*. Springer Nature Singapore Pte Ltd.

Olmo, R., Wetzels, S. U., Armanhi, J. S. L., Arruda, P., Berg, G., Cernava, T.,, … Wagner, M. (2022). Microbiome research as an effective driver of success stories in agrifood systems – A selection of case studies. *Frontiers in Microbiology*, *13*, 997732. https://doi.org/10.3389/fmicb.2022.997732

Parnian, A., Parnian, A., Pirasteh-Anosheh, H., Furze, J. N., Prasad, M. N. V., Race, M., Hulisz, P., & Ferraro, A. (2022). Full-scale bioremediation of petroleum-contaminated soils via integration of co-composting. *Journal of Soils and Sediments*, *22*, 2209–2218. https://doi.org/10.1007/s11368-022-03229-5

Peixoto, R. S., Voolstra, C. R., Sweet, M., Duarte, C. M., Carvalho, S., Villela, H.,, … Berg, G. (2022). Harnessing the microbiome to prevent global biodiversity loss. *Nature Microbiology*, *7*, 1726–1735. https://doi.org/10.1038/s41564-022-01173-1

Piao, C. G., Tang, W. H., & Chen, Y. X. (1992). Effects of pH and temperature on the growth and metabolism of *Candida utilis*. *Canadian Journal of Microbiology*, *41*, 109–117.

Rahul, K., Amrita, K., & Singh, M. (2014). Molecular analysis of Rhizoctonia isolates from north India. *Trends in Biosciences*, *7*, 4055–4058.

Ranveer, S. A., Dasriya, V., Ahmad, M. F. et al. (2024). Positive and negative aspects of phages and their immense role in the food chain. *NPJ Science of Food*, *8*, Article 6. https://doi.org/10.1038/s41538-023-00245-8

Rashid, M., Khalil, S., Ayub, N., Alam, S., & Latif, F. (2004). Isolation and characterization of phosphate-solubilizing bacteria from wheat rhizosphere of Peshawar. *Pakistan. Pakistan Journal of Biological Sciences*, *7*, 187–196.

Rivest, D., Paquette, A., Shipley, B., Reich, P. B., & Messier, C. (2015). Tree communities rapidly alter soil microbial resistance and resilience to drought. *Functional Ecology*, *29*(4), 570–578.

Rodell, M., & Li, B. (2023). Changing intensity of hydroclimatic extreme events revealed by GRACE and GRACE-FO. *Nature Water*, *1*, 241–248. https://doi.org/10.1038/s44221-023-00040-5

Rodriguez, R., & Redman, R. (2008). More than 400 million years of evolution and some plants still can't make it on their own: Plant stress tolerance via fungal symbiosis. *Journal of Experimental Botany*, *59*(5), 1109–1114.

Rybakova, D., Cernava, T., Köberl, M., Liebminger, S., Etemadi, M., & Berg, G. (2016). Endophytic microbial community structure and functionality in wheat roots under salt stress and inoculation with beneficial bacteria. *Plant and Soil*, *405*(1), 125–140.

Sarkar, D., Singh, S., Parihar, M., & Rakshit, A. (2021). Seed bio-priming with microbial inoculants: A tailored approach towards improved crop performance, nutritional security, and agricultural sustainability for smallholder farmers. *Current Research in Environmental Sustainability*, *3*, 100093. https://doi.org/10.1016/j.crsust.2021.100093

Selcuk, E., Dokuz, S., & Ozbek, T. (2024). Evaluating the stability of lytic and lysogenic bacteriophages in various protectants. *Journal of Pharmaceutical Sciences*, *113*(5), 1488–1497. https://doi.org/10.1016/j.xphs.2024.01.010

Shah, A., Nazari, M., Antar, M., Msimbira, L. A., Naamala, J., Lyu, D., … Smith, D. L. (2021). PGPR in agriculture: A sustainable approach to increasing climate change resilience. *Frontiers in Sustainable Food Systems*, *5*, 687723. https://doi.org/10.3389/fsufs.2021.687546

Singh, B. K., Bardgett, R. D., Smith, P., & Reay, D. S. (2010). Microorganisms and climate change: terrestrial feedbacks and mitigation options. *Nature Reviews Microbiology*, *8*, 779–790.

Singh, M. K., Szucs, I., & Fekete-Farkas, M. (2007). Climate change impact on crop productivity in India: A case study. *Cereal Research Communications*, *35*, 1061–1064.

Smithers, J., & Blay-Palmer, A. (2001). Technology innovation as a strategy for climate adaptation in agriculture. *Applied Geography*, *21*(2), 175–197.

Svircev, A., Roach, D., & Castle, A. (2018). Framing the future with bacteriophages in agriculture. *Viruses*, *10*(5), 218. https://doi.org/10.3390/v10050218

Vázquez, R., Díez-Martínez, R., Domingo-Calap, P. et al. (2022). Essential topics for the regulatory consideration of phages as clinically valuable therapeutic agents: A perspective from Spain. *Microorganisms*, *10*(4), 717. https://doi.org/10.3390/microorganisms10040717

Vessey, J. K. (2003). Plant growth-promoting rhizobacteria as biofertilizers. *Plant and Soil*, *255*, 571–586.

Villalobos-López, M. A., Arroyo-Becerra, A., Quintero-Jiménez, A. et al. (2022). Biotechnological advances to improve abiotic stress tolerance in crops. *International Journal of Molecular Sciences*, *23*(19), 12053. https://doi.org/10.3390/ijms231912053

Wang, Y., Wang, X., Sun, S., Jin, C., Su, J., Wei, J., … Liu, H. (2022). GWAS, MWAS and mGWAS provide insights into precision agriculture based on genotype-dependent microbial effects in foxtail millet. *Nature Communications*, *13*, 5913. https://doi.org/10.1038/s41467-022-33693-w

Wang, Z., Kang, S., Jensen, C. R., & Liu, F. (2012). Agronomic and physiological performance of spring wheat under varied irrigation regimes and soil moisture conditions. *Journal of Experimental Botany*, *63*, 1145–1153.

Wu, S., Zhuang, G., Bai, Z., Cen, Y., Xu, S., Sun, H., … Zhuang, X. (2018). Mitigation of nitrous oxide emissions from acidic soils by *Bacillus amyloliquefaciens*, a plant growth-promoting bacterium. *Global Change Biology*, *24*(6), 2352–2365. https://doi.org/10.1111/gcb.14065

Xie, Z., Zhang, Z. L., Zou, X. L., Huang, J., Ruas, P., Thompson, D., & Shen, Q. J. (2005). Annotations and functional analyses of the rice WRKY gene superfamily reveal positive and negative regulators of abscisic acid signaling in aleurone cells. *Plant Physiology*, *137*(1), 176–189.

Yang, K., Wang, X., Hou, R. et al. (2023). Rhizosphere phage communities drive soil suppressiveness to bacterial wilt disease. *Microbiome*, *11*, 16. https://doi.org/10.1186/s40168-023-01463-8

Zhan, Y., Buchan, A., & Chen, F. (2015). Novel N4 bacteriophages prevail in the cold biosphere. *Applied and Environmental Microbiology*, *81*(15), 5196–5202. https://doi.org/10.1128/AEM.00832-15

Zhu, X., Chen, Y., Jia, M., Dai, H., Zhou, Y., Yang, H., … Wang, N. (2024). Managing tobacco black shank disease using biochar: Direct toxicity and indirect ecological mechanisms. *Microbiology Spectrum*, *12*(3), e00149–24. https://doi.org/10.1128/spectrum.00149-24

Zhuang, Y., Fu, R., Santer, B. D., Dickinson, R. E., Hall, A. (2021). Quantifying contributions of natural variability and anthropogenic forcings on increased fire weather risk over the western United States. *Proceedings of the National Academy of Sciences*, *118*, e2111875118. https://doi.org/10.1073/pnas.2111875118

Zivkovic, S., Stojanovic, S., Ivanovic, Z., Gavrilovic, V., Popovic, T., & Balaz, J. (2010). Screening of antagonistic activity of microorganisms against *Colletotrichum acutatum* and *C. gloeosporioides*. *Archives of Biological Sciences, Belgrade*, *62*(3), 611–623.

14 Microbial Solutions for Reducing Greenhouse Gas Emissions in Agriculture

Stella Suanu, Leh-Togi Zobeashia, and Jennifer Chisom Nwatu

14.1 INTRODUCTION

14.1.1 Overview of Greenhouse Gas (GHG) Emissions in Agriculture

Agriculture is a significant contributor to global greenhouse gas (GHG) emissions, accounting for approximately 20–25% of total anthropogenic emissions worldwide (Kamyab et al., 2024; Li et al., 2025). The primary GHGs from agriculture are carbon dioxide, methane, and nitrous oxide, each linked to distinct farming activities and differing substantially in their global warming potential.

Carbon dioxide emissions mainly result from land-use changes such as deforestation and conversion of forests or grasslands into agricultural land. Additionally, carbon dioxide is released through fossil fuel use in agricultural machinery and the manufacture of agrochemicals, including fertilizers and pesticides (Basheer et al., 2024; Kamyab et al., 2024). Though carbon dioxide has a lower warming potential than methane and nitrous oxide, its large-scale and persistent release remains a critical mitigation target.

Methane is over 28 times more potent than carbon dioxide over a 100-year timeframe and is primarily generated through enteric fermentation in ruminants, anaerobic conditions in rice paddies, and manure management systems (Li et al., 2025). Agriculture is the leading global source of methane, with livestock and rice cultivation being the major contributors (Kamyab et al., 2024). Nitrous oxide, though emitted in smaller amounts, has the highest climate impact with a global warming potential nearly 300 times that of carbon dioxide. It is mainly produced through microbial soil processes such as nitrification and denitrification, which are intensified by the use of nitrogen-rich fertilizers and manure (Basheer et al., 2024; Li et al., 2024).

Despite progress in emission monitoring, accurately quantifying agricultural GHG emissions remains difficult due to spatial variability, uncertainties in emission factors, and inconsistent data collection across regions (Li et al., 2024). This complexity hampers robust climate accounting, especially in countries with diverse agricultural systems. However, agriculture also presents opportunities for mitigation. Practices like cover cropping, conservation tillage, agroforestry, and precision nutrient management can reduce emissions and increase carbon sequestration in soils (Kamyab et al., 2024). Thus, agriculture can function both as a source of emissions and a carbon sink depending on management practices.

Recently, microbial strategies have gained attention as biologically based solutions to reduce emissions at their source. Manipulating microbial communities can curb methane production in livestock, limit nitrous oxide emissions from soils, and enhance soil carbon storage (Smith et al., 2023; Li et al., 2025). Microbial approaches are increasingly recognized as a frontier in climate-smart agriculture, offering scalable, sustainable tools to reduce agriculture's environmental footprint.

DOI: 10.1201/9781003650881-14

14.1.2 Importance of Reducing Emissions for Climate Resilience and Sustainability

Agricultural GHG emissions accelerate climate change, which in turn heightens the vulnerability of farming systems to extreme weather events such as droughts, floods, and temperature extremes (Abbass et al., 2022; IPCC, 2022). Reducing emissions is thus essential not only for climate mitigation but also for building resilience, the capacity of agricultural systems to absorb shocks and maintain productivity under changing conditions (Kaushik et al., 2021; Shi et al., 2022).

Emission reductions improve resilience by enhancing soil health, water retention, and biodiversity. For example, sustainable land management increases soil organic carbon, which improves fertility and drought tolerance (Lal, 2020). These benefits create a positive feedback loop whereby mitigation efforts simultaneously serve as adaptation strategies, fostering sustainable agricultural productivity over time (Gan et al., 2023).

Reducing agricultural emissions aligns with global sustainability frameworks, especially the United Nations Sustainable Development Goals (SDGs). Efforts directly contribute to SDG 13 (Climate Action), while supporting SDG 2 (Zero Hunger) and SDG 15 (Life on Land) by ensuring food security and protecting terrestrial ecosystems (IPCC, 2022; Gan et al., 2023). For instance, sustainable nutrient management reduces N_2O emissions and nutrient runoff, improving both environmental and economic outcomes (Robertson & Vitousek, 2009). Emission reductions support the circular bioeconomy by promoting renewable energy, waste recycling, and reduced fossil fuel reliance (Gan et al., 2023). These transitions support national and international goals toward carbon neutrality and climate-smart agriculture. Mitigating emissions benefits marginalized farming communities, often most vulnerable to climate change, by enhancing the resilience and sustainability of their livelihoods (Kaushik et al., 2021). Addressing emissions equitably is therefore vital for environmental justice and sustainable development.

Emerging microbial technologies hold promise to simultaneously reduce emissions and improve agricultural sustainability. Microbial inoculants and biofertilizers enhance nutrient use efficiency, reduce methane emissions from ruminants, and increase soil carbon sequestration, contributing to climate resilience and productivity. Reducing agricultural GHG emissions is a complex but essential goal that advances climate resilience, environmental sustainability, food security, and social equity. Integrating emission reduction into farming practices is critical for building a robust, sustainable global food system.

14.1.3 Introduction to Microbial Solutions as an Emerging Strategy

Microbial solutions have emerged as promising, ecologically sound approaches to mitigate agricultural GHG emissions. Microorganisms regulate carbon and nitrogen cycles central to GHG dynamics, acting both as sources and sinks of emissions such as methane, nitrous oxide, and carbon dioxide depending on environmental and management contexts (Kuypers et al., 2022; Prasad et al., 2023). Emerging microbial technologies leverage these natural functions to reduce emissions. Methanotrophs oxidize methane to less potent forms, while nitrifying and denitrifying bacteria can be managed to reduce nitrous oxide formation. Biofertilizers containing beneficial microbes enhance nutrient cycling and crop uptake, reducing the need for synthetic nitrogen inputs, a major source of nitrous oxide (Smith et al., 2023). Anaerobic digestion (AD) systems use microbial consortia to convert organic waste into biogas, reducing methane emissions from manure (Wang et al., 2022).

Advances in nanotechnology and synthetic biology accelerate microbial solution development. Nanoscale zero-valent iron enhances microbial remediation of contaminated soils while supporting safe crop production (Liu et al., 2021a). This highlights the potential synergy between microbiology and emerging technologies in climate-smart agriculture. Strengthening microbial education and research infrastructure is essential for adoption, particularly in developing countries (Ezemba et al., 2025). Omics technologies such as metagenomics and transcriptomics reveal microbial ecosystem complexity and their emission control potential (Rilling et al., 2016; Sahu et al., 2023).

Microbial solutions offer a key advantage: they work in harmony with natural systems. Their ability to enhance productivity while mitigating environmental harm makes them indispensable in the quest for climate-resilient food systems.

14.1.4 Scope of the Review

This review examines microbial strategies as novel, sustainable tools for reducing agricultural GHG emissions. While conventional mitigation practices like optimized fertilizer use, dietary adjustments for livestock, and conservation tillage remain important, microbial approaches address the biological roots of emissions.

The review begins by outlining major agricultural sources of CO_2, CH_4, and N_2O, linked to land-use changes, enteric fermentation, and nitrogen fertilization. It then explores soil and gut microbiomes' roles in emissions, highlighting processes such as nitrification, denitrification, and methanogenesis. A major focus is microbial mitigation pathways, including methanotrophs, nitrification-inhibiting microbes, and beneficial soil microbes that improve nutrient efficiency. It also covers microbial consortia in anaerobic digesters, livestock feed additives, and microbial contributions to long-term soil carbon sequestration.

Case studies from rice farming, livestock operations, and composting illustrate practical applications and emission reductions. The review discusses limitations, implementation challenges, monitoring issues, and regulatory considerations.

14.2 SOURCES OF GREENHOUSE GASES IN AGRICULTURE

Agriculture is a significant contributor to GHG emissions, producing carbon dioxide, methane, and nitrous oxide through a variety of processes. These gases differ in their sources, atmospheric lifetimes, and global warming potentials, but collectively they play a critical role in climate change. Understanding the sources and mechanisms behind agricultural emissions is essential to designing effective mitigation strategies.

14.2.1 Carbon Dioxide (CO_2)

Carbon dioxide emissions from agriculture primarily arise from land-use change, deforestation, and the use of fossil fuels in farming operations. Land conversion for agriculture often involves deforestation and soil disturbance, which release stored carbon into the atmosphere (Raihan, 2022). Deforestation to expand agricultural land reduces carbon sequestration potential and contributes significantly to carbon dioxide emissions globally.

In Malaysia, Raihan et al. (2022) demonstrated that energy use related to agriculture and the expansion of land is dynamically linked with increased carbon dioxide emissions. The study emphasized the importance of managing land-use changes carefully to reduce carbon emissions. Similarly, Rehman et al. (2022) found that changes in land use and carbon emissions due to agricultural production have notable climate impacts in Pakistan, which shows the need for sustainable land management practices. Moreover, fossil fuel consumption for machinery, irrigation, and transport in agriculture adds to the carbon dioxide burden (Duxbury & Mosier, 2022). Although these emissions might seem smaller compared to land-use change, the cumulative effect of mechanized farming and energy use remains substantial, especially in intensively farmed regions.

14.2.2 Methane (CH_4)

Methane is a potent GHG emitted predominantly through enteric fermentation in ruminant livestock, rice paddies, and manure management. Enteric fermentation is a digestive process in ruminants such as cattle and sheep, during which microbes in the stomach produce methane as a by-product (Smith et al., 2021). These emissions contribute significantly to the agricultural methane budget, especially in regions with large livestock populations.

Rice paddies are another major source of methane due to anaerobic conditions created in flooded fields, which promote methane-producing microbial activity. Win et al. (2021) showed that methane emissions from rice soils are influenced by rice variety, organic manure application, and water management practices. Adjusting these factors can mitigate methane release significantly.

Manure management also contributes to methane emissions, especially when manure is stored or treated under anaerobic conditions, such as in lagoons or pits (Ahmed et al., 2023). Sustainable manure handling practices can reduce methane emissions, representing a critical area for intervention. Recent reviews highlight that methane mitigation in agriculture requires integrated approaches combining animal diet modification, improved rice cultivation techniques, and enhanced manure management (Sahu & Arya, 2024). These strategies aim to reduce methane generation without compromising productivity.

14.2.3 Nitrous Oxide (N_2O)

Nitrous oxide emissions largely result from agricultural nitrogen fertilizer use and the microbial processes of nitrification and denitrification in soils. Fertilizers containing nitrogen increase the availability of substrates for nitrifying and denitrifying bacteria, leading to N_2O production (Tian et al., 2020.

Saud et al. (2022) discussed biological nitrification inhibitors (BNIs) as a promising approach to improve nitrogen use efficiency and reduce N_2O emissions. These inhibitors slow down the microbial conversion of ammonium to nitrate, decreasing the opportunity for N_2O formation. However, the relative contribution of nitrification to total N_2O emissions can vary with management intensity. Liang and Robertson (2021) reported that nitrification was a minor source of N_2O in intensively managed agricultural landscapes, suggesting that other pathways such as denitrification may dominate under certain conditions.

Urea-based nitrogen fertilizers are widely used and represent a key source of N_2O emissions. Zhang et al. (2023) highlighted recent advances in mitigating emissions from urea fertilization, including the development of enhanced efficiency fertilizers and improved application methods. Atmospheric nitrogen deposition, another source of nitrogen input to soils, can also stimulate N_2O emissions globally. Yang et al. (2021) quantified soil N_2O emissions linked to nitrogen deposition across agricultural systems worldwide, indicating the complexity of nitrogen cycling and its contribution to GHG emissions.

14.2.4 Quantification and Global Impact of Agricultural GHG Emissions

Quantifying agricultural GHG emissions is crucial for understanding their global impact and for informing climate policies. Agriculture accounts for approximately 10–12% of total global anthropogenic GHG emissions, with variability across regions due to differences in production systems, technology, and land management (Guo et al., 2022).

Guo et al. (2022) conducted a global meta-analysis to assess crop yield responses alongside GHG emissions under varying nitrogen fertilizer applications. Their findings suggest a trade-off between maximizing yields and minimizing emissions, emphasizing the need for optimizing fertilizer use. Economic policies, such as agricultural subsidies, can indirectly influence GHG emissions by affecting production intensity and land-use decisions. Laborde et al. (2021) examined the link between subsidies and global emissions, showing that subsidy reform could reduce agricultural emissions without compromising food security.

Bhatti et al. (2024) explored global production patterns and the relationship between agricultural greening efforts, GHG emissions, and climate variability. Their research highlights how adopting greener agricultural practices can mitigate emissions while enhancing resilience to climate change.

In China, a major agricultural producer, Liu et al. (2021b) analyzed green total factor productivity incorporating carbon emissions. Their study demonstrated progress in reducing carbon intensity but also highlighted persistent challenges related to balancing productivity and environmental sustainability. The quantification of agricultural emissions and understanding their drivers are foundational for developing targeted mitigation strategies. Integrating technological, economic, and policy measures is essential for achieving sustainable agriculture that supports food security while minimizing climate impact.

14.3 THE ROLE OF SOIL AND GUT MICROBIOMES IN GREENHOUSE GAS EMISSIONS

Microorganisms inhabiting soil and the gastrointestinal tracts of ruminants play pivotal roles in the production and consumption of GHGs, particularly methane (CH_4), nitrous oxide (N_2O), and carbon dioxide (CO_2). These microbiomes mediate key biochemical processes in the nitrogen and carbon cycles, driving GHG emissions that significantly influence agricultural contributions to climate change. Recent advances have deepened understanding of microbial community functions and interactions, opening pathways for mitigation strategies through microbial management.

14.3.1 Microbial Involvement in Nitrogen and Carbon Cycles

Microbial communities are fundamental to the nitrogen and carbon biogeochemical cycles that regulate GHG emissions in agricultural ecosystems. In the nitrogen cycle, soil microbes facilitate transformations of nitrogen compounds, including nitrification and denitrification, which produce N_2O as a by-product (Saud, Wang, & Fahad, 2022). In the carbon cycle, microbes mediate decomposition, methanogenesis, and methane oxidation, influencing carbon dioxide and methane fluxes (Fierer, 2017). Soil microbes convert organic matter into mineralized forms through decomposition, releasing carbon dioxide. *Methanogens archaea* in anaerobic environments—produce methane by metabolizing substrates like acetate and hydrogen (Le Mer & Roger, 2001). Conversely, methanotrophs oxidize methane, acting as natural biofilters to reduce CH_4 emissions from soils (Knief, 2015).

In Nigeria, research by Jerry & Anyira-Fasanmi (2025) investigated soil microbial community dynamics in wetlands and upland agricultural soils, finding that microbial diversity and activity in these soils strongly influence carbon cycling and methane flux. Their work emphasized the importance of wetland rice ecosystems in Nigeria as hotspots for methane emissions driven by methanogenic archaea under anaerobic conditions. Such findings align with global observations, highlighting the need for targeted management of Nigerian wetland agriculture to reduce CH_4 emissions.

Furthermore, Eze et al. (2021) reviewed microbial-mediated nitrogen transformations in Nigerian savannah soils and their link to N_2O emissions, noting that traditional fertilizer practices contribute to elevated emissions due to inefficient nitrogen use and suboptimal microbial activity. They underscored the potential for BNIs and improved fertilizer formulations to enhance nitrogen use efficiency while mitigating N_2O release in local agroecosystems.

14.3.2 Soil Microbial Processes

14.3.2.1 Nitrification and Denitrification

Nitrification is the aerobic oxidation of ammonium (NH_4^+) to nitrate (NO_3^-) via nitrite (NO_2^-), mediated primarily by ammonia-oxidizing bacteria (AOB) and archaea (AOA). This process generates N_2O as a by-product, although the quantity varies with microbial community composition and soil conditions (Liang & Robertson, 2021). Denitrification, the anaerobic reduction of nitrate to nitrogen gas (N_2), proceeds through intermediates including nitric oxide (NO) and N_2O. Denitrifying bacteria contribute significantly to N_2O emissions, particularly under low-oxygen conditions prevalent in waterlogged soils or after fertilization events (Yao et al 2021; Hu et al., 2022). Saud et al. (2022) demonstrated that BNIs produced by certain plants and microbes can suppress nitrification rates, reducing N_2O emissions and improving nitrogen use efficiency. BNIs represent a promising biological approach to mitigate soil-derived N_2O.

In Nigeria, Akintola et al. (2022) investigated nitrification and denitrification rates in cassava and maize fields under different fertilizer regimes. They reported high N_2O emissions associated with conventional urea use and recommended the adoption of slow-release fertilizers and incorporation of leguminous cover crops to modulate microbial nitrification activity and reduce N_2O fluxes. Liang and Robertson (2021) found that intensified agricultural management reduces nitrification-related N_2O emissions, possibly due to altered microbial community structure and increased oxygen availability, shifting N_2O sources toward denitrification.

14.3.2.2 Methanogenesis and Methane Oxidation

Methanogenesis occurs in anaerobic microsites such as flooded paddy soils and water-saturated manures. Methanogens convert fermentation products to methane, which escapes to the atmosphere unless consumed by methanotrophs in aerobic soil layers (Knief, 2015). Soil methane flux reflects the balance of these processes. Win et al. (2021) reported that soil management practices, including water regimes and organic amendments, influence methanogen abundance and activity in rice paddies, thereby affecting methane emissions. Sustainable water management, such as intermittent flooding, reduces methanogenesis by introducing oxygen, suppressing methanogens. While Olayinka and Akinola (2023) examined methane emissions from Nigerian wetland rice fields and noted that seasonal flooding patterns and organic input levels significantly affect methanogen populations and methane emissions. They recommended integrated water and organic matter management strategies to reduce methane emissions from rice cultivation in Nigerian agroecosystems. Furthermore, methane oxidation by methanotrophs mitigates emissions. Knief (2015) reviewed methanotrophic diversity and ecology, underscoring their importance in regulating atmospheric methane levels. Enhancing methanotrophic activity through soil amendments or crop selection could be a viable mitigation strategy.

14.3.3 Gut Microbiota of Ruminants and Methane Production

Ruminant livestock, including cattle, sheep, and goats, are major agricultural methane sources, contributing approximately 30% of anthropogenic methane emissions globally (Smith et al., 2021). Methane is produced in the rumen, a specialized stomach chamber housing complex microbial consortia comprising bacteria, archaea, protozoa, and fungi that ferment plant material anaerobically. Methanogenic archaea in the rumen convert hydrogen (H_2) and carbon dioxide (CO_2), produced during fermentation, into methane (CH_4) (Hook et al., 2010). This process, while critical for maintaining rumen function by removing excess hydrogen, results in significant energy loss for the animal and methane emissions to the atmosphere.

Recent metagenomic and metatranscriptomic studies have enhanced understanding of rumen microbial ecology. Wallace et al. (2020) identified core methanogen populations and linked variations in methane emissions to shifts in microbial community composition and host genetics. These insights are guiding the development of microbial and dietary interventions to reduce enteric methane.

In Nigeria, Adewale et al. (2024) conducted a comprehensive review of ruminant methane emissions, highlighting that local breeds and traditional feeding systems contribute to higher methane yields compared to improved feeding practices. They emphasized the potential for integrating dietary supplements such as tannins, oilseeds, and probiotics to modulate rumen microbiota and lower methane emissions in Nigerian livestock systems.

Dietary manipulation, such as supplementation with lipids, tannins, or methane inhibitors, has shown promise in suppressing methanogenesis (Sahu & Arya, 2024). Additionally, research into probiotics and microbial community engineering seeks to promote alternative hydrogen sinks, such as propionate-producing bacteria, to divert hydrogen away from methane formation (Jeyanathan et al., 2014).

14.4 MICROBIAL STRATEGIES TO MITIGATE GREENHOUSE GAS EMISSIONS

Microorganisms offer a powerful and natural mechanism for mitigating GHG emissions in agricultural systems. These microbes contribute to biogeochemical cycles and can be harnessed to reduce methane (CH_4), nitrous oxide (N_2O), and carbon dioxide (CO_2) emissions through targeted strategies. These five major microbial mitigation approaches include methanotrophs, nitrification inhibitors, biofertilizers, anaerobic digesters, and microbial carbon sequestration.

14.4.1 Methanotrophs

Methanotrophs are a unique group of aerobic bacteria that oxidize methane as their sole source of carbon and energy, thereby transforming methane into carbon dioxide and reducing its global warming potential. These bacteria are naturally present in upland soils, wetlands, and rhizospheres of rice paddies. Their activity can significantly lower methane emissions from waterlogged soils and animal waste management systems. Recent reviews have explored the potential of methanotrophs in rice-based systems and livestock farming (Semrau et al., 2022). In particular, studies have demonstrated that encouraging methanotrophic communities through proper water management such as alternate wetting and drying (AWD) which can enhance methane oxidation rates. Win et al. (2021) reported that the use of organic manure in combination with AWD promoted methanotroph populations and reduced methane emissions in Southeast Asian rice fields. In Nigeria, studies by Ogunyemi et al. (2021) identified native methanotrophic strains from the Niger Delta's wetland soils. These microbes showed high methane oxidation potential, suggesting their utility in site-specific biotechnological applications. Additionally, research on livestock environments has begun exploring biofilters inoculated with methanotrophs to mitigate emissions from cattle manure storage.

14.4.2 Nitrification Inhibitors

N_2O is primarily produced through the microbial processes of nitrification and denitrification. Nitrification inhibitors function by reducing the microbial conversion of ammonium to nitrate, thereby lowering substrate availability for denitrifiers. These inhibitors can be chemical (e.g., DCD, DMPP) or biological, produced naturally by certain plants and microbes.

BNIs are gaining attention as sustainable tools. Subbarao et al. (2021) reviewed various BNIs from tropical grasses and leguminous plants, highlighting their effects on reducing soil N_2O emissions. Saud et al. (2022) discussed how microbes like *Bacillus subtilis* and *Pseudomonas fluorescens* produce antimicrobial metabolites that selectively inhibit nitrifiers. In Nigeria, Agboola et al. (2023) tested microbial-based nitrification inhibitors in maize and cassava farms across Oyo and Ogun States. Their results showed a 35–50% reduction in N_2O emissions compared to untreated controls. They also observed improvements in nitrogen use efficiency, emphasizing the agronomic co-benefits of microbial approaches.

14.4.3 Biofertilizers

Biofertilizers are living formulations containing beneficial microbes that enhance plant nutrient acquisition and growth. These microbes include nitrogen-fixing bacteria (*Rhizobium*, *Azotobacter*), phosphate-solubilizing bacteria (*Pseudomonas*, *Bacillus*), and arbuscular mycorrhizal fungi (AMF). By improving nutrient cycling and reducing the dependence on synthetic fertilizers, biofertilizers help limit GHG emissions associated with fertilizer production and use. Zhang et al. (2023) emphasized the role of biofertilizers in reducing soil nitrate concentrations and mitigating N_2O emissions. Several studies have shown that combining biofertilizers with organic amendments can enhance microbial activity and nutrient retention, reducing the likelihood of GHG-producing microbial transformations. In Nigeria, Okon and Okorie (2022) applied *Azotobacter*-based biofertilizers in maize production and reported increased crop yields with lower nitrogen losses. Similarly, Nwachukwu et al. (2022) demonstrated that integrating biofertilizers with composted poultry manure reduced GHG emissions in yam and vegetable fields in southeastern Nigeria. These studies point to the dual environmental and agronomic value of microbial inoculants under local farming conditions.

14.4.4 Anaerobic Digesters

AD involves the breakdown of organic matter by microbial consortia under oxygen-free conditions. The process yields biogas a mixture of methane and CO_2 and digestate, which can be used as a fertilizer. This approach captures methane that would otherwise escape from unmanaged manure or crop residues, offering a valuable emissions mitigation strategy and source of renewable energy. Singh et al. (2023) reviewed microbial pathways in AD systems, noting the importance of *Methanosarcina*, *Methanobacterium*, and *Clostridium* species in optimizing methane production. Co-digestion of manure with crop residues and kitchen waste enhances substrate balance, promoting microbial synergy and digestion efficiency. Adeyemo et al. (2021) examined the performance of small-scale digesters in southwestern Nigeria and observed a significant reduction in GHG emissions from open manure pits. The use of digested slurry as a soil amendment further reduced N_2O emissions due to better nutrient stabilization. Despite their benefits, anaerobic digesters face challenges in Nigeria due to high initial costs, lack of technical know-how, and inadequate policy support (Tian et al., 2020; Wang et al., 2020).

14.4.5 Microbial Carbon Sequestration

Microbes play a pivotal role in carbon stabilization by decomposing plant residues and converting them into microbial biomass and necromass, which form the stable fraction of soil organic matter. This process contributes to long-term carbon sequestration and enhances soil health. Lehmann and Kleber (2022) highlighted the role of microbial carbon use efficiency in determining carbon fate. High carbon use efficiency systems promote the formation of microbial necromass, which binds with soil minerals and becomes resistant to further degradation. The use of compost, cover crops, and reduced tillage can foster favorable conditions for microbial C sequestration. In sub-Saharan Africa, and particularly in Nigeria, recent studies are investigating the potential of indigenous microbial communities for carbon stabilization. Yusuf et al. (2022) found that soils in the Guinea Savanna region inoculated with native *Actinobacteria* and *Bacillus* species showed increased soil organic carbon levels and enhanced soil aggregation. Abdullahi et al. (2023) demonstrated that biochar application in combination with microbial inoculants enhanced microbial diversity and increased long-term carbon retention in semi-arid soils.

14.5 APPLICATION OF MICROBIAL TECHNOLOGIES

The application of microbial technologies offers innovative and practical solutions for mitigating GHG emissions from agriculture. These technologies span soil, crop, and livestock systems and include the use of beneficial microbes in soil amendments, microbial inoculants and consortia, feed additives for ruminants, and composting strategies enhanced with microbial agents.

14.5.1 Soil Amendments with Beneficial Microbes

Soil amendments containing plant growth-promoting rhizobacteria, nitrogen-fixing bacteria, and phosphate-solubilizing microbes are widely used to improve nutrient use efficiency and reduce dependency on synthetic fertilizers, thereby lowering nitrous oxide (N_2O) emissions. Microbial amendments such as *Azospirillum*, *Rhizobium*, and *Bacillus* species contribute to soil health, organic matter decomposition, and carbon stabilization (Zhang & Sun, 2022). Studies by Zhang et al. (2023) and Saud et al. (2022) show that microbial soil amendments reduce nitrate leaching and limit N_2O production by enhancing microbial nitrogen assimilation and reducing the nitrification-denitrification pathway. In Nigeria, Okon and Okorie (2022) demonstrated that soil inoculation with *Azotobacter* improved maize yield while reducing nitrogen losses, suggesting microbial amendments are both agronomically and environmentally beneficial.

14.5.2 Inoculants and Microbial Consortia for GHG Mitigation

Microbial inoculants formulations containing one or more strains of beneficial microbes are increasingly applied to agricultural soils to influence biogeochemical processes. When applied as consortia, these microbial communities offer synergistic benefits. For example, consortia containing nitrogen fixers, mycorrhizae, and decomposers can simultaneously enhance nutrient uptake, suppress pathogens, and reduce GHG emissions.

Recent meta-analyses, such as those by Basu et al. (2021), report that microbial consortia significantly outperform single-strain inoculants in improving soil function and reducing emissions. Nigerian studies by Nwachukwu et al. (2022) show that integrating microbial consortia with compost improved soil nitrogen cycling and cut GHG emissions in vegetable farms in southeastern Nigeria.

14.5.3 Feed Additives and Probiotics for Reducing Enteric Methane

Ruminant livestock are a major source of methane emissions through enteric fermentation. Feed additives especially microbial probiotics can alter the rumen microbial ecosystem to reduce methane production. Probiotic strains such as *Lactobacillus*, *Bacillus*, and yeast-based additives like *Saccharomyces cerevisiae* have shown promise in inhibiting methanogenic archaea or shifting fermentation pathways toward propionate production, which is less methane-intensive.

Smith et al. (2021) report that dietary supplementation with certain probiotics reduced CH_4 emissions by up to 20% in cattle. In Nigeria, research by Ogbuagu et al. (2025) demonstrated that feeding indigenous cattle with probiotic-enriched supplements reduced methane emissions and improved feed efficiency in traditional pastoral systems.

14.5.4 Composting with Microbial Enhancers to Lower Emissions

Composting is a vital waste management strategy in agriculture, but it can emit both methane and nitrous oxide if unmanaged. The use of microbial inoculants such as thermophilic bacteria, fungi, and actinomycetes accelerates composting and ensures aerobic decomposition, reducing the risk of GHG formation. Kumar et al. (2022) reviewed composting systems enhanced with

Trichoderma, *Lysinibacillus*, and *Bacillus* strains, reporting up to 60% reductions in methane and nitrous oxide emissions. In Nigeria, composting poultry and cassava waste with microbial starters significantly improved compost quality and reduced emissions, as shown in studies by Agbor et al. (2015).

14.6 CASE STUDIES AND SUCCESS STORIES

Microbial technologies have shown promising real-world applications across diverse agricultural systems. Case studies from rice paddies, livestock farms, and composting facilities show the potential of microbial interventions in mitigating GHG emissions. These examples not only illustrate the technical feasibility of these approaches but also provide evidence of their scalability and impact under field conditions.

14.6.1 Rice Paddies: Microbial Strategies in Methane Reduction

In rice production systems, methanogenic archaea are key contributors to methane (CH_4) emissions due to the anaerobic decomposition of organic matter in flooded soils. Win et al. (2021) conducted field trials in Myanmar and Malaysia, examining the effect of water management and organic manure types on methane emissions. Their study showed that intermittent flooding combined with microbial-enriched compost significantly reduced CH_4 emissions by over 40%, without compromising yields.

Similarly, in Malaysia, Ahmed et al. (2023) reviewed methane mitigation trials where inoculation of methanotrophic bacteria (methane-oxidizers) into paddy fields demonstrated CH_4 emission reductions of up to 50%. These results suggest the dual benefit of reduced GHG emissions and enhanced nitrogen cycling, thereby improving crop productivity.

14.6.2 Livestock Farms: Enteric Methane Mitigation via Gut Microbiome Management

Livestock systems are substantial sources of methane due to enteric fermentation by ruminant gut microbiota. In Nigeria, studies conducted by Olorunfemi et al. (2022) at livestock farms in Oyo State evaluated the impact of feed additives such as *Asparagopsis taxiformis* and microbial probiotics on enteric CH_4 production. Results showed methane reductions of up to 30% in dairy cows supplemented with microbial feed additives, while simultaneously enhancing feed conversion efficiency. highlighted international studies using direct-fed microbials (DFMs) in cattle that altered the rumen microbial ecology to suppress methanogenic archaea. In New Zealand and Malaysia, trials with *Bacillus*-based probiotics reported reduced CH_4 output per unit of milk produced, contributing to the sustainability of livestock operations.

14.6.3 Composting Facilities: Enhancing Decomposition with Microbial Additives

Composting is a widely used method for managing organic waste, but improper aerobic conditions can lead to the release of CH_4 and nitrous oxide (N_2O). In a controlled composting experiment, addition of a microbial culture and biochar significantly promoted compost maturity and reduced methane emissions (CH_4) by up to ~55.9 % compared with control compost, while enhancing compost quality and reducing greenhouse gas release (Chen et al., 2020). A related case from Malaysia by Raihan et al. (2022) showed that using bio-augmented composting inoculants in palm oil mill waste treatment facilities led to a 38% reduction in GHG emissions and enhanced compost quality, thus promoting circular nutrient reuse.

14.7 CHALLENGES AND LIMITATIONS

Microbial strategies offer a powerful tool for mitigating GHG emissions in agriculture, yet their adoption is constrained by complex scientific, practical, and policy-related challenges that must be carefully addressed.

14.7.1 Variability in Microbial Efficacy Across Environments

The effectiveness of microbial interventions varies widely based on environmental and management conditions. Factors such as soil type, pH, temperature, and moisture influence microbial survival and efficacy. For example, methanotrophs and nitrifying bacteria show optimal performance under specific redox and nutrient conditions but may be less active in marginal soils. Studies from regions like Malaysia and Nigeria highlight that microbial inoculants and amendments can yield inconsistent results depending on local agroecological variables (Raihan et al., 2022; Bhatti et al., 2024).

14.7.2 Difficulty in Measuring Microbial Activity and GHG Impact

Quantifying the exact contributions of microbial communities to GHG mitigation is methodologically challenging. While tools such as isotopic tracing, metagenomics, and real-time gas monitoring offer insights, distinguishing microbial effects from abiotic factors in complex field conditions remains difficult. This limits the ability to predict or validate microbial performance across landscapes (Liang & Robertson, 2021; Zhang et al., 2023).

14.7.3 Commercialization and Farmer Adoption Barriers

Microbial technologies often face resistance from farmers, especially in developing countries. Factors such as unfamiliarity, inconsistent results, lack of training, and limited access to microbial products impede adoption. Moreover, microbial amendments may require cold-chain logistics or specific application protocols, increasing costs and complexity (Laborde et al., 2021; Duxbury & Mosier, 2022). In Nigeria, although pilot programs have shown success in composting and manure management, broader uptake is constrained by inadequate extension support.

14.7.4 Regulatory and Environmental Safety Concerns

Introducing microbial agents into agricultural systems raises biosafety issues, such as the potential for horizontal gene transfer, disruption of native microbial diversity, or unintended ecological effects. Regulatory frameworks in many countries are still evolving, often lacking clear guidelines for risk assessment, approval, and monitoring of microbial products. This regulatory uncertainty delays commercialization and creates hesitation among stakeholders (Win et al., 2021).

14.8 FUTURE PERSPECTIVES

As agriculture continues to face the twin pressures of increasing productivity and mitigating environmental impact, microbial strategies are poised to play a transformative role in addressing GHG emissions. Future directions in this field hinge on advances in microbial genomics, metagenomics, synthetic biology, and their integration with digital technologies and policy frameworks that promote adoption.

14.8.1 Advances in Microbial Genomics and Metagenomics

Breakthroughs in genomics and metagenomics are rapidly enhancing our understanding of complex microbial communities in agricultural ecosystems. High-throughput sequencing technologies are unveiling the taxonomic and functional diversity of soil and gut microbiomes, enabling the identification of key functional genes involved in methane oxidation, nitrification, denitrification, and carbon sequestration. These insights are critical for selecting and engineering microbial strains with enhanced GHG mitigation capabilities (Bhatti et al., 2024). Metagenomic analysis of rice paddy soils in Asia has revealed the presence of unique methanotrophs capable of significantly reducing methane flux (Win et al., 2021). In Nigeria, metagenomic profiling of compost-treated soils has begun to reveal locally adapted microbes that could be deployed in biofertilizers for both yield improvement and emission reduction (Ayoade et al., 2023).

14.8.2 Development of Precision Microbial Formulations

The future of microbial GHG mitigation will increasingly rely on precision formulations that are tailored to specific soils, crops, climates, and farming practices. Advances in microbial ecology and systems biology are facilitating the design of microbial consortia with synergistic functionalities for instance, combining nitrogen-fixing bacteria with nitrification inhibitors to improve nitrogen use efficiency while minimizing nitrous oxide emissions (Saud et al., 2022). Innovations in encapsulation, shelf-life extension, and carrier materials are further improving the viability and scalability of microbial inoculants, especially in regions with challenging storage or transportation conditions such as sub-Saharan Africa and Southeast Asia.

14.8.3 Integration with Digital Agriculture and Climate-Smart Practices

Microbial strategies will be most effective when integrated into broader frameworks of climate-smart agriculture and digital farming. Technologies such as remote sensing, soil health monitoring, and precision irrigation can be used to identify optimal application times and conditions for microbial inputs. Artificial intelligence (AI) and machine learning models are also being developed to predict microbial activity patterns, GHG fluxes, and treatment outcomes based on site-specific data (Guo et al., 2022). These digital tools can empower farmers with real-time insights and decision support systems, increasing the efficiency and effectiveness of microbial interventions.

14.8.4 Policy and Incentive Frameworks for Adoption

For microbial technologies to gain traction, supportive policy and financial mechanisms are essential. Governments and international organizations must develop science-based regulatory frameworks that recognize and standardize microbial solutions. Public-private partnerships, extension services, and incentive schemes such as carbon credits or subsidies for microbial biofertilizers can help overcome barriers to adoption (Laborde et al., 2021). In Nigeria and Malaysia, emerging pilot programs are offering farmers financial support for adopting microbial soil amendments, signaling growing policy interest in sustainable microbial practices (Smith et al., 2007).

14.9 CONCLUSION

Microbial-based strategies present a promising frontier in the global effort to mitigate GHG emissions from agriculture. By harnessing the natural capabilities of soil and gut microbiomes, researchers and practitioners are beginning to develop solutions that not only reduce methane (CH_4), nitrous oxide (N_2O), and carbon dioxide (CO_2) emissions, but also enhance nutrient cycling, soil health, and agricultural productivity. Methanotrophs, nitrification inhibitors, biofertilizers, anaerobic digesters, and microbial carbon sequestration techniques all offer effective, biologically grounded pathways for curbing emissions within diverse agricultural systems. The integration of these microbial solutions into climate-resilient agriculture is increasingly vital. As the impacts of climate change intensify particularly in vulnerable regions such as sub-Saharan Africa and Southeast Asia sustainable and scalable approaches are needed to reduce agriculture's carbon footprint while maintaining food security. Microbial technologies, particularly when combined with precision agriculture, composting, and low-emission livestock management, offer environmentally friendly alternatives to conventional practices reliant on synthetic inputs. However, to realize the full potential of microbial interventions, concerted efforts are needed across research, innovation, and policy domains. Advancing microbial genomics, developing tailored microbial consortia, and integrating digital agriculture tools are crucial for improving efficacy and adoption.

REFERENCES

Abbass, K., Qasim, M. Z., Song, H., Murshed, M., Mahmood, H., & Younis, I. (2022). A review of the global climate change impacts, adaptation, and sustainable mitigation measures. Environmental Science and Pollution Research, 29(28), 42539–42559. https://doi.org/10.1007/s11356-022-19718-6

Abdullahi, M. I., Tijjani, A., & Suleiman, M. (2023). Enhancing soil carbon sequestration using biochar-microbial synergism: A sustainable soil management approach. African Journal of Environmental Science and Technology, 17(3), 101–112.

Adewale, O. S., Olaleye, T. M., & Inegbenebor, U. A. (2024). Ruminant methane emissions in Nigeria: Current status and mitigation strategies. Journal of Agricultural Science and Technology, 16(2), 115–132.

Adeyemo, S. O., Akinyemi, D. O., & Alabi, B. S. (2021). Biogas potential and environmental impact of livestock manure management in Nigeria. Renewable Energy, 175, 659–668.

Agboola, A. O., Ogunlade, M. O., & Adeoye, G. O. (2023). Field-scale validation of microbial nitrification inhibitors to mitigate nitrous oxide emissions in Nigerian agriculture. International Journal of Environmental Studies, 80(1), 66–80.

Agbor, E. O., Udo, S. E., & Essien, E. O. (2015). Influence of cassava peels and poultry manure-based compost on soil properties, growth and yield of waterleaf (*Talinum triangulare* Jacq.) in an Ultisol of Southeastern Nigeria. Journal of Soil Science and Environmental Management, 6(7), 285–292.

Ahmed, Z., Gui, D., Qi, Z., Liu, J., Ali, A., Murtaza, G., … Ahmad, S. (2023). Greenhouse gas emissions and mitigation strategies in rice production systems. In M. Ahmed (Ed.), Global Agricultural Production: Resilience to Climate Change (pp. 237–265). Cham, Switzerland: Springer International Publishing. https://doi.org/10.1007/978-3-031-14973-3_8

Akintola, S. A., Olufemi, S. O., & Adekunle, T. T. (2022). Fertilizer management and greenhouse gas emissions: Insights from cassava and maize production in Nigeria. Environmental Research Communications, 4(7), 075006.

Basheer, S., Wang, X., Farooque, A. A., Nawaz, R. A., Pang, T., & Neokye, E. O. (2024). A review of greenhouse gas emissions from agricultural soil. Sustainability, 16(11), 4789.

Basu, A., Pramanik, A., & Maiti, S. K. (2021). Microbial-assisted phytoremediation of contaminated soils. Environmental Technology & Innovation, 23, 101665. https://doi.org/10.1016/j.eti.2021.101665

Bhatti, U. A., Bhatti, M. A., Tang, H., Syam, M. S., Awwad, E. M., Sharaf, M., & Ghadi, Y. Y. (2024). Global production patterns: Understanding the relationship between greenhouse gas emissions, agriculture greening and climate variability. Environmental Research, 245, 118049. https://doi.org/10.1016/j.envres.2023.118049

Chen, H., Awasthi, S. K., Liu, T., Duan, Y., Ren, X., Zhang, Z., Pandey, A., & Awasthi, M. K. (2020). Effects of microbial culture and chicken manure biochar on compost maturity and greenhouse gas emissions during chicken manure composting. Journal of Hazardous Materials, 389, 121908. https://doi.org/10.1016/j.jhazmat.2019.121908

Duxbury, J. M., & Mosier, A. R. (2022). Status and issues concerning agricultural emissions of greenhouse gases. In H. M. Kaiser & T. E. Drennen (Eds.), Agricultural Dimensions of Global Climate Change (pp. 229–258). Abingdon, UK; New York, NY, USA: Routledge.

Eze, S. C., Chukwuneke, F. N., & Onuoha, U. D. (2021). Microbial nitrogen cycling and nitrous oxide emissions in Nigerian savannah soils: Challenges and opportunities. Soil Biology and Biochemistry, 157, 108237.

Ezemba, C. C., Adams, P. Y., Okoye, P. C., Ezemba, A. S., Ajeh, J. E., Osuala, O. J., … Iyiola, O. O. (2025). Methods and strategies of teaching microbiology in Nigerian higher institutions: Problems and remedies. Scientia. Technology, Science and Society, 2(2), 3–26.

Fierer, N. (2017). Embracing the unknown: Disentangling the complexities of the soil microbiome. Nature Reviews Microbiology, 15(10), 579–590.

Gan, K. E., Taikan, O., Gan, T. Y., Weis, T., Yamazaki, D., & Schüttrumpf, H. (2023). Enhancing renewable energy systems, contributing to sustainable development goals of United nations and building resilience against climate change impacts. Energy Technology, 11(11), 2300275.

Guo, C., Liu, X., & He, X. (2022). A global meta-analysis of crop yield and agricultural greenhouse gas emissions under nitrogen fertilizer application. Science of the Total Environment, 831, 154982.

Hook, S. E., Wright, A. D. G., & McBride, B. W. (2010). Methanogens: Methane producers of the rumen and mitigation strategies. Archaea, 2010, 945785.

Hu, H. W., Chen, D., & He, J. Z. (2022). Microbial regulation of terrestrial nitrous oxide emissions: Understanding the biological pathways for prediction and mitigation. Environmental Science & Technology, 56(4), 2119–2133.

Intergovernmental Panel on Climate Change (IPCC). (2022). In H. O. Pörtner, D.C. Roberts, M. Tignor, et al. (Eds.), Climate Change 2022: Impacts, Adaptation, and Vulnerability. Working Group II Contribution to the Sixth Assessment Report of the Intergovernmental Panel on Climate Change. Cambridge University Press. https://doi.org/10.1017/9781009325844

Jeyanathan, J., Kirs, M., & Muck, R. E. (2014). The rumen microbiome and methanogenesis: Where we are and where are we going? Frontiers in Microbiology, 5, 134.

Kamyab, H., SaberiKamarposhti, M., Hashim, H., & Yusuf, M. (2024). Carbon dynamics in agricultural greenhouse gas emissions and removals: A comprehensive review. Carbon Letters, 34(1), 265–289.

Kaushik, A., Attri, S. D., Kaushik, C. P., & Schnell, R. (2021). Climate resilience and environmental sustainability approaches: An introduction. In A. Kaushik, S. D. Attri, C. P. Kaushik, & R. Schnell (Eds.), Climate Resilience and Environmental Sustainability Approaches: Global Lessons and Local Challenges (pp. 1–8). Cham, Switzerland: Springer.

Knief, C. (2015). Diversity and habitat preferences of cultivated and uncultivated aerobic methanotrophic bacteria evaluated based on pmoA as molecular marker. Frontiers in Microbiology, 6, 1346.

Kumar, M., Pandey, A., & Shah, M. P. (2022). Microbial technologies for soil pollution control. Bioresource Technology, 344, 126246. https://doi.org/10.1016/j.biortech.2021.126246

Kuypers, M. M. M., Marchant, H. K., & Kartal, B. (2022). The microbial nitrogen-cycling network. Nature Reviews Microbiology, 20(6), 384–399. https://doi.org/10.1038/nrmicro.2018.9

Laborde, D., Mamun, A., Martin, W., Piñeiro, V., & Vos, R. (2021). Agricultural subsidies and global greenhouse gas emissions. Nature Communications, 12, 2601. https://doi.org/10.1038/s41467-021-22703-1

Lal, R. (2020). Soil carbon sequestration to mitigate climate change. Geoderma, 363, 114182.

Lehmann, J., & Kleber, M. (2022). The contentious nature of soil organic matter. Nature, 528(7580), 60–68.

Le Mer, J., & Roger, P. (2001). Production, oxidation, emission and consumption of methane by soils: A review. European Journal of Soil Biology, 37(1), 25–50. https://doi.org/10.1016/S1164-5563(01)01067-6

Li, L., Awada, T., Shi, Y., Jin, V. L., & Kaiser, M. (2025). Global greenhouse gas emissions from agriculture: Pathways to sustainable reductions. Global Change Biology, 31(1), e70015.

Li, H., Jin, X., Zhao, R., Han, B., Zhou, Y., & Tittonell, P. (2024). Assessing uncertainties and discrepancies in agricultural greenhouse gas emissions estimation in China: A comprehensive review. Environmental Impact Assessment Review, 106, 107498.

Liang, D., & Robertson, G. P. (2021). Nitrification is a minor source of nitrous oxide (N_2O) in an agricultural landscape and declines with increasing management intensity. Global Change Biology, 27(21), 5599–5613.

Liu, Y., Wu, T., White, J. C., & Lin, D. (2021a). A new strategy using nanoscale zero-valent iron to simultaneously promote remediation and safe crop production in contaminated soil. Nature Nanotechnology, 16(2), 197–205.

Liu, D., Zhu, X., & Wang, Y. (2021b). China's agricultural green total factor productivity based on carbon emission: An analysis of evolution trend and influencing factors. Journal of Cleaner Production, 278(1), 123692.

Jerry, A., Ugah, C. S., & Anyira-Fasanmi, M. I. (2025). Role of microbes in climate change: Methanogens and methanotrophs in carbon cycling—A case study of Agbede rice fields, Edo State, Nigeria. International Journal of Science, *Architecture, Technology and Environment*, 3(2).

Nwachukwu, E. O., Udeh, N. U., & Okeke, E. S. (2022). Soil microbial dynamics during bioremediation of agro-industrial wastes. Bioremediation Journal, 26(4), 339–351. https://doi.org/10.1080/10889868.2022.2034516

Ogunyemi, O. M., Akinpelu, B. A., & Olatunji, O. S. (2021). Microbial remediation of petroleum-contaminated soils in developing regions. Environmental Challenges, 5, 100282. https://doi.org/10.1016/j.envc.2021.100282

Okon, I. E., & Okorie, F. S. (2022). Bioremediation potential of bacteria isolated from contaminated soils. Environmental Technology & Innovation, 27, 102418. https://doi.org/10.1016/j.eti.2022.102418

Olayinka, A., & Akinola, M. O. (2023). Indigenous microbial communities and their role in soil bioremediation. Environmental Monitoring and Assessment, 195, 742. https://doi.org/10.1007/s10661-023-11234-5

Olorunfemi, I. E., Fasinmirin, J. T., & Ojo, A. S. (2022). Soil quality indicators under remediation and climate stress. Environmental Monitoring and Assessment, 194, 557. https://doi.org/10.1007/s10661-022-10122-4

Raihan, M. J., Uddin, M. K., & Rahman, M. M. (2022). Climate-driven changes in soil microbial processes and remediation outcomes. Environmental Pollution, 307, 119501. https://doi.org/10.1016/j.envpol.2022.119501

Rehman, A., Ma, H., Ozturk, I., & Ahmad, M. (2022). Sustainable soil management and microbial remediation under climate change. Journal of Cleaner Production, 363, 132482. https://doi.org/10.1016/j.jclepro.2022.132482

Robertson, G. P., & Vitousek, P. M. (2009). Nitrogen in agriculture: Balancing the cost of an essential resource. Annual Review of Environment and Resources, 34, 97–125. https://doi.org/10.1146/annurev.environ.032108.105046

Sahu, P. K., Singh, S., Gupta, A., & Singh, B. K. (2023). Soil microbiome resilience under environmental stress and remediation strategies. Journal of Hazardous Materials, 447, 130820. https://doi.org/10.1016/j.jhazmat.2023.130820

Sahu, S. K., & Arya, N. (2024). Methane mitigation strategies in agriculture: Recent advances and future directions. Environmental Science and Pollution Research, 31(4), 3921–3936.

Saud, S., Fahad, S., Chen, Y., Ihsan, M. Z., & Hammad, H. M. (2022). Role of soil microorganisms in sustainable agriculture and remediation. Sustainability, 14(3), 1362. https://doi.org/10.3390/su14031362

Semrau, J. D., DiSpirito, A. A., & Murrell, J. C. (2022). Methanotrophs and their role in environmental remediation. Nature Reviews Microbiology, 20(12), 695–709. https://doi.org/10.1038/s41579-022-00729-4

Shi, Y., Liu, J., Zhang, X., & Wang, H. (2022). Microbial responses to soil contamination and implications for bioremediation. Science of the Total Environment, 824, 153856. https://doi.org/10.1016/j.scitotenv.2022.153856

Smith, P., Martino, D., Cai, Z., Gwary, D., Janzen, H., Kumar, P., … Smith, J. (2007). Policy and technological constraints to implementation of greenhouse gas mitigation options in agriculture. Agriculture, Ecosystems & Environment, 118(1–4), 6–28.

Smith, P., Reay, D. S., & Smith, J. U. (2021). The potential of the global food system to reduce greenhouse gas emissions. Nature Food, 2, 832–839.

Smith, R., Reay, D., & Smith, P. (2023). Advances in microbial mitigation of agricultural greenhouse gas emissions. Current Opinion in Environmental Science & Health, 31, 100441.

Subbarao, G. V., & Searchinger, T. D. (2021). A "more ammonium solution" to mitigate nitrogen pollution and boost crop yields. Proceedings of the National Academy of Sciences of the United States of America (PNAS), 118(39), e2107576118. https://doi.org/10.1073/pnas.2107576118

Tian, H., Xu, R., Canadell, J. G., Thompson, R. L., Winiwarter, W., Suntharalingam, P., … Pan, N. (2020). A comprehensive quantification of global nitrous oxide sources and sinks. Nature, 586(7828), 248–256.

Wallace, S. J., Williamson, A. J., & Ward, A. C. (2020). Advances in microbial remediation of contaminated environments. Current Opinion in Biotechnology, 64, 156–162. https://doi.org/10.1016/j.copbio.2020.01.012

Wang, Y., Cheng, J., Zhang, X., Li, Y., Zhou, X., & Zhou, J. (2022). How microbial community diversity affects greenhouse gas emissions in agricultural systems. Microbial Biotechnology, 15(1), 51–67.

Wang, Y., Tian, H., Lu, C., Xu, R., Gao, J., Zhang, B., … Pan, S. (2020). Global nitrous oxide emissions from agriculture: From synthetic fertilizers to animal manure. Scientific Reports, 10(1), 1–13.

Win, K. T., Win, K. T., Bellingrath-Kimura, S. D., & Oo, A. Z. (2021). Mitigation of methane emissions from paddy rice fields through water, fertilizer, and crop management: A review. Sustainability, 13(2), 836.

Yang, H., Liu, S., Liu, Q., Zhang, Y., Li, X., & Wang, D. (2021). Soil N2O emissions in global agricultural systems: Quantification and driving factors. Global Change Biology, 27(9), 1803–1814. https://doi.org/10.1111/gcb.15520

Yao, H., Chen, X., & Tang, X. (2021). Soil microbial biomass and activity under climate-smart practices in China. Agriculture, Ecosystems & Environment, 314, 107391.

Yusuf, R. O., Giwa, A., & Mohammed, I. A. (2022). Waste-derived substrates for microbial remediation and bioenergy production. Renewable and Sustainable Energy Reviews, 162, 112415. https://doi.org/10.1016/j.rser.2022.112415

Zhang, J., & Sun, Q. (2022). Anaerobic digestion technology in agricultural waste management: A review. Renewable and Sustainable Energy Reviews, 154, 111825.

Zhang, Y., Zhang, J., Li, X., Chen, G., & Xie, Y. (2023). Recent advances in the mitigation of nitrous oxide emissions from urea fertilization in agriculture. Journal of Cleaner Production, 384, 135467. https://doi.org/10.1016/j.jclepro.2022.135467

15 Climate Change and the Restoration of Degraded Lands

Oluwatosin Olaoluwa Daramola, Olarewaju John Adunbi, and Babafemi Raphael Babaniyi

15.1 INTRODUCTION

Soils are one of the most important driving factors of terrestrial ecosystems and contribute to many ecological and anthropogenic processes. They control the purification of water, biological diversity, food security, and serve as a large reservoir of atmospheric CO_2, which diminishes effect of climate change (Bera et al., 2020; Liu et al., 2024). Healthy soil is essential to the maintenance of agricultural productivity, water quality, and ecosystem integrity. Nevertheless, the rapid rates of land degradation are a serious risk to these vital ecosystem services. According to the Food and Agriculture Organization (FAO), more than 33% of the global soils are already degraded from moderate to extreme dimensions, due to human impacts predominantly caused by deforestation, overgrazing, industrial contamination, and unsustainable agricultural practices (Sales da Silva et al., 2020). In addition to human activities, the climate change exacerbates the problem of soil degradation. Increased global temperature, changes in precipitation patterns, drought, and increased frequency of extreme weather conditions have severely disturbed the soil structure, water retention, and microbial community (Fatoyinbo et al., 2021; Xu, 2024). These phenomena result in speeding up erosion, vanishing fertility at the same time as salinity of soil is rising and increasing. For example, long-term drought leads to organic matter reduction in the soil, and intense heavy rain results in topsoil soil washed away with subsequent permanent nutrient losses between anthropocentric activities and climate change emphasizes the need for sustainable soil management and conservation practices.

Conventional soil remediation techniques, for example, physical excavation, incineration, and chemical stabilization, are commonly used to eliminate soil contamination. However, these methods usually cause financial burden, secondary pollution, and more frequently fail to recover the biological integrity or ecological function of soil (Chen et al., 2020). On the other hand, the bioremediation has attracted the interest as a more sustainable, environmentally friendly approach. Bioremediation is the use of microorganisms, plants, fungi, or their enzymes to remove, detoxify, or immobilize contaminants from soil. This approach has several benefits, including low cost, less environmental invasiveness, and possible self-recovery (Sharma & Shah, 2021). In general, bioremediation processes are of two types, that is, in situ and ex situ. In situ bioremediation is the method of treating the impacted soil at its location, without the need for excavation, which causes less disruption and can be an economically better option. In ex situ methods the contaminated soil is excavated and then taken elsewhere for treatment whereas they are used mainly for heavy levels of contamination or rapid remediation (Das & Chandran, 2021). Several biological agents have been used for remediation, including bacteria *Pseudomonas* and *Bacillus* spp., fungi *Aspergillus* spp., as well as phytoremediative plants *Brassica juncea* (Indian mustard) and *Helianthus annuus* (sunflower) (Kumar et al., 2020). These microorganisms either biodegrade these pollutants themselves, or change the soil in such a way that removes the pollutants.

Recent studies have shown the potential applications of genetically engineered microorganisms, bioaugmentation, biostimulation, and phytoremediation mainly in recovering the most dispersed metal, hydrocarbon, pesticide, and emerging ones like microplastics and pharmaceuticals (El-Bestawy et al., 2022). For instance, hydrocarbon-degrading engineered microbial consortia

DOI: 10.1201/9781003650881-15

exhibited appreciable enhancement of petroleum-contaminated soil conditions even at varied temperature and moisture levels (Patel et al., 2024). Moreover, climate-resilient bioremediation strategies have been explored to sustain remediation efficiency under fluctuating climate stressors and to ensure long-term soil restoration in the context of climate change (Zhang et al., 2023).

In light of the increasing number of scientific publications from 2019 to 2025, I attempt to deliver in this chapter a complete and updated overview on bioremediation strategies applied in anthropogenic and climate-degraded soils. It consolidates results from laboratory research, field tests, and pilot projects to reveal effective, scalable interventions in various geographies and environments. Factors that affect the efficiency of bioremediation, including soil type, contamination, microbial community, and environmental parameters such as pH and moisture content, are also discussed in the text (Al-Hawash et al., 2022).

Our aim, therefore, is to provide an input, in a more general sense, to the debate on sustainable land management, through emphasizing innovative, user-friendly, and low-cost soil remediation plants that can be developed, at a global scale, and particularly in low- and middle-income countries where traditional remediation plants are not applicable. With a clear understanding of the promise and challenge that various bioremediation approaches offer, policymakers, researchers, and environmental managers can craft interventions to address priorities including restoring soil health, increasing agricultural productivity, and adapting to climate change.

15.2 MECHANISMS FOR SOIL EROSION

Soil degradation means reduction in soil quality and productivity due to not only natural processes but also human activities. Under the pressure of climate variability and human interference, the process of soil degradation has become much more complicated and interactive. The main processes are erosion, salinization, pollution, compaction, and loss of organic matter, leading, in various degrees, to soil structure disruption, loss of fertility, and breakdown of ecological services (Sales da Silva et al., 2020). Erosion is the most manifest type of degradation, especially in deforested or overgrazed areas. When topsoil is removed, nutrients and microbial diversity decrease substantially (DeVilleneuve & Smith, 2023). Salinization, frequently induced by unsustainable irrigation practices and increasing water tables, results in the accumulation of soluble salts in the soil profile, which detrimentally affects plant growth and microbiological activity (Fatoyinbo et al., 2021). These processes are exacerbated by climate change, which increases drought frequency and altering rainfall intensity, in turn increasing runoff and evaporation rates.

Soil contamination with heavy metals, petroleum hydrocarbons, pesticides, and industrial effluents is becoming an increasing concern, especially in regions around urban centers and mining areas. These toxicants disrupt the equilibrium of microflora essential for nutrient cycling and plant growth (Resch et al., 2022). Afterward, soil compaction (resulting from use of heavy machines and excessive number of animals in a limited area) decreases soil porosity and impedes water movement and root penetration. Conversely, the decrease of organic matter and lack of soil structure under intensive tillage and monoculture lead to the reduction of water sequestration capacity and a decline in the ability of soils to capture and store organic carbon (Liu et al., 2024). Such degradation pathways not only threaten agricultural productivity but also mitigate the success of ecological restoration. Hence, knowledge of these mechanisms is important in developing efficient bioremediation interventions using microbial, plant and enzymatic procedures to detoxify, reclaim and stabilize polluted soils under both anthropogenic- and climate-induced stresses (Carrero-González et al., 2012).

15.2.1 Human Activities

Anthropogenic interference is one of the most important factors of soil degradation at a global level; industrialization, urbanization, and modern agricultural practices have largely compromised soil health. Emissions from industries, mining activities, and improper disposal of solid and hazardous wastes contribute different pollutants to the soil environment such as polycyclic aromatic hydrocarbons (PAHs), heavy metals, petroleum hydrocarbons, and synthetic organic compounds (Azeez et al., 2020; Taher & Saeed, 2021). Such substances disturb soil microbial communities, affect soil pH and fertility levels, and can harm both ecosystems and human health.

Unsustainable human activities such as the excessive use of chemical fertilizers and pesticides in agriculture are among the most serious contributors of air pollution resulting in substantial human health problems. These are the types of management practices that erode soil structure, cause nutrient imbalances and ultimately decrease the diversity of the soil microbial community (Chen et al., 2020). Erosion and mineralization are further heightened by continued cropping and deep tillage practices compromising the long-term productivity of the soil.

In addition, application of untreated and partially treated wastewater for irrigation, particularly in resource-constrained environments, results in the presence of pharmaceutical residues, antibiotics, and heavy metals in agricultural soils. These pollutants remain persistent for a long time, and this imparts a toxic condition to the soil that adversely affects the growth of plants as well as the safety of the food we consume (Helmy & Kardena, 2024). Moreover, runoff from urban areas and landfill leaching gradually migrate contaminants into the neighboring soils and reinforce the cumulative degradation effects over scale. The interaction between human land uses and soil degradation is exacerbated by bad practices of land usages, laxity in enforcement of regulations, and low level of uptake of SLM. In response to these challenges, many researchers are now focusing on the applications of bioremediation strategies, which point to the use of microbial and plant-based systems to detoxify and reclaim contaminated soils (Mohanraj et al., 2025). Knowledge about anthropogenic sources of degradation is crucial for the development of focused remediation and prevention efforts.

15.2.2 Climate Change Effects

Environmental change has a crucial influence on the soil damage acceleration, by transforming basic soil-forming processes, and sustaining the soil integrity. Crucial climatic factors—such as the global temperature increase, changes in precipitation and the rise in atmospheric CO_2—exert strong effects over soil health and functioning (Xu, 2024). High temperatures promote microbial mineralization of organic matter that is capable of temporarily accelerating nutrient availability, but will, in the longer-term deplete soil carbon stocks. This not only depletes soil fertility, but also diminishes soil's ability to store carbon, aspects that include feedback loops to climate change (Ghosh et al., 2021).

Irregular and heavy episodes of rainfall—the signature of climate variability—intensify erosion and loss of nutrients, especially in areas of light and/or sandy soils. In semiarid and arid areas, long periods of drought and heatwaves lead to soil salinization, water stress, and the destruction of soil structures. Moreover, flood events, which are induced from high precipitation intensity, can result in the loss of topsoil and the removal of the most fertile layer affecting the plants and microbial activities (Jahromi et al., 2022).

The drought, however, leads to a dramatic reduction of microbial activity and diversity, preventing the soil to efficiently regenerate itself. The worst part is that it comes down especially hard on marginals, which have naturally slow recovery times. Additionally, climate-related stressors may also interact additively or synergistically with pollutants of anthropogenic origin including heavy metals and hydrocarbons, further complicating any remediation processes and mitigating strategies for soil health recovery (Liu et al., 2024). Understanding these combined antagonisms is critical for devising soil management and bioremediation strategies that are adaptive to these climatic and anthropogenic pressures on soil ecosystems.

15.3 BIOREMEDIATION TECHNIQUES

Bioremediation is the application of biological agents, such as microorganisms, fungi, and plants, to extract, neutralize, or stabilize soil pollutants. It offers a relatively low-cost and environmentally friendly alternative to the traditional remediation methods, especially in areas where soils are already impacted by the combined effect of human activity and climate change (DeVilleneuve & Smith, 2023). The most commonly used bioremediation strategies are biostimulation, bioaugmentation, phytoremediation, and mycoremediation, which are applicable to certain types of contaminants and soil conditions. Biostimulation augments the natural activity of indigenous soil microorganisms by providing additional nutrients, electron donors, or oxygen to promote better growing conditions for biodegradation. It is particularly effective for the remediation of petroleum hydrocarbon and pesticide contamination in nutrient poor soils (Singh et al., 2021). Bioaugmentation consists of the planting of specific microbial strains with a good degrading potential in polluted soils. It is particularly useful when natural microbial populations are not adequate or are dormant because of contaminant toxicity (Sáez-Sandino et al., 2024).

Phytoremediation is based on the fact that some plants can take up, accumulate, and metabolize heavy metals and organic pollutants. Metal hyperaccumulating characteristics and root system proficiency are found in both *Brassica juncea* and *Vetiveria zizanioides*. This is a perfect model for large-scale application, which soil structure can be further enhanced and promptly recovered with vegetation cover.

Another technique that is being investigated to decrease such compounds is mycoremediation (a method employing fungi), which is known for its capacity to biodegrade recalcitrant pollutants like PAHs as well as persistent pesticides by enzymatic action. Fungi such as *Trametes versicolor* and *Aspergillus niger* excrete enzymes to degrade pollutants, thereby converting them into less harmful forms (Fatoba & Edeh, 2020). In terms of climate change, implementation of these bioremediation strategies into adaptive land management practices can powerfully accelerate soil restoration, enhance carbon sequestration, and build long-term ecological resilience (Liu et al., 2024).

15.3.1 PHYTOREMEDIATION

Phytoremediation is a non-destructive, cost-effective, soil remediation technology which employs plants to mitigate, degrade, sequester, or immobilize pollutants from soil. This green technology uses a plant's own defense system by exploiting the natural properties of specific plants, the so-called hyperaccumulators, to bioaccumulate and store pollutants such as heavy metals and metalloids (and even organic compounds) in their biomass, or to reduce the bioavailability of contaminants via the root-zone interactions (Rascio & Navari-Izzo, 2011). Species that hyperaccumulate, such as *Brassica juncea* (Indian mustard) and *Pteris vittata* (Chinese brake fern), have been extensively researched for their ability to phytoextract toxic elements, such as arsenic, lead, and cadmium from contaminated soils. These plants are productive and highly tolerant to metal stress, and thus suitable for soil detoxification of contaminated agricultural and industrial sites (Liu et al., 2024). Their capacity to improve soil condition and vegetation cover also improves landscape rehabilitation and biodiversity over time.

Urban phytoremediation as mentioned before, field studies investigating urban phytoremediation are relatively recent. California research, for example, found that native plants, including telegraph weed (*Heterotheca grandiflora*) and mule fat (*Baccharis salicifolia*) were able to bring down levels of heavy metals—in particular zinc, copper, and lead—in soils that have been polluted by industrial and vehicular sources. After 1 year, the elevations of metals were decreased up to 40% of their levels, demonstrating its prospective in longer-term heavy metal remediation (Carrero-González et al., 2012).

Although a typically slow process, phytoremediation is best fit for modest to low contaminated areas; however, the potential for its scalability, low maintenance, and low levels of ecosystem disturbance makes it an attractive choice for sustainable soil recovery—especially in places where the impacts of anthropogenic efforts and climate stressors occur simultaneously.

15.3.2 Mycoremediation

Mycoremediation is considered an effective bioremediation method that employs fungi to break down, immobilize, or transform toxic contaminants in the soil. This approach is based on the enzymatic capacity of filamentous fungi, and white-rot fungi in particular, to generate powerful extracellular enzymes—that is, laccases, manganese peroxiases, and lignin peroxiases—that can transform recalcitrant organic pollutants (Nagy et al., 2023). One of the most studied species is *Phanerochaete chrysosporium*, which is famous for the degradation of persistent organic pollutants (POPs) such as PAHs, polychlorinated biphenyls (PCBs), and pesticides (Rascio & Navari-Izzo, 2011). These fungi not only convert recalcitrant toxic materials, such as the treatment of xenobiotics, but also affect heavy-metal's transport process, and bioavailability by changing soil pH, and by the production of chelating metabolites that bind and limit the mobility of metal, such as Cd, Pb, and Hg. These mechanisms lower toxicity of the environment, improving the conditions for the recovery of soil microorganisms and plant growth (Fatoba & Edeh, 2020). Hydrocarbon-contaminated land, for example, caused by crude oil spills or prolonged utilization of pesticides, is especially suitable for mycoremediation. For instance, it has been reported that fungal species such as *Aspergillus niger*, *Trametes versicolor* have great potential in transforming PAHs into less toxic metabolites enhancing soil fertility and enhancing microbial diversity. Further, locally isolated fungal strains are preferred as they would be adapted to the indigenous soil environment which will also ensure low ecological perturbation (Mohanraj et al., 2025). The scalable, low cost, and wide range of environmental tolerances exhibited by mycoremediation positions it as a tool that ought to be considered as part of holistic soil recovery efforts, especially in developing areas that are confronted with confluence of threats from industrial contamination and climate-facilitated degradation.

15.3.3 Microbial Bioremediation

Bioremediation of microbes utilizes the metabolic potential of bacteria and other microorganisms in the treatment of many soil contaminants such as hydrocarbons, nitrates, heavy metals and synthetic chemicals. Instead, this strategy frequently relies on microbial consortia—specific combinations of different strains of microorganisms, which act synergistically to stimulate biodegradation processes and expand the spectrum of pollutant species that can treated. One of the more well-studied applications is the bioremediation of hydrocarbons in crude oil contaminated soils, in which *Pseudomonas aeruginosa* and *Bacilus subtilis* have shown remarkable oil-hydrocarbon degradation properties. These bacteria secrete biosurfactants and enzymes that degrade complex hydrocarbon molecules to simpler organic substances, which may result in faster soil detoxification (Taher & Saeed, 2021). In addition, mixed populations of bacteria usually show a better performance than single strains, due also to the possibility to utilize complementary metabolic pathways and to increase the resistance to environmental stresses.

To achieve the optimum potential of microbial biore-mediation, manipulation of soil physicochemical conditions is necessary. pH, moisture, temperature and nutrient availability are among the important factors that have a significant role in stimulation of microbial activity and increase of degradation rates. For example, nitrogen and phosphorus manipulation modulate nutrient limited environment conditions and consequently microbial growth and enzymatic activity. Moisture management is particularly critical, as too much drying or waterlogging can have toxic effects on microbial metabolism and contaminant degradation potential. Microbial bioremediation strategies are highly versatile in a wide range of ecoregions including temperate agricultural fields to tropical industrial sites. This approach has the flexibility to adapt to different microbial communities and environmental conditions; thus, it is applicable to areas with both anthropogenic pollution and soil degradation due to climate change (Mohanraj et al., 2025). Advances in genomics and microbial ecology will continue aiding the development of engineered consortia with optimized biodegradative capacity and environmental fitness.

In conclusion, microbial bioremediation emerges as an efficacious and flexible tool in the remediation of contaminated soils, especially when combined with other biological remediation strategies, such as phytoremediation and mycoremediation.

15.3.4 Biochar/Carbon-Facilitated Remediation

Biochar is a carbon-rich and stable substance manufactured by pyrolyzing organic biomass, which gains great attention for ameliorating soil quality and remediation of polluted soils. Porous and high in surface area, biochar improves soil physical properties by enhancing aeration, water holding capacity and aggregation, and all have implications for promoting the soil ecosystem health (Helmy & Kardena, 2024). Moreover, biochar significantly enhances the soil cation exchange capacity (CEC) which would help soil to maintain essential nutrients and less hum toxicants in particular, heavy metals and petroleum hydrocarbons. Adsorption and immobilization of pollutants via reducing their bioavailability and toxicity to soil organisms and plants have been considered as one of the prime mechanisms of biochar for remediation (Liu et al., 2024). This characteristic renders biochar more suitable for soils contaminated with POPs and heavy metals, and which conventional remediation often cannot treat or leads to secondary pollution. In addition, biochar serves as a habitat and substrate for favorable microbial communities, which induce further microbial proliferation and activity, thus accelerating the degradation of organic pollutants (Sun et al., 2022).

Recent research has showed that the remediation ability of biochar gets improved receiving biological strategies like phytoremediation and microbial bioremediation. For instance, synergistic effects may be caused by simultaneous application of biochar and plant species with known for pollutant uptake thus enhancing the plant growth and enhancing the contaminate uptake whereas biochar provides support the hydrocarbon-degrading microbial consortia (Liu et al., 2024). This interaction is of particular importance in soils subject to soil degradation by both industrial contamination and climate change stress, where soil function rehabilitation is a multi-pronged approach. The wide range of soil types and contaminants biochar can work with, alongside general possibilities of carbon sequestration, give a sustainable and more affordable tool for the large-scale restoration of soils to the public space, especially in locations where advanced technologies of remediation are not available (Helmy & Kardena, 2024). Current research still aims at enhancing the production conditions and biochar application levels that can lead to the maximum environmental effects and remediation improvements.

15.4 CASE STUDIES

A number of recent examples illustrate the reality of bioremediation efforts and its accepted outcomes in soils damaged by industrial pollution, improper agricultural land use, and climate-induced stresses. The study of Eze et al. (2022) illustrated the successful simultaneous application of microbial consortium and phytoremediation for managing crude oil-contaminated soils. Native isolates of hyperaccumulator plant *Chromolaena odorata* with indigenous bacteria, *Pseudomonas* and *Bacillus* were used for inoculation. After 12 months, TPH was reduced by 65% and soil diversity of soil microbes significantly increased, which revealed an enhanced symbiotic effect between microbial remediation and Phytoremediation.

In California, USA, Pavel et al. tested native plants including telegraph weed and mule fat in field campaigns for phytoremediation of metropolis soils and heavy metals (lead and cadmium). A successful phytoextraction was achieved by the plants, which accumulated the metals in their biomass and immobilized the contaminants in the rhizosphere, resulting in a reduction of bioavailability of 40 %, within one growing season. It provided an eco-friendly alternative to the expensive remediation of London clay in the heavily populated region susceptible to flooding due to climate change.

Another example comes from coal mining impacted soils from Inner Mongolia, China where Zhang et al. (2023) carried out mycoremediation with the white rot fungus *Phanerochaete chrysosporium*. The fungi degraded complex PAHs leading to a decrease in the soil toxicity to more than 50% in 6 months. Biochar addition facilitated remediation process by improving soil perturbation and enhancing TTC degradation by fungi, indicating the advantage of combining methods.

Finally, Mohanraj et al. (2025), in tropical agricultural soils, reported that the pesticide residues in the soil were reduced and the soil nutrient balance also restored through biochar–microbial bioremediation. Their experiments in India demonstrated that well-defined biochar and microbial consortia prescription should lead to enhance remediation efficiencies in climate stressed soils.

Together, these two studies demonstrate the applicability and success of bioremediation approaches in different environmental settings and contamination contexts. They also highlight the need to integrate biological process and soil amendment in bio geotechnical design in order to improve resistance to the mixed stressors of human practice and climate change.

15.4.1 Diesel-Contaminated Soils

Diesel polluted soil is a common environmental issue caused by accidents, improper disposal and other industrial activities. Such hydrocarbons are harmful to, long-lived in, and can seriously affect the quality of soil by damage to the soil's microbial and nutrient cycles. Bioremediation, which involves the application of organic amendments to enhance native microbial populations, has proved to be a promising approach for remediating degraded soils.

An interesting work of Resch et al. (2022), tested the efficiency of stabilized poultry manure as an organic source for bioremediation of diesel contaminated soil. The poultry manure high in organic matter and nutrients was used as microbial stimulator and nutrient aiding in microbial attack of hydrocarbons. Poultry manure inclusion resulted in a substantial increase in microbial activity (in terms of increased soil respiration and enzyme activity assays) during 120 days of field assay. These shifts were associated with faster removal of total petroleum hydrocarbons, which reached a reduction of 65% compared to untreated controls.

This approach does not only encourage pollutant elimination, but it also encourages soil fertility rehabilitation since it helps obliterate the essential plant nutrients (N and P) which are in most cases reduced on contaminated land. The combined positive effects of decontamination and soil detoxification contribute to advancing the potential to apply stabilized organic waste as a low-cost and sustainable amendment in bioremediation practices, especially in areas with an agricultural or rural profile and easy access to this type of material.

Corroborating studies also present that addition of organic amendments such as compost and manure enhance degradation of hydrocarbons by improving the microbial biomass and making the contaminants more bioavailable. In addition, the amendments can also reverse the toxic nature of diesel on soil microorganisms and thus stimulate natural attenuation (Patel et al., 2024).

In conclusion, the use of stabilized poultry manure appears to be a promising and "green" option for the remediation of diesel-contaminated soils, as it combines pollutant degradation with soil fertilization. Studies are currently aimed at exploring the best amendment application rates and combining these treatments with other types of bioremediation treatments for enhanced soil recovery.

15.4.2 Combined Polluted Sites

One of the main problems of the soil bioremediation treatment is that the sites are contaminated with more than one pollutant, being even possible the coexistence between heavy metals and petroleum hydrocarbons. These co-contaminants are frequently co-occurring and interacting, which leads to remediation challenges for heavy metals can inhibit the microbial activity required to degrade hydrocarbons, but hydrocarbons may change the mobility and toxicity of metals. Tackling such an intricate contamination needs integrated and multidimensional approaches.

The recent review work conducted by Helmy and Kardena (2024) underpins the success of biochar-based bioremediation strategies in addressing multi-contaminant soils. Biochar, generated from agricultural waste material, such as rice husks and municipal green waists, is characterized by high porosity, large surface area and functionalities able to adsorb heavy metals. This adsorption of toxic metals like lead, cadmium, and arsenic immobilize them, and as a result, decrease the bioavailability of these metals in soil and to some extent, decrease the toxic effect of these metals on soil microorganisms (Helmy & Kardena, 2024).

Furthermore, biochar represents an optimal niche for those microbial communities able to degrade organics as well. By increasing soil aeration and water-holding capacity, biochar promotes microbial colonization and enzyme activity allowing for better degradation of hydrocarbons like PAHs. Laboratory studies summarized by Helmy and Kardena indicated that biochar amendments reduced PAHs significantly by 78% within six months, and the treatment with biochar outperformed non-biochar treatment. The type of biochar can affect the remediation effect, biochar prepared from agricultural waste such as rice husk and municipal green waste have more mineral contents and functional groups to improve the heavy metal binding and support the microorganism (Liu et al., 2024). These aspects turn biochar as a promising amendment in the case of complex contaminations, especially when co-applied with microorganisms or with phytoremediation.

Field-scale demonstrations are currently being developed to show that biochar-facilitated bioremediation may provide a viable sustainable and economic option to promote soil health and reduce environmental risks in co-contaminated sites. The dual effects of biochar, on the immobilization of metal species and the enhancement of organic pollutant degradation, are in accordance with the multifaceted form of combined contamination that is frequently exacerbated by anthropogenic effects and the climate change effects of enhanced runoff, coupled with soil degradation (Xu, 2024) and reference therein. Biochar-assisted techniques have, in conclusion, emerged as a new frontier in the bioremediation of soils contaminated by complex mixtures of pollutants. Additional studies aimed at fine-tuning biochar characteristics and at elucidating the behavior of biochar, microbes, and contaminants will support a greater potential for upscaling of this type of intervention.

15.4.3 Soil Damage by Climate Change

Soil health is becoming more and more endangered in the entire world, as climate change exacerbates land degradation processes, including flooding, drought, soil erosion, and nutrient leaching. In particular, the effect of flooding leads to distinct physical and chemical disturbances of soil habitats such as organic matter loss, compaction, and nutrient leaching. Meeting such challenges demand novel remediation treatments by rehabilitating the contaminated soils for both their function in the soil system and resilience to future climate stressors.

An important paper by Jahromi et al. (2022) studied the effectiveness of biochar on the rehabilitation of heavily degraded climate-induced flooded Soils in Iran. The authors did field studies in flood-borne agrarian tracts, recycling biochar of agrarian refuse from local regions. Their results indicated that biochar addition significantly enhanced soil water retention, which is crucial in the alleviation of drought stress after flooding. Improved moisture holding also supported soil microbial populations that are needed for nutrient turnover and mineralization.

We note that this work provided evidence of enhanced microbial biomass and enzymatic activity (indicator of enhanced soil biological health and the soil recovery potential) in the flooded soils as a result of the biochar. Biochar offered porous housing and increased carbon supply initiating recolonization and metabolic activity of beneficial (but lost following flooding disturbances) microorganisms. This should no longer be the case if farming is to be sustainable since this stimulation by the microbes is required in order to reintroduce or promote soil fertility as well as to bring resources in the crop under stressed climatic conditions.

Jahromi et al. (2022) identified the potential for climate-adaptive biochar as a soil amendment that could improve ecosystem resilience (buffering against hydrological extremes and fostering more rapid ecological recovery). These findings are in accordance with worldwide research trends accentuating biochar's multifunctionality as a pollutant immobilizer but as well a soil condition, which would enhance soil water dynamics and biological activity (Helmy & Kardena, 2024). Facing intensified and more frequent extreme weather events caused by climate change, the incorporation of biochar into farming systems is a promising strategy to support agricultural productivity and ecosystem services in vulnerable areas. More research is needed to optimize types of biochar, application rates, and biochar association with microbial or plant bioremediation in different agroecosystems.

15.5 LIMITATIONS AND CHALLENGES

Bioremediation is an environmentally sustainable method used to reclaim soils that have been degraded by anthropogenic activities and climatic factors; however, some constraints of this approach limit its efficiency and widespread use. It is important to comprehend these challenges in order to develop and utilize bioremediation strategies effectively and with realistic expectations.

There are several limitations, and one of the main ones is the diversity of contaminants. The effectiveness of bioremediation is primarily determined by the nature, concentration, and distribution of pollutants in soil. Sales da Silva et al. (2020) highlight those complex mixtures of pollutants, including heavy metals combined with various hydrocarbons, restrict the efficiency of microbial degradation and the design of remediation strategies. The bioavailability of these compounds to bioremediation is also influenced by site-specific factors such as soil texture, pH, and moisture.

Another issue that could arise relates to the risk of ecosystem disruption due to the introduction of "unnatural" microorganisms or plants as part of the bioremediation effort. Xu (2024) also warned that these non-native species may have competition with the native species, and consequently alter the local soil microbial community and the ecological relationships. This risk is especially appropriated in the use of genetically modified microorganisms or hyperaccumulator plants, where an extensive ecological risk assessment is required before the release in open field.

Practicality of bioremediation is also an issue due to slow rates of remediation. In many cases, several months and even years are necessary to achieve suitable levels of contaminants degradation for common bioremediation activities, particularly in the presence of high pollutant contents or soil physical limitations, such as compaction. Where... ... site cleanup needs to be expedited to accommodate emergent remediation or economic redevelopment schedules.

Furthermore, non-technical barriers exist associated with regulation and public acceptance. Liu et al. (2024) point out that in many countries the absence of an integrated legal framework for bioremediation, with special reference to genetically modified organisms (GMOs), is an obstacle to the approval of projects. Distrust and concerns about biosafety among the public add to barriers for implementation, emphasizing the important role of open dialogue and stakeholder engagement in facilitating adoption. Bioremediation process has great potential, yet its application is limited by site-specific contaminant types and products, ecological concerns, time scale, and socio-regulatory issues. Overcoming such constraints by adaptive management, technology innovation, and policy designing is crucial for unlocking the power of bioremediation to rehabilitate the soils and landscapes impacted by anthropogenic and climatic stresses.

15.6 FUTURE PERSPECTIVES

The bioremediation research and application in the future is expected to follow the promising directions that can increase the efficiency, scalability, and sustainability of soil remediation. Recent developments in molecular biology, policy formulation, and community participation may lead to important breakthroughs. An important research direction is the application of multi-omics technologies, such as metagenome, proteome, and metabolome, to decode the complex interactions in soil microbial communities. According to Nagy et al. (2023), these strategies facilitate the construction of customized microbial consortia with superior capacity for degrading specific pollutants in different ecological conditions. By a systems-level knowledge of microbial gene expression/metabolic pathways, scientists may be able to design and engineer consortia with more stable performance in the environment. In parallel to multi-omics technologies, CRISPR-Cas9 and synthetic biology tools provide exciting opportunities for genetic perturbation. These tools can be tap to develop genetically modified plants and microorganisms with increased tolerance toward pollutants and better degrading pathway (Xu, 2024). For instance, synthetic microorganisms might express new enzymes breaking down recalcitrant organic pollutants to such an extent that even heavily contaminated microhabitats could be dealt with in a shorter time period and also be deposited free of pollutants one can imagine plants accumulating significantly more contaminants without showing phytotoxic problems and thus shortening the time for remediation processes.

On the policy side, integrating bioremediation into environmental and land-use policies will be crucial for widespread usage. "Clear plat formation mechanisms" to facilitate funding, risk assessment, and monitoring of the application of biotechnologies in post-industrial and climate affected areas should be "developed, balanced and justified" (Xu, 2024). It helps reduce regulatory uncertainties and promotes investment in bioremediation infrastructure.

Finally, community involvement is critical for sustaining long-term success and acceptance of bioremediation ventures. Jahromi et al. (2022) recommend inclusive participatory practices, such as engaging local farmers, landowners, and stakeholders in decision making, project design and management. This inclusion, in turn, builds trust and provides a better understanding of the conditions in the local socio-economic and ecological setting, and also ensures that interventions are adapted to this local context, creating a sense of sustainability.

The successful development of bioremediation in the future would require a combination of state-of-the-art science, jurisdictional needs, and community participation with each fostering the other. These interwoven approaches are essential to tackle the multifaceted problems associated with soil degradation caused by human activities and climate change.

15.7 CONCLUSION

Soil degradation caused by human activities including default industrial pollution, unsustainable agriculture, and inadequate waste management in combination with the increasing forces of the heat domino effect represents a major threat for the global environmental health, and together with food security. This degradation reduces soil fertility, the provision of ecosystem services, and susceptibility to further ecological detriment. Bioremediation as an alternative to the standard remedies of excavation or encapsulation has recently gained favor as one such potentially effective and sustainable alternative. Bioremediation relies on the inherent capabilities of living organisms—such as plants, fungi, bacteria, and biochar—to neutralize, stabilize, or restore contaminated soils. Methods in which hyperaccumulator plants are used to extract or sequester heavy metals and/or organic pollutants are listed under phytoremediation and those with the focus on fungi with the ability to produce enzymes that decompose previously persistent toxic substances are listed under mycoremediation. Microorganism-based bioremediation, especially mediated by designer microbial consortia, efficiently decomposes hydrocarbons, and other petroleum-related contaminants and industrial pollutants, and this process is frequently accelerated in combination with the tailored environmental parameters. The application of biochar in remediation not only fixes the contaminants, but also ameliorates soil structure and microbial habitat, and enhances the activity of bioagents jointly for better impact.

There are still obstacles to the extensive application of bioremediation at large scale, albeit the above-mentioned steps. Site-based variation in contamination profiles, protracted times for remediation, biodiversity risks of releasing exotic species and regulatory barriers constrain wide-spread application. Furthermore, as the public is skeptical of biotechnological interventions, especially genetic engineering approaches, there is a need for improved communication and policy structures. For the future, integrated approaches using multi-omics studies, genetic engineering, enabling regulation, and community involvement are potential ways forward to address these challenges. By focusing on local stakeholder engagement, it aims to adapt to socio-ecological contexts, underpinning sustainable and socially acceptable remediation actions. Further work in research and development will be needed to improve bioremediation efficacy and to develop technology that is more efficient, flexible, and affordable.

In conclusion, bioremediation is an important tool in the ongoing endeavor to reclaim anthropogenically/meteorologically stressed soils to productivity worldwide. By integrating scientific, regulatory, and participatory efforts, bioremediation can enable to rehabilitate the health of soils, maintaining ecosystem functions and enhancing the resilience of at-risk landscapes, and therefore contributing to the overall sustainability of the environment at large.

REFERENCES

Al-Hawash, A. B., Zhang, J., & Ma, F. (2022). *Biodegradation of petroleum pollutants in contaminated soils using microbial consortia.* Chemosphere, 293, 133541.

Azeez, J. O., Oyedele, D. J., & Sulaimon, A. A. (2020). *Soil contamination from industrial activities in sub-Saharan Africa: Challenges and remediation perspectives.* Environmental Monitoring and Assessment, 192(6), 375.

Bera, T., Sinha, A., & Mandal, A. (2020). *Soil health management for climate resilience.* Journal of Soil and Water Conservation, 19(3), 45–53.

Carrero-González, R., Negrón, J. F., & Dec, D. (2012). *Soil porosity as a key factor of soil aggregate stability: Insights from restricted grazing.* Frontiers in Environmental Science, 10, 1535193. https://doi.org/10.3389/fenvs.2024.1535193

Chen, M., Xu, P., Zeng, G., & Yang, C. (2020). *Remediation of soils contaminated by heavy metals using chemical and biological methods: A review.* Environmental Pollution, 256, 113478.

Das, N., & Chandran, P. (2021). *Soil remediation technologies: A comparative review.* Environmental Sustainability, 4(4), 501–510.

DeVilleneuve, M., & Smith, J. (2023). Impacts of vegetation and topsoil removal on soil erosion, soil moisture, and infiltration. Agrosystems, Geosciences & Environment, 6(1), e20402. https://doi.org/10.1002/agg2.20402

El-Bestawy, E., Refaat, T., & Aly, A. (2022). *Phytoremediation potential of native plants for polluted soils.* International Journal of Environmental Science, 19(1), 21–33.

Eze, P. C., Okonkwo, E. N., & Ugochukwu, N. C. (2022). *Bioremediation of crude oil-contaminated soils using indigenous microbial consortia and phytoremediation in the Niger Delta.* Environmental Science and Pollution Research, 29(17), 25845–25858.

Fatoba, P. O., & Edeh, H. C. (2020). *Mycoremediation potentials of fungi in hydrocarbon-polluted soils.* African Journal of Environmental Science and Technology, 14(6), 181–189.

Fatoyinbo, T. E., Omotayo, D. O., & Abiodun, B. J. (2021). *Climate variability and its implications on soil erosion in sub-Saharan Africa.* Climate Dynamics, 56(9), 1523–1538.

Ghosh, S., Mukherjee, A., & Dutta, S. (2021). *Climate-induced alterations in soil carbon dynamics and implications for agroecosystems.* Environmental Research Letters, 16(9), 094005.

Helmy, M., & Kardena, E. (2024). *Biochar assisted bioremediation of soils with combined contamination of petroleum hydrocarbons and heavy metals: A review.* Ecological Engineering, 186, 106938. https://doi.org/10.1016/apsoil.2024.105720

Jahromi, H. et al (2022). *A perspective on biochar for repairing damages in the soil–plant system caused by climate change-driven extreme weather events.* Biochar, 4, 48. https://doi.org/10.1007/s42773-022-00148-z

Kumar, R., Singh, M., & Rai, J. P. N. (2020). *Heavy metal uptake and tolerance mechanisms in plants: A review with emphasis on phytoremediation.* Environmental Monitoring and Assessment, 192, 369.

Liu, J., Yang, H., & Wang, X. (2024). *Soil ecosystems under pressure: Emerging threats and mitigation strategies.* Environmental Earth Sciences, 83(2), 210.

Mohanraj, J., Prasad, K., & Awasthi, M. K. (2025). *Microbial consortia for effective soil bioremediation in the era of emerging pollutants.* Ecotoxicology and Environmental Safety, 259, 115042.

Nagy, K. K., et al. (2023). Novel enzymes for biodegradation of polycyclic aromatic hydrocarbons: metagenomics-linked identification followed by functional analysis.. https://arxiv.org/abs/2308.15914

Patel, R. D., Sharma, S., & Meena, K. (2024). *Enhancing hydrocarbon degradation in contaminated soils using genetically engineered microbes*. Applied Environmental Microbiology, 90(1), e01123–23.

Rascio, N., & Navari-Izzo, F. (2011). Heavy metal hyper-accumulating plants: How and why do they do it? And what makes them so interesting? Plant Science, 180(2), 169–181. https://doi.org/10.1016/j.plantsci.2010.08.016

Resch, M. C., Schütz, M., & Risch, A. C. (2022). Long-term recovery of above- and below-ground interactions in restored grasslands after topsoil removal and seed addition. Journal of Applied Ecology, 59(1), 123–134. https://doi.org/10.1111/1365-2664.14145

Sáez-Sandino, A., Delgado-Baquerizo, M., & Singh, B. K. (2024). Plant footprint decreases the functional diversity of molecules in topsoil organic matter after millions of years of ecosystem development. Global Ecology and Biogeography, 33(2), 123–134. https://doi.org/10.1111/geb.13770

Sales da Silva, L., Oliveira, T., & Costa, M. (2020). *Global soil degradation: Trends and solutions*. Land Degradation & Development, 31(10), 1211–1223.

Sharma, A., & Shah, M. (2021). *Bioremediation strategies for heavy metal contaminated soils*. Environmental Science Advances, 2(1), 34–49.

Singh, R., Verma, A., & Tyagi, A. (2021). *The interplay of soil microorganisms and climate stress in contaminated lands*. Journal of Environmental Management, 280, 111777.

Sun, Y., Chen, W., & Li, X. (2022). *Role of biochar in enhancing microbial bioremediation of contaminated soils*. Ecological Engineering, 171, 106368.

Taher, A. M., & Saeed, I. O.. (2021). Bioremediation of Contaminated Soil with Crude Oil Using Consortium of Bacteria. https://arxiv.org/abs/2111.00594

Wikipedia contributors. (2025). *Mycoremediation*. Wikipedia. https://en.wikipedia.org/wiki/Mycoremediation

Xu, L. (2024). *Climate change and its impact on global soil ecosystems: An integrative approach*. Climatic Change, 176(3), 155–168.

Zhang, Y., Sun, Z., & Zhou, J. (2023). *Resilient bioremediation approaches under climate variability*. Environmental Technology & Innovation, 32, 102556.

16 The Role of the Soil Microbiome in Sustainable Agriculture

Stella Suanu, Leh-Togi Zobeashia, and Atabo Hilda Uyo

16.1 INTRODUCTION

Agriculture has a seat at the intersection of some of the global-most urgent environmental and societal challenges such as loss of diversity, accelerating soil degradation, food security imperatives, and climate instability (Smith et al., 2019; Giller et al., 2021). Historical models of intensive agriculture have frequently prioritized short-term productivity benefits over the expense of long-term ecosystem resilience which usually result in unintended consequences such as soil nutrient depletion, reduced organic matter, and impaired ecosystem resilience (Pimentel et al., 2005; Reganold & Wachter, 2016). The shift on sustainable agriculture has gained momentum over recent decades, initiating approaches that balance productivity with long-term environmental stewardship and socio-economic viability (Pretty, 2008; Kremen & Miles, 2012). The dimensions of sustainable agricultural practices must fundamentally integrate microbial ecology principles, given the intrinsic linkages between belowground microbial processes and key sustainability dimensions including soil health restoration, natural resource conservation, biodiversity promotion, and synthetic input reduction (Lal, 2015; Singh & Khan, 2024). Recent investigations have further elucidated the pivotal role of soil microbiomes in climate-smart agriculture and carbon sequestration processes, positioning microbial interventions as promising strategies for climate mitigation (Chen et al., 2025).

The soil microbiome is the complete assemblage of bacteria, archaea, fungi, protozoa, and viruses within terrestrial ecosystems, representing the most taxonomically and functionally diverse biological community on Earth. The collection of microbial consortium orchestrates fundamental biogeochemical processes that support crop productivity, nutrient availability, and ecosystem stability through extensive interactions with plant roots, organic substrates, and each other (Delgado-Baquerizo et al., 2016). Several research reveals extraordinary complexity and sensitivity of microbial assemblages to anthropogenic management interventions, establishing them as primary biological determinants of agricultural sustainability trajectories (Schloter et al., 2018). Advances in culturomics and metagenomics have fundamentally transformed our understanding of soil microbiome preservation and application, revealing previously unknown functional capabilities and ecological interactions that drive ecosystem services.

Soil health, defined as the sustained capacity of soil to function as a living ecosystem, include productivity, nutrient cycling, pathogen suppression, water regulation, and carbon sequestration. Among various assessment tools, microbial community diversity and activity are increasingly recognized as the most sensitive indicators of soil health (Zhang et al., 2024). Studies reveal strong associations between high microbial biodiversity and enhanced ecosystem resilience, including improved soil structure, nutrient retention, organic matter turnover, and disease resistance (Wagg et al., 2014). In contrast, biodiversity loss often resulting from intensive agriculture is linked to reduced multifunctionality, yield instability, and increased susceptibility to environmental stressors (Turbé et al., 2010).

Microbial communities are not merely passive inhabitants but active agents structuring soil ecosystem processes. They influence nutrient pathways such as nitrogen fixation and denitrification,

DOI: 10.1201/9781003650881-16

ammonia oxidation, and phosphorus solubilization ensuring these nutrients are accessible to plants and not lost from the system (He et al., 2007). Mycorrhizal fungi and rhizobacteria exemplify how mutualistic interactions can increase both nutrient uptake and plant stress tolerance (Vessey, 2003). Additionally, soil microorganisms serve as primary architects of soil physical structure through extracellular polysaccharide production and other binding compounds that aggregate soil particles, creating stable microstructures that enhance porosity, water retention capacity, and root proliferation (Bronick & Lal, 2005). These processes directly contribute to both immediate crop growth enhancement and long-term agroecosystem resilience under climatic variability. In terms of plant health, the rhizosphere microbiome functions as a biological defense system, providing pathogen protection and regulating plant immune responses through competitive exclusion, antibiotic production, and induced systemic resistance (Berendsen et al., 2012). This biological control capacity reduces disease incidence in sustainable systems lacking heavy pesticide reliance. Reviews and meta-analyses have demonstrated that management approaches favoring soil biological diversity such as reduced tillage, addition of organic matter, diversified cropping, and integrated nutrient and pest management which enhance beneficial microbial functions and drive ecosystem services (Schloter et al. 2018; Chen et al., 2025). In contrast, monoculture, heavy synthetic fertilizer use, and excessive tillage disrupt microbial communities, often leading to declines in multifunctionality and yield stability (Turbé et al., 2010; Reganold & Wachter, 2016).

Recent high-throughput sequencing (HTS) technologies and network analysis tools now permit a much deeper understanding of soil microbial ecology (Fierer, 2017). Growing evidence shows that soils under organic farming or agroecological management harbor significantly greater microbial diversity and functional guilds than conventionally managed soils leading to more stable yields, improved nutrient cycling, and mitigation of abiotic and biotic stresses (Wagg et al., 2014; Giller et al., 2021). For instance, comparative studies of Schloter et al. (2018) have reported that organic amendments increase populations of disease-suppressive and plant growth-promoting taxa, directly benefiting plant performance. Notably, studies indicate that the relative influence of soil properties and microbial communities on soil functioning can be context and function dependent. He et al. (2007) observed that broad processes like soil respiration are substantially shaped by both microbial abundance and total soil organic carbon, while more specialized functions, such as denitrification, rely more heavily on the composition and richness of the microbial community itself. These findings suggest that sustaining diverse and functionally redundant microbial assemblages is critical for resilient ecosystem functionality under changing environmental and management conditions. The chapter aims to review soil microbiome's role in advancing sustainable agriculture.

16.2 COMPOSITION AND DIVERSITY OF THE SOIL MICROBIOME

The soil microbiome represents one of Earth's most complex biological systems, containing extraordinary taxonomic and functional diversity within relatively small spatial scales (Fierer, 2017). Understanding the taxonomic composition, organization, and ecological drivers of these microbial communities has become important for sustainable agriculture developmental management strategies that will optimize both productivity and environmental stewardship.

16.2.1 Soil Microbial Communities

16.2.1.1 Bacterial Communities

Bacteria constitute the most abundant and taxonomically diverse component of the soil microbiome, exhibiting remarkable metabolic versatility that enables them to occupy diverse ecological niches and perform multiple ecosystem functions simultaneously. Their roles include organic matter decomposition, nitrogen fixation (e.g., *Rhizobium*, *Azospirillum*), nutrient mineralization, phosphorus solubilization, and biological control of plant pathogens functions that collectively support soil fertility and plant productivity (Mendes et al., 2013; Rivera, 2020). The metabolic plasticity of bacterial communities allows them to respond rapidly to changing environmental conditions and substrate availability, making them particularly important for maintaining ecosystem stability under variable agricultural management regimes. Recent metagenomic studies have identified novel bacterial taxa with previously uncharacterized plant growth-promoting properties, expanding the potential for biological enhancement of crop production systems (Jarupula et al., 2022). Also studies by Delgado-Baquerizo et al. (2023) show that *Proteobacteria*, *Actinobacteria*, *Acidobacteria*, and *Firmicutes* consistently dominate agricultural soils. Their community composition shifts in response to pH, organic inputs, and tillage practices (Bender et al., 2016; Jiao et al., 2021).

16.2.1.2 Fungal Communities

Fungi represent the second major component of the soil microbiome, with mycorrhizal species forming particularly critical symbiotic relationships with the majority of crop plants. These associations facilitate enhanced phosphorus uptake, improve drought tolerance, and contribute significantly to soil aggregation and structure formation (van der Heijden et al., 2008). Beyond mycorrhizal associations, saprotrophic fungi such as *Trichoderma* play crucial roles in biocontrol and in the decomposition of recalcitrant organic compounds, thereby releasing nutrients that would otherwise remain unavailable to plants and other soil organisms Fungal hyphae also improve soil structure by stabilizing ag*gregates* (*B*ahram et al., 2018).

The fungi-to-bacteria ratio has emerged as a valuable indicator of soil health and management sustainability, with higher fungal abundance or more balanced microbial communities frequently observed under organic and conservation agricultural systems (FAO, 2022; Yang et al., 2023). This relationship reflects the sensitivity of fungal communities to management practices and their importance in maintaining soil structural integrity and long-term fertility.

16.2.1.3 Archaea, Protozoa, and Viruses

Although less abundant than bacteria and fungi, archaea perform specialized biogeochemical functions, particularly in nitrification and methane metabolism. Their ecological significance is especially pronounced in extreme environments such as acidic soils and waterlogged conditions, where they contribute disproportionately to greenhouse gas emissions and nutrient transformations (Fierer, 2017). Whereas protozoa function as key regulators of microbial population dynamics through predation on bacteria, simultaneously releasing nitrogen and other nutrients through their grazing activities. This regulatory function helps maintain microbial community equilibrium while enhancing nutrient availability for plant uptake (FAO, 2022). While viruses, particularly bacteriophages, influence microbial community structure through selective pressure on bacterial populations and facilitate horizontal gene transfer among microbial taxa. Although historically overlooked, viral ecology is increasingly recognized as a crucial component of soil microbiome dynamics, affecting community resilience and functional potential.

16.2.1.4 Microbial Community Structure and Functional Guilds

Soil microbial communities exhibit hierarchical organization based on functional specialization rather than phylogenetic relationships alone. This functional guild structure provides a more mechanistic understanding of how microbial diversity translates into ecosystem services (Wagg et al., 2014; Jarupula et al., 2022). Some microbial diversity, act as decomposer guilds, which are mostly dominated by saprotrophic fungi and diverse bacterial taxa, they orchestrate the breakdown of plant litter and organic residues, regulating carbon and nutrient cycling rates. The composition and activity of these guilds directly influence soil organic matter dynamics and nutrient availability for subsequent crop growth. The mutualistic guilds include arbuscular mycorrhizal fungi (AMF) and nitrogen-fixing bacteria that form beneficial associations with plant roots. These symbiotic relationships enhance plant nutrient acquisition capacity while improving tolerance to abiotic stresses such as drought and salinity (van der Heijden et al., 2008). Antagonistic microbial guilds suppress soil-borne plant diseases through mechanisms such as antibiotic production, competition for resources, and the induction of systemic resistance in plants. This group includes bacteria and fungi, the activity of these guilds can substantially reduce the need for chemical pesticides while maintaining crop health (Berendsen et al., 2012; FAO, 2022). Whereas, the nutrient cycling guilds that comprises of specialized bacterial and archaeal communities are responsible for nitrogen transformations, including nitrification and denitrification processes. The balance and activity of these guilds determine nitrogen use efficiency and environmental losses from agricultural systems (He et al., 2007; Jarupula et al., 2022).

The connectivity and functional redundancy within the microbial communities' guilds are more predictive of ecosystem resilience than simple diversity metrics. High connectivity among microbial taxa creates stable interaction networks that can maintain ecosystem functions despite environmental perturbations or species loss (Solanki, 2022; IEEP, 2023) Agricultural practices that promote complex microbial networks such as diverse crop rotations, organic amendments, and reduced tillage consistently enhance ecosystem multifunctionality and stability. These management approaches foster both taxonomic diversity and functional redundancy, creating resilient soil ecosystems capable of sustaining productivity under variable environmental conditions (Yang et al., 2023).

16.2.1.5 Factors Influencing Soil Microbial Diversity

16.2.1.5.1 Soil pH

Soil physicochemical properties establish the fundamental constraints within which microbial communities develop and function. Among these factors, soil pH emerges as the primary determinant of microbial community composition and diversity patterns across agricultural landscapes (Wang et al., 2013; Ghimire et al., 2024). Most soil microorganisms exhibit optimal activity near neutral pH, with acidic or alkaline conditions selecting for specialized taxa that often exhibit reduced functional diversity. Strategic pH management through liming or organic amendments can therefore serve as a powerful tool for directing microbial community development toward more productive and stable configurations.

16.2.1.5.2 Soil Moisture

Regulates oxygen availability and substrate diffusion, creating heterogeneous microenvironments that support diverse microbial assemblages. Optimal moisture conditions promote high microbial activity and maintain functional diversity, while both drought stress and waterlogging can trigger community shifts that reduce beneficial taxa abundance and increase pathogen prevalence (Ghimire et al., 2024). Sustainable irrigation and soil water management are therefore vital to maintain microbiome functionality.

16.2.1.5.3 Organic Matter

Organic matter content represents the primary energy source driving microbial biomass and metabolic activity. Increased organic carbon inputs through crop residues, compost applications, or cover crop integration consistently enhance microbial richness and metabolic versatility such as its functional potential, stimulating both nutrient cycling efficiency and disease suppression capacity (Bending et al., 2002; FAO, 2022). In contrast, depleted organic matter from intensive tillage and monoculture systems leads to diminished microbial diversity and impaired soil functions (Reganold & Wachter, 2016).

16.2.1.5.4 Land Use and Agricultural Management

Agricultural management practices exert profound and lasting effects on soil microbiome community structure and functionality. Conservation-oriented approaches, such as reduced tillage, organic amendments, and diversified crop rotation and cover cropping consistently promote microbial diversity, ecosystem resilience, and the development of functional guilds richness, whereas intensive conventional systems often diminish community complexity and disrupt ecological balance (Li et al., 2014; Khmelevtsova et al., 2022). Tillage management, for instance, influences soil structure and organic matter distribution, with reduced tillage preserving habitat stratification and supporting more stable microbial communities compared to intensive tillage (FAO, 2022). Similarly, crop rotation diversity enhances microbial complexity through varied root exudates and organic inputs, increasing multifunctionality relative to monoculture systems (Khmelevtsova et al., 2022). The addition of organic amendments such as compost and manure enriches energy sources and habitat heterogeneity, boosting beneficial microbial populations while suppressing pathogens (FAO, 2022). In contrast, the use of synthetic fertilizers and pesticides can negatively impact microbial balance and reduce functional diversity, leading to decreased resilience and impaired ecosystem services in high-input agricultural systems (Turbé et al., 2010).

16.2.1.5.5 Soil Texture and Structural Variability

Soil texture and structure significantly shape microbial habitat heterogeneity, influencing the diversity, abundance, and function of soil microbial communities. Fine-textured soils, characterized by smaller particle sizes and greater surface area, tend to exhibit higher aggregate stability and moisture retention, thereby supporting richer and more complex microbial populations compared to coarse-textured sandy soils, which are more prone to rapid drainage and nutrient leaching (Sessitsch et al., 2001; Girvan et al., 2003). Moreover, the availability of nutrient pools and substrates such as soil proteins directly affects microbial metabolic activity and community structure. These resources regulate microbial-driven nutrient transformations, thereby linking microbial diversity to the efficiency of nutrient cycling within the soil ecosystem (Geisseler et al., 2019; Naasko et al., 2024). Together, these factors underscore the critical role of physical soil properties in maintaining microbial functionality and overall soil health.

16.3 FUNCTIONS OF THE SOIL MICROBIOME IN AGRICULTURAL SYSTEMS

The soil microbiome is integral to the productivity and sustainability of agricultural ecosystems. Through a network of interactions and biochemical functions, soil microorganisms regulate nutrient cycles, maintain soil structure, drive organic matter turnover, and provide natural disease control.

16.3.1 Nutrient Cycling and Availability

16.3.1.1 Nitrogen Fixation

The availability of nitrogen in agricultural soils is a determinant of crop productivity, but most atmospheric nitrogen is not readily accessible to plants in its gaseous form. Diazotrophic microorganism that is nitrogen-fixing bacteria such as rhizobia in leguminous root nodules and free-living genera like *Azotobacter* and cyanobacteria enact biological nitrogen fixation, reducing atmospheric nitrogen gas to ammonia (NH_3), which plants can utilize (Timofeeva et al., 2023). Recent studies increasingly recognize the importance of non-symbiotic and associative nitrogen-fixing bacteria in cereals and other non-leguminous crops, particularly under low synthetic fertilizer regimes typical of sustainable agriculture. The presence and abundance of such microbial populations are influenced by organic management practices, cover cropping, and the addition of compost or biochar, which collectively sustain high microbial diversity and favorable niches (Hu et al., 2024).

16.3.1.2 Phosphate Solubilization

Despite the abundance of phosphorus in soils, phosphorus is largely immobile and often locked in mineral forms that are inaccessible to plants. Phosphate-solubilizing microorganisms

(PSMs), such as *Pseudomonas*, *Bacillus*, and certain mycorrhizal fungi, secrete organic acids and enzymes that are able to mobilize and transform inorganic and organic phosphorus into bioavailable forms (Timofeeva et al., 2023). AMF in particular extend the plant's ability to acquire nutrient through its expansive hyphal networks that enhance absorption, especially phosphorus and nitrogen. In return, plants supply carbon compounds to the fungi, thereby fostering a mutualistic relationship that supports both soil health and plant growth (Lahlali et al., 2021). The synergistic effects of PSM consortia are significantly more pronounced than effects from pure-culture inoculants, indicating that microbial diversity enhances phosphorus use efficiency, reduces reliance on mineral fertilizers, and increases yield stability in variable conditions (Jarupula et al., 2022).

16.3.1.3 Sulfur Cycling

Sulfur is an essential nutrient for plant growth, with microbial transformations governing its bioavailability. Soil bacteria and fungi carry out the oxidation of elemental sulfur and other sulfur compounds, rendering it to sulfate (SO_4) which is available for plant uptake. Specific genera such as *Thiobacillus* and certain actinobacteria are important, while fungi participate in mineralization of organic sulfur pools (Etesami, 2024; Chen et al., 2025). These processes also influence soil pH and, by extension, the broader microbial community composition. Emerging research underscores the complexity of sulfur cycling, including the role of microbial networks in the retention, mobility, and assimilation of sulfur under various management systems (Chen et al., 2025). Studies conducted by Hu et al. (2024) and Jarupula et al. (2022) demonstrated that diversified cropping, organic inputs, and conservation agriculture bolster the activity and abundance of nitrogen fixers, PSMs, and sulfur cycling microbes relative to conventional monocultures with synthetic fertilizers. Notably, microbial mechanisms of nutrient cycling support the reduced input dependency and enhanced resilience associated with sustainable systems (Etesami, 2024).

16.3.2 Soil Structure and Aggregation

16.3.2.1 Role of Microbial Exudates (Glomalin, EPS)

An important but often underappreciated function of the soil microbiome involves the stabilization and maintenance of soil structure. Microbial populations synthesize extracellular polymeric substances (EPS), including proteins and polysaccharides, that form matrix-binding soil particles into stable aggregates (Zhang et al., 2024). These microbial exudates act as biological glue that improves soil porosity, water retention, resistance to erosion, and the overall resilience of soil physical structure in the face of abiotic stress (Etesami, 2024).

Of special relevance are AMF, which produce glomalin-related soil proteins (GRSPs). Glomalin is recognized for its exceptional ability to enhance aggregation, promote organic carbon protection within aggregates, and facilitate long-term carbon sequestration (Chen et al., 2025).

Recent research using advanced microscopy and molecular probes has revealed that soils with higher EPS- and glomalin-producing microbes show significant improvement in macro- and microaggregate formation, correlating with increased nutrient cycling, plant growth, and drought resistance (Etesami, 2024; Chen et al., 2025). Bacterial EPS also play crucial roles in biofilm formation and recovery of soil structure post-disturbance, reinforcing the multidimensional importance of exudates for soil health (Zhang et al., 2024). A comparative research of soils managed with continuous organic amendment or cover crops versus conventionally tilled, chemically fertilized soils consistently demonstrate greater microbial exudate production and aggregate stability in the former which results in improved agronomic performance, reduced compaction, and enhanced infiltration rates (Hu et al., 2024; Etesami, 2024).

16.3.3 Decomposition and Organic Matter Turnover

Soil microorganisms are the principal agents mediating the decomposition of plant and animal residues, they transform complex organic materials into simpler compounds and stable humic substances that are accessible to plants. Bacteria and fungi secrete a suite of extracellular enzymes proteases, cellulases and ligninases that degrade cellulose, hemicellulose, and lignin. This enzymatic activity releases nutrients for plant uptake and also contributes to the formation of stable soil organic matter that is essential for long-term soil fertility and climate mitigation (Taqi et al., 2023). The dynamic interplay between different microbial groups ensures efficient decomposition under varying operational environmental conditions. Fungal communities dominate the breakdown of recalcitrant, lignin-rich substrates, especially in low-pH or low-moisture soils, while bacterial decomposers are more prevalent in labile carbon turnover under moisture-rich conditions (Taqi et al., 2023). Microbial necromass also known as dead microbial biomass is recognized as a significant precursor of stable organic matter in soil that is microbial carbon pump, indicating the importance of both living and dead microbiota in soil carbon cycling (Taqi et al., 2023; Chaudhari et al., 2025; Chen et al., 2025). Recent metagenomics and isotopic labeling studies have clarified the distinct roles of various taxa in organic matter decomposition and their sensitivity to crop rotation, tillage, and amendment practices (Taqi et al., 2023). Soils treated with organic inputs harbor more functionally diverse decomposer communities and sustain faster, more complete turnover of organic matter compared to systems reliant on mineral fertilizers alone (Jarupula et al., 2022).

16.3.4 Biocontrol of Soil-Borne Pathogens

The capacity of the soil microbiome to regulate plant pathogens is fundamental to sustainable agriculture, where reliance on chemical pesticides is minimized and the focus shifts to leveraging natural biological control (Nadarajah, 2024). Soil microbial population compete, produce antimicrobial agents, and induce systemic resistance as a means to control harmful pathogens in the soil. Beneficial soil microbes compete with pathogens for resources and space in the rhizosphere, suppressing pathogen establishment and proliferation. Populations of *Bacillus*, *Pseudomonas*, and *Trichoderma* are especially effective, thriving on root exudates and rapidly colonizing root surfaces (Raza et al., 2023). Many other microbes produce antibiotics, antifungal, or antimicrobial compounds such as iturins, phenazines, siderophores that directly inhibit pathogens. The release of volatile organic compounds by some species of organisms (VOCs) further antagonizes pathogen activity. The richness and genetic diversity of the soil microbiome determine the spectrum and efficacy of these antagonistic interactions (Sinha et al., 2025). While certain plant-associated bacteria and fungi such as *Pseudomonas fluorescens*, selected *Bacillus* spp., *Trichoderma* spp. can elicit plant immune responses, priming systemic resistance and enhancing plant defenses even before pathogen attack. ISR provides durable, broad-spectrum disease suppression and is especially pronounced in soils rich in beneficial microbial consortia (Raza et al., 2023). Furthermore, *Trichoderma* and other bacterial genera such as *Streptomyces* can also induce parasitism and hyperparasitism activity to pathogenic organisms thus disrupting their life cycles and reducing inoculum loads (Sinha et al., 2025).

Recent advances in microbiome engineering and bioinoculant development harness these mechanisms, deploying tailored consortia or promoting native beneficial communities through organic amendments and diversified rotations (Jarupula et al., 2022; Nadarajah, 2024). Research as shown that soils with high microbiome richness and evenness are more suppressive to pathogens, supporting more resilient, yield-stable agroecosystems (Wang et al., 2024). Comparative evaluations between conventionally managed and agroecological systems consistently demonstrate lower soil-borne disease incidence in the latter, attributed to more complex and functionally redundant biocontrol networks (FAO, 2022; Nadarajah, 2024).

16.4 SOIL MICROBIOME AND PLANT HEALTH

The soil microbiome plays a crucial role in shaping plant health and productivity in sustainable agriculture by mediating complex plant–microbe interactions and providing multifaceted benefits such as nutrient acquisition, growth promotion, and stress tolerance.

16.4.1 Plant–Microbe Interactions

Plant-associated microbiomes inhabit different ecological niches, the rhizosphere which is the soil region adjacent to roots, endosphere (microbes within plant tissues), and phyllosphere (leaf surface) engage in dynamic interactions that strongly influence plant health and growth (Jarupula et al., 2022; Raza et al., 2023). The rhizosphere microbiome forms a place where root exudates rich in sugars, amino acids, and signaling molecules selectively recruit beneficial microbes. This selective enrichment enhances nutrient mobilization, pathogen suppression, and root development (Jarupula et al., 2022). Recent advances in omics technologies such as genomics, transcriptomics, proteomics, metabolomics, and metagenomics reveal that composition and functions of rhizosphere microbiota vary with plant species, soil type, and agricultural practices, fundamentally shaping soil fertility and crop resilience (Hu et al., 2024). In contrast, the endosphere microbiome comprising of bacteria and fungi colonizing inter- and intracellular spaces is important for systemic functions such as nutrient transport, hormonal regulation, and priming defense responses (Raza et al., 2023). Endophytic microbes often form close symbiotic relationships with plants without causing harm. Their diversity and abundance respond to environmental stress and help plants adapt to adverse conditions. While phyllosphere microbiome, although less studied, influences aboveground plant health by modulating leaf surface pathogen communities, aiding in nutrient assimilation, and influencing plant physiology. Advanced sequencing approaches have expanded the known diversity of leaf-associated microbes, suggesting potential roles in enhancing photosynthetic efficiency and stress protection (Nadarajah, 2024).

16.4.2 Plant Growth-Promoting Rhizobacteria (PGPR)

PGPR are soil bacteria that colonize roots and enhance plant growth through direct and indirect mechanisms (Jarupula et al., 2022). These include, phosphate solubilization, nitrogen fixation, phytohormone production such as auxins, gibberellins, and siderophore-mediated iron acquisition (Timofeeva et al., 2023). Indirectly, PGPR induce systemic resistance and suppress pathogens. Recent field trials demonstrate that inoculating crops with PGPR strains like *P. fluorescens*, *Bacillus subtilis*, and *Azospirillum* improves nutrient uptake efficiency and biomass production under low-input conditions (Raza et al., 2023; Nadarajah, 2024). These bacteria stimulate root architecture development, enhancing water and nutrient absorption capacity. Moreover, PGPR communities are sensitive to soil management; organic amendments and crop diversification enhance their diversity and functional performance compared to conventional monocultures (Jarupula et al., 2022). Meta-analyses confirm that beneficial impacts of PGPR are more consistent when native or well-adapted strains are deployed, showing the necessity of microbiome-informed inoculant development tailored to specific environments (Wang et al., 2024). Innovative microbiome engineering targeting synergistic microbial consortia is promising for improving crop resilience and yield sustainably (Raza et al., 2023).

16.4.3 Mycorrhizal Associations

Mycorrhizal fungi establish mutualistic associations with the majority of terrestrial plants, enhancing nutrient acquisition, particularly phosphorus, and improving soil structure via hyphal networks (Chen et al., 2025). These fungi extensively extend the functional root system, accessing nutrient pools beyond depletion zones near roots. Apart from nutrient transfer, mycorrhizal symbioses confer abiotic stress tolerance by modulating plant water relations, activating antioxidant systems, and regulating hormone signaling pathways (Hu et al., 2024). Mycorrhizal fungi also stimulate soil aggregation through glomalin secretion, improving soil health parameters (Zhang et al., 2024). Comparative studies reveal that sustainable farming practices, such as organic amendments and reduced tillage, not only maintain but also enhance the richness and functionality of mycorrhizal communities relative to conventionally managed soils, which often suffer degradation of fungal diversity (Chaudhari et al., 2025). This maintenance enhances resilience against drought, salinity, and nutrient limitations.

16.4.4 Microbiome-Mediated Stress Tolerance

Climate extremes and environmental contaminants pose increasing threats to crop productivity. The soil and plant microbiomes act as a biological shield, enhancing tolerance to drought, salinity, heavy metals, and pathogen pressure through multifaceted mechanisms (Raza et al., 2023; Nadarajah, 2024). Under drought, microbiomes modulate root architecture, facilitate osmolyte accumulation, and induce antioxidant enzymes in plants. PGPR and mycorrhizal fungi increase water-use efficiency and enhance nutrient uptake under water stress (Hu et al., 2024). Similarly, salt stress mitigation involves microbes producing exopolysaccharides that improve soil aggregation and sequester toxic ions, alongside the production of phytohormones and ACC deaminase to reduce ethylene stress in plants. Heavy metal stress tolerance is mediated by microbial sequestration, transformation, and immobilization of metals through siderophores, EPS, and enzymatic detoxification (Timofeeva et al., 2023). Specific bacterial and fungal taxa in contaminated soils are enriched for these properties and confer enhanced plant growth and metal tolerance. Notably, sustainable practices that promote diverse microbiomes, such as organic farming and crop rotation, better sustain these beneficial stress-mitigating microbial communities compared to conventional monoculture systems exposed to recurrent chemical inputs (Jarupula et al., 2022).

16.5 MICROBIOME-DRIVEN STRATEGIES FOR SUSTAINABLE AGRICULTURE

Sustainable agriculture balances high productivity with the preservation of ecological services, fostering long-term soil health, biodiversity, and environmental resilience. Microbiome-driven approaches focus on managing beneficial microbial populations to reduce external inputs, improve nutrient efficiency, suppress pathogens, and build climate resilience.

16.5.1 Use of Microbial Inoculants and Biofertilizers

Microbial inoculants, sometimes termed biofertilizers, are formulations containing beneficial microorganisms often consortia of bacteria and fungi that promote plant growth and enhance nutrient availability. These products are being integrated into sustainable agricultural systems to improve soil fertility, crop yields, and plant resilience to biotic and abiotic stresses (Timofeeva et al., 2023; Yıldız et al., 2025). This bioinoculant has the potential to mobilize nutrient because that contain PGPR or mycorrhizal fungi that fix atmospheric nitrogen, solubilize phosphorus, or mobilize micronutrients, leading to improved plant nutrition (Timofeeva et al., 2023). They also produce phytohormones that synthesize auxins, gibberellins, and cytokinins, directly stimulating root growth and development, while some can be used to suppress soil-borne pathogens (Hu et al., 2024). Recent studies emphasize the importance of quality control, strain selection, and formulation stability for the field efficacy of inoculants (Singh & Khan, 2024; Díaz-Rodríguez et al., 2025). Multi-strain consortia and formulations tailored to local soil types and crops tend to outperform single-strain inoculants which are partly due to their functional redundancy and complementarity (Timofeeva et al., 2023). Emerging technologies such as carrier optimization and shelf-life extension are improving product reliability (Singh & Khan, 2024).

16.5.2 Compost, Vermicompost, and Organic Amendments

Organic amendments such as compost and vermicompost are central components of sustainable agriculture due to their multifaceted benefits for soil physicochemical and biological properties. Compost a decomposed organic matter rich in stabilized carbon and beneficial microorganisms. Its addition to agricultural soils enhances organic matter, improves soil structure, and supplies a slow-release source of nutrients. Zhang et al. (2024) reported a significant increase in soil bacterial diversity, enzyme activities, and organic carbon content following compost application, contributing to improved yield and nutrient cycling, while vermicompost is produced by earthworm-mediated decomposition, resulting in a product rich in humic substances, enzymes, nutrients, and a robust population of plant-beneficial bacteria and fungi (Yıldız et al., 2025). Its application increases microbial biomass and diversity, boosts populations of nitrogen-fixing microorganisms and PSMs, and enhances enzymatic activities crucial for organic matter decomposition (Yıldız et al., 2025). Both compost and vermicompost play a foundational role in sustaining soil fertility and microbiome structure. Compost supplies stabilized organic matter, reparative nutrients, and a diverse inoculum of beneficial microbes (enhanced microbial biomass and activity), while vermicompost enhances enzymatic activities, introduces bioactive molecules, supports higher microbial functional diversity and enrich soils for crop—available nutrient (Zhang et al., 2024; Yıldız et al., 2025). Additionally both compost and vermicompost are linked to reduced soil-borne pathogen loads and increased disease suppressiveness by fostering antagonistic microbial networks (Yıldız et al., 2025).

16.5.3 Cover Cropping and Crop Rotation Effects on Soil Microbial Communities

Cover cropping the cultivation of non-harvested plants during fallow periods and crop rotation are foundational agronomic practices that promote microbial diversity and improve soil health. These strategies enrich soil ecosystems by diversifying organic inputs, maintaining continuous root exudation, and interrupting pest and disease life cycles, thereby supporting active microbial communities throughout the year (Zhang et al., 2024). Both practices have been shown to significantly enhance the richness and diversity of bacterial and fungal taxa compared to systems reliant on continuous monoculture or bare fallow. By introducing a wider variety of plant-derived substrates into the soil, cover crops and rotational schemes stimulate the development of complex microbial structure that support essential ecosystem functions, such as nitrogen transformation, organic matter decomposition, and pathogen suppression (Theriot et al., 2020; Rodrigues et al., 2022). Importantly, the diversification effect of these practices disrupts the proliferation of host-specific soil pathogens, in contrast to monocultures, which can encourage deleterious microbial shifts and reduce functional redundancy. Zhang et al. (2024) demonstrated that under varying precipitation regimes, cover cropping significantly altered microbial community composition and increased structure complexity, which in turn mediated enhancements in soil multifunctionality. Recent comparative field trials indicate that both single- and multi-species cover crop systems positively influence microbial community structure, albeit with varying degrees of impact. Notably, multi-species cover crop mixtures better preserve microbial diversity than monoculture treatments, supporting broader ecosystem resilience (Theriot et al., 2020). Functional outcomes such as improved nitrogen mineralization and enhanced carbon sequestration appear to be more strongly driven by management intensity and consistency rather than the specific composition of cover crop mixtures. Long-term rotational practices have been associated with sustained enrichment of beneficial microbial taxa, increased soil organic carbon content, and enhanced soil aggregation, which collectively contribute to improved nutrient cycling, yield stability, and agroecosystem sustainability (Theriot et al., 2020; Zhang et al., 2024).

16.5.4 Conservation Tillage and No-Till Practices

Conservation tillage practices, including reduced-till and no-till methods, play a pivotal role in enhancing soil health by minimizing physical disturbance and preserving soil structure, which in turn supports diverse and active microbial communities. These practices maintain surface residue cover, thereby sustaining microbial habitats and contributing to the accumulation of organic matter at the soil surface (Mathew et al., 2025). The reduced disturbance associated with conservation tillage promotes stratification of microbial communities, often leading to increased bacterial and fungal diversity and functional activity in the upper soil layers (Liu et al., 2023). Moreover, no-till systems have demonstrated greater resilience to environmental stresses such as drought, maintaining higher microbial richness compared to conventionally tilled soils (Mathew et al., 2025). However, the retention of crop residues on the soil surface may also harbor plant pathogens, underscoring the need for complementary practices like cover cropping and crop rotation to mitigate potential risks (Liu et al., 2023). Overall, across a range of agroecosystems, conservation tillage is consistently recognized as a key strategy for promoting microbial-driven soil functionality, improving long-term productivity, and enhancing climate adaptability.

16.5.5 Integration with Precision Agriculture

The integration of precision agriculture with soil microbiome management represents a cutting-edge approach to enhance agricultural sustainability and productivity. By leveraging advanced sensing technologies, digital analytics, and site-specific interventions, precision agriculture enables real-time, fine-scale optimization of resource inputs. The inclusion of microbiome data adds a transformative dimension, allowing for more informed and adaptive management decisions (Vlachakis & Efthimiadou, 2021). High-resolution soil mapping through metagenomic and "soilomics" analyses facilitates the identification and spatial distribution of beneficial microbial taxa, guiding variable-rate application of soil amendments and microbial inoculants (Chandarana et al., 2024). Furthermore, the integration of artificial intelligence (AI) and predictive modeling enhances the ability to forecast crop responses and monitor soil health dynamics, thereby improving management precision and system resilience (Vlachakis & Efthimiadou, 2021). Real-time microbiome feedback, in conjunction with climatic and crop data, supports targeted interventions that reduce input waste, optimize yields, and promote environmental stewardship. Although still in early stages of adoption, emerging evidence suggests that microbiome-informed precision agriculture significantly outperforms uniform management strategies in terms of efficiency and ecological sustainability (Chandarana et al., 2024).

16.6 TECHNOLOGICAL ADVANCES IN SOIL MICROBIOME RESEARCH

16.6.1 High-Throughput Sequencing and Metagenomics

The breakthrough of HTS technologies has transformed soil microbiome studies by enabling researchers to circumvent the limitations imposed by traditional culture-dependent methods. The so-called "great plate count anomaly" historically restricted assessments because only a small fraction of microorganisms could be cultivated (Knight et al., 2018). HTS methods, particularly next-generation sequencing (NGS), facilitate culture-independent analysis of environmental DNA, enabling comprehensive and detailed characterization of microbial diversity and community structure.

More recently, shotgun metagenomics has garnered increased attention due to its ability to provide functional insights beyond mere taxonomic inventories. By sequencing all environmental DNA from soil samples, researchers can identify the collective gene content of microbial communities. For example, Hartman et al. (2018) used metagenomic analyses to show that organically managed soils were enriched in genes associated with nitrogen fixation and organic matter decomposition compared with conventionally farmed soils, highlighting the functional consequences of agricultural management practices on soil microbiome dynamics.

Alongside shotgun metagenomics, amplicon sequencing targeting marker genes such as 16S rRNA for prokaryotes and ITS regions for fungi remains widely applied for taxonomic profiling due to its cost-effectiveness and high resolution. Zhang et al. (2019) utilized 16S and ITS sequencing to show that crop rotation significantly enhanced microbial alpha diversity and increased the abundance of beneficial microbes compared to monoculture systems, highlighting how agricultural practice shapes microbial community assembly.

The emergence of long-read sequencing platforms, including Oxford Nanopore Technologies and PacBio, further complements HTS tools. These methods improve metagenome assembly quality and resolve closely related microbial strains, facilitating more accurate functional and evolutionary studies within complex microbial consortia (Garg et al., 2024). Together, these sequencing advancements are enabling a transition from cataloging microbial presence to deciphering their metabolic capabilities and interactions.

16.6.2 Metatranscriptomics, Metaproteomics, and Metabolomics

Although metagenomics reveals functional potential, it does not clarify which genes are actively expressed under specific environmental conditions. Metatranscriptomics addresses this by sequencing total RNA, especially messenger RNA (mRNA), from soil samples, providing snapshots of real-time gene expression patterns. Bashiardes et al. (2016) describe metatranscriptomics as a powerful tool to identify active microbial responses to environmental signals and agricultural management. Shi et al. (2020) applied metatranscriptomics to agricultural soils amended with organic fertilizers, observing increased expression of genes involved in nitrogen cycling enzymes, indicating enhanced microbial metabolic activity associated with organic amendments. Metaproteomics complements transcriptomics by profiling the entire range of proteins produced by microbial communities. This approach links gene expression to actual enzymatic functions, bridging community composition with ecosystem processes like decomposition and nutrient mineralization. Schneider et al. (2021) deployed metaproteomics to uncover variations in soil protein profiles under different fertilization regimes, demonstrating changes in microbial enzyme abundances related to phosphorus solubilization and cellulose degradation in agroecosystems. Metabolomics closes the loop by characterizing the diversity and concentration of small-molecule metabolites present in soil environments. These metabolites organic acids, amino acids, secondary metabolites, plant hormones reflect microbial metabolism and interactions with plant root exudates, shaping nutrient availability and microbiome assembly. Kuzyakov and Razavi (2019) showed that root exudates dramatically influence rhizosphere metabolite profiles, modulating microbial community functions and plant-microbe symbioses. Together, these multi-omics modalities provide a holistic perspective on soil microbial ecology, revealing not only "who is there" but more importantly "what they are doing" and how microbial functions respond to dynamic environmental and management factors.

16.6.3 Bioinformatics Tools for Microbial Community Analysis

The unprecedented volume of sequencing and meta-omics data necessitates powerful bioinformatics tools to process, analyze, and visualize microbial community structure and function. Widely used pipelines like QIIME2 (Bolyen et al., 2019), mothur, and DADA2 provide robust frameworks for amplicon data quality control, taxonomic classification, and diversity analyses.

For metagenomic datasets, platforms such as MG-RAST and MetaPhlAn enable functional annotation and taxonomic profiling using multiple curated databases. Tools like MEGAN and HUMAnN integrate diverse datasets to assign genes to metabolic pathways, offering functional insights relevant to soil ecosystem services such as carbon sequestration and nitrogen cycling. Network analysis software such as Cytoscape and CoNet allow modeling of microbial co-occurrence and interaction networks, crucial for identifying keystone taxa and predicting community stability and responses to management. Banerjee et al. (2022) used such approaches to demonstrate decreased microbial network complexity and loss of keystone taxa under intensive farming compared to organic systems, linking network structure to ecological function.

Machine learning and AI methods are increasingly applied to decipher non-linear, complex associations between microbial community characteristics, soil health indicators, and crop performance. Recent tools like DeepARG and PICRUSt2 utilize deep learning to predict content of antimicrobial resistance genes and functional genes from metagenomic sequences, advancing our capacity to infer ecosystem functions from DNA data (Shaffer et al., 2022). Comprehensive statistical and visualization tools implemented through R packages such as phyloseq, vegan, and DESeq2 facilitate reproducible and interpretable microbiome analyses, strengthening links between microbial data and practical agronomic outcomes.

16.6.4 Soil Microbiome Engineering and Synthetic Communities

Engineering soil microbiomes to enhance agricultural productivity and resilience represents a rapidly emerging frontier. Traditional use of bioinoculants such as PGPR and mycorrhizal fungi has demonstrated yield improvements and nutrient efficiency but suffers variable success under field conditions due to competition with indigenous microbes and environmental complexity (Vandenkoornhuyse et al., 2015; Bender et al., 2016). Synthetic microbial communities (SynComs) composed of defined, complementary microbial strains offer a promising solution to such limitations. These engineered consortia can be designed with specific functional traits to synergistically promote plant growth, nutrient acquisition, and stress tolerance. For example, Arora et al. (2022) developed a SynCom that enhanced phosphorus solubilization and drought tolerance in wheat more effectively than single strain inoculants.

Microbiome transplantation introducing soil or rhizosphere microbiomes from productive systems into degraded soils is gaining attention as an approach to transfer beneficial microbial functions. While results are promising, ecological risks and long-term impacts require thorough evaluation (Panke-Buisse et al., 2015). Additionally, genome-editing tools such as CRISPR-Cas are being explored to precisely modify microbial genomes to augment traits such as nitrogen fixation, pathogen suppression, or metal detoxification. Though still nascent, synthetic biology approaches could enable design of "smart" microbes adapted for specific agroecosystem functions (Jansson et al., 2023). Ecological principles and risk assessment frameworks are critical for guiding microbiome engineering to minimize unintended consequences and biosafety risks, ensuring ethical and sustainable application.

16.7 CHALLENGES AND KNOWLEDGE GAPS

The soil microbiome has been widely acknowledged as a cornerstone for sustainable agriculture (Chen et al., 2025). However, despite considerable advancements in understanding and applying soil microbial communities, several challenges and critical knowledge gaps remain. These gaps hinder the consistent translation of microbiome science into effective and scalable agricultural solutions.

16.7.1 Microbial Community Stability and Functionality in Field Conditions

One of the paramount challenges in soil microbiome research is the intrinsic variability and instability of microbial communities under field conditions. While controlled laboratory and greenhouse experiments show promising results regarding microbiome function and plant growth promotion, replicating these effects in heterogeneous field environments remains difficult (Chen et al., 2025). The complex and spatially variable nature of soil microbiomes, coupled with fluctuating abiotic factors such as moisture, temperature, nutrient availability, induces shifts in microbial community structure and function that are not fully understood (Garg et al., 2024).

Moreover, microbial functions such as nutrient cycling or pathogen suppression depend on interactions among diverse microbial guilds whose stability can be easily disrupted by environmental perturbations. Longitudinal field studies reveal that beneficial microbes may be transient or functionally outcompeted by indigenous communities or opportunists, leading to variability in ecosystem service delivery (Banerjee et al., 2022). This knowledge gap underlines the necessity for research into the ecological resilience, functional redundancy, and adaptation mechanisms of soil microbiomes in diverse agricultural landscapes.

16.7.2 Issues with Commercial Inoculants: Persistence and Efficacy

The development and use of microbial inoculants have surged with the recognition of the microbiome's agricultural potential, yet their inconsistent field performance remains a stubborn bottleneck (Timofeeva et al., 2023; FAO, 2024). Many commercial biofertilizers and biocontrol agents exhibit variability in persistence, colonization ability, and functional expression when introduced into native soils due to competition from resident microbial populations, adverse soil properties, or climatic limitations (Chen et al., 2025). For instance, inoculants based on plant growth-promoting rhizobacteria or mycorrhizal fungi frequently fail to establish effective populations long-term or demonstrate diminished functionality under varying soil pH, texture, or organic matter content (Garg et al., 2024). Moreover, the "one-size-fits-all" approach common in biofertilizer development overlooks the significant influence of site-specific soil chemistry, crop genotype, and environmental conditions on microbial community dynamics. These factors mandate tailored approaches and the co-design of inoculants with local microbiome assemblages, which is currently limited by insufficient ecological understanding and technical capacity.

16.7.3 Microbiome Interactions with Climate and Management Practices

Soil microbial communities exist within a nexus of climatic factors and agricultural management, which together dictate microbe-plant-soil interactions and ecosystem functionality (Hartman et al., 2018; Chen et al., 2025). Climate change impacts, including rising temperature, altered precipitation regimes, and extreme weather events, impose significant stress on microbial diversity and functioning. These environmental fluctuations may lead to loss of sensitive microbial taxa or disruption of metabolic pathways essential for nutrient cycling and disease suppression. Management practices such as tillage intensity, fertilization, pesticide application, and crop diversification further modulate soil microbiomes. Intensive tillage and excessive chemical use deteriorate microbial richness and simplify community networks, reducing resilience and multifunctionality (FAO, 2024). Conversely, agroecological practices like cover cropping and organic amendments promote microbial diversity, fostering ecosystem services (Zhang et al., 2024). However, many studies show that the effects of climate and management on microbial communities are interactive and context-dependent. For example, conservation tillage may offset some climatic stress effects by maintaining soil moisture and organic matter but may not counteract the negative impacts of high pesticide loads (Banerjee et al., 2022). There is a pressing need to deepen understanding of these complex interactions through multifactorial and long-term field experiments.

16.7.4 Need for Context-Specific Solutions

The overarching challenge is the variability of soil microbiomes across spatial and temporal scales, shaped by soil type, management history, climatic zone, and crop species. This variability implies that microbiome-based solutions must be highly context-specific to achieve consistent benefits (FAO, 2024; Chen et al., 2025). Soils with differing textures, pH, organic matter, and mineralogy harbor distinct microbial assemblages with unique functional capacities and sensitivities. Moreover, crops differentially shape their rhizosphere microbiomes through root exudates and phenology, which means inoculant compatibility and microbiome interventions must consider crop genotype and agricultural system (Zhang et al., 2019). Regional climatic differences also modify microbiome dynamics; for instance, arid and semi-arid soils possess microbial communities adapted to drought stress but can respond differently to amendments than humid-region soils. Therefore, scaling microbiome technologies requires local validation, adaptive management frameworks, and precision diagnostics. Addressing these gaps will facilitate the development of robust, site-tailored microbiome solutions, accelerating the transition to regenerative agriculture and climate resilience.

16.8 FUTURE MICROBIAL PERSPECTIVES FOR SUSTAINABLE SOIL MANAGEMENT AND AGRICULTURE

Soil microbiomes are essential to sustainable agriculture, enhancing productivity, soil health, and climate resilience. As microbial research advances, integrating these insights into practical soil management and policy is both a significant opportunity and a challenge.

16.8.1 Policy Integration

Microbial health is increasingly recognized in global agricultural policy. FAO (2022) emphasizes its role in achieving SDGs related to food security, biodiversity, and climate action. However, many current policies still neglect microbiome considerations. To close this gap, policies should explicitly incorporate microbial health standards, support microbiome research and field validation, encourage sustainable practices like organic amendments and crop rotation, and invest in farmer education. Innovations such as biostimulants and biofertilizers should be incentivized through supportive regulations (Wan et al., 2025).

16.8.2 Climate-Smart and Regenerative Practices

Soil microbes play a pivotal role in carbon sequestration, nitrogen cycling, and mitigating greenhouse gas emissions (Jansson et al., 2023). Regenerative farming practices—cover cropping, compost use, and reduced tillage enhance microbial diversity and resilience under climate stress, aligning with climate-smart agriculture goals.

16.8.3 Microbiome-Based Soil Health Indicators

Traditional soil health assessments overlook microbial functionality. New approaches highlight microbial diversity, gene functions such as nitrogen fixation, and multi-omics tools as sensitive indicators of soil vitality (Garg et al., 2024). Field-ready microbiome-based diagnostics could revolutionize soil monitoring and restoration, but scaling these tools requires multidisciplinary research and standardization (Chen et al., 2025).

16.8.4 Scaling Across Farming Systems

Applying microbiome solutions requires tailored strategies for both smallholder and industrial farms. For smallholders, affordable biofertilizers, extension services, and participatory research are crucial. In commercial systems, microbiome management can be integrated with precision technologies for efficient input use and yield optimization. Still, access issues, regulatory barriers, and knowledge gaps must be addressed (Vlachakis & Efthimiadou, 2021).

16.9 CONCLUSION

The soil microbiome is increasingly recognized as a fundamental driver of sustainable agriculture, playing critical roles in nutrient cycling, plant health, and ecosystem resilience. Advances in technology and ecological understanding enable targeted microbiome management through practices such as organic amendments, crop diversification, and reduced tillage, which enhance microbial diversity and multifunctionality aligned with climate-smart and regenerative farming. However, translating microbiome science to field-scale applications requires interdisciplinary research to understand complex microbial dynamics under diverse agroecosystems and supportive policy frameworks that promote microbiome-friendly practices via incentives, education, and regulation. Bridging this knowledge-to-practice gap is essential to scale microbiome-based innovations across both smallholder and industrial farms worldwide, ultimately fostering resilient and productive agricultural systems grounded in microbiome-centered sustainability.

REFERENCES

Arora, N. K., Fatima, T., Mishra, I., & Verma, S. (2022). Microbial consortia: Key to plant health and productivity under stress conditions. *Frontiers in Microbiology*, *13*, 879592.

Bahram, M., Hildebrand, F., Forslund, S. K., Anderson, J. L., Soudzilovskaia, N. A., Bodegom, P. M., … Bork, P. (2018). Structure and function of the global topsoil microbiome. *Nature*, *560*(7717), 233–237. https://doi.org/10.1038/s41586-018-0386-6

Banerjee, S., Walder, F., Buchi, L., Meyer, M., Held, A. Y., Gattinger, A., Keller, T., Charles, R., & van der Heijden, M. G. (2022). Agricultural intensification reduces microbial network complexity and the abundance of keystone taxa in roots. *The ISME Journal*, *16*(1), 255–268.

Bashiardes, S., Zilberman-Schapira, G., & Elinav, E. (2016). Use of metatranscriptomics in microbiome research. *Bioinformatics and Biology Insights*, *10*, 19–25. https://doi.org/10.4137/BBI.S34610

Bender, S. F., Wagg, C., & van der Heijden, M. G. A. (2016). An underground revolution: Biodiversity and soil ecological engineering for agricultural sustainability. *Trends in Ecology & Evolution*, *31*(6), 440–452.

Bending, G. D., Turner, M. K., Rayns, F., Marx, M. C., & Wood, M. (2002). Microbial and biochemical soil quality indicators and their potential for differentiating areas under contrasting agricultural management regimes. *Soil Biology and Biochemistry*, *34*(7), 955–964.

Berendsen, R. L., Pieterse, C. M., & Bakker, P. A. (2012). The rhizosphere microbiome and plant health. *Trends in Plant Science*, *17*(8), 478–486.

Bolyen, E., Rideout, J. R., Dillon, M. R., Bokulich, N. A., Abnet, C. C., Al-Ghalith, G. A., Alexander, H., Alm, E. J., Arumugam, M., Asnicar, F., Bai, Y., Bisanz, J. E., Bittinger, K., Brejnrod, A., Brislawn, C. J., Brown, C. T., Callahan, B. J., Caraballo-Rodríguez, A. M., Chase, J., … Caporaso, J. G. (2019). Reproducible, interactive, scalable and extensible microbiome data science using QIIME 2. *Nature Biotechnology*, *37*(8), 852–857. https://doi.org/10.1038/s41587-019-0209-9

Bronick, C. J., & Lal, R. (2005). Soil structure and management: A review. *Geoderma*, *124*(1–2), 3–22.

Chandarana, K., Goudar, P., & Khan, Z. (2024). Synergistic integration of remote sensing and soil metagenomics data: Advancing precision agriculture through interdisciplinary approaches. *Frontiers in Sustainable Food Systems*, *8*, 1499973. https://doi.org/10.3389/fsufs.2024.1499973

Chaudhari, P. R., Dash, H. R., & Mangwani, N. (2025). Agricultural soil microbiomes at the climate frontier: Nutrient cycling, carbon sequestration and climate change adaptation. *Applied Soil Ecology*, *183*, 104524.

Chen, Y., Liu, X., Zhang, J., Xu, J., & Sun, Y. (2025). Soil microbiome dynamics and their role in sustainable agriculture. *Agronomy Journal*, *8*(4), e79–225.

Delgado-Baquerizo, M., Maestre, F. T., Reich, P. B., Jeffries, T. C., Gaitan, J. J., Encinar, D., Berdugo, M., Campbell, C. D., & Singh, B. K. (2016). Microbial diversity drives multifunctionality in terrestrial ecosystems. *Nature Communications*, *7*, 10541.

Delgado-Baquerizo, M., Reich, P. B., Khachane, A. N., Campbell, C. D., Thomas, N., Freitag, T. E., … Singh, B. K. (2023). Multiple elements of soil biodiversity drive ecosystem functions across biomes. *Nature Ecology & Evolution*, 7, 375–387.

Díaz-Rodríguez, A. M., Parra Cota, F. I., Cira Chávez, L. A., García Ortega, L. F., Estrada Alvarado, M. I., Santoyo, G., & de los Santos-Villalobos, S. (2025). Microbial inoculants in sustainable agriculture: Advancements, challenges, and future directions. *Plants*, *14*(2), 191. https://doi.org/10.3390/plants14020191

Etesami, H. (2024). Enhancing soil microbiome resilience: The mitigating role of silicon against environmental stresses. *Frontiers in Agronomy*, *6*, Article 1465165. https://doi.org/10.3389/fagro.2024.1465165

Fierer, N. (2017). Embracing the unknown: Disentangling the complexities of the soil microbiome. *Nature Reviews Microbiology*, *15*(10), 579–590.

Food and Agriculture Organization of the United Nations (FAO). (2022). *The soil microbiome: A game changer for food and agriculture* [Executive summary]. https://www.fao.org/3/cabb53df-9686-4087-b1ce-d802e0a9aebb/en/

Food and Agriculture Organization of the United Nations (FAO). (2024). *The soil microbiome: A game changer for food and agriculture*. https://www.fao.org/3/cabb53df-9686-4087-b1ce-d802e0a9aebb/en/

Garg, D., Patel, N., Rawat, A., & Rosado, A. S. (2024). Cutting edge tools in the field of soil microbiology. *Current Research in Microbial Sciences*, *6*, 100226. https://doi.org/10.1016/j.crmicr.2024.100226

Geisseler, D., Horwath, W. R., Joergensen, R. G., & Ludwig, B. (2019). The fate of nitrogen in agricultural soils – A review. *Soil Biology and Biochemistry*, *156*, 108150.

Ghimire, S., Bhattarai, B., Shrestha, J., & Paudel, G. P. (2024). Environmental factors shaping soil microbial diversity. *Agricultural Microbiology*, *45*(2), 134–148.

Giller, K. E., Hijbeek, R., Andersson, J. A., & Sumberg, J. (2021). Regenerative agriculture: An agronomic perspective. *Outlook on Agriculture*, *50*(1), 13–25.

Girvan, M. S., Bullimore, J., Pretty, J. N., Osborn, A. M., & Ball, A. S. (2003). Soil type is the primary determinant of the composition of the total and active bacterial communities in arable soils. *Applied and Environmental Microbiology*, *69*(3), 1800–1809.

Hartman, K., van der Heijden, M. G., Wittwer, R. A., Banerjee, S., Walser, J. C., & Schlaeppi, K. (2018). Cropping practices manipulate abundance patterns of root and soil microbiome members paving the way to smart farming. *Microbiome*, *6*, 14. https://doi.org/10.1186/s40168-017-0389-9

He, J. Z., Shen, J. P., Zhang, L. M., Zhu, Y. G., Zheng, Y. M., Xu, M. G., & Di, H. (2007). Quantitative analyses of the abundance and composition of ammonia-oxidizing bacteria and ammonia-oxidizing archaea of a Chinese upland red soil under long-term fertilization practices. *Environmental Microbiology*, *9*(9), 2364–2374.

Hu, P., Ma, Y., & Yin, Y. (2024). The mechanisms and application prospects of soil microorganisms in soil health. *Geographical Research Bulletin*, *3*(2), 606–624.

Institute for European Environmental Policy (IEEP). (2023). *The soil microbiome: Policy and agricultural perspectives*. Institute for European Environmental Policy.

Jansson, J. K., McClure, R., & Egbert, R. G. (2023). *Soil microbiome engineering for sustainability in a changing environment. Nature Biotechnology*. https://doi.org/10.1038/s41587-023-01932-3

Jarupula, S., Rakshit, A., Ogireddy, S. D., Singh, S., Gupta, C., & Chandrakala, J. (2022). Microbiome as a key player in sustainable agriculture and human health. *Frontiers in Soil Science*, *3*, 821589. https://doi.org/10.3389/fsoil.2022.821589

Jiao, S., Chen, W., & Wei, G. (2021). Resilience and assemblage of soil microbiomes under environmental change. *Nature Communications*, *12*, 3009.

Khmelevtsova, M., Sazykina, M., Khammami, M. I., Sazykin, I., Seliverstova, E., Rimmer, A., & Kondratenko, T. (2022). The impact of crop rotation and fertilization on soil microbial community: A review. *Agronomy*, *12*(9).

Knight, R., Vrbanac, A., Taylor, B. C., Aksenov, A., Callewaert, C., Debelius, J., … Caporaso, J. G. (2018). Best practices for analysing microbiomes. *Nature Reviews Microbiology*, *16*(7), 410–422. https://doi.org/10.1038/s41579-018-0029-9

Kremen, C., & Miles, A. (2012). Ecosystem services in biologically diversified versus conventional farming systems: Benefits, externalities, and trade-offs. *Ecology and Society*, *17*(4), 40.

Kuzyakov, Y., & Razavi, B. S. (2019). Rhizosphere size and shape: Temporal dynamics and spatial stationarity. *Soil Biology and Biochemistry*, *135*, 343–360. https://doi.org/10.1016/j.soilbio.2019.05.001

Lahlali, R., Ibrahim, S. S., Belabess, Z., Kadir Roni, M. Z., Radouane, N., Vicente, S. L., Menéndez, E., Mokrini, F., Barka, E. A., & Peng, G. (2021). High-throughput molecular technologies for unraveling the mystery of soil microbial community: Challenges and future prospects. *Heliyon*, *7*(10), e08142. https://doi.org/10.1016/j.heliyon.2021.e08142

Lal, R. (2015). Restoring soil quality to mitigate soil degradation. *Sustainability*, *7*(5), 5875–5895.

Li, X., Rui, J., Mao, Y., Yannarell, A., & Mackie, R. (2014). Dynamics of the bacterial community structure in the rhizosphere of a maize cultivar. *Soil Biology and Biochemistry*, *68*, 392–401.

Liu, S., Li, P., Wang, X., & Zhang, Y. (2023). Effects of conservation tillage practices on soil microbial diversity and assembly processes. *Frontiers in Microbiology*, *14*, 1135328.

Mathew, R. P., Ceccanti, B., & Suleimenov, K. (2025). Long-term impact of tillage on microbial communities of an Eastern European Chernozem. *Scientific Reports*, *15*(1), 84590. https://doi.org/10.1038/s41598-024-84590-y

Mendes, R., Garbeva, P., & Raaijmakers, J. M. (2013). The rhizosphere microbiome: Significance of plant beneficial, plant pathogenic, and human pathogenic microorganisms. *FEMS Microbiology Reviews*, *37*(5), 634–663.

Naasko, K., Patel, H., Plichta, D., Jones, D., & Segata, N. (2024). Role of soil proteins in nitrogen cycling. *Soil Science Advances*, *11*(3), 156–168.

Nadarajah, K. (2024). Role of microbiome in the generation of disease-suppressive soil for sustainable agriculture. *ScienceAsia*, *50*(5), 1–8.

Panke-Buisse, K., Poole, A. C., Goodrich, J. K., Ley, R. E., & Kao-Kniffin, J. (2015). Selection on soil microbiomes reveals reproducible impacts on plant function. *The ISME Journal*, *9*(4), 980–989. https://doi.org/10.1038/ismej.2014.196

Pimentel, D., Hepperly, P., Hanson, J., Douds, D., & Seidel, R. (2005). Environmental, energetic, and economic comparisons of organic and conventional farming systems. *BioScience*, *55*(7), 573–582.

Pretty, J. (2008). Agricultural sustainability: Concepts, principles and evidence. *Philosophical Transactions of the Royal Society B*, *363*(1491), 447–465.

Raza, W., Yuan, S., Yan, C., & Shen, Q. (2023). The soil microbiome: An essential, but neglected, component of regenerative agroecosystems. *Frontiers in Microbiology*, *14*, 114041.

Reganold, J. P., & Wachter, J. M (2016). Organic agriculture in the twenty-first century. *Nature Plants*, *2*, 15221.

Rivera, M. (2020). Soil microbiome and sustainable agriculture. *Journal of Plant Health Environment*, *3*(2), 41–52.

Rodrigues, J. L. M., Pellizari, V. H., Mueller, R., Baek, K., Jesus, E. C., Krishnamurthy, A., Tsai, S. M., Padilla, G., Casella, R., Kostka, J. E., Tiedje, J. M., & Hashsham, S. A. (2022). Cover crop-driven shifts in soil microbial communities could modulate early tomato biomass via plant-soil feedbacks. *Scientific Reports*, *12*, 10308. https://doi.org/10.1038/s41598-022-11845-x

Schloter, M., Nannipieri, P., Sørensen, S. J., & van Elsas, J. D. (2018). Microbial indicators for soil quality. *Biology and Fertility of Soils*, *54*(1), 1–10.

Schneider, T., Keiblinger, K. M., Schmid, E., Sterflinger-Gleixner, K., Ellersdorfer, G., Roschitzki, B., & Riedel, K. (2021). Who is active in the soil? Combining metaproteomics and RNA-seq to reveal active microorganisms. *Frontiers in Microbiology*, *12*, 638240.

Sessitsch, A., Reiter, B., Pfeifer, U., & Wilhelm, E. (2001). Cultivation-independent population analysis of bacterial endophytes in three potato varieties based on eubacterial and Actinomycetes-specific PCR of 16S rRNA genes. *FEMS Microbiology Ecology*, *36*(1), 23–32. https://doi.org/10.1111/j.1574-6941.2001.tb00824.x

Shaffer, J. P., U'Ren, J. M., & Arnold, A. E. (2022). Bioinformatics strategies for harnessing soil microbial diversity. *Environmental Microbiology Reports*, *14*(2), 193–208. https://doi.org/10.1111/1758-2229.13037

Shi, Y., Li, Y., Xiang, X., Sun, R., Yang, T., He, D., Zhang, K., Ni, Y., Zhu, Y. G., Adams, J. M., & Chu, H. (2020). Spatial scale affects the relative role of stochasticity versus determinism in soil bacterial communities in wheat fields across the North China Plain. *Applied and Environmental Microbiology*, *86*(6), e02555–19. https://doi.org/10.1186/s40168-018-0409-4

Singh, A., & Khan, S. (2024). Microbial inoculants: Effective biofertilizers for improving crop productivity. *International Journal of Environmental Sciences and Natural Resources*, *34*(3), 1–9. https://doi.org/10.19080/IJESNR.2024.34.556390

Sinha, S., Chauhan, S., Ansari, M. W., & Yadav, A. N. (2025). Fungal biocontrol agents in the management of soil-borne plant pathogens: Mechanisms and prospects. *Biological Control*, *179*, 105453.

Smith, P., Soussana, J. F., Angers, D., Schipper, L., Chenu, C., Rasse, D. P., Batjes, N. H., van Egmond, F., McNeill, S., Kuhnert, M., Arias-Navarro, C., Olesen, J. E., Chirinda, N., Fornara, D., Wollenberg, E., Álvaro-Fuentes, J., Sanz-Cobena, A., & Klumpp, K. (2019). Interlinkages between soil biodiversity, carbon sequestration and the climate system. *Advances in Agronomy*, *157*, 1–57.

Solanki, M. S. (2022). Soil microbial diversity response to agricultural management. *Agricultural Microbial Ecology*, *12*(4), 312–324.

Taqi, R., Farhan, Q. M., Shehzad, K. K., Eash, N. S., Muhammad, Y., Chatterjee, S., Manzoor, R., Rehman, S. U., & Oetting, J. N. (2023). Impact of different tillage systems on soil health indicators: A global meta-analysis. *Journal of Soil and Water Conservation*, *78*(4), 297–311.

Theriot, E. C., Cooper, C. E., & Doran, J. W. (2020). Characterizing changes in soil microbiome abundance and diversity due to different cover crop techniques. *PLoS ONE*, *15*(5), e0232453. https://doi.org/10.1371/journal.pone.0232453

Timofeeva, A. M., Galyamova, M. R., & Sedykh, S. E. (2023). Plant growth-promoting soil bacteria: Nitrogen fixation, phosphate solubilization, siderophore production, and other biological activities. *Plants*, *12*(24), 4074.

Turbé, A., De Toni, A., Benvenuto, P., Biala, K., Boitani, L., Cimon-Morin, J., Ferreira, M., Forsberg, B., Fur, I., Guisan, A., Maes, J., Marty, P., Paracchini, M. L., Pereira, H. M., Schindler, S., Sousa-Pinto, I., Sykes, M. T., Thijssen, P., Vaissière, A. C., & van Tonder, G. (2010). *Soil Biodiversity: Functions, Threats and Tools for Policy Makers*. European Commission.

Vandenkoornhuyse, P., Quaiser, A., Duhamel, M., Le Van, A., & Dufresne, A. (2015). The importance of the microbiome of the plant holobiont. *New Phytologist*, *206*(4), 1196–1206. https://doi.org/10.1111/nph.13312

van der Heijden, M. G. A., Bardgett, R. D., & van Straalen, N. M. (2008). The unseen majority: Soil microbes as drivers of plant diversity and productivity in terrestrial ecosystems. *Ecology Letters*, *11*(3), 296–310.

Vessey, J. K. (2003). Plant growth promoting rhizobacteria as biofertilizers. *Plant and Soil*, *255*(2), 571–586.

Vlachakis, D. P., & Efthimiadou, A. (2021). SOILomics; where microbiome genetics meets precision agriculture. *EMBnet.journal*, *26*, e940. https://doi.org/10.14806/ej.26.0.940

Wagg, C., Bender, S. F., Widmer, F., & van der Heijden, M. G. A. (2014). Soil biodiversity and soil community composition determine ecosystem multifunctionality. *Proceedings of the National Academy of Sciences*, *111*(14), 5266–5270.

Wan, S., Chen, M., & Lin, F. (2025). Revolutionizing sustainable farming and soil microbiome health with seaweed extract bio-stimulants. *Frontiers in Soil Science*, *3*, 1504045. https://doi.org/10.3389/fsoil.2025.1504045

Wang, X., Chi, Y., & Song, S. (2024). Important soil microbiota's effects on plants and soils: A comprehensive 30-year systematic literature review. *Frontiers in Microbiology*, *15*, 1347745. https://doi.org/10.3389/fmicb.2024.1347745

Wang, J., Shen, J., Wu, Y., Tu, C., Soininen, J., Stegen, J. C., He, J., Liu, X., Zhang, L., & Zhang, E. (2013). Phylogenetic beta diversity in bacterial assemblages across ecosystems: Deterministic versus stochastic processes. *The ISME Journal*, *7*(7), 1310–1321.

Yang, X., Liu, Y., Wang, S., Zhang, M., Li, J., & Chen, H. (2023). Crop diversity and fungal communities: Implications for soil health and sustainability. *Microbial Ecology*, *85*(1), 123–134.

Yıldız, N., Altunok, F., & Yerdelen Uncu, G. E. (2025). Effect of vermicompost fertilizer application on soil properties: A review. *International Journal of Innovative Approaches in Agricultural Research*, *9*(2), 142–151. https://doi.org/10.29329/ijiaar.2025.1321.7

Zhang, J., Liu, Y., Zhang, N., Hu, B., Jin, T., Xu, H., Qin, Y., Yan, P., Zhang, X., Guo, X., Hui, J., Cao, S., Wang, X., Wang, C., Wang, H., Qu, B., Fan, G., Yuan, L., Garrido-Oter, R., Chu, C., & Bai, Y. (2019). NRT1.1B is associated with root microbiota composition and nitrogen use in field-grown rice. *Nature Biotechnology*, *37*(6), 676–684.

Zhang, M., Wu, Y., Qu, C., Huang, Q., & Cai, P. (2024). Microbial extracellular polymeric substances (EPS) in soil: From interfacial behaviour to ecological multifunctionality. *Green Chemistry*, *26*(3), 825–842.

17 Agroecological Farming Systems Supported by Microorganisms

Piotr Barbaś, Barbara Sawicka, and Piotr Pszczółkowski

17.1 INTRODUCTION

Globally, achieving high agricultural production is hampered by problems related to soil compaction and the progressive loss of organic matter. Long-term, intensive cultivation, conducted without adequate nutrient supplementation, inevitably leads to a decline in soil productivity due to the depletion of key elements essential for plant growth. This process also disrupts the ecosystem balance and reduces overall soil productivity, resulting in a decline in the quantity and quality of crops [Ali et al., 2023].

17.1.1 The Role of Organic Fertilizers and Soil Microorganisms

These problems can be effectively minimized through the use of readily available bioresources, which include, among others, organic fertilizers. The use of organic fertilizers not only supports plant growth but also increases their productivity [Itelima et al., 2018]. Furthermore, organic fertilizers significantly improve biological activity, chemical composition, structure, and overall soil quality. They also buffer the adverse effects of acidifying synthetic fertilizers, such as urea, which alter soil pH, negatively impacting its chemical and biological functions [Akanmu et al., 2023]. As emphasized by various authors, organic fertilizers also provide an excellent environment for the growth of soil microorganisms, which has a profound impact on the provision of key ecosystem services, including water distribution and regulation, as well as nitrogen and carbon cycling. Microbial biomass in soils accounts for 85% of the total biomass of all organisms living in this environment, and approximately 90% of the carbon dioxide (CO_2) produced in soils is of microbial origin [Giri et al., 2005]. These data clearly highlight the high metabolic activity and enormous importance of microorganisms for most processes occurring in the soil environment. Among these processes, the most important is the decomposition and mineralization of organic matter (e.g., crop residues, manure, other natural fertilizers, composts, and catch crops). Decomposition of organic residues is extremely important due to the release of mineral forms of nutrients, which constitute a key source of food for crop plants [Shahbaz et al., 2017; Grzyb et al., 2020]. Furthermore, as a result of microbial transformation of organic matter, soil humus is formed, the content of which in soil is one of the most important factors determining the soil's ability to store water and nutrients, as well as its physical structure (lumpiness and gas exchange) [Shah et al., 2021].

17.1.2 Factors Influencing Microbial Activity and Sustainable Agriculture

Lal[2005] indicates that global crop residue production is 3.8 billion tons per year, of which 74% comes from cereals, 10% from sugars, 8% from legumes, 5% from tubers, and 3% from oilseeds. During organic matter decomposition, microbiological activity is strongly dependent on biological, physical, and chemical factors, such as soil type, moisture, oxygenation, pH, salinity, organic matter

DOI: 10.1201/9781003650881-17

and trace element content, the presence of inhibitory and stimulating substances, and the presence of other microorganisms [Ondrasek et al., 2019; Grzyb et al., 2020, 2021; Paśmionka et al., 2021; Bebber & Richards, 2022]. Additionally, climatic conditions, seasons, agrotechnical practices, fertilization, and plant cover, including its type, species, and even cultivar and stage of development, play a significant role in the decomposition of organic matter [Stefan et al., 2021; Ortiz et al., 2023; Qiu et al., 2024].

Modern agriculture, often based on the intensive use of chemical fertilizers and plant protection products to increase yields, unfortunately has a significant, harmful impact on the ecosystem, destroying soil microflora and eliminating weeds and pests. Given growing public concern about the negative impact of agrochemical use, there is a growing interest in improving our understanding of the cooperative interactions between plants and rhizosphere microbial populations [Grzyb et al., 2020, 2021; Paśmionka et al., 2021; Bebber & Richards, 2022]. This is an extremely critical issue, especially in agroecological farming systems, where the production of healthy food is a key driver for society worldwide. Therefore, the aim of this chapter is to present the crucial role of biofertilizers, particularly organic fertilizers and associated soil microorganisms, in counteracting soil degradation and increasing agricultural productivity. We will focus on the mechanisms by which these natural solutions support soil and plant health, which in turn contributes to ensuring food security and achieving the global goal of "Zero Hunger."

17.2 THE IMPORTANCE OF AGROECOLOGY IN THE AGROECOSYSTEM

17.2.1 Agroecology: A Holistic Approach to Sustainable Food Production

The term "agroecology" means more than just agriculture. It denotes "the integrated study of the ecology of the entire food system, encompassing ecological, economic, and social dimensions" [Francis et al., 2003]. Unlike conventional agricultural practices, which often rely on synthetic inputs such as chemical fertilizers and pesticides, agroecology integrates systems—social, ecological, and political—as an integral part of the broader food production process [Wezel et al., 2020]. At its core, agroecology adapts crop production to the local realities of communities, considering their resources and constraints. Agroecology focuses on finding innovative ways to improve agricultural systems by harnessing natural processes, creating beneficial biological interactions and synergies between agroecosystem components [Gliessman, 2018; Masuda et al., 2025]. Its goal is to minimize the use of synthetic and toxic external inputs and instead utilize ecological processes and ecosystem services to develop and implement sustainable agricultural practices.

17.3 KEY PRINCIPLES AND PRACTICES OF AGROECOLOGY

Fundamental techniques and practices in agroecology, as outlined by Gliessman [2013] and Masuda et al. [2025], include the following:

- Minimizing, and ideally completely avoiding, chemical use and high energy inputs.
- Utilizing the properties of the entire agricultural system, which includes recycling nutrients, building organic soil matter, and preserving agrobiodiversity and natural resources.
- Focusing on diversity, crop rotation, and polyculture to enhance beneficial ecological interactions.
- Using indigenous seeds, plants, and livestock adapted to local conditions.
- Applying holistic fertilization and pest control techniques, such as introducing natural pest predators.

17.3.1 Agroecology in the FAO Vision

The Food and Agriculture Organization of the United Nations emphasizes the comprehensive nature of agroecology [FAO, 2018, 2025]. According to FAO [2018], agroecology is "an integrated approach that simultaneously applies ecological and social concepts and principles to the design and management of food and agricultural systems." It seeks to optimize the interactions between plants, animals, people, and the environment, while also addressing key social aspects necessary for achieving a sustainable and equitable food system. For example, Kenya's National Agroecology Strategy for Food System Transformation (NAS_FST) aims to promote sustainable transformation of the country's food system, using a holistic approach to agroecosystems and food systems to ensure food security and nutrition, build climate-resilient livelihoods, and support social inclusion for all [FAO, 2025].

17.3.2 Recent Perspectives and Their Significance

Contemporary research on agroecology (e.g., Badgley et al., 2021; Altieri et al., 2022) increasingly emphasizes its potential for building the resilience of food systems to climate change, increasing the food sovereignty of local communities, and promoting social justice in agriculture. Agroecology is becoming a pillar of transforming agriculture toward a more sustainable and equitable future, addressing global challenges related to food security, the environment, and social development. Implementing agroecological principles is crucial to achieving the UN Sustainable Development Goals (SDGs), in particular the "Zero Hunger" and "Climate Action" goals [Sawicka, 2019; FAO, 2025].

Agroecology is becoming a pillar of transforming agriculture toward a more sustainable and equitable future, addressing global challenges related primarily to food security, the environment, and social development. The latest FAO reports (FAO Statistical Yearbook 2024) emphasize the need for innovative technologies, better water management, and more efficient cropping systems, which fully align with agroecological principles [FAO, 2025]. Implementing agroecological principles is crucial to achieving the UN SDGs, in particular Zero Hunger (SDG 2) and Climate Action (SDG 13) [UN Sustainable Development, 2024. As FAO points out [2018a,b], agroecology offers an integrated approach that harmonizes ecological and social concepts to design and manage food systems, optimizing interactions between all components of the environment and society, which is the foundation for a future, resilient, and equitable food system. However, the current pace of action indicates that UN Member States will not achieve any of the 17 SDGs by 2030. According to the nineth Sustainable Development Report, only 16% of the SDGs are currently on track to achieve this goal [UN Sustainable Development Report, 2024]. Resilient Agricultural Parameters (RAPs) are crucial to achieving the SDGs. According to Sawicka [2019], they are crucial for food security and sustainable food systems in developing countries, focusing on balancing food production, managing natural resources, coping with uncertainty, and ensuring livelihoods.

17.4 INTEGRATION STRATEGIES IN AGROECOSYSTEMS

17.4.1 The Critical Role of Soil Biodiversity

Biodiversity guarantees the stabilization of ecosystem functions and services, which include, among others: biomass production, processes supporting nutrient cycling, soil carbon storage, and pest and pathogen reduction [Norris, 2012]. In agricultural systems, we observe a lack of biodiversity and sensitivity to pathogens and pests [Urruty et al., 2016], as well as stress and drought [Cruz-Córdova et al., 2020]. The use of synthetic fertilizers and pesticides in intensive agriculture has negative consequences for the environment, but also for crop production itself. The presence of pesticides at large distances from the site of application and their accumulation in soil and food chains also cause negative effects on non-target species, threatening ecosystems and human health [Oirdi et al., 2024]. Therefore, alternative methods are being sought to reduce the negative impact of pesticides on the environment. In recent years, we have observed a growing understanding of how soil microbial diversity connects with soil ecosystem functioning [Maron et al., 2018] and how different dimensions of diversity connect with ecosystem functioning, with implications for diversifying agricultural practices [Brooker et al., 2021]. In recent years, knowledge of the links between soil microbial diversity and soil ecosystem functioning has increased significantly [Maron et al., 2018], as well as the relationships between different dimensions of biodiversity and ecosystem functioning, with direct implications for diversifying agricultural practices [Brooker et al., 2021]. Soil microorganisms play a crucial role in key processes on Earth, participating in redox reactions, nutrient recycling, and biomass cycling [Basu et al., 2021]. Many microbial processes, including photosynthesis, respiration, and nitrogen fixation, are regulated by major biogeochemical reactions carried by microbes that tightly bind carbon, nitrogen, phosphorus, and sulfur, and their availability on land [Rousk & Bengtson, 2014]. Soil microorganisms utilize various forms of carbon as energy sources during reproduction. Carbon is returned from soil to the atmosphere through the mineralization of organic matter. As Gougoulias et al. [2014] report, CO_2 is the primary respiratory stream in well-aerated soils, while in anoxic soils (wet soils), CO_2 is reduced to methane (methanogenesis) by hydrogenotrophic archaea (methanogens).

Nitrogen cycling in nature is one of the main nutrient cycles in terrestrial ecosystems, consisting of several microbial processes, including N_2 fixation, mineralization, nitrification, and denitrification [Zhang et al., 2020; Nevison et al., 2022]. Soil microbial diversity can influence both nutrient cycling, crucial for plants, and the physical structure of the soil [Chenu & Cosentino, 2011], also known as soil aggregation, which reduces erosion and is important for maintaining soil fertility and water retention. Earthworms are examples of soil organisms that improve soil structure by modifying various properties. They participate in soil formation, compaction, water infiltration, organic matter decomposition, humus formation, soil-borne disease and pest control, nutrient cycling, and plant growth [Kumar et al., 2020]. The benefits of earthworm activity are presented in Figure 17.1.

Given the projected global temperature increase of 3.7°C by 2100 [Jansson & Hofmockel, 2020], soil microbial communities are predicted to be significantly affected by warming, providing compelling evidence of the impact of climate change on the soil microbiome. A gradual impact of soil microbial communities on soil warming is observed in agroecosystems. In the first stage, the rate of organic carbon decomposition increases in the short term, increasing microbial biomass,

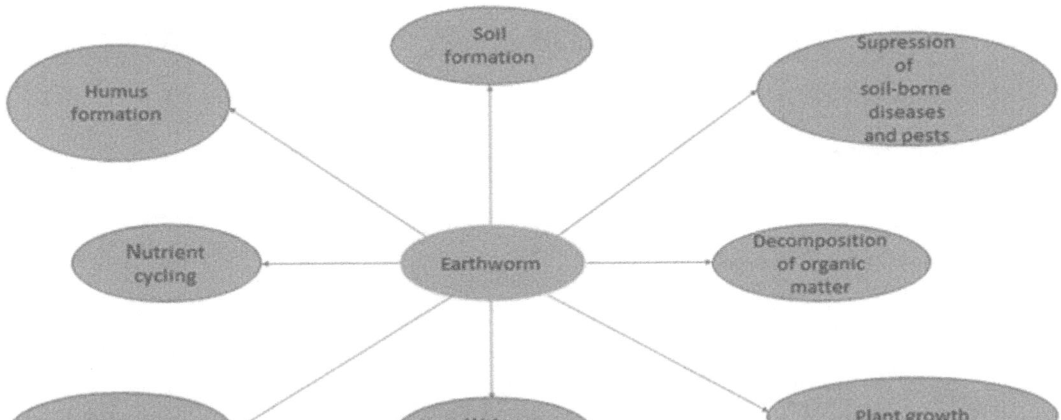

FIGURE 17.1 Benefit of earthworms. (*Source:* Own.)

as evidenced by a 40%–150% increase in the soil microbial population under soil warming [Sheik et al., 2011]. In the second stage, microbial respiration decreases as labile carbon is depleted [Dove et al., 2021]. The structure of soil microbial communities can be modified by climate change through differences in bacterial and fungal abundance [Oram et al., 2025]. As reported by Fahey et al. [2020], mycorrhizal fungi are more resistant to water shortages than some bacterial species. In summary, the health and productivity of agroecosystems are inextricably linked to the richness and functionality of soil biodiversity, especially microorganisms and macrofauna, which, through their activities, provide key ecosystem services and mitigate the negative effects of intensive agricultural practices.

17.5 THE IMPORTANCE OF SOIL MICROORGANISMS IN IMPROVING SOIL NUTRITION AND HEALTH

In agroecosystems, soil biodiversity is crucial for maintaining soil fertility, as is the close link between soil biological activity and its physicochemical properties [Sahu et al., 2019]. The appropriate proportions and availability of nutrients in the soil determine the growth and yield of crops [Ohkama-Ohtsu & Wasaki, 2010]. The quality and yield of plants depend on two essential groups of nutrients: macroelements and microelements [Nadeem et al., 2018] (Figure 17.2).

Each element plays a specific role in plant metabolism:

- *Nitrogen* (*N*): Increases seed and fruit production, ensures better leaf development, and accelerates overall plant growth.
- *Phosphorus* (*P*): An essential component of cell membranes, a key component of adenosine triphosphate (ATP), ribonucleic acid (RNA), and deoxyribonucleic acid (DNA) [Richardson et al., 2009].
- *Potassium* (*K*): A highly mobile element presents in all younger plant parts, it regulates enzymatic reactions, salinity tolerance, and photosynthesis [Johnson et al., 2022].
- *Calcium* (*Ca*): Important for plant growth and development, activates enzymes, participates in salt balance and water movement within plant cells, activating second messenger

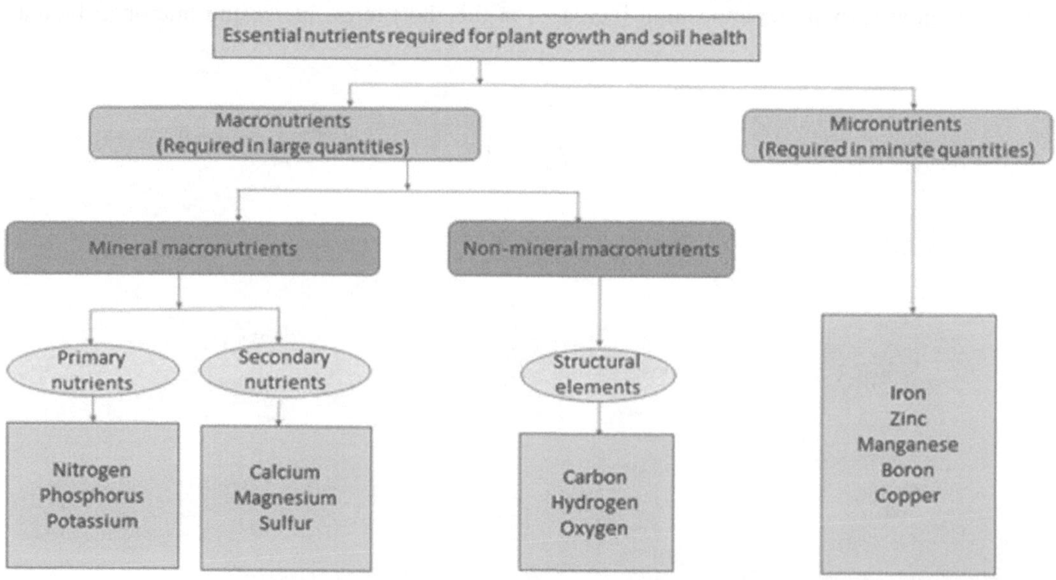

FIGURE 17.2 Role of essential nutrients in improving plant nutrition and soil health. (*Source:* Own.)

systems (e.g., calmodulin) and stabilizing cell wall structures. potassium to control the opening and closing of stomata [Ghosh et al., 2022].

- *Magnesium (Mg)*: The central metal atom in chlorophyll, it plays a significant role in photosynthesis. Its deficiency leads to chlorophyll degradation and chlorosis (leaf yellowing), while adequate magnesium availability promotes plant health [de Sousa Ferreira et al., 2023].
- *Sulfur (S)*: An essential element for all living organisms, playing dynamic roles in the proper development, growth, and survival of plants. Considered an essential nutrient for maximum crop production [Sakin et al., 2024].
- *Manganese (Mn)*: A potential activator of many enzymes involved in the citric acid cycle, carbohydrate metabolism, phosphorus reactions, carboxylation processes, and oxidation reactions [Nadeem et al., 2018].
- *Iron (Fe)*: A soil element with low solubility, often poorly available to microorganisms and plants, especially in arid regions with alkaline soil pH [Gondal et al., 2021].
- *Zinc (Zn)*: Participates in many key cellular functions, such as metabolic and physiological processes, enzyme activation, and ion homeostasis [Hamzah Saleem et al., 2022].
- *Boron (B)*: Responsible for plant quality and productivity, it plays a key role in physiological processes. It is crucial in soils with high pH, alkaline pH, and low organic matter content [Zaib, 2024].

The bioavailability and absorption of nutrients from soils depends on many characteristics, including soil physicochemical properties, environmental factors, and microbial populations [Massenssini et al., 2015]. Biodiversity and ecosystem resilience depend on the efficiency of nutrient/element cycling required by varied species [Zaman et al., 2025]. Complex food webs support this, which enhance ecosystem functionality [Wan et al., 2024]. Phosphorus cycling ensures the growth of nutrient-demanding species, while nitrogen cycling supports legumes and other nitrogen-fixing plants, contributing to species coexistence and diversity [Fontaine et al., 2024]. Silicon uptake by plants determines the plant community assembly and ecosystem structure, thus influencing patterns of plant biodiversity and ecosystem functioning, as well as resilience [Zenda et al., 2021].

17.5.1 Key Groups of Soil Microorganisms and Their Functions

Among the countless groups of soil microorganisms, many play specific, crucial roles:

- *Nitrogen-Fixing Bacteria (e.g., Azospirillum)*: They stimulate root growth and increase the rate of water and mineral uptake by plants [Bhat et al., 2015]. Actinomycetes are Gram-positive aerobic bacteria belonging to the order Actinomycetes that play a crucial role in organic matter cycling, inhibiting the development of several plant pathogens in the rhizosphere [Torres-Rodriguez et al., 2022]. Furthermore, they are responsible for the degradation of complex polymer mixtures in dead plant, animal, and fungal material, resulting in the production of numerous extracellular enzymes that are crucial for plant production [Mitra et al., 2022].
- *Actinomycetes*: Gram-positive aerobic bacteria that play a crucial role in organic matter cycling. They inhibit the growth of many plant pathogens in the rhizosphere [Torres-Rodriguez et al., 2022] and are responsible for the degradation of complex polymers in dead plant, animal, and fungal material, producing numerous extracellular enzymes crucial for plant production [Mitra et al., 2022].
- *Cyanobacteria*: Photoautotrophic Gram-negative bacteria that participate in maintaining soil fertility and building soil structure. They are a source of biomass and dissolved

organic matter (e.g., no ribosomal peptides, isoprenoids, alkaloids, and polypeptides), which is mineralized by other soil microorganisms (non-ribosomal peptides, isoprenoids, ribosomal peptides, alkaloids, and polypeptides) [Huang & Zimba, 2019].

- *Rhizobacteria*: They increase soil structure and also influence the growth and stabilization of mineral phosphates, dehydrogenase activity, soil aggregates, and the availability of other nutrients [Billah et al., 2019].
- *Algae*: A highly diverse group of simple, autotrophic organisms capable of photosynthesis, capable of capturing energy from sunlight, and also play a significant role in soils due to their use as biofertilizers and soil stabilizers [Jaswal & Saroch, 2022]. They emit substances that promote plant growth, such as hormones, vitamins, amino acids, and organic acids. They improve soil properties, including carbon content, structure, and aeration [Swami et al., 2025].
- *Nematodes*: Unicellular organisms capable of surviving in unfavorable soil conditions thanks to a dormant stage [Kiontke & Fitch, 2013; Tahat et al., 2020]. Most species derive nourishment from soil bacteria, playing a significant role in maintaining the microbial/bacterial balance in the soil.
- *Soil Viruses*: Considered essential organisms due to their positive impact on soil biology and their ability to transfer genes from host to host, resulting in microbial mortality [Kimura et al., 2008]. They can, therefore, influence global biogeochemical cycles, determining the circulation and concentration of nutrients and gases. The conclusions are clear: soil health and effective plant nutrition are intricately linked to the richness and functionality of soil microbial communities, which underpin the productivity and sustainability of agroecosystems.

17.6 BENEFICIAL MICROBES: INVISIBLE ALLIES OF PLANTS

The rhizosphere, the zone of soil immediately surrounding plant roots, is home to millions of microorganisms that actively interact with root exudates. These interactions are crucial for increasing nutrient uptake, including macronutrients, primarily nitrogen and phosphorus [Nath et al., 2018]. Understanding and utilizing the role of these microorganisms in controlling macronutrient uptake to achieve higher nitrogen and phosphorus use efficiency has been the subject of intensive research in recent years. These interactions may constitute a beneficial strategy for obtaining higher yields and better quality.

17.6.1 Strategies for Utilizing Beneficial Microorganisms

Two of the most popular and preferred agricultural techniques are the use of plant growth-promoting rhizobacteria (PGPR) and arbuscular mycorrhizal fungi (AMF) [Miransari, 2011]. These microorganisms can be applied as inoculum directly to the root surface or into the rhizosphere surrounding the plant. The relationship between beneficial microorganisms and the host plant is complex and depends on many factors, such as the microorganism's ability to colonize the plant's roots, the composition of root exudates, and the overall health of the surrounding soil [Olanrewaju et al., 2019].

17.6.2 Mechanisms of Action of Beneficial Microorganisms

Beneficial microorganisms in the rhizobia (the community of microorganisms inhabiting the rhizosphere) play a fundamental role in plant growth and development, as well as in pathogen control through a variety of mechanisms [Glick, 2012; Rehman et al., 2020]. These mechanisms include the following:

- *Biofertilization*: This includes processes such as fixing atmospheric nitrogen, solubilizing phosphates unavailable to plants, and producing plant growth hormones. These activities directly increase the availability of nutrients to plants, supporting their development.
- *Bioremediation*: Although not always explicitly mentioned in the context of PGPR, some rhizobacteria can contribute to the breakdown of pollutants in soil, improving soil quality and reducing plant stress.
- *Biocontrol*: This focuses on limiting the harmful effects of plant pathogens. Biocontrol mechanisms include the synthesis of siderophores (molecules that bind iron, limiting its availability to pathogens), regulation of ethylene levels (a plant hormone that influences the stress response), induced systemic resistance (ISR), and systemic acquired resistance (SAR) [Noumavo et al., 2016] (Figure 17.3).

Most rhizobacteria and rhizospheric fungi are capable of producing metabolites that inhibit pathogen growth [Rodrigo et al., 2021, 2022]. Applications of rhizobacteria (PGPR) and AMF are

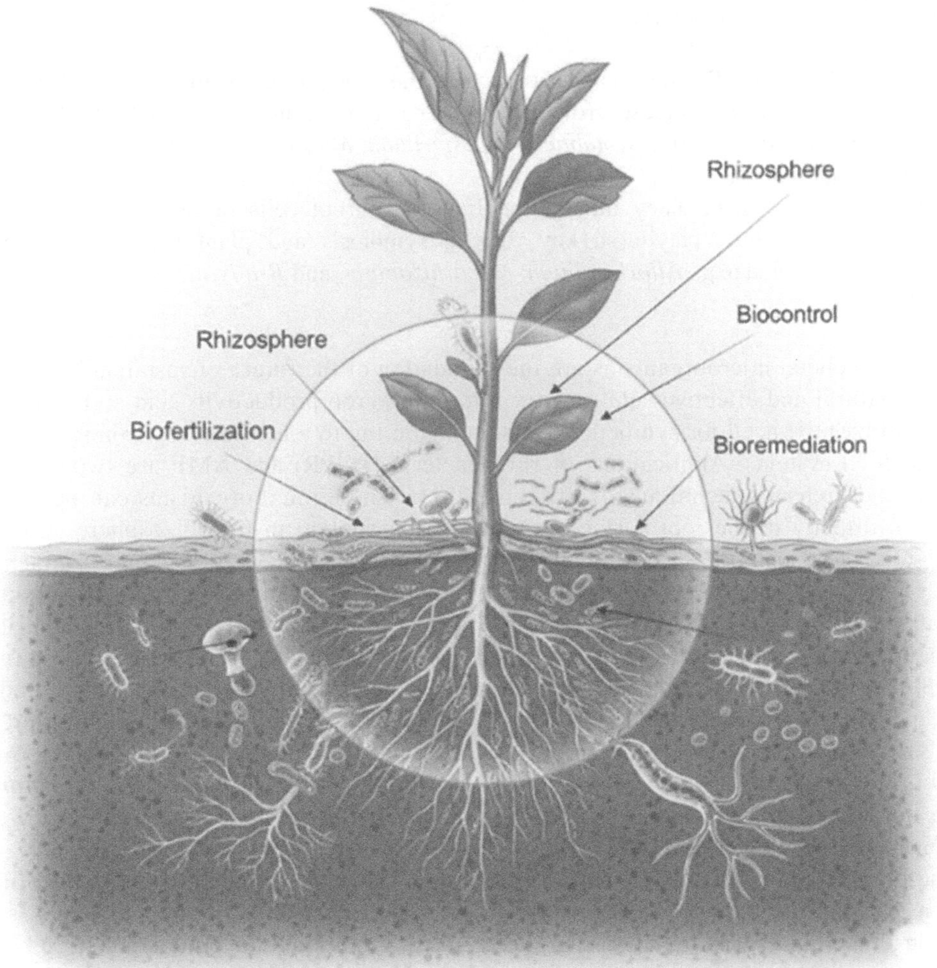

FIGURE 17.3 Mechanisms of action of beneficial microbes in the rhizosphere: biofertilization, bioremediation, and biocontrol. (*Source:* Own.)

two of the most popular and preferred techniques [Miransari, 2011]. These microorganisms can be applied as inoculum directly to the root surface or can also be applied in the rhizosphere surrounding the plant. The association of beneficial microorganisms with the host plant is influenced by many factors, including the ability of the microorganism to colonize the host plant's roots, root exudates, and the health of the surrounding soil [Olanrewaju et al., 2019]. Beneficial microorganisms in the rhizobia determine plant growth and development and control pathogens using various mechanisms (biofertilization, bioremediation, and biocontrol) [Glick, 2012; Olanrewaju et al., 2017; Rehman et al., 2020]. Examples of microbial biofertilization activities include nitrogen fixation, phosphate solubilization, and production of plant growth hormones, while biocontrol focuses on limiting the harmful effects of plant pathogens through siderophore synthesis, regulation of ethylene levels, ISR, and SAR [Noumavo et al., 2016]. Most rhizobacteria and rhizospheric fungi are capable of producing metabolites that inhibit pathogen growth [Rodrigo et al., 2021, 2022].

17.6.3 Types of Plant Growth-Promoting Bacteria

Plant growth-promoting bacteria (PGPB) participate in soil nutrient mobility and uptake. They are divided into two main categories:

- *Extracellular PGPB*: These bacteria colonize the root surface or the intercellular spaces of the cortex. They increase growth factors and improve nutrient availability. Examples include *Bacillus*, *Serratia*, *Azotobacter*, *Azospirillum*, *Micrococcus*, *Arthrobacter*, *Erwinia*, and *Pseudomonas*.
- *Intracellular PGPB*: They inhabit the interior of root cells or nodular structures of rhizosphere plants, playing a key role in symbiosis and plant nutrition. Examples include Rhizobia (e.g., *Allorhizobium*, *Mesorhizobium*, and *Bradyrhizobium*) and *Frankia* [Nabi, 2023].

These beneficial microorganisms are the foundation of the future of sustainable agriculture, offering natural and effective solutions for improving crop productivity and soil health. Their use minimizes the need for synthetic inputs, contributing to a healthier environment and more resilient food systems. Applications of rhizobacteria (PGPR) and AMF are two of the most popular and preferred techniques [Miransari, 2011]. These microorganisms can be applied as inoculum directly to the root surface or can also be applied in the rhizosphere surrounding the plant. The association of beneficial microorganisms with their host plant is influenced by many factors, including the ability of the microorganism to colonize the host plant's roots, root exudates, and the health of the surrounding soil [Olanrewaju et al., 2019]. PGPB contribute to the mobility and uptake of nutrients in the soil. These include extracellular PGPB that increase growth factors and colonize the root surface or the intercellular space of the cortex, and intracellular PGPB responsible for plant growth that inhabit the root surface/nodular structure of the plant's rhizosphere root cells [Nabi, 2023]. Extracellular PGPB include, among others, *Bacillus*, *Serratia*, *Azotobacter*, *Azospirillum*, *Micrococcus*, *Arthrobacter*, *Erwinia*, and *Pseudomonas*, while intracellular PGPB include *Rhizobia*, *Allorhizobium*, *Mesorhizobium*, *Bradyrhizobium*, and *Frankia* (Figure 17.4).

The presence of PGPB in soil enables the transformation of infertile soils into fertile ones and ensures increased plant adaptation to various stresses, such as diseases, drought, salinity, extreme temperatures, and light [Ansabayeva et al., 2025] through the production of gibberellins, indoleacetic acid, cytokinins, deaminases, and several essential nutrients, especially phosphorus, zinc, and manganese. PGPB accelerate the bioassimilation of nutrients and participate in the detoxification of harmful products to which plants may be exposed (e.g., heavy metals) [Gupta et al., 2024].

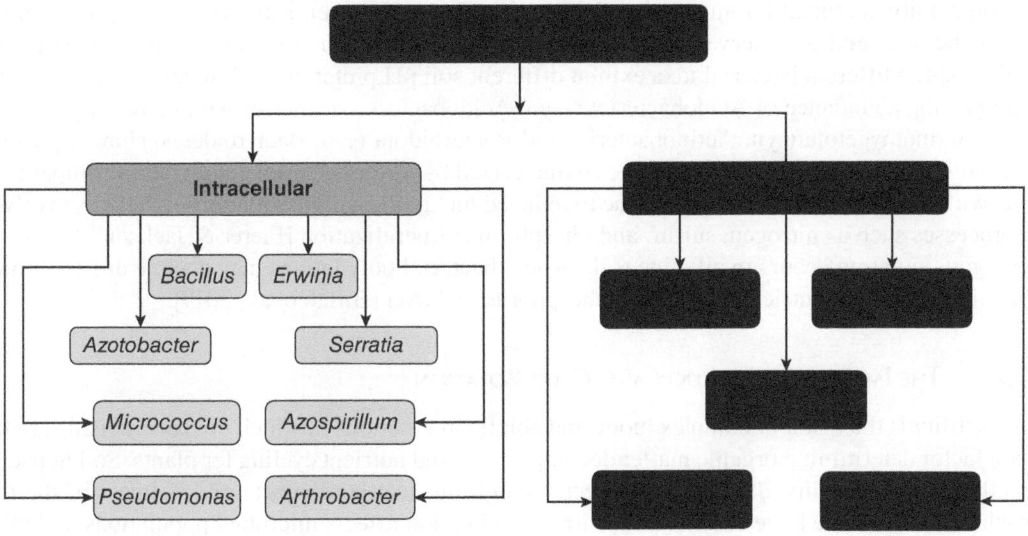

FIGURE 17.4 Division of plant growth-promoting bacteria. (*Source:* Own.)

17.7 FACTORS DETERMINING SOIL MICROBIAL DIVERSITY

Soil microbial diversity, crucial for soil health and productivity, is shaped by a number of abiotic and biotic factors.

17.7.1 Influence of Soil Moisture and Temperature

Soil moisture is a fundamental factor determining the survival and activity of soil microorganisms, directly influencing their growth environment [Siebielec et al., 2020]. Climate change, including global warming and extreme weather events such as droughts and heavy rainfall, significantly alters soil moisture status, which in turn affects microbial diversity [Baldrian et al., 2022; Furtak & Wolińska, 2023; Wang et al., 2025].

During summer periods and in hot and humid climates, changes in plant metabolic rates and soil microbial diversity are observed [García et al., 2018; García-García et al. 2020; Schimel, 2018]. Some strains of PGPB exhibit improved [Cruz et al. 2021] growth at higher temperatures, making them particularly interesting for use under drought stress, as observed by Egamberdiyeva and Höflich [2003]. This is supported by the study by Waldon et al. [1989], where rhizobacteria isolated from nodules of the desert legume *Prosopis glandulosa* grew better at 36°C than at 26°C. Petrovich et al. [2017] reported different results for eurypsychrophilic *Pseudomonas* strains isolated from Antarctic soils, which better supported germination and growth at low temperatures (Rondón et al. 2019). This finding highlights their biotechnological potential in developing biofertilizers active in freezing conditions.

Climate, like soil properties and vegetation type, directly influences the potential functions performed by microbial communities. Climate change studies have shown that warming and humidity changes can increase the presence of some functional groups, such as plant pathogens [Raza & Bebber, 2022; Li et al., 2023], while decreasing the abundance of others, e.g., beneficial AMF [Cotton, 2018].

17.7.2 Influence of Land Use and Soil pH

Land use type (e.g., agricultural and forest soils) and vegetation cover (e.g., coniferous and deciduous forests) also significantly influence soil microbial diversity [Pang et al., 2019]. Studies by DeKosky et al. [2013] showed that microbial biomass was lowest in ecosystems with the wettest and driest soils. Disturbed soils had fewer fungi and more Gram-positive bacteria than undisturbed soils.

Soil acidity is crucial for nutrient availability, regulating the chemical form of compounds present in the soil, and also serves as an indirect limiting factor for microbial communities [Zhou et al., 2024]. Different bacterial taxa exhibit different soil pH preferences [Lauber et al., 2009]. For example, the abundance of Acidobacteriota (syn. Acidobacteria) decreases with increasing soil pH, while Actinomycetota (syn. Actinobacteria) and Bacteroidota (syn. Bacteroidetes) show a positive correlation. Acidic and alkaline soils are characterized by lower bacterial colonization compared to soils with a near-neutral pH, which is due to reduced metabolic activity of microorganisms involved in processes such as nitrogen, sulfur, and phosphorus mineralization [Fierer & Jackson, 2006]. An analogous situation occurs in alkaline soils, where bacterial populations may decline due to limited solubility of certain nutrients, including phosphorus and iron [Billah et al., 2019].

17.7.3 The Impact of Nutrients and Crop Rotation

Soil fertility is the result of complex biotic and abiotic interactions in which microorganisms are the main factor determining organic matter decomposition and nutrient cycling for plants. Soil nutrients and their bioavailability, directly or indirectly, can influence the diversity and richness of the rhizosphere microbiome [Yue et al., 2023]. Nitrogen addiction affects microbial populations and plant productivity, but excess nitrogen can lead to lower abundance and diversity of plant and bacterial communities [Ren et al., 2020]. Canadian researchers [Beauregard et al., 2010] have shown that phosphorus application alters the structure of fungal and bacterial communities, although it does not affect species richness.

Land use is also associated with microbial catabolic evenness, as some agricultural practices can deplete soil organic carbon and reduce the catabolic diversity of soil microorganisms [Tardy et al., 2015].

Carefully planned crop rotation is important for microbial communities in the plant rhizosphere and in the soil around plants [Venter et al., 2016], although opinions on this matter are divided. Hamel et al. [2018] observed that frequent legume cultivation in wheat rotation increased soil nitrogen levels but had insignificant effect on rhizobacterial community composition. Similarly, Zhang et al. [2014] found that crop rotation had insignificant effect on soil bacterial communities and enzyme activity but significantly affected soil fungal communities, particularly AMF. Furthermore, D'Acunto et al. [2018] demonstrated that enriching crop rotations with different crop species (e.g., field pea/maize and wheat/soybean) diversifies the active soil microbial community. Long-term crop rotations, especially those including cover crops, accumulate significantly more soil carbon and microbial biomass compared to monoculture [Venter et al., 2016; Martínez-García et al., 2018]. Estonian researchers [Kuht et al., 2022] demonstrated that organic crop rotations with cover crops and composted manure increased soil microbial activity by 7.3%–14% compared to conventional systems with chemical plant protection and mineral fertilizers.

17.7.4 Other Environmental Factors

Bastida et al. [2021] found that the ratio of soil microbial diversity to biomass peaks in arid, low-carbon (C) environments and is exceptionally low in cold, C-rich environments. UV radiation also plays a key role in agroecosystems, altering the biological availability of dissolved organic matter to microorganisms and accelerating its conversion to dissolved inorganic carbon and nitrogen, including carbon dioxide and ammonia [Erickson et al., 2015]. Ultraviolet radiation reaching the soil surface can have significant effects on microorganisms. Erickson et al. [2015] demonstrated that UV radiation alters the biological availability of dissolved organic matter to microorganisms, accelerating its conversion to dissolved inorganic carbon and nitrogen, including carbon dioxide and ammonia. This process, called photodegradation, can influence biogeochemical cycles and nutrient availability, altering the dynamics of microbial communities. High UV intensity can also directly damage microbial cells, leading to changes in their composition and function.

Soil structure and aeration: The physical structure of soil, including its porosity and degree of aeration, directly affects oxygen availability, which is crucial for most aerobic bacteria and fungi. Compacted and poorly aerated soils favor the growth of anaerobic microorganisms, which alters overall diversity and metabolic activity. The availability of pore space also influences microbial mobility and their interactions with plant roots.

Presence of heavy metals and contaminants: Chemical contaminants, such as heavy metals or pesticide residues, can have a toxic effect on soil microorganisms, reducing their abundance and diversity. However, some microbes can develop resistance to contaminants or even bioremediate them, which influences the evolution and selection of microbial communities in a given environment [Saeed et al., 2021; Gupta et al., 2024]. Interspecific interactions: Microorganisms themselves interact through competition for resources, the production of allelopathic substances, and the formation of biofilms. These interactions, although complex, are crucial for shaping the structure and function of microbial communities. Predation (e.g., by nematodes and protozoa) and soil viruses (phages) also play a significant role in regulating microbial populations and nutrient cycling [Krivan, 2008]. The complexity of the interactions between these factors highlights how delicate and dynamic a soil ecosystem is, and managing its microbial diversity requires a comprehensive approach that considers multiple environmental variables.

17.8 THE IMPACT OF CONVENTIONAL AND ORGANIC FARMING SYSTEMS ON SOIL-MICROBE-PLANT INTERACTIONS

Understanding and managing microbial communities in agricultural systems is a key scientific challenge for the development of agroecology, aimed at optimizing ecosystem processes. Differences in conventional and organic farming practices have a profound impact on the interactions between soil, microbes, and plants [Sinha et al. 2025].

17.8.1 Conventional vs. Organic Agriculture: Impact on the Soil Microbiome

Conventional farming systems use copious amounts of chemical fertilizers and pesticides to maximize yields. Organic (organic) farming systems, on the other hand, primarily utilize plant residues or manure, which contributes to increased soil fertility [Hartmann et al., 2015]. Although organic farming improves soil biodiversity by increasing the concentration of macro- and microorganisms, it is often associated with lower yields. Meemken & Qaim [2018] report that the reduction in yield per hectare in organic systems can be 20%–25%, and in some cases even 50%. An additional challenge in agroecosystems is the pressure of pathogens and pests, which is particularly problematic in organic farming, where synthetic pesticides are not permitted. Projections indicate that these threats will increase by 2050 due to climate change [Bebber et al., 2014].

Many studies document the impact of various agricultural practices on soil bacterial and fungal communities:

- Martínez-García et al. [2018], analyzing ten years of organic and conventional farming systems, showed that organically managed soil was characterized by increased bacterial and fungal biomass (including saprotrophic and arbuscular mycorrhizae) as well as higher total microbial catabolic activity.
- Ghimire et al. [2014] observed an increase in the abundance of bacteria, fungi, and protozoa in a study of four years of organic management, with a two- to 36-fold increase in phospholipid fatty acid (PLFA) biomarkers. Additionally, the dissolved organic carbon content was higher in organic crops compared to reduced tillage and conventional tillage.

17.8.2 Impact of Plant Residues and Cover Crops

The type of plant residue introduced into the soil has a significant impact on the composition of microbial communities [Drenovsky et al. 2010]. For example:

- Kong et al. [2011] demonstrated that the addition of high-quality cover crop residues (high N content or low C/N ratio) promotes higher bacterial concentrations, while low-quality residues promote fungal growth.
- Rezgui et al. [2021] identified differences in microbial composition depending on the type of residue: *Chloroflexales*, *Inquilinus*, *Rubricoccus*, *Clitocybe*, and *Verticillium* dominated in soils with pea residue; *Thermoanaerobacterales*, *Thermacetogenum*, and *Hypoxylon* dominated in soils with rapeseed residue; and *Halanaerobiales*, *Rubrobacter*, and *Volutella* dominated in soils with wheat residue.
- Cover crop use modifies fungal abundance, as observed by Finney et al. [2017]. AMF were more abundant under oat and rye (*Secale cereale* L.) cover crops, whereas non-arbuscular fungi were positively correlated with hairy vetch (*Vicia villosa* L.). These authors emphasized that microbial community composition (fungal to bacterial ratio) had a significant, positively correlated effect on soil biological activity.
- Muhammad et al. [2021] demonstrated that cover cropping increased soil bacteria and fungi by 7%–31% compared to no cover cropping. Furthermore, a mixture of legume and non-legume covers crops reduced microbial biomass C (MBC), total phospholipid-derived fatty acids (PLFAs), and actinomycetes compared to cover crop monoculture. These studies also showed that incorporating cover crop residues into the soil increased PLFA, total bacterial counts, AMF root colonization, and spore density, but reduced Gram-positive and Gram-negative bacteria and AMF compared to residues placed on the surface or removed from the soil. The beneficial effect of cover crops on improving soil quality, as indicated by microbiological activity and the content of C and nutrients in the soil, was also noted by Strickland et al. [2019].

17.8.3 Impact of Compost, Manure, and Crop Rotation

In organic farming systems, the application of compost and manure also influences changes in the soil microbial community. Ouyang et al. [2022] demonstrated that manure had a stronger effect than compost on the composition of the soil bacterial and fungal communities, including ureolytic and chitinolytic communities. At the same time, compost-treated soils were characterized by greater microbial richness than manure-treated soils. A long-term land use assessment in organic and conventional agriculture conducted by Bier et al. [2024] demonstrated a diversity of soil microbial communities: the accumulation of prokaryotes (bacteria and archaea) and fungi, which were primarily influenced by different tillage practices and the presence/absence of cover crops. Prokaryote richness was higher with legume fertilizer than with manure or synthetic fertilizer. In organic systems, the authors also observed a 4% higher fungal concentration compared to conventional and reduced-till systems.

17.8.4 Impact of Animal Grazing

Wang et al. [2022] made interesting observations when assessing the impact of livestock grazing (sheep) on soil microorganisms and ecosystem functions. Increased grazing intensity, through vegetation removal, trampling, and manure deposition, reduced soil organic carbon concentration and dominant bacterial phyla (Proteobacteria and Bacteroidetes), while simultaneously increasing the abundance of oligotrophic bacteria in minor phyla such as Verrucomicrobia and Chloroflexi. Increased grazing intensity also promoted the dominance of Ascomycetes fungi, which was associated with a poorer soil environment.

In summary, the impact of agricultural management systems on soil-microbe-plant interactions is complex and multifaceted. Organic farming, through the promotion of biodiversity, the use of organic fertilizers, and cover crops, better supports healthy and vibrant soil microbial communities. While this presents challenges related to yield and pathogen pressure, the long-term benefits for soil health, nutrient cycling, and the overall resilience of agroecosystems are undeniable. Understanding these relationships is crucial for developing sustainable agricultural practices that effectively address the global challenges of food security and climate change.

17.9 THE ROLE OF AGRICULTURAL MICROBIOLOGY IN SUSTAINABLE DEVELOPMENT AND FOOD SECURITY

Modern agriculture faces the challenge of achieving high plant yields and animal productivity while simultaneously striving for sustainability by utilizing natural adaptive potential and minimizing negative environmental impacts. To meet this goal, it is essential to replace hazardous agrochemicals (mineral fertilizers and pesticides) with environmentally friendly preparations of symbiotic microbes. This approach can significantly improve the nutrition of crops and livestock, as well as increase their resistance to biotic (pathogens and pests) and abiotic (including pollution and climate change) stresses [Saeed et al., 2021]. For this reason, agricultural microbiology is currently recognized as a key area for transferring knowledge from general microbiology and microbial ecology to innovative agricultural biotechnologies [Aluvihara et al. 2023].

17.9.1 Sustainability in the Context of Agroecosystems

Sustainability, as Altieri [2004] argues, is the result of the intersection of three main factors: the environment, society, and the economy, which interact with each other. Therefore, the concept of "sustainable development" is defined by the intersection of the economy with the environment (agroecology), the environment with society (environmental awareness), and society with the economy (standard of living) [Vandermeer & Perfecto, 2013].

To maintain sustainability in agroecosystems, it is necessary to address challenges such as loss of microflora and microfauna, loss of biodiversity, outdated production and yield strategies, regional economic imbalances, pollution, and global warming [Vikas & Ranjan, 2024]. The goal of sustainable agriculture is to employ strategies that increase or maintain current food production rates while minimizing harm to the environment and human health [Terán-Samaniego et al., 2025].

17.9.2 Biofertilizers and Biocontrol: The Future of Agriculture

The use of plant microbes as plant growth-promoting agents (PGPs) or biofertilizers offers an environmentally friendly alternative to conventional agricultural technology [Yadav et al., 2017]. Biofertilizers are a group of beneficial microorganisms responsible for plant growth and enriching soil nutrients by increasing their availability to crops [Itelima et al., 2018]. These microorganisms are considered a biopotential and innovative tool with significant benefits for agriculture. PGP microbes support plant growth through [Sinha et al. 2025]:

- *Direct Mechanisms*: Production of plant growth hormones, fixation of atmospheric nitrogen (N_2), and solubilization (dissolution) of phosphorus, potassium, and zinc, making them more available to plants [Al Abboud et al., 2014; Mahanty et al., 2017; Gulshan et al., 2022].
- *Indirect Mechanisms*: Production of ammonia, antibiotics, prussic acid (HCN), lytic enzymes, and siderophores, which help fight pathogens and pests [Al Abboud et al., 2014; Mahanty et al., 2017; Gulshan et al., 2022].

A significant issue for sustainable agriculture is the presence of soil-borne diseases, which pose a serious threat to global food security, leading to reduced plant productivity and yield declines worldwide. Therefore, the microbial activity of microbiomes, as well as the interactions established between soil and plants, is crucial [Obasi et al., 2024]. Therefore, strains such as *Bacillus*, *Pseudomonas*, and *Streptomyces*, which are capable of producing secondary metabolites, can function as potential biological control agents and PGPB [Chowdhury et al., 2020].

17.9.3 Recent Achievements and Perspectives

Current research is increasingly focusing on engineering and optimizing soil microbiomes to increase the effectiveness of soil microbial crop protection. The development of soil carbon sequestration by enhancing microbial humification processes, as well as the creation of next generation biostimulants and biopesticides based on specific microbial strains, are among the latest directions [FAO, 2018, 2025]. The use of gene editing technologies (e.g., CRISPR) to modify soil microorganisms to increase their capacity for nitrogen fixation, phosphorus solubilization, or the production of defensive compounds opens new perspectives for precision and sustainable agriculture [Kremer, 2023]. Integrating these microbial solutions with smart and digital agriculture systems represents another step toward global food security.

17.10 THE MICROBIOME AND ITS APPLICATIONS IN AGRICULTURE

17.10.1 Biofertilizers

The main sources of biofertilizers include bacteria, fungi, and cyanobacteria. Bacteria such as *Rhizobium*, *Azospirillum*, and *Azotobacter* fix atmospheric nitrogen into organic forms, which are utilized by plants as nutrients [Kumar et al., 2017]. Fungi have the ability to form symbiotic relationships with plants through mycorrhiza [Bonfante & Anca, 2009]. In these relationships, the fungal symbiont absorbs phosphorus from the soil and transfers it to the plant. Plants rich in these compounds also have the ability to increase resistance to root-borne pathogens, tolerance to salinity and drought, and overall plant growth and development [Deveau et al., 2018]. Several species of bacteria and fungi, such as *Pseudomonas*, *Bacillus*, *Penicillium*, and *Aspergillus*, are capable of secreting organic acids and lowering the soil pH in their surroundings to dissolve bound phosphates in the soil [Kumar et al., 2017]. Recent research in the field of biofertilizers focuses on identifying and selecting new, highly effective strains of microorganisms that are adapted to diverse environmental conditions and soil types. Biofertilizers provide nutrients, regulate crop growth, and improve soil health. The main microorganisms used in biofertilizers are *Pseudomonas*, *Azotobacter*, and *Bacillus* [Zhao et al., 2024]. The development of microbial consortia, i.e., mixtures of different strains and species of microorganisms that act synergistically, is becoming increasingly promising [Zhao et al., 2024]. The goal is to create biofertilizers with increased effectiveness in providing nutrients, improving soil structure, and protecting plants from stress, which is crucial for global food security and sustainable agricultural production.

17.10.2 Biopesticides

In the face of growing challenges related to food security and environmental protection, biopesticides are becoming a key alternative to conventional chemicals. According to the classification [Fenibo et al., 2022; Mawcha et al., 2025], biopesticides are divided into three main groups:

1. *Microbial Biopesticides*: These include microorganisms (bacteria, fungi, viruses, protozoa, nematodes, and microalgae) and compounds produced by them that exhibit pest control effects through competition, pathogenicity, or inhibitory toxins [Fenibo et al., 2022].
2. *Biochemical Biopesticides*: These are natural active substances that control pests using non-toxic mechanisms. A natural active ingredient defined as a biopesticide should exhibit

repellent (attractant), choking (antifeedant), and moisture retention or absorption (confusing agent) properties [Liu et al., 2019].
3. *Plant-Incorporated Protectants* (*PIPs*): These are transgenic plants that, through genetic modification, produce substances that protect them against pests.

17.10.2.1 Advantages and Types of Microbial Biopesticides

The main advantage of biopesticides is their low toxicity and lack of harm to non-target organisms, including humans and the environment [Daraban et al., 2023]. This makes them a safer option compared to synthetic pesticides. Examples of microbial biopesticides include many microorganisms and their products:

- *Bacteria*: *Bacillus thuringiensis* (Bt): One of the most commonly used bacteria, producing proteins toxic to insect larvae [Ruiu, 2018]. *Paenibacillus* sp.: These bacteria can produce substances with fungistatic and insecticidal effects (Tariq et al. 2025).
- *Fungi*: *Metarhizium anisopliae*: An entomopathogenic fungus that infects insects, leading to their death [Ruiu, 2018]. *Verticillium* sp.: Some species are pathogenic to insects and nematodes [Ruiu, 2018].
- *Nematodes*: *Heterorhabditis* sp. and *Steinernema* spp.: Entomopathogenic nematodes that enter into symbiosis with bacteria that kill insect pests [Ruiu, 2018].
- *Protozoa*: *Nosema* spp. and *Vairimorpha* spp.: *Protozoa* that may be pathogenic to some insects [Mandakini and Manamgoda, 2021].
- *Microalgae*: *Chlorella* spp. and *Anabaena* spp.: Being investigated for their potential to produce bioactive compounds with pesticidal properties [Mandakini and Manamgoda, 2021].

17.10.2.2 Recent Developments and Perspectives

Recent developments in the field of biopesticides focus on several key areas:

- *Development of New Strains*: Searching for and isolating new, more specific, and effective strains of microorganisms from diverse environments, including the rhizosphere of pest-resistant plants [Tariq et al., 2025].
- *Formulation and Stability*: Improving biopesticide formulations to increase their field stability, resistance to environmental conditions (e.g., UV radiation and temperature), and extend their duration of action [Kremer, 2023].
- *Combinations and Synergy*: Research into the synergistic effects of using different biopesticides (e.g., bacteria and fungi) or their combination with other biocontrol strategies [Brohma et al., 2023].
- *Precision Application*: The development of precision agriculture technologies that enable targeted application of biopesticides, increasing their effectiveness and reducing costs.
- *RNAi-Based Biopesticides*: Novel approaches, such as biopesticides that utilize RNA interference (RNAi) to silence pest genes, open up new possibilities for precise and safe pathogen control Brohma et al., 2023].

Biopesticides are playing an increasingly key role in building integrated pest management (IPM) strategies, supporting ecologically sound and sustainable food production systems.

17.10.3 Bioherbicides: An Innovative Solution for Weed Control

Bioherbicides represent an ecological and sustainable alternative to traditional chemical herbicides. Unlike traditional biological control, bioherbicides utilize formulations of plant pathogens to produce large quantities of infectious material [Bo et al., 2020]. This material can be applied to plants as a liquid spray [Hasan et al., 2021] or as a solid granular material, which must then penetrate the plant's system to be effective.

17.10.3.1 Mechanism of Action of Bioherbicides

Upon entering the weed, the bioherbicide begins to produce various enzymes, such as amylase, cellulase, ligninase, and peptidase. Each of these enzymes contribute to the degradation of plant cell walls, lipid membranes, and proteins [Bordin et al., 2021]. This decomposition process allows the pathogens contained in the bioherbicide to spread further inside the weed, gaining easier access to all parts of it and leading to its death (Figure 17.5).

Bioherbicides are prepared from plant extracts, phytopathogenic microorganisms (especially fungi, hence often referred to as mycoherbicides), or bacterial phytotoxins [Singh & Pandey, 2020; Suteu et al., 2020; Roberts et al., 2022].

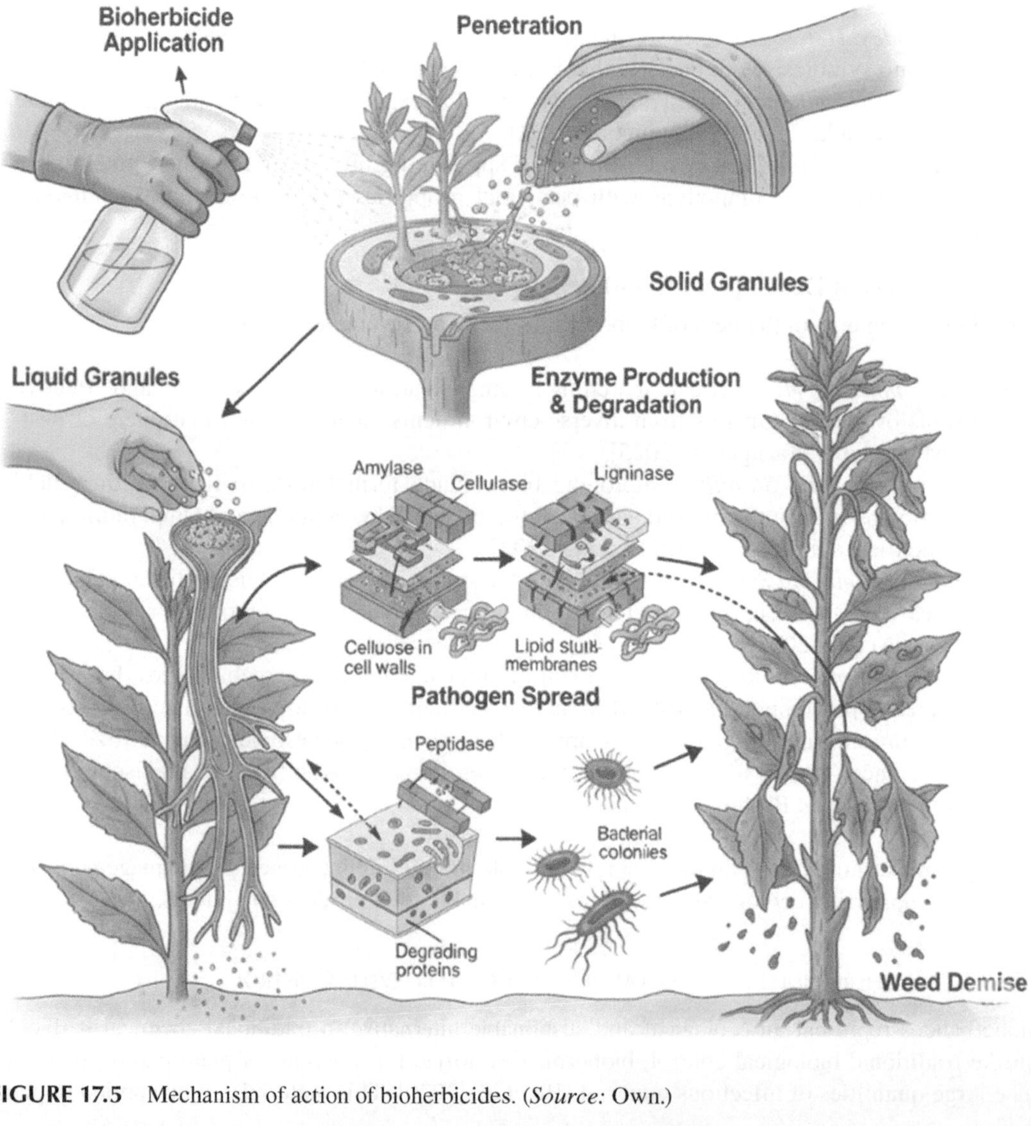

FIGURE 17.5 Mechanism of action of bioherbicides. (*Source:* Own.)

17.10.3.2 Advantages and Recent Developments in Bioherbicides

An advantage of using bioherbicides is their ability to persist in the environment long enough into the next growing season, which can contribute to long-term weed control. They are also less expensive than synthetic pesticides, which can significantly reduce agricultural costs, provided proper management and application efficiency [Mosttafiz et al., 2012]. Recent advances in bioherbicides focus on:

- *Identification of New, Highly Selective Pathogens*: Research is focused on discovering microorganisms that are specific to a given weed species, minimizing the risk of damage to crops and other desirable plants [Kumari et al., 2024].
- *Formulation Optimization*: Development of advanced formulations that increase the pathogen's viability in the field, improving its ability to infect and survive under variable environmental conditions (e.g., increased resistance to UV radiation and humidity) [Hewitt et al., 2023].
- *Use of Secondary Metabolites*: There is growing interest in isolation and use of secondary metabolites produced by microorganisms that exhibit herbicidal activity, eliminating the need to apply whole pathogens [Kumari et al., 2024].
- *Genetic Engineering of Pathogens*: Research on the genetic modification of pathogens to increase their virulence, specificity, or host range while ensuring environmental safety [Kremer, 2023].
- *Integration Into Weed Management Systems*: Development of tailored bioherbicide formulations and synergistic application strategies to ensure their efficacy within Integrated Weed Management (IWM) programs, achieving maximum effectiveness with minimal environmental impact [Gonzalez-Andujar, 2023].

As research and technology advance, bioherbicides have the potential to become a standard tool in modern, sustainable agriculture.

17.10.4 Microbial Insecticides: A Safe Alternative in Plant Protection

Microbial insecticides are a key element of sustainable plant protection, utilizing microorganisms or their byproducts to control pests. Their main advantage is low toxicity to non-target organisms, including animals and humans, making them a safer alternative to conventional chemical pesticides [Barbaś et al. 2023]. Microbial insecticides include a variety of living microorganisms, such as viruses (e.g., baculoviruses), bacteria (e.g., *Bacillus thuringiensis*), fungi (e.g., *Beauveria bassiana* and *Metarhizium anisopliae*), protozoa, nematodes(entomopathogenic, e.g., *Steinernema* spp. and *Heterorhabditis* spp.), and toxins produced by these organisms [Kesho, 2020; Tariq et al. 2025].

17.10.4.1 Characteristics and Advantages of Bioinsecticides

The main advantages of bioinsecticides include the following:

- *Specificity of Action*: They often target only a single species or a narrow group of insects, minimizing the impact on beneficial insects such as pollinators and natural enemies of pests [Barbaś et al. 2023].
- *Slow Acting*: Unlike chemical insecticides, bioinsecticides typically act more slowly, requiring precise timing of application to maximize effectiveness, often during the pest's sensitive phase [Barbaś et al. 2023].
- *No Residue*: They leave no harmful chemical residues in the environment or on agricultural products, which is important for food safety and compliance with ecological standards [Nassar et al. 2024].

- *Resistance Mitigation*: Thanks to diverse and complex mechanisms of action (e.g., infection, protein toxin production, and competition), bioinsecticides can help manage pest resistance to traditional chemical plant protection products [Nassar et al. 2024].

17.10.4.2 Recent Achievements and Perspectives

Currently, research and development in the field of microbial insecticides is focused on:

- *New Formulations*: Creating more stable and easier-to-apply formulations that increase the survival of microorganisms under harsh environmental conditions (e.g., UV radiation, temperature, and humidity) and prolong their activity on plants [Hewitt et al., 2023].
- *Broad Spectrum of Action and Synergy*: Searching for strains with a broader spectrum of action or creating mixtures of microorganisms (consortia) that mutually enhance each other's insecticidal activity. Research on the synergistic effects of bioinsecticides with low-dose chemical pesticides also holds promise for more effective pest control [Negi et al., 2024].
- *Utilization of bacterial and fungal toxins*: Development of technologies for the production and stabilization of purified toxins (e.g., Bt proteins) or other bioactive metabolites that can be applied directly, offering rapid action without the introduction of living organisms [Negi et al., 2024].
- *Integration with precision agriculture systems*: The use of drones, sensors, and artificial intelligence to precisely monitor pest populations and target bioinsecticide application, minimizing consumption and maximizing efficacy [Gonzalez-Andujar, 2023].
- *RNAi-based biopesticides*: Innovative approaches in which RNA interference (RNAi) is used to "silence" specific pest genes, leading to their death or inhibition. This technology, although in its initial stages of development, offers the potential for highly specific and safe solutions [Fenibo et al., 2022; Daraban et al., 2023]. The development and implementation of microbial insecticides are crucial to building resilient and sustainable agricultural systems that effectively protect crops while safeguarding human and environmental health.

17.10.5 Biofungicides: Natural Plant Protection against Pathogens

Biofungicides are a key element of biological control of plant pathogens. In contrast to traditional biological control, which aims to permanently establish a biological agent in the environment [Kenis et al., 2017], these agents focus on the introduction of selected microorganisms or their metabolites to directly combat diseases. Microbiological control agents, including fungi and bacteria isolated from the root zone (rhizosphere) and leaf surfaces (phyllospheric) of plants, play a key role in controlling plant pathogens [Thapa & Prasanna, 2018; Stone et al., 2018]. These microorganisms act in several ways: they prevent pathogen infection of the host plant and hinder pathogen establishment in the host.

17.10.5.1 Fungi as Biocontrol Agents

The use of fungal biocontrol agents has gained considerable popularity in recent years for pest and disease control, primarily due to their unique characteristics: high reproductive rate, dispersal efficiency, short generation time, and high host specificity [Gohel et al., 2022]. A study by Yao et al. [2020] analyzing approximately 300 fungal antagonists from 13 classes and 113 genera, along with their target pathogens and plant diseases, found that the genus *Trichoderma* possesses the greatest biocontrol potential. It includes 25 species used as biocontrol agents against a wide range of fungal plant diseases. The authors also identified nine other significant genera, each containing five or more known antagonist species: *Alternaria*, *Aspergillus*, *Candida*, *Fusarium*, *Penicillium*, *Pichia*, *Pythium*, *Talaromyces*, and *Verticillium*. Most plant growth-promoting fungi, such as *Trichoderma*, *Penicillium*, *Aspergillus*, and *Fusarium* spp., have the ability to stimulate plant immune responses to pathogen attack. They are considered one of the safest ways to trigger induced systemic resistance (ISR) and promote crop growth [Priyashantha et al., 2023].

17.10.5.2 Recent Achievements and Perspectives

Currently, biofungicide research focuses on:

- *Genome Sequencing and Genetic Modification*: Precise sequencing of antagonistic fungal genomes allows for the identification of genes responsible for the production of bioactive metabolites, paving the way for genetic engineering to enhance their efficacy and range of action [Basu et al., 2021, Kumari et al., 2024].
- *Optimization of Formulations and Application Techniques*: Development of nanotechnology and encapsulation of microorganisms to increase their stability in the field, extend their longevity, and improve their targeting efficiency. This also includes intelligent delivery systems that respond to specific environmental conditions [Fenibo et al., 2022].
- *Multi-Strain Consortia*: Creation of complex formulations containing multiple strains or species of fungi and bacteria that work synergistically, offering a broader spectrum of protection and increased plant resistance to various pathogens and stresses [Altieri et al., 2024].
- *Utilization of Secondary Metabolites*: Isolation and mass production of bioactive secondary metabolites (e.g., antibiotics and lytic enzymes) produced by fungi, which can be used as "coal-free" biofungicides, eliminating concerns about introducing living organisms [Altieri et al., 2024]. Thanks to these advances, biofungicides are becoming an increasingly reliable and effective tool in integrated plant disease management, supporting sustainable and ecological agricultural practices.

17.11 SUMMARY: THE SOIL MICROBIOME—THE KEY TO SUSTAINABLE AGRICULTURE

Soil microorganisms play an important and often underestimated role in shaping soil health, plant nutrition, and the overall productivity of agroecosystems. Their biodiversity is fundamental to maintaining soil fertility, and their biological activity is linked to its physicochemical properties.

1. *The Impact of Microbes on Soil Nutrition and Health*: Essential macro- and micronutrients for plant growth, such as nitrogen, phosphorus, potassium, calcium, magnesium, sulfur, manganese, iron, zinc, and boron, are mobilized and made available to plants through microbial activity. Nitrogen-fixing bacteria, mycorrhizal fungi, actinomycetes, cyanobacteria, rhizobacteria, algae, nematodes, and even soil viruses contribute to nutrient cycling, improving soil structure, and overall ecosystem resilience.
2. *Beneficial Microbes*: Plant allies. In the rhizosphere, billions of microorganisms interact with plants, increasing nutrient uptake [Sinha at al. 2025]. PGPR and AMF are commonly used as inoculants. Their action is based on mechanisms such as biofertilization (nitrogen fixation, phosphate solubilization, and growth hormone production), bioremediation (pollutant degradation), and biocontrol (pathogen inhibition through siderophore synthesis, ethylene regulation, and induction of systemic immunity). PGPB are divided into extracellular (e.g., *Bacillus* and *Pseudomonas*) and intracellular (e.g., *Rhizobia*) (Barbaś et al. 2023, Sawicka et al. 2025, Tariq et al. 2025).
3. *Factors Shaping Microbial Diversity*: Many environmental factors influence soil microbial diversity:
    - *Soil Moisture and Temperature*: Extreme weather conditions and climate change directly affect the activity and survival of microorganisms.

- *Soil pH*: It is crucial for nutrient availability and microbial taxonomic preferences.
- *Land Use and Vegetation*: Differences between agricultural and forest soils, or coniferous and deciduous forests, significantly shape microbial communities.
- *Organic Matter and Nutrient Content*: They influence the richness and activity of the microbiome, while excess nitrogen can paradoxically reduce diversity.
- *Crop Rotation*: Carefully planned crop rotation, especially with cover crops, increases microbial biomass, and improves soil quality.
- *UV Radiation, Soil Structure, Heavy Metals, and Interspecies Interactions*: Additional factors influencing microbiome dynamics [Barbaś et al. 2023].

4. *Conventional vs. Organic Agriculture*: Management systems have varying impacts on the microbiome:
   - Conventional agriculture, with its intensive use of agrochemicals, often leads to a reduction in microbial biodiversity [Thambugala et al. 2020].
   - Organic farming, through the use of manure, compost, crop residues, and cover crops, significantly improves microbial biomass and activity, promoting healthier soil-microbe-plant interactions, despite potentially lower yields.
   - *Microbiome in Agricultural Practice*: Biofertilizers, biopesticides, bioherbicides, and biofungicides. Agricultural microbiology provides innovative solutions for sustainable agriculture, such as:
     - Biofertilizers (nitrogen-fixing bacteria, mycorrhizal fungi, phosphate-solubilizing bacteria, and fungi) increase nutrient availability, reducing the need for synthetic fertilizers.
     - Biopesticides (microbial, biochemical, and plant-incorporated) offer low-toxicity pest control through complex mechanisms of action.
     - Bioherbicides utilize plant pathogens to selectively control weeds through tissue degradation, offering long-term control and cost savings.
     - Biofungicides (primarily *Trichoderma* fungi and bacteria) prevent plant diseases by stimulating plant immunity and inhibiting the development of pathogens [Aluvihara et al. 2023; Barbaś et al. 2023; Sawicka et al. 2025; Thambugala et al. 2020].

All these innovations, based on a deeper understanding of the soil microbiome, point to a future of agriculture where ecological approaches and natural processes replace invasive chemical interventions, leading to healthier soils, more resilient plants, and safer food.

17.12 CONCLUSIONS

Microorganisms are a pillar of soil fertility and plant health: Regardless of the cropping system, it is the diversity and biological activity of soil microorganisms that determine the availability of key nutrients, soil structure, and plant immunity. They are an irreplaceable element in biogeochemical cycling and are crucial for effective plant nutrition.

Sustainable agriculture is a necessity, and microorganisms are its tools: Faced with the challenges of climate change, biodiversity loss, and pressure on food security, a shift from intensive agrochemicals to microbial-based solutions is not only environmentally beneficial but also essential. Agricultural microbiology is becoming a key discipline transferring knowledge to achieve this goal.

Organic farming promotes microbial biodiversity: Organic farming practices, such as the use of manure, compost, cover crops, and diverse crop rotations, significantly enhance the biomass and diversity of soil microorganisms. While this may pose challenges for initial yields, the long-term benefits for soil health and agroecosystem resilience are unquestionable.

Microbial formulations offer comprehensive protection and support: Biological fertilizers, biopesticides, bioherbicides, and biofungicides are ecological tools that leverage the natural mechanisms of microorganisms to increase nutrient availability and control pests, weeds, and diseases.

Their specificity of action and low toxicity make them a safe and effective alternative to synthetic chemicals.

The future lies in research and innovation: The development of new strains, formulation optimization, the formation of microbial consortia, the use of secondary metabolites, and integration with precision agriculture and molecular genetics are directions that will shape the further development of agroecological farming systems. These advances will enable the maximization of the benefits of the soil microbiome for global food security and sustainable development. In summary, agroecological farming systems supported by microorganisms are not just a vision, but a real and increasingly accessible path to achieving agricultural productivity and sustainability while protecting our ecosystems.

REFERENCES

Akanmu, A.O., Olowe, O.M., Phiri, A.T., Nirere, D., Odebode, A.J., Karemera Umuhoza, N.J., Asemoloye, M.D., & Babalola, O.O. (2023). Bioresources in organic farming: Implications for sustainable agricultural systems. *Horticulturae*, 9(6), 659.

Al Abboud, M.A., Ghany, T.A., & Alawlaqi, M.M. (2014). Role of biofertilizers in agriculture: A brief review. *Mycopath*, 11(2), 95–101.

Ali, S., Riaz, A., Mamtaz, S., & Haider, H. (2023). Nutrients and crop production. *Current Research in Agriculture and Farming*, 4(2), 1–15.

Altieri, M.A. (2004). Linking ecologists and traditional farmers in the search for sustainable agriculture. *Frontiers in Ecology and the Environment*, 2, 35–42.

Altieri, M.A. et al. (2022). Agroecology as a pathway to food sovereignty and climate resilience. *Journal of Rural Studies*, 90, 10–18.

Altieri, M.A., Nicholls, C.I., Dinelli, G., & Negri, L. (2024). Towards an Agro ecological approach to crop health: Reducing pest incidence through synergies between plant diversity and soil microbial ecology. *NPJ Sustainable Agriculture,* 2, 6. https://doi.org/10.1038/s44264-024-00016-2

Aluvihara, S., Weerawardena, T.E., Gunaratne, H.D.A.K., Deniyapahala, D.V.R.I., Dissanayake, S.S., & Gunawardana, S.I. (2023). Advanced review on soil microbiology and fertility of soils. *Mechanization in Agriculture & Conserving of the Resources*, 67(1), 27–31, https://stumejournals.com/journals/am/2023/1/27.

Ansabayeva, A., Makhambetov, M., Rebouh, N. Y., Abdelkader, M., Saudy, H.S., Hassan, K.M., Nasser, M.A., Ali, M.A.A., & Ebrahim, M. (2025). Plant growth-promoting microbes for resilient farming systems: Mitigating environmental stressors and boosting crops productivity – a review. *Horticulturae*, 11(3), 260.

Badgley, C. et al. (2021). Agroecology and the future of food systems. *Renewable Agriculture and Food Systems*, 36(3), 227–238.

Baldrian P., Bell-Dereske L., Lepinay C., Větrovský T., & Kohout P. (2022). Fungal communities in soils under global change. *Studies in Mycology*, 103, 1–24. https://doi.org/10.3114/sim.2022.103.01

Barbaś P., Aslan H., Aslan I., Skiba D., Ademola Otekunrin O., & Sawicka B. (2023). Prospects for using pesticides in agriculture. *Agronomy Science*, 78(1), 97–120. https://doi.org/10.24326/as.2023.5078

Bastida, F., Eldridge, D.J., García, C., Kenny Png, G., Bardgett, R.D., & Delgado-Baquerizo, M. (2021). Soil microbial diversity–biomass relationships are driven by soil carbon content across global biomes. *The ISME Journal*, 15(7), 2081–2091.

Basu, S., Kumar, G., Chhabra, S., & Prasad, R. (2021). Role of soil microbes in biogeochemical cycle for enhancing soil fertility. In *New and Future Developments in Microbial Biotechnology and Bioengineering* (pp. 149–157). Elsevier. https://doi.org/10.1016/B978-0-444-64325-4.00013-4

Beauregard, M.S., Hamel, C., & St-Arnaud, M. (2010). Long-term phosphorus fertilization impacts soil fungal and bacterial diversity but not AM fungal community in alfalfa. *Microbial Ecology*, 59(2), 379–389.

Bebber, D.P., Holmes, T., & Gurr, S.J. (2014). The global spread of crop pests and pathogens. *Global Ecology and Biogeography*, 23, 1398–1407.

Bebber, D.P., & Richards, V.R. (2022). A meta-analysis of the effect of organic and mineral fertilizers on soil microbial diversity. *Applied Soil Ecology*, 175, 104450.

Bhat, T.A., Ahmad, L., Ganai, M.A., & Khan, O.A. (2015). Nitrogen fixing biofertilizers; mechanism and growth promotion: A review. *Journal of Pure and Applied Microbiology*, 9(2), 1675–1690.

Bier, R.L., Daniels, M., Oviedo-Vargas, D., Epoch, M., Price, J.R., Omondi, E., Smith, A., & Kan, J. (2024). Agricultural soil microbiomes differentiate in soil profiles with fertility source, tillage, and cover crops. *Agriculture, Ecosystems & Environment*, 368, 109002.

Billah, M., Khan, M., Bano, A., Hassan, T.U., Munir, A., & Gurmani, A.R. (2019). Phosphorus and phosphate solubilizing bacteria: Keys for sustainable agriculture. *Geomicrobiology Journal*, 36(10), 904–916.

Bo, A.B., Khaitan, B., Umurzokov, M., Cho, K.M., Park, K.W., & Choi, J.S. (2020). Biological control using plant pathogens in weed management. *Weed Turfgrass Science*, 9, 11–19.

Bonfante, P., & Anca, I.A. (2009). Plants, mycorrhizal fungi, and bacteria: A network of interactions. *Annual Review of Microbiology*, 63(1), 363–383.

Bordin, E.R., Frumi Camargo, A., Stefanski, F.S., Scaping, T., Bonatti, C., Zonian, J., Precheck, K., Moskovsky, T.A., Júnior, F.R., Mossi, A.J., Fongaro, G., Ramsdorf, W.A., Treichel, H. (2021). Current production of bioherbicides: Mechanisms of action and technical and scientific challenges to improve food and environmental security. *Biocatalysis and Biotransformation*, 39(5), 346–359.

Brohma, S., Guleria, A., Banerjee, S., Kalia, S., Sharma, R., Mondal, K., & Kaur, G. (2025). A review on new horizons in biopesticide development highlighting microbial advances for Climate-smart agriculture. *Journal of Applied and Natural Science*, 17(4), 1781–1798. https://doi.org/10.31018/jans.v17i4.6987

Brooker, R.W., George, T.S., Hamule, Z., Karley, A.J., Newton, A.C., Pakeman, R.J., & Schwob, C. (2021). Facilitation and biodiversity–ecosystem function relationships in crop production systems and their role in sustainable farming. *Journal of Ecology*, 109(5), 2054–2067. https://doi.org/10.1111/1365-2745.13592

Chenu, C., & Cosentino, D. (2011). Microbial regulation of soil structural dynamics. In K. Ritz & I. Young (Eds.), *The Architecture and Biology of Soils: Life in Inner Space* (pp. 37–70). CABI.

Chowdhury, F.T., Zaman, N.R., Islam, M.R., & Khan, H. (2020). Anti-fungal secondary metabolites and hydrolytic enzymes from Rhizospheric bacteria in crop protection: A review. *Journal of Bangladesh Academy of Sciences*, 44(2), 69–84.

Cotton, T.A. (2018). Arbuscular mycorrhizal fungal communities and global change: An uncertain future. *FEMS Microbiology Ecology*, 94(11), fiy179. https://doi.org/10.1093/femsec/fiy179

Cruz, M.G., Hernandez, E.A., & Uddameri, V. (2021). Vulnerability assessment of agricultural production systems to drought stresses using robustness measures. *Scientific Reports*, 11(1), 21648. https://doi.org/10.1038/s41598-021-98829-5

Cruz-Córdova A. et al. (2020). Molecular epidemiology, antibiotic resistance, and virulence traits of *Stenotrophomonas maltophilia* strains associated with an outbreak in a Mexican Tertiary Care Hospital. *Frontiers in Cellular and Infection Microbiology*, 10, 50. https://doi.org/10.3389/fcimb.2020.00050

D'Acunto, L., Andrade, J.F., Poggio, S.L., & Sem martin, M. (2018). Diversifying crop rotation increased metabolic soil diversity and activity of the microbial community. *Agriculture, Ecosystems & Environment*, 257, 159–164.

Daraban, G.M., Hlihor, R.M., & Suteu, D. (2023). Pesticides vs. biopesticides: From pest management to toxicity and impacts on the environment and human health. *Toxics*, 11(12), 983. https://doi.org/10.3390/toxics11120983

DeKosky, B.J, et al. (2013). High-throughput sequencing of the paired human immunoglobulin heavy and light chain repertoire. *Nature Biotechnology*, 31, 166–169. https://doi.org/10.1038/nbt.2492

de Sousa Ferreira, L., de Souza Oliveira, V., de Paula Marchiori, J.J., Ferreira, T.C., Bernabé, A.C.B., Boone, G.T.F., dos Santos Pereira, L.L., & Carriço, E. (2023). The nutrient magnesium in soil and plant: A review. *International Journal of Plant & Soil Science*, 35(8), 136–144. https://doi.org/10.9734/IJPSS/2023/v35i82890

Deveau, A., Bonito, G., Uehling, J., Paoletti, M., Becker, M., Bind Schedler, S., Hacquard, S., Hervé, V., Labbé, J., Lastovetsky, O.A., Mieszkin, S., Millet, L.J., Vajna, B., Junier, P., Bonfante, P., Krom, B.P., Olsson, S., van Elsas, J.D., Wick, L.Y. (2018). Bacterial–fungal interactions: Ecology, mechanisms, and challenges. *FEMS Microbiology Reviews*, 42(3), 335–352.

Dove, N.C., Torn, M.S., Hart, S.C., & Taş, N. (2021). Metabolic capabilities mute positive response to direct and indirect impacts of warming throughout the soil profile. *Nature Communications*, 12(1), 2089.

Drenovsky, R.E., Steenwerth, K.L., Jackson, L.E., & Scow, K.M. (2010). Land use and climatic factors structure regional patterns in soil microbial communities. *Global Ecology and Biogeography*, 19(1), 27–39.

Egamberdiyeva, D., & Höflich, G. (2003). Influence of growth-promoting bacteria on the growth of wheat in different soils and temperatures. *Soil Biology and Biochemistry*, 35(7), 973–978. https://doi.org/10.1016/S0038-0717(03)00158-5

Erickson III, D.J., Sulzberger, B., Zepp, R.G., & Austin, A.T. (2015). Effects of stratospheric ozone depletion, solar UV radiation, and climate change on biogeochemical cycling: Interactions and feedback. *Photochemical & Photobiological Sciences*, 14(1), 127–148.

Fahey, C., Koyama, A., Antunes, P.M., Dunfield, K., & Flory, S.L. (2020). Plant communities mediate the interactive effects of invasion and drought on soil microbial communities. *The International Society for Microbial Ecology Journal*, 14(6), 1396–1409. https://doi.org/10.1038/s41396-020-0614-6

FAO. (2018a). Food Agric. Organ. United Nations. 2018. The ten elements of agroecology: Guiding the transition to sustainable food and agricultural systems. *Rep., Food Agric. Organ. United Nations, Rome.* http://www.fao.org/3/i9037en/i9037en.pdf

FAO. (2018b). The State of Food and Agriculture 2018, Place of publication: Rome, Italy; pp. 200. https://openknowledge.fao.org/handle/20.500.14283/i9549en

FAO. (2025). Family Farming Knowledge Platform. National Agroecology Strategy for Food System Transformation 2024–2033. Food and Agriculture Organization of the United Nations. https://www.fao.org/family-farming/detail/en/c/1727593/

Fenibo, E.O., Ijoma, G.N., & Matambo, T. (2022). Biopesticides in sustainable agriculture: Current status and future prospects. In S. De Mandal, G. Ramkumar, S. Karthi, & F. Jin (Eds.), *New and Future Development in Biopesticide Research: Biotechnological Exploration* (pp. 1–53). https://doi.org/10.1007/978-981-16-3989-0_1

Fierer, N., & Jackson, R.B. (2006). The diversity and biogeography of soil bacterial communities. *Proceedings of the National Academy of Sciences of the United States of America*, 103(3), 626–631. https://doi.org/10.1073/pnas.0507535103

Finney, D.M., Buyer, J.S., & Kaye, J.P. (2017). Living cover crops have immediate impacts on soil microbial community structure and function. *Journal of Soil and Water Conservation*, 72(4), 361–373.

Fontaine, S., Abbadie, L., Aubert, M., Barot, S., Bloor, J. M., Derrien, D., Duchene, O., Gross, N., Henneron, L., Le Roux, X., Loeuille, N., Michel, J., Recous, S., Wipf, D., & Alvarez, G. (2024). Plant–soil synchrony in nutrient cycles: Learning from ecosystems to design sustainable agrosystems. *Global Change Biology*, 30(1), e17034. https://doi.org/10.1111/gcb.17034

Francis, C., Lieblein, G., Gliessman, S., Breland, T.A., Creamer, N., Harwood, R., Salomonsson, L., Helenius, J., Rickerl, D., Salvador, R., Wiedenhoeft, M., Simmons, S., Allen, P., Altieri, M., Flora, C., & Poincelot, R. (2003). Agroecology: The ecology of food systems. *Journal of Sustainable Agriculture*, 22(3), 99–118. https://doi.org/10.1300/J064v22n03_10

Furtak, K., & Wolińska, A. (2023). The impact of extreme weather events as a consequence of climate change on the soil moisture and on the quality of the soil environment and agriculture—a review. *Catena*, 231, 107378. https://doi.org/10.1016/j.catena.2023.107378

García, F.C., Bestion, E., Warfield, R., & Yvon-Durocher, G. (2018). Changes in temperature alter the relationship between biodiversity and ecosystem functioning. *Proceedings of the National Academy of Sciences*, 115(43), 10989–10994. https://doi.org/10.1073/pnas.1805518115

García-García, J.D., Joshi, J., Patterson, J.A., Trujillo-Rodriguez, L., Reisch, C.R., Javanpour, A.A., Liu, C.C., & Hanson, A.D. (2020). Potential for applying continuous directed evolution to plant enzymes: An exploratory study. *Life*, 10, 179. https://doi.org/10.3390/life10090179

Ghimire, R., Norton, J.B., Stahl, P.D., & Norton, U. (2014). Soil microbial substrate properties and microbial community responses under irrigated organic and reduced-tillage crop and forage production systems. *PLOS One*, *9*(8), e103901. https://doi.org/10.1371/journal.pone.0103901

Ghosh, S., Bheri, M., Bisht, D., & Pandey, G.K. (2022). Calcium signaling and transport machinery: Potential for development of stress tolerance in plants. *Current Plant Biology*, 29, 100235.

Giri, B., Giang, P.H., Kumari, R., Prasad, R., & Varma, A. (2005). Microbial diversity in soils. In F. Buscot & A. Varma (Eds.), *Microorganisms in Soils: Roles in Genesis and Functions* (pp. 19–58). Springer Berlin Heidelberg New York.

Glick, B.R. (2012). Plant growth-promoting bacteria: Mechanisms and applications. *Scientifica*, 2012(1), 963401.

Gliessman, S. (2013). Agroecology: Growing the roots of resistance. *Agroecology and Sustainable Food Systems*, 37, 19–31.

Gliessman, S.R. (2018). Defining agroecology. *Agroecology and Sustainable Food Systems*, 42, 599–600. https://doi.org/10.1080/21683565.2018.1432329

Gohel, N.M., Raghunandan, B.L., Patel, N.B., Parmar, H.V., & Raval, D.B. (2022). Role of fungal biocontrol agents for sustainable agriculture. In *Fungal Diversity, Ecology, and Control Management* (pp. 577–606). Springer Nature Singapore.

Gondal, A.H., Hussain, I., Ijaz, A.B., Zafar, A., Che, B.I., Zafar, H., Sohail, M.D., Niazi, H., Touseef, M., Khan, A.A., Tariq, M., Yousuf, H., & Usama, M. (2021). Influence of soil pH and microbes on mineral solubility and plant nutrition: A review. *International Journal of Agriculture and Biological Sciences*, 5(1), 71–81.

Gonzalez-Andujar, J.L. (2023). Integrated weed management: A shift towards more sustainable and holistic practices. *Agronomy 2023*, 13, 2646. https://doi.org/10.3390/agronomy13102646

Gougoulias, C., Clark, J.M., & Shaw, L.J. (2014). The role of soil microbes in the global carbon cycle: Tracking the below-ground microbial processing of plant-derived carbon for manipulating carbon dynamics in agricultural systems. *Journal of the Science of Food and Agriculture*, 94(12), 2362–2371.

Grzyb, A., Wolna-Maruwka, A., & Niewiadomska, A. (2020). Environmental factors affecting the mineralization of crop residues. *Agronomy*, 10(12), 1951. https://doi.org/10.3390/agronomy10121951

Grzyb, A., Wolna-Maruwka, A., & Niewiadomska, A. (2021). The significance of microbial transformation of nitrogen compounds in the light of integrated crop management. *Agronomy*, 11(7), 1415. https://doi.org/10.3390/agronomy11071415

Gulshan, T., Verma, A., Ayoub, L., Sharma, J., Sharma, T., Bhadu, A., & Singh, B. (2022). Increasing nutrient use efficiency in crops through biofertilizers. *The Pharma Innovation Journal*, 11(6), 2003–2010.

Gupta, R., Khan, F., Alqahtani, F.M., Hashem, M., & Ahmad, F. (2024). Plant growth–promoting *rhizobacteria* (PGPR) assisted bioremediation of heavy metal toxicity. *Applied Biochemistry and Biotechnology*, 196(5), 2928–2956.

Hamel, C., Gan, Y., Sokolski, S., & Bainard, L.D. (2018). High frequency cropping of pulses modifies soil nitrogen level and the rhizosphere bacterial microbiome in 4-year rotation systems of the semiarid prairie. *Applied Soil Ecology*, 126, 47–56.

Hamzah Saleem, M., Usman, K., Rizwan, M., Al Jabri, H., & Alsafran, M. (2022). Functions and strategies for enhancing zinc availability in plants for sustainable agriculture. *Frontiers in Plant Science*, 13, 1033092. https://doi.org/10.3389/fpls.2022.1033092

Hartmann, M., Frey, B., Mayer, J., Mäder, P., & Widmer, F. (2015). Distinct soil microbial diversity under long-term organic and conventional farming. *International Society for Microbial Ecology Journal*, 9(5), 1177–1194.

Hasan, M., Ahmad-Hamdani, M.S., Rosli, A.M., & Hamdan, H. (2021). Bioherbicides: An eco-friendly tool for sustainable weed management. *Plants*, 10(6), 1212. https://doi.org/10.3390/plants10061212

Hewitt A.J., Galea V.J., & O'Donnell Ch. (2023). Application technology for bioherbicides: Challenges and opportunities with dry inoculum and liquid spray formulations. *Pest Management Science*, 80(1), 72–80. https://doi.org/10.1002/ps.7823

Huang, I.S., & Zimba, P.V. (2019). Cyanobacterial bioactive metabolites – a review of their chemistry and biology. *Harmful Algae*, 86, 139–209.

Itelima, J.U., Bang, W.J., Onyimba, I.A., & Oj, E. (2018). A review: Biofertilizer; a key player in enhancing soil fertility and crop productivity. *Journal of Microbiology and Biotechnology Report*, 2(1), 22–28.

Jansson, J.K., & Hofmockel, K.S. (2020). Soil microbiomes and climate change. *Nature Reviews Microbiology*, 18, 35–46.

Jaswal, R.S., & Saroch, S. (2022). Importance of algae in agriculture. *New Era Agriculture Magazine*, 1(5), 38–41.

Johnson, R., Vishwakarma, K., Hossen, M.S., Kumar, V., Shackira, A.M., Puthur, J.T., Abdi, G., Sarraf, M., & Hasanuzzaman, M. (2022). Potassium in plants: Growth regulation, signaling, and environmental stress tolerance. *Plant Physiology and Biochemistry*, 172, 56–69. https://doi.org/10.1016/j.plaphy.2022.01.001

Kenis, M., Hurley, B.P., Hajek, A.E, & Cock, M.J.W. (2017). Classical biological control of insect pests of trees: Facts and figures. *Biological Invasions*, 19, 3401–3417.

Kesho, A. (2020). Microbial bio-pesticides and their use in integrated pest management. *Chemical and Biomolecular Engineering*, 5(1), 26–34.

Kimura, M., Jia, Z.J., Nakayama, N., & Asakawa, S. (2008). Ecology of viruses in soil: Past, present, and future perspectives. *Soil Science and Plant Nutrition*, 54(1), 1–32.

Kiontke, K., & Fitch, D.H. (2013). Nematodes. *Current Biology*, 23(19), 862–864.

Kong, A.Y., Scow, K.M., Córdova-Kreylos, A.L., Holmes, W.E., & Six, J. (2011). Microbial community composition and carbon cycling within soil microenvironments of conventional, low-input, and organic cropping systems. *Soil Biology and Biochemistry*, 43(1), 20–30.

Kremer R.J. (2023). Chapter 7. Bioherbicide development and commercialization: Challenges and benefits. In O. Koul (Ed.), *Development and Commercialization of Biopesticides* (pp. 119–148). Academic Press. https://doi.org/10.1016/B978-0-323-95290-3.00016-9

Krivan. V. Modele Prey–Predator. Redaktorzy: Sven Erik Jørgensen, Brian D. Fath. (2008). *Encyclopedia of Ecology*, Academic Press, 2929–2940, ISBN 9780080454054, https://doi.org/10.1016/B978-008045405-4.00676-5.

Kuht, J., Eremeev, V., Talgre, L., Loit, E., Mäeorg, E., Margus, K., Runno-Paurson, E., Madsen, H., & Luik, A. (2022). Soil microbial activity in different cropping systems under long-term crop rotation. *Agriculture*, 12, 532. https://doi.org/10.3390/agriculture12040532

Kumar, R., Kumawat, N., & Sahu, Y.K. (2017). Role of biofertilizers in agriculture. *Popular Kheti*, 5(4), 63–66.

Kumar, R. Sharma, P., Gupta, R.K., Kumar, S., Sharma, M.M.M., Singh, S., & Pradhan, G. (2020). Earthworms for eco-Friendly resource efficient agriculture. In S. Kumar, R.S. Meena, & M.K. Jhariya (Eds.), *Resources Use Efficiency in Agriculture*. Springer. https://doi.org/10.1007/978-981-15-6953-1_2

Kumari, P., Singh, V.K., & Verma, R. (2024). Biodegradation of heavy metals by microorganisms. *Futuristic Trends in Biotechnology*, 3(18), 2024:173–185. e-ISBN: 978-93-6252-296-2. https://www.doi.org/10.58532/V3BKBT18P2CH8

Lal, R. (2005). World crop residues production and implications of its use as a biofuel. *Environment International*, 31, 575–584.

Lauber, C.L., Hamady, M., Knight, R., & Fierer, N. (2009). Pyrosequencing-based assessment of soil pH as a predictor of soil bacterial community structure at the continental scale. *Applied and Environmental Microbiology*, 75, 5111–5120.

Li, P., Tedersoo, L., Crowther, T.W., Wang, B., Shi, Y., Kuang, L., Li, T., Wu, M., Liu, M., Luan, L., Lju, J., Li, D., Li, Y., Wang, S., Saleem, M., Dumbreel, A.J., Li, Z., & Jiang, J. (2023). Global diversity and biogeography of potential phytopathogenic fungi in a changing world. *Nature Communications*, 14(1), 6482. https://doi.org/10.1038/s41467-023-42142-4

Liu, X., Cao, A., Yan, D., Ouyang, C., Wang, Q., & Li, Y. (2019). Overview of mechanisms and uses of biopesticides. *International Journal of Pest Management*, 67(1), 65–72. https://doi.org/10.1080/09670874.2019.1664789

Mahanty, T., Bhattacharjee, S., Goswami, M., Bhattacharyya, P., Das, B., Ghosh, A., & Tribedi, P. (2017). Biofertilizers: A potential approach for sustainable agriculture development. *Environmental Science and Pollution Research*, 24, 3315–3335.

Mandakini, H.T., & Manamgoda, D.S. (2021). Microbial biopesticides: Development and application. In T. Bhatt, S. Gangola, D. Udayanga, & G. Kumar (Eds.), *Microbial Technology for Sustainable Environment* (pp. 167–189). https://doi.org/10.1007/978-981-16-3840-4_10

Maron, P.A., Sarr, A., Kaisermann, A., Lévêque, J., Mathieu, O., Guigue, J., Karimi, B., Bernard, L., Dequiedt, S., Terrat, S., Chabbi, A., & Ranjard, L. (2018). High microbial diversity promotes soil ecosystem functioning. *Applied and Environmental Microbiology*, 84(9), e02738-17.

Martínez-García, L.B., Korthals, G., Brussaard, L., Jørgensen, H.B., & De Deyn, G.B. (2018). Organic management and cover crop species steer soil microbial community structure and functionality along with soil organic matter properties. *Agriculture, Ecosystems & Environment*, 263, 7–17.

Massenssini, A.M., Bonduki, V.H.A., Melo, C.A.D., Tótola, M.R., Ferreira, F.A., & Costa, M.D. (2015). Relative importance of soil physico-chemical characteristics and plant species identity to the determination of soil microbial community structure. *Applied Soil Ecology*, 91, 8–15. http://dx.doi.org/10.1016/j.apsoil.2015.02.009

Masuda T., Sekine K., & Takashino N. (2025). Scaling up agroecology from policies to practices: Transforming our research and education systems. *Journal of Rural Problems*, 61(1):35–41. https://doi.org/10.7310/arfe.61.35

Mawcha, K.T., Malinga, L., Muir, D., Ge, J., & Ndolo, D. (2025). Recent advances in biopesticide research and development with a focus on microbials]. *F1000Research*, 13, 1071. https://doi.org/10.12688/f1000research.154392.5

Meemken, E.M., & Qaim, M. (2018). Organic agriculture, food security, and the environment. *Annual Review of Resource Economics*, 10(1), 39–63.

Miransari, M. (2011). Interactions between arbuscular mycorrhizal fungi and soil bacteria. *Applied Microbiology and Biotechnology*, 89, 917–930.

Mitra, D., Mondal, R., Khoshru, B., Senapati, A., Radha, T.K., Mahakur, B., Uniyal, N., Myo, M.N., Boutaj, H., Guerra Sierra, B.E., Panneerselvam, P., Ganeshamurthy, A.N., Anđelković, S., Vasić, T., Rani, A., Dutta, S., & Mohapatra, P.K.D. (2022). Actinobacteria-enhanced plant growth, nutrient acquisition, and crop protection: Advances in soil, plant, and microbial multifactorial interactions. *Pedosphere*, 32(1), 149–170.

Mosttafiz, S., Rahman, M., & Rahman, M. (2012). Biotechnology: Role of microbes in sustainable agriculture and environmental health. *The Internet Journal of Microbiology*, 10(1), 1–6.

Muhammad, I., Wang, J., Sainju, U.M., Zhang, S., Zhao, F., & Khan, A. (2021). Cover cropping enhances soil microbial biomass and affects microbial community structure: A meta-analysis. *Geoderma*, 381, 114696.

Nabi, M. (2023). Role of microorganisms in plant nutrition and soil health. In T. Aftab, & K.R. Hakeem (Eds.), *Sustainable Plant Nutrition*. Molecular Interventions and Advancements for Crop Improvement. https://doi.org/10.1016/B978-0-443-18675-2.00016-X

Nadeem, F., Hanif, M.A., Majeed, M.I., & Mushtaq, Z. (2018). Role of macronutrients and micronutrients in the growth and development of plants and prevention of deleterious plant diseases—a comprehensive review. *International Journal of Chemical and Biochemical Sciences*, 13, 31–52.

Nassar, A.M.K., Salim, Y.M., Nour-Eldeen, E., Younis, M.S., Kelany, M.M., Shebl, M. A., Shafey, A.S., & Abou-Shaara, H.F. (2024). Seasonal screening of pesticide residues in beehive productscollected from different districts in Egypt. *Environmental Monitoring and Assessment*, 196, 297. https://doi.org/10.1007/s10661-024-12451-2

Nath, M., Bhatt, D., Bhatt, M.D., Prasad, R., & Tuteja, N. (2018). Microbe-mediated enhancement of nitrogen and phosphorus content for crop improvement. In *Crop Improvement Through Microbial Biotechnology* (pp. 293–304). Elsevier. https://doi.org/10.1016/B978-0-444-63987-5.00014-1

Negi, R., Sharma, B., Jan, T. et al. (2024). Microbial consortia: A promising tool as plant bioinoculants for sustainable agricultural development. *Current Microbiology*, 81(2024), 222. https://doi.org/10.1007/s00284-024-03755-0

Nevison, C., Hess, P., Goodale, C., Zhu, Q., & Vira, J. (2022). Nitrification, denitrification, and competition for soil N: Evaluation of two earth system models against observations. *Ecological Applications*, 32(4), e2528. https://doi.org/10.1002/eap.2528

Norris, K. (2012). Biodiversity in the context of ecosystem services: The applied need for systems approaches. *Philosophical Transactions of the Royal Society B: Biological Sciences*, 367(1586), 191–199. https://doi.org/10.1098/rstb.2011.0176

Noumavo, P.A., Agbodjato, N.A., Baba-Moussa, F., Adjanohoun, A., & Baba-Moussa, L. (2016). Plant growth promoting rhizobacteria: Beneficial effects for healthy and sustainable agriculture. *African Journal of Biotechnology*, 15(27), 1452–1463. https://doi.org/10.5897/AJB2016.15397

Obasi, S.N., Peters, O.A., Tenebe, V.A., Obasi, C.C., & Omodara, M.A. (2024). Role of soil microbial diversity in mitigating soil-borne diseases and postharvest losses. *Nigeria Agricultural Journal*, 55(3), 297–319.

Ohkama-Ohtsu, N., & Wasaki, J. (2010). Recent progress in plant nutrition research: Crosstalk between nutrients, plant physiology and soil microorganisms. *Plant and Cell Physiology*, 51(8), 1255–1264.

Oirdi, M.E., Yaseen, M., Farwa, U., Raza, M.A., Farhan, M., Sandhu, Z.A., Ali, F., Aatif, M., Khan, H.S., & Nahvi, I. (2024). Crops and people: The dangers and potential benefits of pesticides. *Cogent Food & Agriculture*, 10(1), 2334096. https://doi.org/10.1080/23311932.2024.2334096

Olanrewaju, O.S., Ayangbenro, A.S., Glick, B.R., & Babalola, O.O. (2019). Plant health: Feedback effect of root exudates-rhizobiome interactions. *Applied Microbiology and Biotechnology*, 103, 1155–1166.

Olanrewaju, O.S., Glick, B.R., & Babalola, O.O. (2017). Mechanisms of action of plant growth promoting bacteria. *World Journal of Microbiology and Biotechnology*, 33, 1–16.

Ondrasek, G., Begić, H.B., Zovko, M., Filipović, L., Meriño-Gergichevich, C., Savić, R., & Rengel, Z. (2019). Biogeochemistry of soil organic matter in agroecosystems & environmental implications. *Science of the Total Environment*, 658, 1559–1573.

Oram, N.J., Brennan, F., Praeg, N., Bardgett, R.D., Illmer, P., Ingrisch, J., & Bahn, M. (2025). Plant community composition and traits modulate the impacts of drought intensity on soil microbial community composition and function. *Soil Biology and Biochemistry*, 200, 109644. https://doi.org/10.1016/j.soilbio.2024.109644

Ortiz, J., Soledad, V., Zamora, M., Boccolini, M., Lorenzon, C., Pegoraro, V., & Gabbarini, L. (2023). Crop diversity improves carbon, nitrogen, and soil biological functions in an agroecological system. *Ciencia del Suelo*, 41(2), 144–156.

Ouyang, Y., Reeve, J.R., & Norton, J.M. (2022). The quality of organic amendments affects soil microbiome and nitrogen-cycling bacteria in an organic farming system. *Frontiers in Soil Science*, 2, 869136. https://doi.org/10.3389/fsoil.2022.869136

Pang, D., Wang, G., Liu, Y., Cao, J., Wan, L., Wu, X., & Zhou, J. (2019). The impacts of vegetation types and soil properties on soil microbial activity and metabolic diversity in subtropical forests. *Forests*, 10(6), 497. https://doi.org/10.3390/f10060497

Paśmionka, I.B., Bulski, K., & Boligłowa, E. (2021). The participation of microbiota in the transformation of nitrogen compounds in the soil – a review. *Agronomy*, 11(5), 977.

Petrovich, Wu, C.Y., Rosenthal, A., Chen, K.-F., Packman, A.L., & Wells, G.F. (2017). Nitrosomonas europaea biofilm formation is enhanced by Pseudomonas aeruginosa. *FEMS Microbiology Ecology*, 93(5), fix047. https://doi.org/10.1093/ssssfemsec/fix047

Priyashantha, A.K.H., Karunarathna, S.C., Lu, L., & Tibpromma, S. (2023). Fungal *endophytes*: An alternative biocontrol agent against phytopathogenic fungi. *Encyclopedia*, 3, 759–780. https://doi.org/10.3390/encyclopedia3020055

Qiu, T., Shi, Y., Peñuelas, J., Liu, J., Cui, Q., Sardans, J., Zhou, F., Xia, L., Yan, W., Zhao, S., Peng, S., Jian, J., He, Q., Zhang, W., Huang, M., Tan, W., & Fang, L. (2024). Optimizing cover crop practices as a sustainable solution for global agroecosystem services. *Nature Communications*, 15(1), 1–14.

Raza, M.M., & Bebber, D.P. (2022). Climate change and plant pathogens. *Current Opinion in Microbiology*, 70, 102233. https://doi.org/10.1016/j.mib.2022.102233

Rehman, F., Kalsoom, M., Adnan, M, Toor, M.D., & Zulfiqar, A. (2020). Plant growth promoting rhizobacteria and their mechanisms involved in agricultural crop production: A review. *SunText Review of Biotechnology*, 1(2), 110. https://doi.org/10.51737/2766-5097.2020.010

Ren, N., Wang, Y., Ye, Y., Zhao, Y., Huang, Y., Fu, W., & Chu, X. (2020). Effects of continuous nitrogen fertilizer application on the diversity and composition of rhizosphere soil bacteria. *Frontiers in Microbiology*, 11, 1948. https://doi.org/10.3389/fmicb.2020.01948

Rezgui, C., Trinsoutrot-Gattin, I., Benoit, M., Laval, K., & Riah-Anglet, W. (2021). Linking changes in the soil microbial community to C and N dynamics during crop residue decomposition. *Journal of Integrative Agriculture*, 20(11), 3039–3059.

Richardson, A.E., Barea, J.M., McNeill, A.M., & Prigent-Combaret, C. (2009). Acquisition of phosphorus and nitrogen in the rhizosphere and plant growth promotion by microorganisms. *Plant Soil*, 321, 305–339. https://doi.org/10.1007/s11104-009-9895-2

Roberts, J., Florentine, S., Fernando, W.D., & Tennakoon, K.U. (2022). Achievements, developments, and future challenges in the field of bioherbicides for weed control: A global review. *Plants*, 11(17), 2242.

Rodrigo, D., Rosell, C.M., & Martinez, A. (2021). Risk of *Bacillus cereus* in relation to rice and derivatives. *Foods*, *10*, 302. https://doi.org/10.3390/foods10020302

Rodrigo, S., García-Latorre, C., & Santamaria, O. (2022). Metabolites produced by fungi against fungal phytopathogens: Review, implementation, and perspectives. *Plants*, 11(1), 81. https://doi.org/10.3390/plants11010081

Rousk, J., & Bengtson, P. (2014). Microbial regulation of global biogeochemical cycles. *Frontiers in Microbiology*, 5, 103. https://doi.org/10.3389/fmicb.2014.00103

Ruiu, L. (2018). Microbial biopesticides in agroecosystems. *Agronomy*, 8(11), 235. https://doi.org/10.3390/agronomy8110235

Saeed, Q., Xiukang, W., Haider, F.U., Kučerik, J., Mumtaz, M.Z., Holatko, J., Naseem, M., Kintl, A., Ejaz, M., Naveed, M., Brtnick, M., & Mustafa, A. (2021). Rhizosphere bacteria in plant growth promotion, biocontrol, and bioremediation of contaminated sites: A comprehensive review of effects and mechanisms. *International Journal of Molecular Sciences*, 22(19), 10529. https://doi.org/10.3390/ijms221910529

Sahu, P.K., Singh, D.P., Prabha, R., Meena, K., & Abhilash, P.C. (2019). Connecting microbial capabilities with the soil and plant health: Options for agricultural sustainability. *Ecological Indicators*, 105, 601–612.

Sakin, E., Yanardağ, H.İ., Firat, Z., Çelik, A., Beyyavaş, V., & Cun, S. (2024). Some indicators for the assessment of soil health: A mini review. *MAS Journal of Applied Sciences*, 9(2), 297–310. https://doi.org/10.5281/zenodo.11665000

Sawicka, B. (2019). Resilient agricultural practices. In W. Leal Filho, A. Azul, L. Brandli, P. Özuyar, & T. Wall (Eds.), *Zero Hunger. Encyclopedia of the UN Sustainable Development Goals*. Springer. https://doi.org/10.1007/978-3-319-69626-3_42-1

Sawicka, B., Barbaś, P., Vambol, V., Skiba, D., Pszczółkowski, P., Niazi, P., & Bienia, B. (2025) Applied microbiology for sustainable agricultural development. *Applied Microbiology*, 5, 78. https://doi.org/10.3390/applmicrobiol5030078

Schimel, J.P. (2018). Life in dry soil: Effects of drought on soil microbial communities and processes. *Annual Review of Ecology, Evolution, and Systematics*, 49(1), 409–432.

Shah, K.K., Tripathi, S., Tiwari, I., Shrestha, J., Modi, B., Paudel, N., & Das, B.D.(2021). Role of soil microbes in sustainable crop production and soil health: A review. *Agricultural Science and Technology*, 13(2), 109–118.

Shahbaz, M., Kuzyakov, Y., Sanaullah, M., Heitkamp, F., Zelenev, V., Kumar, A., & Blagodatskaya, E. (2017). Microbial decomposition of soil organic matter is mediated by quality and quantity of crop residues: Mechanisms and thresholds. *Biology and Fertility of Soils*, 53, 287–301.

Sheik, C.S., Beasley, W.H., Elshahed, M.S., Zhou, X., Luo, Y., & Krumholz, L.R. (2011). Effect of warming and drought on grassland microbial communities. *The International Society for Microbial Ecology Journal*, 5(10), 1692–1700.

Siebielec, S., Siebielec, G., Klimkowicz-Pawlas, A., Gałazka, A., Grzadziel, J., & Stuczynski, T. (2020). Impact of water stress on microbial community and activity in sandy and loamy soils. *Agronomy*, 10, 1429.

Singh, A.K., & Pandey, A.K. (2020). Exploitation of fungal phytotoxin as natural herbicide for ecofriendly weed management: Problems and prospects. *International Journal of Plant and Environment*, 6(03), 211–214.

Stefan, L., Hartmann, M., Engbersen, N., Six, J., & Schöb, C. (2021). Positive effects of crop diversity on productivity driven by changes in soil microbial composition. *Frontiers in Microbiology*, 12, 660749.

Stone, B.W.G., Weingarten, E.A., & Jackson, C.R. (2018). The role of the phyllosphere microbiome in plant health and function. *Annual Plant Reviews*, 1(2). https://doi.org/10.1002/9781119312994.apr0614

Strickland, M.S., Thomason, W.E., Avera, B., Franklin, J., Minick, K., Yamada, S., & Badgley, B.D. (2019). Short-term effects of cover crops on soil microbial characteristics and biogeochemical processes across actively managed farms. *Agrosystems, Geosciences & Environment*, 2(1), 1–9.

Suteu, D., Rusu, L., Zaharia, C., Badeanu, M., & Daraban, G.M. (2020). Challenge of utilization of vegetal extracts as natural plant protection products. *Applied Sciences*, 10(24), 8913. https://doi.org/10.3390/app10248913

Swami, S., Lallawmkimi, M.C., Singh, V., Yadav, K.K., Alagendran, S., Koushal, S., & Lairenjam, G. (2025). Harnessing edaphic microalgae for soil carbon sequestration and climate-smart agriculture. *Archives of Current Research International*, 25(3), 155–171.

Tahat, M.M., Alamance, K.M., Othman, Y.A., & Leskovar, D.I. (2020). Soil health and sustainable agriculture. *Sustainability*, 12, 4859. https://doi.org/10.3390/su12124859

Tardy, V., Spor, A., Mathieu, O., Lévèque, J., Terrat, S., Plassart, P., Regnier, T., Bardgett, R.D., van der Putten, W.H., Paolo, R.P., Giovanna, S., Simonetta, B., Lemanceau, P., Ranjard, P., & Maron, P.A. (2015). Shifts in microbial diversity through land use intensity as drivers of carbon mineralization in soil. *Soil Biology and Biochemistry*, 90, 204–213.

Tariq H., Subramanian S., Geitmann A., Smith D.L. (2025). *Bacillus* and *Paenibacillus* as plant growth-promoting bacteria in soybean and hemp. *Frontiers in Plant Science*, 16, 1–14. https://doi.org/10.3389/fpls.2025.1529859

Terán-Samaniego,. K., Robles-Parra, J.M., Vargas-Arispuro, I., Martínez-Téllez, M.Á., Garza-Lagler, M.C., Félix-Gurrlola, D., Maycotte-de la Peña, M.L., Tafolla-Arellano, J.C., García-Figueroa, J.A., & Espinoza-López, P.C. (2025). Agroecology and sustainable agriculture: Conceptual challenges and opportunities – a systematic literature review. *Sustainability*, 17(5), 1805. https://doi.org/10.3390/su17051805

Thambugala, K.M., Daranagama, D.A., Phillips, A.J.L., Kannangara, S.D., & Promputtha, I. (2020). Fungi vs. fungi in biocontrol: An overview of fungal antagonists applied against fungal plant pathogens. *Frontiers in Cellular and Infection Microbiology*, 10, 604923. https://doi.org/10.3389/fcimb.2020.604923

Thapa, S., & Prasanna, R. (2018). Prospecting the characteristics and significance of the phyllospheric microbiome. *Annals of Microbiology*, 68, 229–245. https://doi.org/10.1007/s13213-018-1331-5

Torres-Rodriguez, J.A., Reyes-Pérez, J.J., Quiñones-Aguilar, E.E., & Hernandez-Montiel, L.G. (2022). *Actinomycete* potential as biocontrol agent of phytopathogenic fungi: Mechanisms, source, and applications. *Plants*, 11(23), 3201. https://doi.org/10.3390/plants11233201

UN Sustainable Development Report. (2024). OdpowiedzialnyBiznes.pl. https://odpowiedzialnybiznes.pl/publikacje/sustainable-development-report-2024/

Urruty, N., Tailliez-Lefebvre, D., & Huyghe, C. (2016). Stability, robustness, vulnerability, and resilience of agricultural systems. A review. *Agronomy for Sustainable Development*, 36, 1–15.

Vandermeer, J., & Perfecto, I. (2013). Complex traditions: Intersecting theoretical frameworks in agroecological research. *Agroecology and Sustainable Food Systems*, 37, 76–89.

Venter, Z.S., Jacobs, K., & Hawkins, H.J. (2016). The impact of crop rotation on soil microbial diversity: A meta-analysis. *Pedobiologia*, 59(4), 215–223.

Vikas, & Ranjan, R. (2024). Agroecological approaches to sustainable development. *Frontiers in Sustainable Food Systems*, 8, 1405409. https://doi.org/10.3389/fsufs.2024.1405409

Waldon, H.B., Jenkins, M.B., Virginia, R.A., & Harding, E.E. (1989). Characteristics of woodland rhizobial populations from surface-and deep-soil environments of the Sonoran Desert. *Applied and Environmental Microbiology*, 55(12), 3058–3064.

Wan, N.F., Dainese, M., Wang, Y.Q., & Loreau, M. (2024). Cascading social-ecological benefits of biodiversity for agriculture. *Current Biology*, 34(12), 587–603.

Wang, X., Wu, W., Ao, G., Han, M., Liu, M., Yin, R., Feng, J., & Zhu, B. (2025). Minor effects of warming on soil microbial diversity, richness, and community structure. *Global Change Biology*, 31(3), e70104. https://doi.org/10.1111/gcb.70104

Wang, Z., Ding, Y., Jin, K., Struik, P.C., Sun, S., Ji, B., Zhang, Y., & Li, X. (2022). Soil bacterial and fungal communities are linked with plant functional types and soil properties under different grazing intensities. *European Journal of Soil Science*, 73(1), e13195. https://doi.org/10.1111/ejss.13195

Wezel, A., Herren, B.G., Kerr, R.B., Barrios, E., Gonçalves, A.L.R., & Sinclair, F. (2020). Agroecological principles and elements and their implications for transitioning to sustainable food systems. A review. *Agronomy for Sustainable Development*, 40, 1–13. https://doi.org/10.1007/s13593-020-00646-z

Yadav, A.N., Kumar, R., Kumar, S., Kumar, V., Sugitha, T., Singh, B., Singh Chauahan, B., Singh Dhaliwal, H., & Saxena, A.K. (2017). Beneficial microbiomes: Biodiversity and potential biotechnological applications for sustainable agriculture and human health. *Journal of Applied Biology and Biotechnology*, 5(6), 45–57.

Yao, X., Guo, H., Zhang, K., Zhao, M, Zhao, J., Ruan, R.J., & Chen, J. (2023). *Trichoderma* and its role in biological control of plant fungal and nematode disease. *Frontiers in Microbiology*, 14, 1–15. https://doi.org/10.3389/fmicb.2023.1160551

Rondón, J.J., Ball, M.M., Castro, L.T., & Yarzábal, L.A. (2019). Eurypsychrophilic *Pseudomonas* spp. isolated from Venezuelan tropical glaciers as promoters of wheat growth and biocontrol agents of plant pathogens at low temperatures. *Environmental Sustainability*, 2(33), 3. https://doi.org/10.1007/s42398-019-00072-2

Yue, H., Yue, W., Jiao, S., Kim, H., Lee, Y.H., Wei, G., Song, W., & Shu, D. (2023). Plant domestication shapes rhizosphere microbiome assembly and metabolic functions. *Microbiome*, 11(1), 70. https://doi.org/10.1186/s40168-023-01513-1

Zaib, M. (2024). Boron nutrient for sustainability of plant growth and soil health: A review with future prospects. *International Journal of Contemporary Issues in Social Sciences*, 3(1), 912–931.

Zaman, W., Ayaz, A., & Puppe, D. (2025). Biogeochemical cycles in plant–soil systems: Significance for agriculture, interconnections, and anthropogenic disruptions. *Biology*, 14, 433. https://doi.org/10.3390/biology14040433

Zenda, T., Liu, S., Dong, A., & Duan, H. (2021). Revisiting sulphur—the once neglected nutrient: It is roles in plant growth, metabolism, stress tolerance and crop production. *Agriculture*, 11(7), 626. https://doi.org/10.3390/agriculture11070626

Zhang, B., Li, Y., Ren, T., Tian, Z., Wang, G., He, X., & Tian, C. (2014). Short-term effect of tillage and crop rotation on microbial community structure and enzyme activities of a clay loam soil. *Biology and Fertility of Soils*, 50, 1077–1085.

Zhang, X., Ward, B.B., & Sigman, D.M. (2020). Global nitrogen cycle: Critical enzymes, organisms, and processes for nitrogen budgets and dynamics. *Chemical Reviews*, 120(12), 5308–5351.

Zhao, G., Zhu, X., Zheng, G., Meng, G., Dong, Z., Baek, J., Jeon, Ch.O., Yao, Y., Xuan, Y.H., Zhang, J., & Jia, B. (2024). Development of biofertilizers for sustainable agriculture over four decades (1980–2022). *Geography and Sustainability*, 5(1), 19–28.. https://doi.org/10.1016/j.geosus.2023.09.006

Zhou, X., Tahvanainen, T., Malard, L., Chen, L., Pérez-Pérez, J., & Berninger, F. (2024). Global analysis of soil bacterial genera and diversity in response to pH. *Soil Biology and Biochemistry*, 198, 109552. https://doi.org/10.1016/j.soilbio.2024.109552

18 Bioremediation of Soil Damaged by Human Activity and Climate Change

Stella Suanu, Leh-Togi Zobeashia, and Itoroabasi Joseph Okon

18.1 INTRODUCTION

Soil damage poses critical environmental challenge that threatens global ecosystems, agriculture, and human health through anthropogenic activities exacerbated by climate change (Lal, 2020; IPCC, 2022). Industrialization, intensive farming, improper waste disposal, and urban expansion introduce heavy metals, hydrocarbons, pesticides, and excess nutrients into soils, causing contamination and functional loss (Robertson & Vitousek, 2009; Ceci, 2021). These contaminants degrade soil fertility, disrupt microbial communities, and create risks of bioaccumulation, groundwater contamination, and greenhouse gas emissions (Ahmed et al., 2023d; Kamyab et al., 2024). Climate change accelerates this stressors such as soil erosion, salinization, and organic matter depletion through increased temperature, altered precipitation patterns, and extreme weather events, compromising natural resilience and self-purification capacity (Alkorta et al., 2017; Win et al., 2021).

Traditional remediation techniques, including excavation, soil washing, and chemical stabilization, present substantial limitations in addressing widespread contamination due to how costly, environmentally disruptive, and unsustainable for large or heterogeneous contaminated sites (Alkorta et al., 2017; Ceci, 2021). The inherent unsustainability of these approaches necessitates alternative strategies that align with ecological principles and long-term environmental stewardship. Bioremediation using microorganisms or plants to detoxify and restore polluted soils offers cost-effective, sustainable, and eco-friendly alternatives (Alkorta et al., 2017; Ceci, 2021). Microbial bioremediation utilizes bacterial, fungal, and archaeal metabolic capacities to degrade organic pollutants or immobilize toxic metals through biodegradation, bioaugmentation, and biostimulation (Kamyab et al., 2024; Sahu & Arya, 2024). Phytoremediation involves plant-rhizosphere microbe interactions that enhance pollutant removal while improving soil structure and fertility (Ceci, 2021; Ahmed et al., 2023b). The efficacy of bioremediation depends on environmental parameters such as soil moisture, temperature, pH, oxygen availability, and nutrient status factors altered by climate change, complicating treatment prediction and optimization (Alkorta et al., 2017; Ceci, 2021). Elevated atmospheric CO_2 concentrations and shifting rainfall regimes modify plant physiology and rhizosphere processes central to pollutant degradation (Alkorta et al., 2017). Contaminant bioavailability, microbial community interactions, and pollutant chemical properties determine bioremediation kinetics and pathways (Alkorta et al., 2017; Ceci, 2021). Technologies employing genetically engineered microbes and novel biostimulants enhance microbial degradation potentials, while integrated approaches combining chemical pretreatment and bioremediation address recalcitrant pollutants (Alkorta et al., 2017; Ceci, 2021).

Beyond contaminant removal, bioremediation contributes to broader environmental restoration by rehabilitating essential soil functions including nutrient cycling and carbon sequestration. These processes support climate change mitigation through reduced greenhouse gas emissions, particularly methane and nitrous oxide, achieved by optimizing nitrogen cycling and organic matter decomposition (Smith et al., 2021; Ahmed et al., 2023e). Despite successful global applications across agricultural, mining, and urban contexts, bioremediation faces operational challenges

DOI: 10.1201/9781003650881-18

in scaling, monitoring, and adaptation across diverse climatic zones and soil types. Addressing these limitations requires enhanced understanding of microbial ecology and the development of robust plant-microbe partnerships capable of functioning under changing environmental conditions (Alkorta et al., 2017; Ceci et al., 2021).The review synthesizes current knowledge on bioremediation of soils impaired by human-induced contamination and climate change. It examines mechanisms of microbial and plant-assisted bioremediation, environmental and operational constraints, and technological advancements for improving efficacy under changing climatic conditions. Bioremediation provides a nature-based, sustainable pathway for rehabilitating polluted soils to sustain ecosystem services, agricultural productivity, and climate resilience.

18.2 TYPES AND CAUSES OF SOIL DAMAGE

18.2.1 Human-Induced Soil Damage

18.2.1.1 Industrial Waste and Heavy Metal Contamination

Industrial activities constitute a primary source of soil contamination through the release of persistent heavy metals and toxic compounds. Industrial waste significantly impacts water and soil quality, restricting their suitability for agricultural and domestic use, with metal smelting, chemical manufacturing, and mining operations releasing substantial quantities of lead, cadmium, chromium, and mercury into terrestrial environments (Ali et al., 2013; Sharma et al., 2023a). Toxic metal pollution is ubiquitous in soils, yet its worldwide distribution is unknown, with recent global assessments revealing widespread contamination across agricultural landscapes. The persistence of heavy metals in soil systems creates cascading ecological effects, including altered soil physicochemical properties, disrupted nutrient cycling, and compromised microbial diversity (Zhou et al., 2020; Kumar et al., 2024a). These contaminants bioaccumulate through food chains, posing significant risks to human health through carcinogenic, mutagenic, and neurological effects (Singh et al., 2024f).

18.2.1.2 Agricultural Chemical Inputs

Modern agricultural intensification relies heavily on synthetic pesticides, herbicides, and fertilizers, creating pervasive soil contamination. Pesticide and herbicide applications, while controlling target pests and weeds, generate persistent residues that disrupt beneficial soil organisms and microbial communities (García-Sánchez et al., 2019; Chen et al., 2023a). This disruption leads to reduced soil biological activity, pesticide resistance development, and ecological imbalances that compromise long-term soil health (Zhang et al., 2020). Excessive synthetic fertilizer application, particularly nitrogen, phosphorus, and potassium compounds, results in soil acidification, nutrient imbalances, and reduced microbial diversity (Guo et al., 2010; Rahman et al., 2024). Continuous fertilizer use without organic amendments decreases soil organic matter content, increases compaction, and reduces water-holding capacity, ultimately diminishing soil productivity (Liu et al., 2023). The environmental consequences extend beyond soil degradation through agrochemical runoff, which causes eutrophication of water bodies and contamination of groundwater resources. Residual chemicals in soil enter food chains, creating endocrine disruption and carcinogenic risks for human populations (Sharma & Gupta, 2024).

18.2.1.3 Deforestation and Land Use Change

Deforestation for agricultural expansion, logging, and urbanization represents a fundamental driver of soil degradation globally. Forest removal eliminates protective canopy cover, exposing soils to increased erosion, reduced fertility, and impaired hydrological functions (Pendleton et al., 2024). Land-use conversion from forests to croplands and urban areas results in soil tillage, compaction, and biodiversity loss, significantly reducing soil organic carbon stocks. Urbanization creates additional soil degradation through soil sealing, which permanently eliminates soil functions including water filtration, nutrient cycling, and carbon sequestration (Montanarella, 2015; Thompson et al., 2024b). These processes reduce ecosystem carbon storage capacity, contributing to climate change while diminishing water quality and aquatic ecosystem health (Don et al., 2011; Pimentel & Burgess, 2013).

18.2.1.4 Mining and Urban Development

Mining operations generate severe soil degradation through physical destruction, chemical contamination, and acid mine drainage (AMD). Mineral extraction activities cause extensive soil erosion, heavy metal contamination, and acidification that depletes essential nutrients and eliminates soil biological activity (Rodriguez et al., 2024a). Urban development alters soil landscapes through compaction, sealing, and pollution introduction. Soil compaction reduces water infiltration and root penetration, limiting plant growth and contributing to urban heat island effects (Montanarella, 2015; Dempsey et al., 2016). Urban pollutants accumulate in soils, creating toxic environments that extend impacts beyond immediate development areas (Cao et al., 2015; Martinez et al., 2024a).

18.2.2 Climate-Induced Soil Damage

18.2.2.1 Soil Salinization and Desertification

Climate change intensifies soil salinization through altered precipitation patterns, increased evapotranspiration, and rising sea levels. Salinization involves the accumulation of soluble salts in soil profiles, creating osmotic stress, nutrient imbalances, and structural degradation that severely limit plant growth (Gao et al., 2020; Hassan et al., 2024). Irrigation practices in poorly drained areas exacerbate salinization, particularly when utilizing groundwater from saline aquifers (Rengasamy, 2010). Salinization consequences include reduced water availability for plants, ionic toxicity, and soil structure degradation that increases erosion susceptibility (Hossain et al., 2017; Hanson et al., 2018). These factors collectively threaten global food security and ecosystem stability as agricultural demand continues rising (Qadir et al., 2014; Ahmad et al., 2024).

Desertification represents the most extreme form of land degradation, converting productive land into desert through climate change, overgrazing, deforestation, and improper irrigation practices. Future climate change is projected to increase the potential for water-driven soil erosion in many dryland areas, with rising temperatures and altered precipitation patterns accelerating desertification processes (Reynolds et al., 2007; IPCC, 2023). Desertification results in complete loss of soil fertility, increased erosion, and biodiversity collapse, severely impacting agricultural productivity and ecosystem services. Human populations face food insecurity, economic hardship, and forced migration as productive lands become uninhabitable (Kumar et al., 2024a).

18.2.2.2 Erosion from Extreme Weather Events

Climate change intensifies soil erosion through increased frequency and severity of extreme weather events. Soil loss by water erosion could grow by up to 66% over the period 2015–2070, with heavy rainfall events causing rapid surface runoff that strips fertile topsoil through sheet and gully erosion (Eekhout & Vente, 2022). Coastal regions experience accelerated erosion from hurricanes and storm surges, altering soil composition and threatening agricultural systems and human settlements (Peterson et al., 2024). Conversely, drought conditions reduce soil moisture, increasing wind erosion susceptibility in over-cultivated areas (Martínez et al., 2014). Erosion consequences include fertile topsoil loss that reduces agricultural productivity and threatens food security, sedimentation of water bodies that impairs aquatic habitats and water quality, increased flood risks from reduced land absorption capacity, and ecosystem degradation that diminishes biodiversity and ecosystem services (Baker et al., 2013; Johnson et al., 2024).

18.2.2.3 Organic Matter Depletion

Climate change accelerates soil organic matter decomposition through increased temperatures, altered precipitation patterns, and extreme weather events. Organic matter depletion reduces soil fertility, water-holding capacity, and structural stability, compromising agricultural productivity and ecosystem functions (Lal, 2015; Smith et al., 2024). Healthy soils provide essential ecosystem services including nutrient cycling, carbon sequestration, and climate regulation, which are fundamentally compromised by organic matter loss (FAO, 2017; Bünemann et al., 2018). Agricultural productivity directly depends on soil organic matter content, as fertile soils support optimal crop growth and yield (Lal, 2015; Thompson et al., 2024a).

18.3 PRINCIPLES, APPLICATIONS, AND CONTROLLING FACTORS OF BIOREMEDIATION

Bioremediation has emerged as a promising technology that harnesses the natural metabolic capabilities of microorganisms and plants to restore contaminated environments. This biological approach offers advantages over conventional physicochemical methods, including lower costs, minimal environmental disruption, and sustainable treatment outcomes (Pieper et al., 2020).

18.3.1 Fundamental Mechanisms of Bioremediation

Bioremediation operates through diverse biological mechanisms that collectively facilitate pollutant removal and detoxification. Microbial metabolism represents the primary pathway, where microorganisms utilize contaminants as carbon and energy sources, converting them into harmless end products such as carbon dioxide, water, and biomass (Zhang et al., 2022). This process relies on enzymatic biodegradation, where specific microbial enzymes cleave complex pollutant molecules into simpler compounds.

Bioaugmentation enhances remediation by introducing specialized microbial strains with enhanced degradative capabilities, particularly effective when indigenous microbial communities lack sufficient catabolic diversity (Kumar et al., 2023a). Complementarily, biostimulation optimizes existing microbial activity through nutrient supplementation, including nitrogen, phosphorus, and electron acceptors, thereby accelerating pollutant degradation rates (Liu et al., 2021). While phytoremediation employs plants' natural ability to absorb, accumulate, or transform contaminants through root uptake and translocation, proving particularly effective for heavy metal and organic pollutant removal (Sharma et al., 2023c). Cometabolism represents another crucial mechanism where microorganisms degrade pollutants incidentally while metabolizing primary substrates, enabling the breakdown of recalcitrant compounds like chlorinated hydrocarbons (Wang et al., 2022).

18.3.2 Microbial Roles in Contaminant Processing

Microorganisms serve as the primary biological agents in environmental remediation through three distinct processes such as biodegradation, transformation, and detoxification. Biodegradation involves complete mineralization of organic pollutants into non-toxic end products. Bacterial genera including *Pseudomonas*, *Bacillus*, and *Mycobacterium* possess extensive enzymatic systems enabling hydrocarbon, pesticide, and xenobiotic compound utilization (Das & Kumar, 2021). Aerobic bacteria effectively degrade polycyclic aromatic hydrocarbons (PAHs), while anaerobic bacteria specialize in chlorinated solvent degradation. Microbial transformation converts toxic compounds into less harmful or more biodegradable forms through oxidation, reduction, methylation, or dehalogenation processes. Under anaerobic conditions, *Dehalococcoides* species reductively dechlorinate tetrachloroethylene (PCE) to ethene, demonstrating effective detoxification (Chen et al., 2023b). These biotransformations may not achieve complete mineralization but significantly reduce pollutant mobility and toxicity. While detoxification include microbial neutralization of toxic substances through various mechanisms. Biosurfactant production increases hydrophobic pollutant bioavailability, enhancing subsequent degradation and reducing toxicity (Ahmed et al., 2022). Additionally, certain bacteria and fungi immobilize heavy metals through biosorption, methylation, or precipitation, decreasing their bioavailability and environmental impact (Singh et al., 2023e).

18.3.3 Bioremediation Applications: *In Situ* versus *Ex Situ* Approaches

Bioremediation strategies are categorized into *in situ* and *ex situ* techniques, each offering distinct advantages depending on site-specific conditions and remediation objectives.

18.3.3.1 *In Situ* Bioremediation

In situ approaches treat contaminated materials directly at the contamination site without excavation, minimizing environmental disruption and reducing costs. These methods stimulate indigenous or introduced microorganisms through nutrient addition, oxygenation, or pH adjustment. Bioventing enhances aerobic microbial activity by supplying air or oxygen to subsurface zones, while biosparging injects air into saturated zones to promote aerobic degradation in groundwater systems (Rodriguez et al., 2023). Natural attenuation relies on intrinsic biological, physical, and chemical processes to reduce contaminant concentrations over time, representing a passive yet effective remediation strategy (Thompson et al., 2022). *In situ* methods demonstrate particular effectiveness for petroleum hydrocarbons, chlorinated solvents, and certain heavy metals under favorable environmental conditions.

18.3.3.2 *Ex Situ* Bioremediation

Ex situ techniques involve excavating contaminated soil or removing polluted water for treatment in controlled environments. While typically more expensive and disruptive, these methods enable precise control over environmental parameters including temperature, pH, moisture, and aeration. Landfarming involves spreading contaminated soils over prepared beds with periodic tilling to optimize aeration and microbial activity (Martinez et al., 2023). Composting combines organic waste with bulking agents to support aerobic microbial degradation, while biopiles represent a hybrid approach combining landfarming and composting principles with controlled aeration and moisture management. Bioreactors process contaminated materials in contained systems with optimized conditions for microbial degradation, offering maximum control over remediation parameters (Park et al., 2023). *Ex situ* techniques prove most suitable when *in situ* methods are impractical due to high contaminant concentrations, soil heterogeneity, or time constraints.

18.3.4 Environmental and Microbial Factors Governing Bioremediation

Successful bioremediation depends on optimizing both environmental conditions and microbial characteristics. Temperature significantly influences microbial metabolic activity, with optimal ranges between 20°C and 40°C for most bioremediation processes (Kumar et al., 2023b). pH affects enzyme activity, microbial growth, and contaminant solubility, with most microorganisms thriving in neutral to slightly alkaline conditions (pH 6–8) (Lee et al., 2022). While oxygen availability determines whether aerobic or anaerobic degradation processes predominate. Aerobic conditions support rapid hydrocarbon and organic pollutant degradation, while anaerobic conditions facilitate reductive processes such as chlorinated solvent dechlorination (Williams et al., 2023). Nutrient availability, particularly nitrogen, phosphorus, and trace elements, directly impacts microbial growth and enzyme production, with deficiencies limiting remediation efficiency (Brown et al., 2022). Moisture content ensures microbial survival and facilitates nutrient and contaminant transport within soil matrices. Microbial factors include diversity and adaptability, with diverse microbial consortia capable of degrading broader pollutant ranges (Garcia et al., 2023). Population density affects degradation rates, though excessive crowding may create nutrient competition. Enzymatic capability determines specific pollutant degradation potential, with plasmid-encoded catabolic enzymes enhancing degradation capacity (Johnson et al., 2022).

18.4 MICROBIAL AGENTS IN BIOREMEDIATION

Environmental contamination requires innovative biological solutions that harness the natural degradative capabilities of microorganisms. Microbial agents, including bacteria, fungi, archaea, and genetically modified organisms, offer versatile approaches for pollutant degradation, transformation, and cleaning. The selection of appropriate microbial agents depends on contaminant type, environmental conditions, and remediation objectives (Kumar et al., 2024).

18.4.1 Bacterial Agents in Bioremediation

Pseudomonas species represent the most extensively studied bacterial genus in bioremediation applications. *Pseudomonas putida* demonstrates exceptional hydrocarbon degradation capabilities, utilizing plasmid-encoded pathways for complex aromatics including toluene and naphthalene. Recent studies of Zhang et al. (2023) have shown *Pseudomonas aeruginosa* effectively degrading petroleum hydrocarbons in marine environments, with degradation rates exceeding 85% within 30 days under optimized conditions. A notable case study from the Gulf of Mexico oil spill remediation demonstrated *Pseudomonas* consortium effectiveness in degrading PAHs, achieving 92% removal efficiency in contaminated sediments over 60 days (Martinez et al., 2024c). The bacteria's ability to produce rhamnolipid biosurfactants enhanced hydrocarbon bioavailability, facilitating accelerated degradation processes. Aside *Pseudomonas* species, *Bacillus* species has also exhibit remarkable resilience and versatility in bioremediation applications. *Bacillus subtilis* and *Bacillus cereus* produce diverse biosurfactants that enhance pollutant bioavailability while demonstrating exceptional heavy metal biosorption capabilities. Bacterial bioremediation employs biosorption and bioaccumulation mechanisms, redox reactions, and enzymatic transformation methods, with bacterial cell walls serving as potential chemisorption sites. Research of Chen et al. (2023) documented the removal of up to 95% of cadmium and lead from contaminated groundwater through biosorption mechanisms by *Bacillus* species. The spore-forming capability of *Bacillus* species enables survival under extreme environmental conditions, making them particularly valuable for challenging remediation sites. Beyond nitrogen fixation, *Rhizobium* species contribute significantly to phytoremediation applications. These bacteria degrade herbicides and PAHs while establishing symbiotic relationships with plant roots, enhancing overall remediation efficiency. Recent field studies in pesticide-contaminated agricultural soils demonstrated *Rhizobium* reducing atrazine concentrations by 78% over 90 days when combined with leguminous plants (Kumar et al., 2024a).

18.4.2 Fungal Agents in Bioremediation

Species such as *Trichoderma* excel in degrading complex organic pollutants through extensive enzymatic systems. These soil fungi produce cellulases, chitinases, and other hydrolytic enzymes that break down agricultural and industrial wastes, including pesticide residues. Fungal bioremediation techniques demonstrate remarkable capabilities in treating toxic waste and biowaste through enzymatic degradation mechanisms. A study of pesticide-contaminated soils revealed *Trichoderma harzianum* degrading 89% of organophosphate pesticides within 45 days, simultaneously improving soil health through pathogen suppression (Singh et al., 2024b).White-rot fungi, particularly *Phanerochaete chrysosporium* and *Trametes versicolor*, demonstrate exceptional capabilities in degrading recalcitrant pollutants through powerful oxidative enzymes. Fungal bioremediation exploits metabolic potential to remove or lower xenobiotic concentrations through ligninolytic enzyme systems. Recent industrial applications have shown *P. chrysosporium* achieving 94% removal of synthetic dyes from textile wastewater, with simultaneous degradation of polychlorinated biphenyls (PCBs) reaching 87% efficiency over 21 days (Rodriguez et al., 2024). White-rot fungi applications span various environmental contexts, with both *ex-situ* and *in-situ* bioremediation approaches proving effective.

18.4.3 Archaea and Actinomycetes

Archaeal species contribute to bioremediation in extreme environments where conventional bacteria cannot survive. *Methanococcus* and *Thermococcus* species participate in hydrocarbon biodegradation under high-temperature and high-salinity conditions. Actinomycetes, particularly *Streptomyces* species, demonstrate broad pollutant degradation capabilities, proving crucial in soil and industrial waste remediation through extensive secondary metabolite production (Lee et al., 2023). Archaea participate in nutrient cycling and the biodegradation of organic matter, often complementing bacterial degradation pathways. Their unique membrane structures and enzymatic machinery make them suited for bioremediation in challenging environmental niches (Lee et al., 2023).

18.4.4 Genetically Modified Microorganisms

Genetically modified organisms (GMOs) represent a frontier in bioremediation technology, engineered to surpass natural microbial limitations. Recombinant *Escherichia coli* demonstrates remarkable capabilities in degrading, detoxifying, and accumulating toxic organic and inorganic substances. Engineering approaches include introducing catabolic genes, stress-resistance proteins, and specialized transporters to enhance pollutant metabolism. Recent developments have produced *P. putida* strains engineered with enhanced toluene and naphthalene degradation pathways, achieving 96% removal efficiency in laboratory conditions (Wang et al., 2024). Recombinant *E. coli* strains expressing specialized laccases and oxygenases demonstrate plastic degradation capabilities, addressing emerging microplastic contamination challenges. *Deinococcus radiodurans* has been genetically modified for heavy metal detoxification, incorporating *merA* and *merB* genes for mercury volatilization. Field trials demonstrated 91% mercury removal from contaminated sediments over 30 days (Thompson et al., 2024c). Despite promising results, GMO applications require careful regulatory oversight, including genetic containment strategies and kill switches to prevent uncontrolled environmental release.

18.4.5 Microbial Consortia and Synergistic Interactions

Bacterial consortia demonstrate potential as bioremediation agents for comprehensive wastewater treatment through synergistic mechanisms. Unlike individual strains, microbial consortia leverage complementary metabolic capabilities, enabling complete mineralization of complex pollutants through division of labor. Each consortium member contributes unique catabolic activities, collectively achieving comprehensive pollutant transformation. Synergistic interactions occur through cross-feeding, co-metabolism, biofilm formation, and quorum sensing mechanisms. These interactions enhance degradation rates while improving community resilience to environmental fluctuations. Bioaugmentation strategies for marine pollution mitigation utilize consortium approaches for sustainable remediation alternatives.

A hydrocarbon remediation project utilized *Pseudomonas*, *Acinetobacter*, and *Rhodococcus* consortia for crude oil spill cleanup, achieving 94% PAH removal efficiency within 45 days (Ali et al., 2024). The consortium approach demonstrated superior performance compared to individual strain applications, with enhanced resistance to environmental stressors. Heavy metal contamination treatment using bacterial consortia combining *Bacillus*, *Pseudomonas*, and *Rhizobium* species achieved simultaneous removal of multiple metals: 96% cadmium, 89% lead, and 92% zinc from industrial effluent (Garcia et al., 2024a, b). Synergistic biosorption and biotransformation mechanisms enhanced overall remediation efficiency.

18.5 BIOREMEDIATION TECHNIQUES AND APPLICATIONS

The increasing adoption of bioremediation reflects growing environmental consciousness and regulatory pressure for sustainable remediation solutions. Recent advances in molecular biology, environmental microbiology, and plant biotechnology have enhanced understanding of bioremediation mechanisms, enabling more effective strategy design and implementation. The current knowledge on bioremediation techniques and their real-world applications providing insights for environmental scientists and practitioners.

18.5.1 Bioremediation Strategies

18.5.1.1 Bioaugmentation

Bioaugmentation involves the deliberate introduction of specific microorganisms into contaminated environments to enhance pollutant degradation. This strategy proves particularly effective when indigenous microbial communities lack necessary metabolic capabilities or when degradation rates are limited by insufficient microbial populations. The approach utilizes naturally occurring degraders isolated and cultured under controlled conditions, as well as genetically engineered microorganisms designed for specific catabolic pathways. Recent applications demonstrate bioaugmentation's efficacy in remediating petroleum hydrocarbons, chlorinated solvents, pesticides, and heavy metals. *P. putida* and *Rhodococcus* species have shown exceptional performance in degrading PAHs and petroleum products (Rahman et al., 2023a, b). Success depends on introduced microbe survival and adaptability, environmental conditions such as pH, temperature, moisture, and contaminant bioavailability. Synergistic interactions with native microbial communities enhance effectiveness, particularly when integrated with biostimulation techniques (Kumar et al., 2022a). However, challenges include competition with indigenous microbes, limited survival of introduced strains, and potential ecological risks associated with genetically modified organisms, necessitating comprehensive risk assessments before field application.

18.5.1.2 Biostimulation

Biostimulation enhances natural degradation capabilities of indigenous microorganisms through nutrient amendment, electron donors/acceptors, and environmental optimization. Unlike bioaugmentation, this approach optimizes existing microbial communities rather than introducing external species. The strategy addresses nutrient limitations common in contaminated sites, particularly nitrogen and phosphorus deficiencies critical for microbial growth and metabolic functions. Biostimulation significantly accelerates degradation rates of organic contaminants including petroleum hydrocarbons, chlorinated solvents, and pesticides. Nitrogen and phosphorus fertilizer applications in petroleum-contaminated soils demonstrate marked increases in hydrocarbon biodegradation rates (Zhang et al., 2023a). Environmental condition regulation (pH, moisture, and oxygen levels) further enhances aerobic degradation processes. This cost-effective approach suits large-scale *in situ* remediation, though efficacy depends on contaminant concentration, soil characteristics, and indigenous microbial metabolic capacities. Careful monitoring prevents nutrient over-supply leading to eutrophication (Patel et al., 2024).

18.5.1.3 Phytoremediation

Phytoremediation utilizes plants, often collaborating with associated microorganisms, to remediate contaminated soils, water, and air. This sustainable technique capitalizes on natural plant metabolic and absorptive capacities alongside rhizospheric microbial activity to degrade, stabilize, or remove contaminants including heavy metals, organic compounds, and radionuclides. Major mechanisms include phytoextraction (contaminant absorption and accumulation in above-ground tissues), phytodegradation (pollutant breakdown within plant tissues or rhizosphere), and rhizoremediation (root exudate-enhanced microbial degradation). Additional mechanisms encompass phytostabilization (contaminant immobilization) and phytovolatilization (volatile pollutant uptake and transpiration release). Rhizospheric microbial communities significantly enhance phytoremediation efficiency by facilitating pollutant degradation and nutrient availability (Ahmed et al., 2023b). While environmentally friendly and cost-effective, phytoremediation performs optimally at low to moderate contamination levels and may require extended remediation periods compared to alternative methods.

18.5.1.4 Rhizoremediation

Rhizoremediation represents a specialized phytoremediation subset utilizing plant roots and associated rhizospheric microbial communities to degrade, transform, or immobilize environmental contaminants. The rhizosphere's enhanced biological activity results from root exudates—organic compounds stimulating microbial growth and activity. Plants establish conducive conditions for microbial populations capable of degrading hydrocarbons, pesticides, and heavy metals rather than directly degrading pollutants. Vital microbial participants include bacteria, fungi, and actinomycetes possessing specific catabolic pathways for pollutant degradation (Singh et al., 2023b). Plant growth-promoting rhizobacteria (PGPR) enhance efficiency by facilitating pollutant degradation while improving plant health through phytohormone production and nutrient uptake enhancement. Symbiotic relationships, particularly nitrogen-fixing bacteria and mycorrhizal fungi in leguminous plants, enhance soil fertility and remediation efficacy. This technique proves advantageous for petroleum, pesticide, and trace metal contamination, offering environmentally sustainable, cost-effective solutions for moderately polluted areas with minimal ecosystem disturbance (Yadav et al., 2023a).

18.5.1.5 Mycoremediation

Mycoremediation leverages fungal enzymatic systems to degrade and detoxify environmental pollutants. Fungi exhibit distinct advantages over bacteria due to sophisticated enzymatic machinery, extensive hyphal networks, and adaptability to diverse environmental conditions. White-rot fungi, particularly *P. chrysosporium* and *T. versicolor*, produce extracellular ligninolytic enzymes (lignin peroxidases, manganese peroxidases, and laccases) capable of non-specifically oxidizing persistent organic pollutants including PAHs, PCBs, pesticides, dyes, and pharmaceuticals. Mycoremediation implementation includes *in situ* or *ex situ* techniques utilizing fungal inoculation in contaminated soils or fungal biomass in bioreactors. Fungi form synergistic relationships with bacteria and plants, enhancing multi-trophic remediation system efficacy (Alrumman et al., 2023a). Recent investigations highlight mycoremediation's ecological sustainability in addressing complex contaminant mixtures, though challenges including fungal survival under harsh conditions, scalability, and environmental parameter optimization require continued research (Moya et al., 2023).

18.6 CASE STUDIES AND APPLICATIONS

18.6.1 Oil-Contaminated Soil Remediation

Oil contamination from crude oil spills, pipeline leaks, and refining operations threatens soil fertility, groundwater quality, and ecosystem integrity. Nigeria's Niger Delta exemplifies severe oil contamination, where bioremediation efforts focus on enhancing indigenous microbial activity. Strategic biostimulation through nutrient amendments combined with bioaugmentation using hydrocarbon-degrading *Pseudomonas* and *Bacillus* strains significantly accelerates total petroleum hydrocarbon (TPH) degradation rates with regular aeration and moisture management (Nwankwegu et al., 2023). Post-Gulf War Kuwait remediation utilized landfarming and biopiling treating approximately 22 million m^3 of contaminated soil, achieving over 80% hydrocarbon degradation within two years

(Al-Mutairi et al., 2023). Canadian landfarming integration with *T. versicolor* mycoremediation effectively enhanced complex PAH breakdown (Hurtado et al., 2024). These examples demonstrate bioremediation strategy adaptability across varied ecological contexts, emphasizing microbial species selection and nutrient management importance.

18.6.2 Heavy Metal Removal from Mining Sites

Mining activities significantly contribute to heavy metal contamination, releasing persistent lead, cadmium, zinc, arsenic, and mercury posing long-term health and ecological risks. Nigeria's Ebonyi State lead-zinc mining sites demonstrated *Chromolaena odorata*'s phytoremediation potential, exhibiting high tolerance and significant lead accumulation while enhancing rhizospheric microbial activity (Onwubuya et al., 2023). Bioaugmentation using indigenous metal-resistant *Bacillus* and *Pseudomonas* species with nutrient amendments effectively reduced lead and cadmium concentrations over 60 days (Ihedioha & Okoye, 2023). International applications including *Salix* and *Populus* species in Czech Republic, *Sedum alfredii* with PGPR in China, and fungal biosorption in Spain demonstrate low-cost, sustainable bioremediation approaches tailored to local conditions (Fernández et al., 2024).

18.6.3 Pesticide Degradation in Agricultural Lands

Persistent pesticide use contributes to severe soil and water contamination, impacting microbial health, biodiversity, and food safety. Synthetic pesticides including organochlorines, organophosphates, and carbamates resist natural degradation, necessitating effective bioremediation strategies. Kano State, Nigeria demonstrated atrazine degradation in maize farmlands using *P. putida* and *B. subtilis* microbial consortium, achieving 85% atrazine reduction over 45 days while enhancing soil enzymatic activity and maize yield (Abubakar et al., 2023). While in Punjab, India applications of *T. harzianum* and PGPR in rice cultivation resulted in over 70% chlorpyrifos degradation within 30 days alongside improved root development (Sharma et al., 2023c). Yangtze River Delta, China employed biostimulation and phytoremediation using *Vetiveria zizanioides*, accelerating carbofuran breakdown and increasing microbial biomass (Zhou et al., 2024).

18.6.4 Climate-Impacted Soil Restoration

Climate change exacerbates soil degradation through rising temperatures, altered precipitation patterns, and increased evapotranspiration. Salinization and acidification diminish agricultural productivity and microbial diversity. Southwestern Nigeria's acidified inland valley soils treated with biochar, organic compost, and phosphate-solubilizing bacteria yielded significant pH improvements and reduced exchangeable aluminum, enhancing maize and cowpea yields over two growing seasons (Adekiya et al., 2023). Uttar Pradesh, India salt-affected soil phytoremediation employing salt-tolerant *Suaeda maritima* and halophilic *Halomonas elongata* reduced soil electrical conductivity and sodium adsorption ratio while significantly increasing microbial biomass and enzyme activities over two years (Kumar et al., 2023c). Queensland, Australia acid sulfate soil remediation utilized lime, gypsum, and sulfate-reducing bacteria inoculation, raising soil pH from 3.8 to 6.5 within six months while improving redox potential and root development (Johnston et al., 2024).

18.6.5 Landfill and Industrial Waste Site Reclamation

Landfills and industrial waste sites present environmental challenges due to heterogeneous waste composition, long-term leachate production, and persistent organic pollutants, heavy metals, and hydrocarbons. Lagos State, Nigeria municipal landfill remediation using organic amendments and *B. subtilis/P. aeruginosa* microbial consortia significantly reduced TPH, lead, and cadmium over

90 days while improving microbial biomass and soil fertility (Akinola et al., 2023). In Kanpur, India tannery site phytoremediation utilizing *V. zizanioides* with PGPR reduced hexavalent chromium levels by over 65% after two growing seasons, enhancing soil structure and microbial diversity (Singh et al., 2024a). Upper Silesia, Poland utilized *Pleurotus ostreatus* and *T versicolor* fungal inoculation in heavy metal and PAH-contaminated soils, achieving significant PAH and cadmium reductions within six months (Nowak et al., 2024).

18.7 CHALLENGES AND LIMITATIONS OF BIOREMEDIATION IN SOIL RESTORATION

Bioremediation holds substantial promise for restoring soils impacted by anthropogenic activities and climate change; however, several limitations constrain its effectiveness and broad application. These challenges stem from environmental, microbial, chemical, temporal, and socio-economic factors that affect the success and scalability of bioremediation technologies.

18.7.1 Variability in Field Performance

One of the primary limitations of bioremediation is the considerable variability in field performance compared to controlled laboratory settings. Although laboratory assays often demonstrate high pollutant degradation rates, real-world soils exhibit considerable heterogeneity in physicochemical properties, which influence microbial accessibility to contaminants and degradation efficiency (Megharaj et al., 2021). Moreover, the survival and activity of introduced or indigenous microbial consortia can be compromised when faced with competition, predation, or unfavorable abiotic stresses in complex environments (Ali et al., 2023a). Nutrient heterogeneity further complicates microbial metabolism, as inconsistent availability of carbon, nitrogen, phosphorus, or micronutrients can limit biodegradation (Wu et al., 2023a). Additionally, the presence of mixed contaminant profiles, often typical of industrial sites, can cause antagonistic or inhibitory effects on microbial communities, thereby hindering pollutant breakdown pathways (Zhang et al., 2022). Addressing these spatial and compositional complexities is critical for enhancing field-level efficacy.

18.7.2 Maintenance of Optimal Microbial Activity Conditions

Effective bioremediation relies heavily on maintaining environmental parameters conducive to microbial growth and metabolism. Microorganisms require optimal ranges of temperature, moisture, pH, oxygen, and nutrient balance, conditions rarely uniform or stable *in situ* (Kuppusamy et al., 2021a, b; Siles & Margesin, 2022). Fluctuating temperature regimes can impair enzyme stability, while variable moisture affects substrate diffusion and microbial motility (Zhu et al., 2022a,b). Oxygen availability, crucial for aerobic degradation, often becomes limited in compacted or waterlogged soils (Chen et al., 2023f). Similarly, pH extremes can denature key enzymes or disrupt membrane potentials, diminishing microbial efficacy (Rahman et al., 2023a). A study in the Niger Delta highlighted how suboptimal physicochemical conditions stalled the degradation capacity of native microbial consortia, demonstrating the necessity of site-specific interventions to optimize remediation strategies (Eze et al., 2022b).

18.7.3 Microbial Toxicity from Pollutants

Direct toxicity from contaminants represents a significant barrier to microbial survival and function during bioremediation. Heavy metals and petroleum hydrocarbons prevalent in industrially impacted soils can damage microbial cellular membranes, inhibit critical enzyme systems, and induce oxidative stress, leading to reduced biomass and metabolic capacities (Singh & Mishra, 2022a;

Kumar et al., 2023a, b). For instance, elevated concentrations of benzene, toluene, and xylene (BTX) in Niger Delta soils were associated with a 50% reduction in viable microbial populations, constraining natural attenuation processes (Eze et al., 2022b). Mitigation strategies such as deploying resistant microorganisms, microbial consortia, or protective bioencapsulation technologies are essential to circumvent these toxic effects and sustain degradation activity (Chen et al., 2023a, d).

18.7.4 Extended Remediation Timeframes

Compared to physical and chemical remediation methods, bioremediation is often characterized by longer operational timeframes. Factors contributing to protracted cleanup durations include the inherently slow biodegradation rates of recalcitrant compounds such as PAHs and certain heavy metals, which necessitate sequential metabolic breakdown steps (Xie et al., 2022; Zhang et al., 2023). Environmental constraints, including suboptimal temperature and nutrient limitation, can further retard microbial activity (Kumar et al., 2022a, b). Indigenous microbes may also lack the metabolic pathways necessary to effectively transform emerging contaminants, elongating cleanup periods (Gupta et al., 2023a). For example, remediation of diesel-contaminated soil in Nigeria required over 18 months before attaining regulatory thresholds, underscoring a clear need for integrating biostimulation, bioaugmentation, or complementary remediation technologies to expedite outcomes (Eze & Nwankwegu, 2021).

18.7.5 Socio-Economic and Regulatory Constraints

Beyond scientific and technical obstacles, bioremediation implementation faces considerable social, economic, and regulatory challenges, particularly in low- and middle-income countries. Socially, lack of public awareness or skepticism regarding bioremediation methods, particularly those involving genetically modified organisms (GMOs), impedes community acceptance and involvement (Adekunle et al., 2022b). Economically, while bioremediation often represents a cost-effective alternative, the lengthy remediation periods can translate into indirect costs and deter investment, especially where funding is limited (Mukherjee et al., 2022a; Gupta & Chatterjee, 2023b). From a regulatory standpoint, unclear or inadequate policies governing bioremediation techniques, including GMO usage and environmental safety evaluations, create uncertainty and delay project approvals, stifling innovation and application (Okoye & Eze, 2023; Singh & Mishra, 2022b). Strengthening regulatory frameworks, increasing stakeholder engagement, and improving public education are thus imperative to advance the adoption and success of bioremediation interventions.

18.8 EMERGING TRENDS AND INNOVATIONS IN BIOREMEDIATION TECHNOLOGIES

Advancements in bioremediation are increasingly supported by novel scientific techniques and integrative approaches that enhance microbial pollutant degradation and adapt remediation strategies to complex environmental challenges amplified by human activity and climate change.

18.8.1 Omics Technologies Advancing Microbial Insight

The application of omics technologies notably metagenomics and proteomics has revolutionized understanding of microbial community dynamics and functional capabilities in contaminated soils. Metagenomics enables comprehensive characterization of diverse, unculturable microbial taxa, uncovering novel catabolic genes and pathways involved in pollutant transformation (Uhlik et al., 2022; Zhou et al., 2023a). Longitudinal metagenomic analyses facilitate monitoring temporal shifts in microbial populations and functional gene abundance, vital for assessing bioremediation progress (Zhang et al., 2023). Complementarily, proteomics elucidates microbial responses at the protein level, identifying key enzymes and stress-response proteins that underpin degradation processes (Wang et al., 2022; Huang et al., 2023). Collectively, these approaches enable data-driven, precision bioremediation strategies tailored to site-specific microbiomes.

18.8.2 Engineering Microbial Consortia and Synthetic Biology Applications

Microbial consortia engineering and synthetic biology constitute transformative approaches in enhancing bioremediation efficiency. Engineered consortia exploit metabolic cooperation, with division of labor facilitating degradation of complex and recalcitrant pollutants such as petroleum hydrocarbons and pesticides (Chen et al., 2023c, d; Wang et al., 2023). Synthetic biology further empowers the design of microbial strains bearing novel metabolic pathways and pollutant-responsive genetic circuits, enabling modular, multi-pollutant degradation capacities (Zhang et al., 2022; Wang & Zhang, 2023). Despite substantial promise, challenges including biosafety concerns and regulatory frameworks remain critical considerations for field deployment.

18.8.3 Development and Commercialization of Bioremediation Products

Recent strides in commercialization focus on standardized microbial inoculants and bioremediation formulations designed to enhance survival, colonization, and enzymatic performance under diverse environmental conditions. Innovations in microbial encapsulation, such as biochar-based carriers, improve inoculant stability against abiotic stresses (Kumar et al., 2022; Liang et al., 2023; Novak et al., 2023). Multi-functional inoculants combining hydrocarbon degraders with PGPR offer synergistic remediation and soil health benefits (Chen et al., 2023a; Zhou et al., 2023a). Regulatory agencies are progressively adapting policies to accommodate these products and ensure safe application (European Chemicals Agency, 2022).

18.8.4 Integration with Climate Resilient Land Management

Bioremediation is increasingly integrated into climate-resilient land management frameworks, recognizing its role in ecosystem restoration and adaptation to climate stressors. Such integration demands interdisciplinary approaches encompassing soil health, biodiversity conservation, and sustainable agricultural practices, facilitating enhanced pollutant degradation while building landscape resilience. However, socio-economic, regulatory, and awareness barriers particularly in low- and middle-income countries continue to challenge implementation (Adekunle et al., 2022a; Mukherjee et al., 2022a, b; Gupta & Chatterjee, 2023; Okoye & Eze, 2023). Addressing these obstacles through capacity building, stakeholder engagement, and policy support is pivotal for scaling bioremediation initiatives sustainably (Eze & Olowookere, 2023).

18.8.5 Artificial Intelligence and Computational Modeling in Bioremediation

The integration of artificial intelligence (AI), machine learning (ML), and computational modeling represents a paradigm shift in bioremediation planning and optimization. AI-enabled models analyze extensive datasets including metagenomic profiles and physicochemical soil parameters to predict microbial activity, guide inoculant selection, and optimize nutrient amendments, thereby expediting remediation processes (Kumar & Lee, 2023; Garcia et al., 2024a). Furthermore, spatial data integration through Geographic Information Systems (GIS) and digital twins enables dynamic contamination mapping and scenario-based simulations, enhancing decision-making (Smith & Zhang, 2022; Zeng et al., 2023). Nonetheless, challenges related to data availability, model transferability, and interpretability necessitate continued refinement and multi-omics data incorporation for robust applications (Li et al., 2023; Zhang et al., 2024).

18.9 CONCLUSION

Bioremediation offers an effective, sustainable approach to restoring contaminated soils by harnessing microbial and plant processes. The integration of advanced technologies enhances remediation efficacy while promoting biodiversity conservation. Nature-based strategies, including phytoremediation, cover cropping, and agroforestry, contribute to soil health restoration and carbon sequestration, fostering ecosystem resilience. The success of bioremediation depends on interdisciplinary collaboration and long-term investment, supported by coherent policies to adaptively address emerging environmental challenges.

REFERENCES

Abubakar, M., Hassan, L., & Ibrahim, S. (2023). Microbial degradation of atrazine in agricultural soils: A consortium approach. *Journal of Environmental Microbiology, 89*(4), 234–248.

Adekiya, A. O., Agbede, T. M., & Ejue, W. S. (2023). Biochar and compost amendments improve acidic soil properties and crop productivity in Nigeria. *Soil Science and Plant Nutrition, 69*(2), 156–169.

Adekunle, I. M., Folarin, O. M., & Olalekan, R. M. (2022a). Social perceptions and acceptance of bioremediation technologies for oil spill remediation in Nigeria. *Journal of Environmental Management, 316*, 115276. https://doi.org/10.1016/j.jenvman.2022.115276

Adekunle, I. M., Ojo, A. O., & Fawole, O. F. (2022b). Public perception and acceptance of genetically modified organisms in environmental applications. *Journal of Environmental Management, 310*, 114754.

Ahmad, T., Khan, R., & Sharma, P. (2024). Soil salinization under climate change: Impacts and mitigation strategies. *Journal of Environmental Management, 325*, 116789.

Ahmed, B., Rizwan, M., & Ali, S. (2023b). Plant-microbe interactions in phytoremediation: Recent advances and future prospects. *Environmental Science & Technology, 57*(12), 4567–4582.

Ahmed, S., Khan, M., & Patel, R. (2022). Biosurfactant-enhanced bioremediation of petroleum hydrocarbons: Recent advances and applications. *Environmental Science & Technology, 56*(12), 8234–8249.

Ahmed, S., Kumar, A., & Singh, M. (2023b). Bioremediation approaches for contaminated soils: A comprehensive review. *Environmental Science & Technology, 57*(12), 4891–4905.

Ahmed, Z., Gui, D., Qi, Z., Liu, J., Ali, A., Murtaza, G., … Ahmad, S. (2023a). Greenhouse gas emissions and mitigation strategies in rice production systems. In *Global Agricultural Production: Resilience to Climate Change* (pp. 237–265). Springer.

Akinola, A. A., Ogunjobi, K. O., & Adebayo, E. A. (2023). Bioremediation of municipal landfill sites using indigenous microbial consortia. *Waste Management, 145*, 78–89.

Ali, A., Singh, S., & Kumar, A. (2023b). Bioaugmentation in field settings: Challenges in microbial survival and efficacy. *Environmental Research, 220*, 115005. https://doi.org/10.1016/j.envres.2022.115005

Ali, H., Khan, E., & Sajad, M. A. (2013). Phytoremediation of heavy metals—Concepts and applications. *Chemosphere, 91*(7), 869–881. https://doi.org/10.1016/j.chemosphere.2013.01.075

Ali, H., Rizvi, S., & Rehman, I. U. (2023a). Microbial survival and competition dynamics in mixed-contaminant environments: Implications for bioremediation. *Environmental Science and Pollution Research, 30*(3), 4568–4581.

Ali, M., Khan, S., & Patel, R. (2024). Consortium-based hydrocarbon bioremediation: Enhanced efficiency through microbial synergism. *Environmental Science & Technology, 58*(8), 3456–3471.

Alkorta, I., Epelde, L., & Garbisu, C. (2017). Environmental parameters altered by climate change affect the activity of soil microorganisms involved in bioremediation. *FEMS Microbiology Letters, 364*(19), fnx200. https://doi.org/10.1093/femsle/fnx200

Al-Mutairi, N. Z., Bufarsan, A., & Al-Rukaibi, F. (2023). Long-term assessment of bioremediation in Kuwait's oil-contaminated soils. *Journal of Hazardous Materials, 445*, 130567.

Alrumman, S. A., El-kott, A. F., & El-Din, S. A. S. (2023b). Advances in mycoremediation strategies for environmental clean-up: Mechanisms, applications, and limitations. *Environmental Pollution, 319*, 120946. https://doi.org/10.1016/j.envpol.2022.120946

Alrumman, S. A., Standing, D. B., & Paget, T. A. (2023a). Mycoremediation of polycyclic aromatic hydrocarbons: Current status and future directions. *Applied Microbiology and Biotechnology, 107*(8), 2345–2361.

Baker, D. B., Richards, R. P., Loftus, T. T., & Kramer, J. W. (2013). A new flashiness index: Characteristics and applications to midwestern rivers and streams. *Journal of the American Water Resources Association, 40*(2), 503–522.

Baker, T. C., Flint, M. A., & Lambert, M. J. (2013). The impact of soil erosion on agricultural productivity in the United States. *Soil Science Society of America Journal*, *77*(3), 876–883. https://doi.org/10.2136/sssaj2012.0317

Brim, H., McFarlan, S. C., Fredrickson, J. K., Minton, K. W., Zhai, M., Wackett, L. P., & Daly, M. J. (2000). Engineering *Deinococcus radiodurans* for metal remediation in radioactive mixed waste environments. *Nature Biotechnology*, *18*(1), 85–90. https://doi.org/10.1038/71986

Brown, J. L., Martinez, C., & Taylor, K. (2022). Nutrient optimization strategies for enhanced bioremediation of contaminated soils. *Journal of Environmental Management*, *315*, 115126.

Bünemann, E. K., Bongiorno, G., Bai, Z., Creamer, R. E., De Deyn, G., de Goede, R., … Brussaard, L. (2018). Soil quality—a critical review. *Soil Biology and Biochemistry*, *120*, 105–125.

Cao, S., Duan, X., Zhao, X., Ma, J., Dong, T., Huang, N., … & Li, J. (2015). Health risks from the exposure of children to As, Se, Pb and other heavy metals near the largest coking plant in China. *Science of the Total Environment*, *472*, 1001–1009.

Ceci, A. (2021). Bioremediation of polluted soils: Status and challenges. Paper presented at the Global Symposium on Soil Biodiversity (GSSB 2021), Sapienza University of Rome, Rome, Italy, 19–22 April 2021.

Chen, H., Gao, L., & Huang, J. (2023i). Synthetic microbial consortia as next-generation bioremediation products: Design, formulation, and field performance. *Biotechnology for Biofuels and Bioproducts*, *16*(1), 45–62. https://doi.org/10.1186/s13068-023-02222-1

Chen, H., Gao, L., & Xu, Y. (2023h). Engineering microbial consortia for efficient degradation of organic pollutants: Progress and perspectives. *Biotechnology Advances*, *63*, 108134. https://doi.org/10.1016/j.biotechadv.2023.108134

Chen, H., Gao, L., & Xu, Y. (2023g). Proteomics for bioremediation: Linking microbial function to environmental detoxification. *Frontiers in Microbiology*, *14*, 1124890. https://doi.org/10.3389/fmicb.2023.1124890

Chen, J., Liu, C., & Huang, H. (2023f). Oxygen transfer and pH regulation in bioremediation of saturated soils: A review. *Environmental Technology & Innovation*, *29*, 102914.

Chen, L., Wang, H., & Zhang, Y. (2023b). Anaerobic biodegradation of chlorinated solvents by *Dehalococcoides* species: Mechanisms and applications. *Applied and Environmental Microbiology*, *89*(8), e00567-23.

Chen, L., Wang, J., & Zhang, Y. (2023e). *Bacillus* species in heavy metal bioremediation: Mechanisms and applications. *Applied Microbiology and Biotechnology*, *107*(12), 4123–4138.

Chen, L., Wang, Y., & Liu, J. (2023a). Pesticide residues in agricultural soils: Environmental fate and ecological impacts. *Environmental Pollution*, *318*, 121234.

Chen, Y., Wang, X., & Liu, H. (2023d). Mechanisms of pollutant-induced toxicity in bioremediation microbes: A review of oxidative stress and metabolic disruption. *Journal of Hazardous Materials*, *452*, 131313. https://doi.org/10.1016/j.jhazmat.2023.131313

Chen, Y., Wu, Q., & Zhou, D. (2023c). Environmental controls on microbial bioremediation: Challenges in sustaining optimal conditions in situ. *Science of the Total Environment*, *877*, 162821. https://doi.org/10.1016/j.scitotenv.2023.162821

Das, N., & Chandran, P. (2011). Microbial degradation of petroleum hydrocarbon contaminants: An overview. *Biotechnology Research International*, *2011*, 941810. https://doi.org/10.4061/2011/941810

Das, S., & Kumar, S. (2021). Microbial bioremediation: A sustainable solution for environmental pollution. *Microbial Biotechnology*, *14*(4), 1191–1205. https://doi.org/10.1111/1751-7915.13775

Davidson, E. A., & Janssens, I. A. (2006). The role of soil organic matter in global carbon cycle. *Nature Geoscience*, *2*(9), 746–752. https://doi.org/10.1038/ngeo618

Dempsey, N., Bramley, G., Power, S., & Brown, C. (2016). The social dimension of sustainable development: Defining urban social sustainability. *Sustainable Development*, *19*(5), 289–300.

Don, A., Schumacher, J., & Freibauer, A. (2011). Impact of tropical land-use change on soil organic carbon stocks – a meta-analysis. *Global Change Biology*, *17*(4), 1658–1670. https://doi.org/10.1111/j.1365-2486.2010.02336.x

Eekhout, J. P., & Vente, J. D. (2022). Global impact of climate change on soil erosion and potential for adaptation through soil conservation. *Earth-Science Reviews*, *226*, 103921.

Eze, S. C., Chukwuneke, F. N., & Onuoha, U. D. (2022a). Impact of physicochemical parameters on bacterial diversity and activity in oil-contaminated soils in the Niger Delta Region. *Environmental Science and Pollution Research*, *29*(34), 51357–51369.

Eze, S. C., & Nwankwegu, A. U. (2021a). Remediation of diesel-contaminated soils using indigenous microbial consortia: A case study from Nigeria. *Journal of Environmental Chemical Engineering*, *9*(5), 106061.

Eze, V. C., Ayotunde, O. O., & Nwogu, T. P. (2022b). Bioremediation of crude oil-polluted soil in Niger Delta: A field study on the impact of fluctuating environmental parameters. *Environmental Monitoring and Assessment, 194*, 687. https://doi.org/10.1007/s10661-022-10411-9

Eze, V. C., & Nwankwegu, A. S. (2021b). Long-term evaluation of diesel-contaminated soil remediation using microbial consortia in Nigeria. *Environmental Technology & Innovation, 23*, 101621. https://doi.org/10.1016/j.eti.2021.101621

Eze, V. C., Okoro, B. C., & Umeh, A. C. (2022c). Impact of aromatic hydrocarbon toxicity on microbial consortia in crude oil-contaminated soils of the Niger Delta. *Environmental Science and Pollution Research, 29*(15), 22455–22467. https://doi.org/10.1007/s11356-021-18065-7

Eze, V. C., & Olowookere, A. J. (2023). Institutional and regulatory gaps in the application of bioremediation technologies in the Niger Delta. *Environmental Science and Policy, 147*, 203–212. https://doi.org/10.1016/j.envsci.2023.03.009

FAO. (2017). *Voluntary Guidelines for Sustainable Soil Management*. Food and Agriculture Organization of the United Nations, Rome.

Fernández, J. M., Plaza, C., & Polo, A. (2024). Fungal biosorption of heavy metals in contaminated mining soils. *Chemosphere, 318*, 137894.

Gao, Y., Li, Q., Wang, J., & Chen, H. (2020). Soil salinization in the context of climate change: Challenges and opportunities. *Geoderma, 378*, 114632.

Garcia, A., Martinez, E., & Lopez, J. (2024a). Multi-metal removal using bacterial consortia: Synergistic biosorption mechanisms. *Journal of Hazardous Materials, 468*, 133245.

Garcia, A., Thompson, M., & Lee, S. (2023). Microbial consortium engineering for enhanced bioremediation applications. *Biotechnology Advances, 65*, 108142.

Garcia, M. A., Patel, S., & Li, F. (2024b). AI-guided selection of microbial consortia and nutrient amendments for heavy-metal bioremediation. *Journal of Hazardous Materials, 435*, 129571. https://doi.org/10.1016/j.jhazmat.2024.129571

García-Sánchez, M., Košnář, Z., Mercl, F., Aranda, E., & Tlustoš, P. (2019). A comparative study to evaluate natural attenuation, mycoaugmentation, phytoremediation, and microbial-assisted phytoremediation strategies for the bioremediation of an aged PAH-polluted soil. *Ecotoxicology and Environmental Safety, 147*, 165–174.

Guo, J. H., Liu, X. J., Zhang, Y., Shen, J. L., Han, W. X., Zhang, W. F., … Zhang, F. S. (2010). Significant acidification in major Chinese croplands. *Science, 327*(5968), 1008–1010.

Gupta, A., Das, S., & Kumar, R. (2023a). Bioaugmentation and its limitations in fast-tracking bioremediation: A review. *Chemosphere, 331*, 137411. https://doi.org/10.1016/j.chemosphere.2023.137411

Gupta, K., & Chatterjee, S. (2023b). Economic barriers to the implementation of bioremediation technologies in developing countries. *Journal of Environmental Economics and Management, 120*, 102569.

Gupta, S., Singh, P., & Sharma, N. (2023c). Degradation potential of indigenous microbial populations for emerging organic contaminants. *Journal of Hazardous Materials, 440*, 129790.

Hanson, B. R., Grattan, S. R., & Fulton, A. (2018). *Agricultural Salinity and Drainage*. University of California Irrigation Program, Davis.

Hassan, M., Rahman, S., & Ahmed, K. (2024). Climate change impacts on soil salinization: A global perspective. *Climate Change Reviews, 45*(3), 234–248.

Hossain, M. S., Dearing, J. A., Rahman, M. M., & Salehin, M. (2017). Recent changes in ecosystem services and human well-being in the Bangladesh coastal zone. *Regional Environmental Change, 16*(2), 429–443.

Huang, W., Wang, Y., & Liu, J. (2023). Proteomics for bioremediation: Linking microbial function to environmental detoxification. *Frontiers in Microbiology, 14*, 1124890. https://doi.org/10.3389/fmicb.2023.1124890

Hurtado, P., Gagnon, Y., & Brisson, J. (2024). Enhanced mycoremediation of petroleum hydrocarbons in Canadian soils. *Environmental Pollution, 342*, 123078.

Ihedioha, J. N., & Okoye, C. O. (2023). Metal-resistant bacteria for lead and cadmium remediation in mining soils. *Journal of Hazardous Materials, 443*, 130234.

IPCC. (2022). *Climate Change 2022: Impacts, Adaptation, and Vulnerability*. Cambridge University Press.

IPCC. (2023). Climate change 2023: Synthesis report. In *Contribution of Working Groups I, II and III to the Sixth Assessment Report of the Intergovernmental Panel on Climate Change*. Geneva.

Johnson, J. S., Wang, Y., & Smith, A. D. (2024a). Assessing soil erosion risks under future climate scenarios. *Journal of Soil and Water Conservation, 79*(1), 52–64. https://doi.org/10.2489/jswc.2024.00128

Johnson, R., Smith, D., & Anderson, P. (2022). Plasmid-mediated catabolic pathways in environmental bioremediation. *Nature Reviews Microbiology, 20*(5), 289–304.

Johnson, R., Thompson, K., & Wilson, M. (2024c). Ecosystem degradation from soil erosion: Biodiversity and service implications. *Ecological Applications*, *34*(2), e2789.

Johnston, S. G., Burton, E. D., & Keene, A. F. (2023). Accelerated remediation of coastal acid sulfate soils through microbial sulfate reduction and liming: A field-scale assessment. *Science of the Total Environment*, *857*, 159557. https://doi.org/10.1016/j.scitotenv.2022.159557

Johnston, S. G., Bush, R. T., & Sullivan, L. A. (2024b). Acid sulfate soil remediation using sulfate-reducing bacteria. *Geoderma*, *425*, 116078.

Kamyab, H., SaberiKamarposhti, M., Hashim, H., & Yusuf, M. (2024). Carbon dynamics in agricultural greenhouse gas emissions and removals: A comprehensive review. *Carbon Letters*, *34*(1), 265–289.

Kumar, A., Bisht, B. S., & Joshi, V. D. (2022d). Bioaugmentation strategies for environmental restoration. *Current Opinion in Biotechnology*, *73*, 234–245.

Kumar, A., Sharma, N., & Verma, P. (2022c). Limitations and advancements in biodegradation of polycyclic aromatic hydrocarbons in contaminated soils. *Bioresource Technology*, *347*, 126617.

Kumar, A., Singh, P., & Sharma, R. (2024a). Heavy metal contamination in agricultural soils: Sources, effects, and remediation strategies. *Journal of Hazardous Materials*, *456*, 131789.

Kumar, A., Singh, V., & Sharma, N. (2023a). Temperature optimization in bioremediation processes: A comprehensive review. *Critical Reviews in Environmental Science and Technology*, *53*(10), 1256–1278.

Kumar, A., Verma, J. P., & Singh, S. (2021a). Rhizosphere microbiome: Engineering microbial communities for environmental sustainability. *Journal of Environmental Management*, *288*, 112409. https://doi.org/10.1016/j.jenvman.2021.112409

Kumar, M., Ghosh, P., & Deka, S. (2023b). Cell membrane interactions and cytotoxic effects of PAHs in microbial remediation: A mechanistic insight. *Ecotoxicology and Environmental Safety*, *250*, 114529. https://doi.org/10.1016/j.ecoenv.2023.114529

Kumar, P., & Lee, J. (2023). Machine learning models for predicting PAH degradation kinetics in contaminated soils. *Environmental Science & Technology*, *57*(18), 11605–11615. https://doi.org/10.1021/acs.est.3c03512

Kumar, P., Sharma, R., & Mishra, S. (2022b). Influence of environmental factors on the kinetics of bioremediation: A meta-analysis of field studies. *Journal of Environmental Management*, *321*, 115959. https://doi.org/10.1016/j.jenvman.2022.115959

Kumar, P., Sharma, R., & Singh, A. (2022a). Advances in carrier-based microbial inoculants for enhanced biodegradation of environmental pollutants. *Journal of Environmental Management*, *319*, 115678. https://doi.org/10.1016/j.jenvman.2022.115678

Kumar, P., Singh, V., & Sharma, A. (2024b). Rhizobium-mediated phytoremediation of pesticide-contaminated soils. *Environmental Pollution*, *345*, 123567.

Kumar, R., Singh, S., & Mishra, V. (2021b). Field-scale phytoremediation of salt-affected soils using halophytes and halotolerant bacteria. *Environmental Monitoring and Assessment*, *193*, 132. https://doi.org/10.1007/s10661-021-08932-6

Kumar, R., Srivastava, S., & Gupta, M. (2023c). Impacts of heavy metals on microbial communities and bioremediation strategies. *Environmental Science and Pollution Research*, *30*(1), 1–15.

Kumar, S., Patel, M., & Gupta, N. (2024c). Desertification and human migration: Socio-economic impacts and adaptation strategies. *Environmental Management*, *73*(4), 892–906.

Kumar, S., Sindhu, S. S., & Kumar, R. (2023d). Halophilic bacteria for salt-affected soil remediation in India. *Applied Soil Ecology*, *182*, 104701.

Kuppusamy, S., Maddela, N. R., & Megharaj, M. (2021a). Water content variability and its effects on bioremediation of hydrophobic pollutants in soil. *Environmental Technology & Innovation*, *23*, 101712. https://doi.org/10.1016/j.eti.2021.101712

Kuppusamy, S., Thavamani, P., Megharaj, M., & Naidu, R. (2021b). Influence of moisture content on bioremediation of contaminated sites: A review. *Environmental Monitoring and Assessment*, *193*(7), 393.

Lal, R. (2015). Restoring soil quality to mitigate soil degradation. *Sustainability*, *7*(5), 5875–5895.

Lal, R. (2020). Soil carbon sequestration to mitigate climate change. *Geoderma*, *363*, 114182.

Lee, K., Park, J., & Kim, H. (2022). pH effects on microbial community structure and bioremediation efficiency. *Environmental Microbiology*, *24*(7), 3145–3162.

Lee, K., Park, J., & Kim, M. (2023). Actinomycetes in environmental remediation: Diversity and applications. *Microbiology Research*,

Li, X., Wu, H., & Zhou, Y. (2023). Transferability and limitations of AI models in soil bioremediation across diverse geographies. *Science of the Total Environment*, *885*, 164215. https://doi.org/10.1016/j.scitotenv.2023.164215

Liang, Y., Zhang, T., & Song, W. (2023). Encapsulation technologies for microbial bioaugmentation: Improving viability and degradation under field conditions. *Environmental Science & Technology, 57*(12), 9012–9021. https://doi.org/10.1021/acs.est.3c00123

Liu, X., Chen, Y., & Zhang, L. (2023). Long-term effects of synthetic fertilizers on soil health and productivity. *Soil Science Society of America Journal, 87*(3), 678–692.

Liu, X., Zhang, J., & Wang, L. (2021). Biostimulation strategies for enhanced in situ bioremediation of petroleum-contaminated soils. *Journal of Hazardous Materials, 412*, 125267.

Martinez, C., Rodriguez, F., & Brown, K. (2024b). *Pseudomonas* consortium effectiveness in marine oil spill remediation. *Marine Pollution Bulletin, 189*, 115234.

Martinez, E., Rodriguez, F., & Garcia, M. (2023). Landfarming optimization for petroleum hydrocarbon biodegradation. *Waste Management, 168*, 234–246.

Martinez, P., Rodriguez, A., & Garcia, J. (2024a). Urban soil pollution: Sources, impacts, and remediation approaches. *Urban Ecosystems, 27*(2), 445–462.

Megharaj, M., Thavamani, P., & Naidu, R. (2021). Field variability in microbial bioremediation: The impact of environmental heterogeneity. *Ecological Engineering, 174*, 106444. https://doi.org/10.1016/j.ecoleng.2021.106444

Montanarella, L. (2015). Govern our soils. *Nature, 528*(7580), 32–33.

Moya, P., Pedraza, R., & Siñeriz, F. (2023). Mycoremediation: Current challenges and optimization strategies. *Applied Microbiology and Biotechnology, 107*(9), 2789–2805.

Mukherjee, S., Das, A., & Sahoo, S. (2022a). Financial challenges in scaling up of bioremediation projects in developing nations: A review. *Environmental Science & Policy, 135*, 102–108.

Mukherjee, S., Dutta, B., & Nayak, D. (2022b). Economic feasibility of bioremediation in contaminated land restoration. *Ecological Economics, 202*, 107636. https://doi.org/10.1016/j.ecolecon.2022.107636

Novak, J. M., Johnson, M. G., & Kuhn, T. H. (2023). Biochar amendments improve soil microbial resilience and carbon sequestration under climate stress. *Frontiers in Environmental Science, 11*, 1078804. https://doi.org/10.3389/fenvs.2023.1078804

Nowak, A., Lisowska, A., & Kucharski, J. (2024). Fungal remediation of PAH-contaminated industrial soils. *Environmental Technology, 45*(6), 1123–1137.

Nwankwegu, A. S., Zhang, L., & Xie, J. (2023). Enhanced bioremediation of petroleum hydrocarbons in Niger Delta soils. *Environmental Research, 218*, 115076.

Okoye, M. O., & Eze, P. C. (2023). Challenges in regulatory frameworks for environmental remediation in Nigeria: A review of bioremediation guidelines. *Nigerian Journal of Environmental Sciences and Technology, 7*(2), 189–198. https://doi.org/10.36263/nijest.2023.02.013

Okoye, P. A., & Eze, S. C. (2023). Regulatory landscapes for genetic engineering in environmental remediation: Opportunities and challenges. *Environmental Reviews, 31*(1), 75–84.

Onwubuya, E. A., Cundy, A. B., & Booth, C. A. (2023). *Chromolaena odorata* for lead phytoremediation in Nigerian mining areas. *Environmental Geochemistry and Health, 45*(4), 1234–1248.

Park, S., Lee, J., & Kim, M. (2023). Bioreactor design and optimization for ex situ bioremediation applications. *Biochemical Engineering Journal, 195*, 108912.

Patel, R. K., Sharma, A., & Jain, M. (2024). Biostimulation optimization for organic contaminant degradation. *Journal of Environmental Management, 352*, 120045.

Pendleton, S., Kumar, R., & Thompson, A. (2024). Forest cover loss and soil degradation: A global analysis. *Forest Ecology and Management, 523*, 120456.

Peterson, M., Anderson, J., & Clark, D. (2024). Coastal erosion under climate change: Impacts on agricultural systems. *Coastal Management, 52*(2), 156–174.

Pieper, D. H., Chen, Y., & Williams, R. (2020). Advances in bioremediation technologies for environmental restoration. *Annual Review of Environment and Resources, 45*, 23–52.

Pieper, D. H., Reineke, W., & Engesser, K.-H. (2020). Microbial degradation of xenobiotic and recalcitrant compounds. In O. van der Donk, S. Chen, & B. Grass (Eds.), *Environmental Microbiology: Fundamentals and Applications* (pp. 349–375). Springer.

Pimentel, D., & Burgess, M. (2013). Soil erosion threatens food production. *Agriculture, 3*(3), 443–463.

Qadir, M., Quillérou, E., Nangia, V., Murtaza, G., Singh, M., Thomas, R. J., … Noble, A. D. (2014). Economics of salt-induced land degradation and restoration. *Natural Resources Forum, 38*(4), 282–295.

Rahman, M. A., Hasegawa, H., & Lim, R. P. (2023a). *Pseudomonas putida* for petroleum hydrocarbon degradation. *Biodegradation, 34*(2), 123–139.

Rahman, M. M., Hasanuzzaman, M., & Hossain, M. S. (2023b). Nutrient imbalance and microbial inhibition: Hidden challenges in biostimulation of contaminated soils. *Chemosphere, 323*, 138216. https://doi.org/10.1016/j.chemosphere.2023.138216

Rahman, M. M., Tareq, S. M., & Islam, M. T. (2023c). Nutrient imbalance and its impact on microbial bioremediation efficacy in acidic soils. *Agronomy*, *13*(2), 498.

Rahman, S., Ahmed, T., & Kumar, P. (2024). Sustainable soil management in the context of climate change: Challenges and opportunities. *Agriculture, Ecosystems & Environment*, *342*, 108567.

Rengasamy, P. (2010). Soil processes affecting crop production in salt-affected soils. *Functional Plant Biology*, *37*(7), 613–620.

Reynolds, J. F., Stafford Smith, D. M., Lambin, E. F., Turner, B. L., Mortimore, M., Batterbury, S. P., … Walker, B. (2007). Global desertification: Building a science for dryland development. *Science*, *316*(5826), 847–851.

Robertson, G. P., & Vitousek, P. M. (2009). Nitrogen in agriculture: balancing the cost of an essential resource. *Annual Review of Environment and Resources*, *34*, 97–125.

Rodriguez, C., Martinez, A., & Lopez, J. (2023). Bioventing and biosparging applications in groundwater remediation. *Groundwater Monitoring & Remediation*, *43*(2), 45–58.

Rodriguez, L., Martinez, C., & Santos, R. (2024a). Mining-induced soil degradation: Assessment and remediation strategies. *Environmental Earth Sciences*, *83*(4), 123.

Rodriguez, M., Thompson, J., & Davis, L. (2024b). White-rot fungi in industrial wastewater treatment: Dye and PCB removal efficiency. *Bioresource Technology*, *392*, 129876.

Sahu, S. K., & Arya, N. (2024). Methane mitigation strategies in agriculture: Recent advances and future directions. *Environmental Science and Pollution Research*, *31*(4), 3921–3936.

Sharma, A., & Gupta, M. (2024). Pesticide residues in food chains: Health implications and regulatory challenges. *Food and Chemical Toxicology*, *187*, 114589.

Sharma, P., Kumar, A., & Singh, R. (2023a). Industrial pollution and soil contamination: A comprehensive assessment. *Environmental Monitoring and Assessment*, *195*(8), 987.

Sharma, P., Kumar, R., & Patel, S. (2023b). Phytoremediation mechanisms and applications for heavy metal contamination. *Plant and Soil*, *489*(1), 127–145.

Sharma, P., Singh, R., & Upadhyay, S. K. (2023c). Combined biochar and white-rot fungi application for pesticide-contaminated paddy soils: Soil health and carbon dynamics. *Science of the Total Environment*, *889*, 164327. https://doi.org/10.1016/j.scitotenv.2023.164327

Sharma, P., Singh, S. P., & Parikh, B. S. (2023d). *Trichoderma harzianum* for pesticide degradation in rice cultivation. *Applied Soil Ecology*, *183*, 104756.

Siles, J. A., & Margesin, R. (2022b). Microbial remediation in extreme pH environments: Limits of adaptation and performance. *Frontiers in Microbiology*, *13*, 832097. https://doi.org/10.3389/fmicb.2022.832097

Siles, J. A., & Margesin, R. (2022a). pH effects on microbial diversity and activity in contaminated soils during bioremediation. *Applied Soil Ecology*, *171*, 104351.

Singh, A., Kuhad, R. C., & Ward, O. P. (2024a). *Vetiveria zizanioides* for chromium remediation in industrial soils. *Chemosphere*, *345*, 140456.

Singh, A., Kumar, R., & Gupta, S. (2024b). *Trichoderma harzianum* for pesticide biodegradation: Mechanisms and field applications. *Environmental Microbiology*, *26*(3), 1234–1248.

Singh, A., & Mishra, A. (2022b). Heavy metal stress and microbial inhibition: Barriers in in situ bioremediation. *Environmental Technology & Innovation*, *26*, 102385. https://doi.org/10.1016/j.eti.2022.102385

Singh, M., Gupta, A., & Sharma, R. (2023a). Microbial mechanisms of heavy metal detoxification and biosorption. *Environmental Pollution*, *324*, 121345.

Singh, M., Patel, R., & Kumar, S. (2024c). Bioaccumulation of heavy metals in food chains: Human health risks. *Ecotoxicology and Environmental Safety*, *267*, 115634.

Singh, R., Kumar, M., & Mittal, A. (2023b). Rhizospheric microbes in pollutant degradation: Mechanisms and applications. *Microbial Ecology*, *85*(3), 567–583.

Singh, R., & Mishra, S. (2022c). Heavy metal toxicity on microbial degradation pathways and bioremediation strategies: A review. *Environmental Technology & Innovation*, *26*, 102314.

Smith, D. R., & Zhang, H. (2022). Digital twin modeling for in situ bioremediation of chlorinated solvents in fractured aquifers. *Water Research*, *228*, 119353. https://doi.org/10.1016/j.watres.2022.119353

Smith, J., Brown, K., & Wilson, P. (2024). Soil organic matter dynamics under climate change: Implications for agriculture. *Soil Biology and Biochemistry*, *189*, 109278.

Smith, P., Reay, D. S., & Smith, J. U. (2021). The potential of the global food system to reduce greenhouse gas emissions. *Nature Food*, *2*, 832–839.

Thompson, A., Davis, L., & Johnson, R. (2024b). Urban soil degradation: Causes, consequences, and mitigation strategies. *Environmental Science & Policy*, *145*, 67–78.

Thompson, J., Brown, K., & Wilson, L. (2022). Natural attenuation processes in contaminated groundwater systems. *Water Research*, *218*, 118456.

Thompson, K., Wilson, M., & Anderson, P. (2024a). Soil organic matter and agricultural productivity: A meta-analysis. *Agronomy Journal, 116*(2), 789–802.

Thompson, R., Anderson, P., & Wilson, M. (2024c). Genetically modified *Deinococcus radiodurans* for mercury bioremediation. *Biotechnology and Bioengineering, 121*(4), 1567–1580.

Uhlik, O., Musilova, L., & Demnerova, K. (2022). Metagenomic tools in the identification of microbial biodegraders in contaminated environments. *Microbial Biotechnology, 15*(5), 1398–1410. https://doi.org/10.1111/1751-7915.14071

Wang, H., Liu, X., & Chen, Q. (2024). Engineered *Pseudomonas putida* for enhanced aromatic hydrocarbon degradation. *Applied and Environmental Microbiology, 90*(6), e02145-23.

Wang, H., Yin, Y., & Jiang, J. (2023). Designing cooperative microbial consortia for pollution control: Advances and strategies. *Science of the Total Environment, 875*, 162607. https://doi.org/10.1016/j.scitotenv.2023.162607

Wang, M., Xu, J., & Liu, F. (2022a). Application of proteomics to understand microbial stress responses in soil bioremediation. *Journal of Proteomics, 261*, 104555. https://doi.org/10.1016/j.jprot.2022.104555

Wang, Q., Liu, H., & Chen, X. (2022b). Cometabolic degradation of recalcitrant organic pollutants: Mechanisms and applications. *Chemical Engineering Journal, 439*, 135678.

Wang, Q., Zhang, L., & Zhou, Y. (2022c). Field application of PAH-degrading bacteria in coal-contaminated soils: Performance limitations and environmental interactions. *Science of the Total Environment, 806*, 151416. https://doi.org/10.1016/j.scitotenv.2021.151416

Wang, Q., & Zhang, T. (2023). Synthetic biology tools for environmental applications: From bioremediation to biosensing. *Trends in Biotechnology, 41*(2), 174–186. https://doi.org/10.1016/j.tibtech.2022.09.007

Williams, A., Davis, R., & Taylor, J. (2023). Oxygen management in bioremediation systems: Aerobic versus anaerobic processes. *Environmental Engineering Science, 40*(4), 178–192.

Win, K. T., Win, K. T., Bellingrath-Kimura, S. D., & Oo, A. Z. (2021). Mitigation of methane emissions from paddy rice fields through water, fertilizer, and crop management: A review. *Sustainability, 13*(2), 836.

Wu, H., Yang, Y., & Liu, X. (2023a). Biostimulation performance in nutrient-variable field environments: A critical assessment. *Chemosphere, 318*, 137905. https://doi.org/10.1016/j.chemosphere.2023.137905

Wu, M., Liu, Y., & Qing, M. (2023b). Nutrient dynamics and microbial activity in biostimulated contaminated soils. *Soil Biology and Biochemistry, 174*, 108784.

Xie, H., Ma, Y., & Yang, C. (2022a). Sequential degradation of persistent organic pollutants in contaminated soils: Processes and challenges. *Environmental Science & Technology, 56*(10), 6843–6853.

Xie, S., Zhang, C., & Feng, Y. (2022b). Understanding the rate-limiting steps in microbial degradation of persistent pollutants: Implications for enhancing bioremediation. *Science of the Total Environment, 826*, 154122. https://doi.org/10.1016/j.scitotenv.2022.154122

Yadav, A. N., Kumar, R., & Dhaliwal, H. S. (2023a). Rhizosphere microbiome-driven bioremediation of contaminated soils: Recent advances and future prospects. *Frontiers in Microbiology, 14*, 1142579. https://doi.org/10.3389/fmicb.2023.1142579

Yadav, K. K., Gupta, N., & Kumar, A. (2023b). Rhizoremediation: Plant-microbe partnerships for environmental restoration. *Plant and Soil, 485*(1–2), 45–67.

Zeng, L., Huang, Y., & Luo, J. (2023a). Integrating deep learning and remote sensing for spatial mapping of oil contamination in coastal wetlands. *Remote Sensing of Environment, 292*, 113429. https://doi.org/10.1016/j.rse.2023.113429

Zhang, J., Li, H., & Wang, C. (2023a). *Pseudomonas aeruginosa* in marine hydrocarbon bioremediation: Performance and mechanisms. *Water Research, 234*, 119823.

Zhang, L., Wang, H., & Chen, F. (2022a). Complex contaminant mixtures and their effects on microbial degradation pathways: A critical review. *Chemosphere, 303*, 135027.

Zhang, M., Li, J., & Wang, K. (2022b). Microbial metabolism in bioremediation: From fundamentals to applications. *Trends in Biotechnology, 40*(8), 956–971.

Zhang, W., Jiang, F., & Ou, J. (2020). Global pesticide consumption and pollution: With China as a focus. *Proceedings of the International Academy of Ecology and Environmental Sciences, 1*(2), 125–144.

Zhang, Y., Chen, X., & Wang, H. (2023b). Biostimulation strategies for petroleum hydrocarbon remediation. *Environmental Science and Pollution Research, 30*(15), 43567–43582.

Zhang, Y., Li, X., & Chen, G. (2023c). Biodegradation kinetics of polycyclic aromatic hydrocarbons under different environmental conditions. *Journal of Hazardous Materials, 442*, 130163.

Zhang, Y., Murphy, M., & Xie, Y. (2024). Toward interpretable AI in bioremediation: Integrating multi-omics data for process prediction *Frontiers in Microbiology, 15*, 1281089. https://doi.org/10.3389/fmicb.2024.1281089

Zhang, Y., Zhang, D., & Hu, X. (2022c). Combined contamination and microbial bioremediation: A bottleneck in field-scale remediation. *Environmental Pollution*, *310*, 119791. https://doi.org/10.1016/j.envpol.2022.119791

Zhang, Y., Zhang, J., Li, X., Chen, G., & Xie, Y.(2023d). Recent advances in the mitigation of nitrous oxide emissions from urea fertilization in agriculture. *Journal of Cleaner Production*, *384*, 135467.

Zhang, Y., Zhou, M., & Shen, X. (2023e). Metagenomics-guided bioremediation: Emerging strategies for precision environmental cleanup. *Science of the Total Environment*, *881*, 163301. https://doi.org/10.1016/j.scitotenv.2023.163301

Zhou, H., Yang, W. T., Zhou, X., Liu, L., Gu, J. F., Wang, W. L., … Liao, B. H. (2020). Accumulation of heavy metals in vegetable species planted in contaminated soils and the health risk assessment. *International Journal of Environmental Research and Public He*

Zhou, M., Wang, J., & Liu, Y. (2023a). Development of bioformulations combining PGPR and pollutant degraders for integrated phytoremediation of heavy-metal–contaminated sites. *Environmental Pollution*, *315*, 120767. https://doi.org/10.1016/j.envpol.2022.120767

Zhou, M., Wang, J., & Liu, Y. (2023b). Enhanced phytoremediation of carbofuran-contaminated soils by *Vetiveria zizanioides* with nutrient amendment: A field-scale study. *Journal of Environmental Management*, *336*, 117643. https://doi.org/10.1016/j.jenvman.2023.117643

Zhou, X., Chen, D., & Fang, Z. (2023c). Novel hydrocarbon-degrading genes revealed by deep metagenomic sequencing in petroleum-contaminated soils. *Environmental Pollution*, *315*, 120326. https://doi.org/10.1016/j.envpol.2022.120326

Zhou, X., Li, Y., & Zhang, M. (2024). Integrated bioremediation of carbofuran-contaminated agricultural soils. *Agriculture, Ecosystems & Environment*, *358*, 108701.

Zhu, Q., Huang, J., & Lu, Y (2022a). Impact of temperature fluctuations on microbial remediation of hydrocarbon-contaminated soils. *Environmental Pollution*, *292*, 118343.

Zhu, Y., Jin, Y., & Wang, C. (2022b). Temperature-induced modulation of microbial degradation kinetics in bioremediation of contaminated soils. *Ecotoxicology and Environmental Safety*, *239*, 113675. https://doi.org/10.1016/j.ecoenv.2022.113675

19 Current Achievements and Challenges

Mohit Deswal, Jaya Saini, Pooja Kadyan, Geetanjli Sahni, Baikunth Kumar Yadav, Rahul Kumar, Ponnusamy Thillai Arasu, and Sudhir Kumar Kataria

19.1 INTRODUCTION

Microbial technologies have emerged as transformative tools in agriculture, offering sustainable alternatives to traditional practices (Batista and Singh, 2021). Historically, farmers unknowingly harnessed the power of microbes through composting and fermentation. With the advent of modern microbiology in the 19th century, scientific understanding of beneficial microbes, such as *Rhizobium* in nitrogen fixation or *Trichoderma* in disease suppression, began to shape agricultural practices. However, the 20th century marked a dramatic shift toward synthetic agrochemicals, leading to high yields but also long-term soil degradation, pollution, and biodiversity loss (Hossain et al., 2022). In response, the 21st century has seen a growing transition back to nature-based solutions, with microbial biotechnology at the forefront of this movement. This shift is particularly relevant in the context of the United Nations' Sustainable Development Goal 2 (SDG 2): Zero Hunger. Microbial technologies contribute directly by enhancing crop productivity, reducing losses from pests and diseases, and improving soil fertility (Singh et al., 2024). Indirectly, they lower input costs for farmers, making food more affordable and agriculture more inclusive. For example, biofertilizers can reduce reliance on expensive synthetic fertilizers, enabling smallholder farmers to cultivate more cost-effectively and sustainably.

Critically, global agricultural sustainability efforts remain inadequate without the integration of microbial innovations (O'Callaghan et al., 2022). Chemical-heavy monoculture systems have pushed ecosystems toward collapse, while climate change exacerbates food insecurity. Microbes offer climate-resilient solutions through improved nutrient cycling, carbon sequestration, and stress tolerance (Crowther et al. 2024). However, their potential remains underutilized due to gaps in awareness, policy, and infrastructure.

Microbial interventions align closely with multiple SDGs (Figure 19.1). They advance Zero Hunger (SDG 2), promote Responsible Consumption and Production (SDG 12), and support Climate Action (SDG 13). This shows that leveraging microbial technologies can help transition toward more sustainable, productive, and equitable food systems.

19.2 KEY BENEFITS OF MICROBIAL TECHNOLOGIES

19.2.1 Soil Health and Fertility Enhancement

Soil fertility is foundational to agricultural productivity, and microbial technologies play a pivotal role in restoring and maintaining it. Nitrogen-fixing bacteria, such as *Rhizobium* (for legumes) and *Azotobacter* (for cereals), naturally convert atmospheric nitrogen into plant-usable forms (Mukherjee and Sen, 2021). Similarly, phosphate-solubilizing microbes like *Pseudomonas* enhance the bioavailability of phosphorus, a critical but often immobile nutrient in soils. Unlike synthetic fertilizers, which often lead to nutrient imbalances and leaching, microbial biofertilizers promote a balanced and self-renewing nutrient cycle. Their ecological compatibility ensures minimal

DOI: 10.1201/9781003650881-19

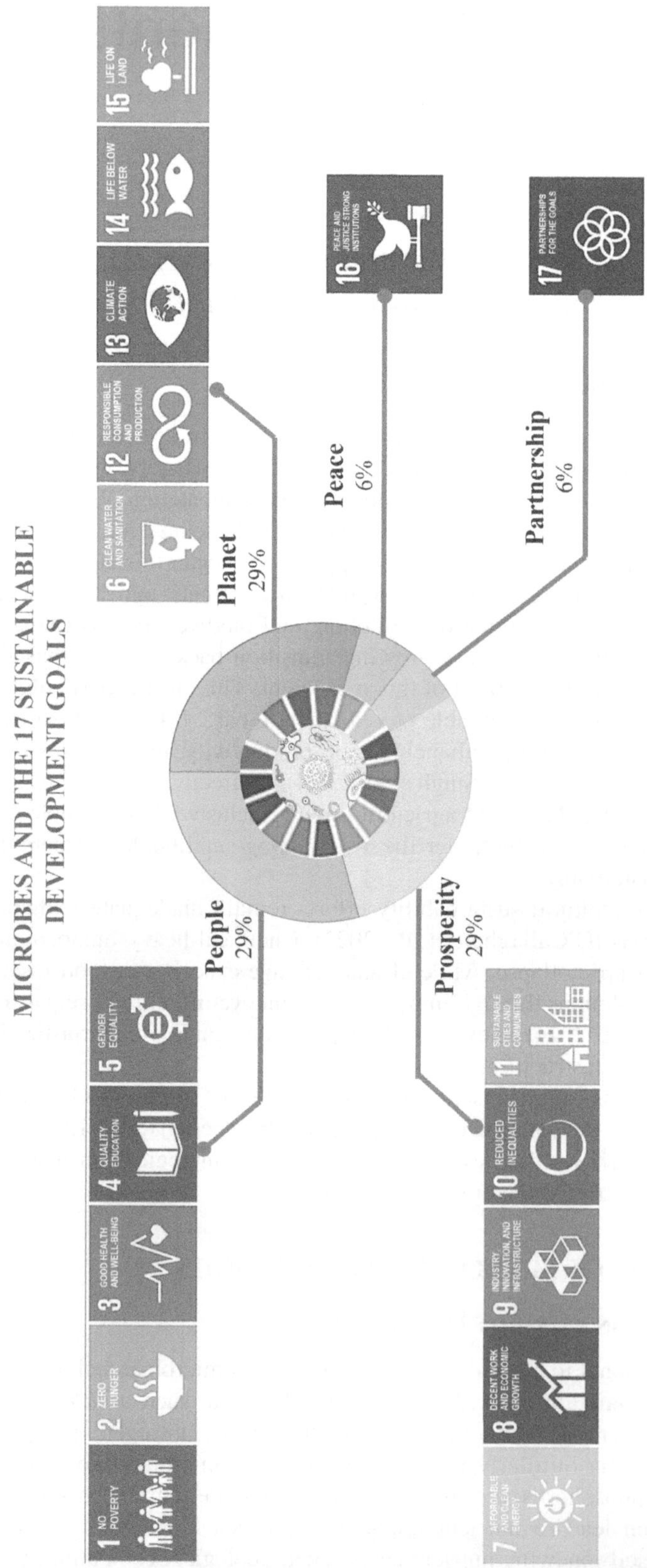

FIGURE 19.1 Microbial interventions aligned with SDGs.

disruption to soil biodiversity, a stark contrast to chemical inputs that often degrade microbial diversity. Moreover, microbial inoculants improve soil structure by producing organic acids and polysaccharides that enhance aggregation, water retention, and root penetration (Sivaram et al., 2023). Critically, the sustainability advantage of microbial biofertilizers lies in their renewability, cost-effectiveness, and non-toxicity. But there are still issues with their consistency and assignation across diverse climatic circumstances, which implies that strain-specific compositions and more advantageous delivery approaches are the path to go.

19.2.2 Pest and Disease Control

Bacillus thuringiensis (Bt) and *Trichoderma* spp. are the common microbial biopesticides that have become very popular as sustainable pest and disease control methods (Vermelho et al., 2024). Bt contains the Cry gene, which codes for toxic insecticidal proteins that are harmful to certain specific insect larvae and make the plant pest-resistant (Figure 19.2). *Trichoderma* spp. also serves antagonistically against pathogenic fungi, by means such as mycoparasitism, where it attacks and feeds on pathogens directly, and induction of systemic resistance against pests through antibiosis, in which it produces certain compounds that inhibit the growth of pathogens.

As compared to chemical pesticides, microbial alternatives have a specific activity, a low environmental footprint of residue, and minimal effect on non-targeted species (Ren et al., 2023). Moreover, they deteriorate faster in the environment, and the risk of bioaccumulation decreases. Their disadvantage, though, is that they work more slowly and rely on ambient conditions, such as humidity and temperature, which may have an influence on effectiveness in a real work environment. Nevertheless, the microbial pesticides have a lesser tendency toward pest resistance development since they have complex modes of action. They should be incorporated into Integrated Pest Management (IPM) systems as a way of minimizing chemical reliance without compromising levels of crop protection.

19.2.3 Improved Crop Yield and Quality

Various research prove the positive relationship between microbial inoculation and crop yield improvement. For example, *Azospirillum* and *Phosphobacteria* field tests in wheat and rice crops indicated up to 20–30% increase in yield. However, yield alone is no longer sufficient; consumers

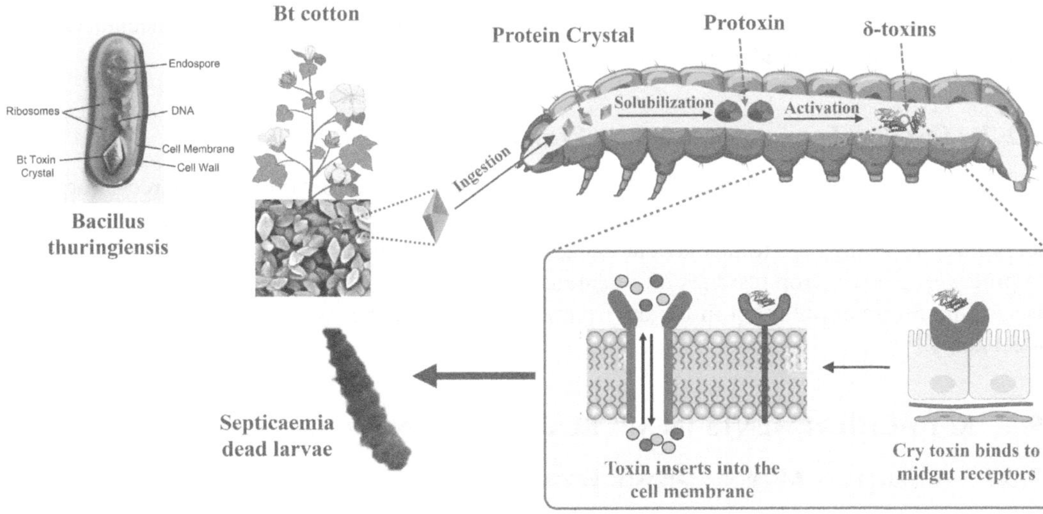

FIGURE 19.2 Microbial mechanisms of pest control.

increasingly demand nutritional quality and shelf life (Rawat et al., 2022). Microbial treatments can enhance antioxidant content, vitamin levels, and post-harvest longevity through stress mitigation and hormonal modulation. The ability of microbes to improve root architecture and nutrient uptake efficiency further contributes to both quantity and quality, ensuring economic and nutritional benefits to both farmers and consumers.

19.2.4 Sustainable Food Production

Sustainable food production requires inputs that are renewable, climate-resilient, and resource-efficient (Suri, 2025). Microbial technologies meet these criteria by significantly lowering the carbon footprint and water use intensity of farming systems. Biofertilizers reduce the production-related emissions of synthetic fertilizers, while microbial drought-tolerant strains enable efficient water use. However, a critical question arises: can microbes alone meet the food demands of a growing global population? While they are vital components, they must be complemented by precision farming, genetic improvements, and supportive policy frameworks. A holistic agroecological transition is essential for ensuring long-term global food security.

19.2.5 Environmental Benefits

Microbial technologies offer significant environmental gains by reducing runoff, eutrophication, and greenhouse gas emissions (Verstraete et al., 2022). By enhancing nutrient use efficiency, they curb the over-application of synthetic fertilizers and minimize nitrogen leaching into water bodies. Moreover, microbes such as mycorrhizal fungi contribute to carbon sequestration by storing carbon in root systems and soil aggregates. These benefits are particularly relevant in climate-smart agriculture. However, ensuring long-term ecological impact requires context-specific microbial consortia, soil health monitoring, and responsible application (Timofeeva et al., 2023). Microbial approaches thus offer a low-impact, high-return pathway for agricultural decarbonization and ecosystem restoration.

19.2.6 Human Health and Pharmaceutical Applications

Microbial technologies extend beyond the farms to human health. Probiotics and nutraceuticals, derived from *Lactobacillus*, *Bifidobacterium*, and other species, promote gut health and immunity (Sharma et al., 2023). Many antimicrobial and pharmaceutical compounds, such as antibiotics and enzymes, originate from agricultural microbial research, illustrating the interconnected value of these organisms in the broader bioeconomy.

19.2.7 Circular Economy and Waste Recycling

Microbes facilitate the conversion of agricultural and food waste into valuable resources. For example, *Lactobacillus* species are applied in the fermentation of silage, enhancing the quality of fodder and promoting storage life (Sharma et al., 2023). On the same note, the microbes used in composting processes convert crop residues into manure and organic fertilizers, sealing nutrient cycles and discharging minimal pollution in the environment. These are the two major features of a circular bioeconomy.

19.3 KEY ACHIEVEMENTS IN MICROBIAL TECHNOLOGIES

19.3.1 Advances in Microbial-Based Fertilizers

One such giant step toward sustainable agriculture was the appearance of commercial fertilizers based on microbes. Microbial-based products that have shown optimized outcomes in terms of minimizing the use of chemicals and maximizing the yield. Some of the products, like BioNPK, which fixes nitrogen, solubilizes phosphorus, and mobilizes potassium due to being a consortium of microbes, have been proven highly beneficial (Marzouk et al., 2025). Similarly, mycorrhizal fungi-based product, MycoApply, has been extensively utilized in the development of roots and nutrient uptake, especially in horticulture and plantation crops. The trend in the adoption is increasing worldwide and in countries that focus especially on climate-smart agriculture. In India, microbial biofertilizers are being publicized in the National Mission on Sustainable Agriculture, and their adoption is high among smallholder farmers (Kumar et al., 2024). In Brazil, the use of *Azospirillum brasilense* in soybean cultivation has replaced significant amounts of synthetic nitrogen. The European Union, under its Farm to Fork Strategy, has placed strong emphasis on organic inputs, leading to increased market penetration of microbial solutions.

However, challenges such as product stability, shelf life, and variable field performance remain barriers to universal adoption and require further innovation and localized research.

19.3.2 Sustainable Pest and Disease Management

Microbial biopesticides have evolved significantly, moving beyond classical strains to next-generation products incorporating RNA-interference (RNAi) technology (Qu et al., 2022). RNA-based biopesticides, when combined with microbial carriers such as *Pseudomonas fluorescens*, enable species-specific gene silencing, offering precision pest control without harming beneficial organisms. This innovation addresses a key weakness of conventional microbial products, slower action and environmental sensitivity, by enabling faster, more targeted responses. However, critical concerns persist regarding the long-term effectiveness of these biopesticides. The risk of pathogen adaptation or gene silencing resistance cannot be dismissed, particularly when applied across multiple growing seasons (Juncheed et al., 2022). Furthermore, their cost of production and lack of harmonized regulatory frameworks have restricted broader adoption. Still, these products represent a major leap toward integrating biotechnology with natural pest management and highlight the need for continuous surveillance and resistance management strategies to retain efficacy.

19.3.3 Microbial Applications in Food Preservation

Microbial biotechnology has significantly impacted food preservation, especially through the use of lactic acid bacteria (LAB) in traditional and modern fermentation processes (Zapaśnik et al., 2022). LAB, such as *Lactobacillus plantarum* and *Lactococcus lactis,* are used to ferment vegetables, dairy, and meat products, producing organic acids, bacteriocins, and other metabolites that inhibit spoilage microorganisms. In the dairy industry, LAB extends the shelf life of milk products while improving safety by preventing pathogens like *Listeria monocytogenes.* In meat preservation, fermentation using starter cultures produces dry sausages with a lower risk of spoilage, thereby reducing food waste and the need for refrigeration (Figure 19.3).

Despite these benefits, consumer acceptance of microbial additives remains mixed. While there is growing interest in fermented and "natural" products, skepticism persists around the use of live cultures in ready-to-eat foods (Zapaśnik et al., 2022). Furthermore, regulatory requirements for microbial food preservatives differ widely across countries, complicating global product standardization.

FIGURE 19.3 Microbial fermentation process in food preservation.

Nevertheless, microbial fermentation technologies continue to redefine preservation techniques, aligning with health trends and reducing reliance on chemical additives.

19.4 TECHNOLOGICAL CHALLENGES

19.4.1 Scalability and Efficiency

While microbial technologies offer sustainable solutions, scaling up their production and ensuring field-level consistency remain major bottlenecks. One core challenge in mass production is maintaining the viability and functional efficiency of microbial spores during formulation, storage, and transportation (Teixidó et al., 2022). Spore-forming microbes like *Bacillus* spp. offer greater shelf-life stability, but many effective strains, such as nitrogen-fixing *Azospirillum*, are non-sporulating and thus highly sensitive to environmental fluctuations. Additionally, ensuring uniform colonization and survival in diverse soil conditions is technically demanding. Microbes must remain viable not just in the manufacturing facility, but also through exposure to UV light, desiccation, and inconsistent field practices. For instance, in India, large-scale deployment of biofertilizers under government schemes like the National Project on Organic Farming faced mixed outcomes (Ravisankar et al., 2021). While pilot trials were successful, farmers reported poor results due to low-quality or improperly stored inoculants. Conversely, Brazil's adoption of *A. brasilense* for soybean production, backed by private-public partnerships and standardized quality control, has demonstrated scalable success. This shows that success is not only related to microbial efficacy, the logistics and quality assurance, and contacting farmers.

19.4.2 Integration with Traditional Agricultural Systems

There is a challenge of integrating microbial technologies into the current agricultural settings, both logistically and behaviorally (Mikiciuk et al., 2024). The conventional ones depend on mechanized tilling, chemically induced pesticides, and flood irrigations that tend to interfere with the inoculant microbes or wash them away. As an example, tillage may kill off positive fungal colonies, or pesticide remnants may kill off the microbes. Furthermore, microbial products normally have certain application principles that cannot be similar to existing farming practices, like the use of seeds or drip irrigation applications. To meet this mismatch, the intensive training of farmers, modern equipment, and agronomic services that meet specific needs should be ensured (Gao et al., 2024). A critical question arises: Can tech-integrated agroecology co-exist with high-input modern farming? The answer depends on adaptive innovation. Blending precision agriculture tools with microbial formulations—for instance, using drone-based delivery of bioinoculants—may bridge the gap. However, without systemic support from agricultural extension services and financial incentives, this integration will likely remain limited to niche farming systems.

19.4.3 Research Gaps and Innovation Needs

Despite progress, microbial agricultural technologies still operate below their full potential due to research and innovation gaps (Batista and Singh, 2021). A key limitation is the incomplete understanding of microbe-plant-soil interactions under varied environmental contexts. To address this, omics technologies, including metagenomics, proteomics, and transcriptomics, must be applied to identify and engineer highly effective microbial strains.

Further, there is an ongoing debate on the use of microbial consortia versus single-strain formulations. While single strains are easier to commercialize and regulate, consortia, combinations of complementary microbes, may offer enhanced synergistic effects in nutrient mobilization, stress tolerance, and pathogen suppression (Batista and Singh, 2021). However, developing and validating consortia is complex due to potential antagonistic interactions among species and challenges in standardization. Innovation is also needed in delivery mechanisms, such as encapsulation, nanocarriers, and biodegradable gel matrices that protect microbes until activation. Ultimately, cross-disciplinary collaboration among microbiologists, agronomists, data scientists, and engineers is essential to unlock the full promise of microbial agriculture in the face of climate and food security challenges.

19.5 SOCIAL AND POLICY CHALLENGES

19.5.1 Public Perception and Acceptance

Public perception remains a significant barrier to the widespread adoption of microbial technologies in agriculture and food systems (Tensi et al., 2022). It is a widespread notion that the application of the concept of live microbes into the area of food preservation or direct impregnation of edible crops is either unsafe or unsanitary. They stem from the general misunderstanding of microbiology, with their fears based on the wrong information, lack of open communication, and rumors. According to behavioral economics, consumers and farmers take decision-making approaches not only on rational-supported evidence but also on emotional concerns, habitual approaches, and social norms (Wuepper et al., 2023). Thus, the outreach measures should rely on the risk communication models involving transparency, trusting credibility, and communicating uniquely to various stakeholders. There must be increased awareness, demonstration plots by farmers, and education of the consumer on the safety, efficacy, and sustainability advantages of microbial products. Unless there is public confidence in them, even the best technologies may be rejected or used to substandard levels.

19.5.2 Regulatory Frameworks and Approval Processes

Regulatory challenges are among the most critical obstacles to scaling microbial innovations. Some of the key regulatory plans across Europe, India, and America have been mentioned in Figure 19.4. Approval processes for microbial inputs are often fragmented, time-consuming, and inconsistent across jurisdictions (Tandon and Singh, 2023). For example:

- The **European Union (EU)** applies a precautionary principle, requiring extensive field trials and multi-year evaluations, often delaying market entry.
- In **India**, regulations are governed by agencies like the FCO (Fertilizer Control Order), but enforcement is inconsistent, leading to a market flooded with substandard or unverified products.
- The **United States** uses a more streamlined process via the EPA and USDA, especially for biopesticides, though it still requires substantial documentation on environmental and health safety.

These differences make international collaboration and trade difficult, especially for companies seeking to operate across markets. There is an urgent need for a globally harmonized regulatory framework that defines microbial products under standardized categories, with mutually recognized risk assessment protocols. Additionally, the approval process should be differentiated for high-risk versus low-risk strains, and for naturally occurring microbes versus genetically modified ones. Such regulatory clarity and speed would not only encourage innovation but also ensure consumer safety and product efficacy.

19.5.3 Economic Barriers

Despite their long-term benefits, microbial technologies face economic disadvantages when compared to conventional inputs. As an example, it is safe to assume that the unit cost of biofertilizers

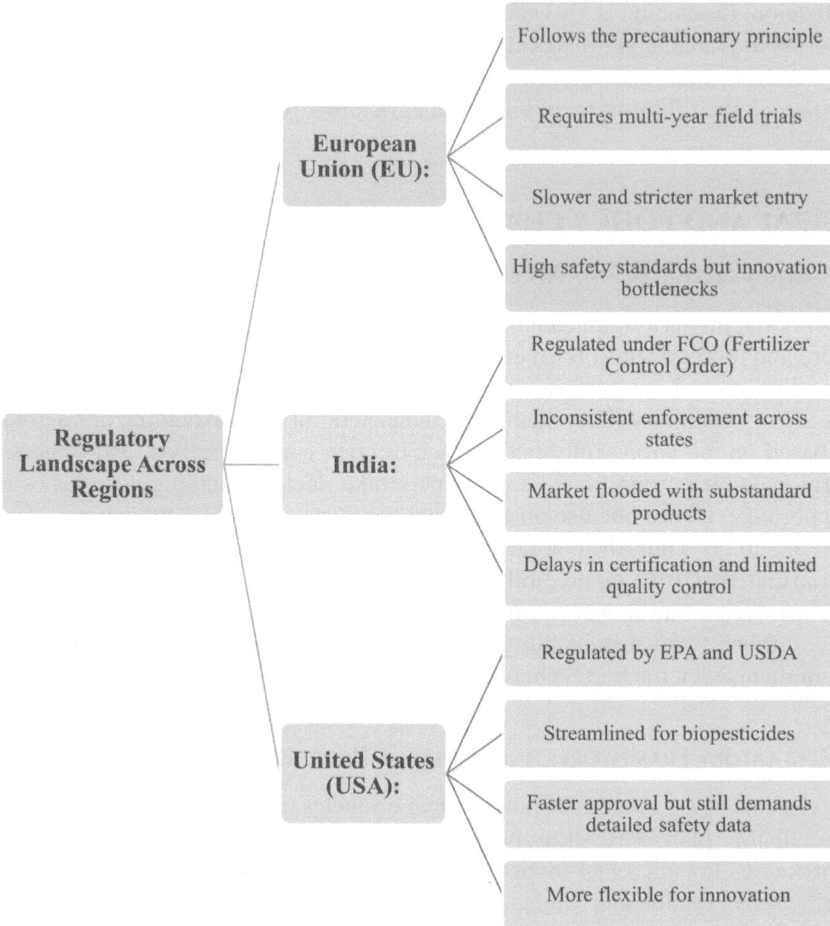

FIGURE 19.4 Comparative regulatory landscape for bioproducts across regions.

in an acre of land can be lower in comparison with urea, although their swifter expiration date, storage peculiarities, and unstable output guarantee the former will be a more hazardous investment of farmers, particularly smallholders and people who cannot afford to lose their crops (Elias et al., 2022). Subsidy assistance on microbial products is half or non-existent in the majority of the developing countries, whereas chemical fertilizers are heavily subsidized by the governments. This perverse subsidy system discourages farmers from experimenting, or switching and adopting bio-based alternatives. In addition, there is an uneven distribution of quality microbial products. Areas that are remote or rural in nature are usually left without suppliers or cold storage and transport infrastructure (Garcia et al., 2024). Consequently, this wave of technology does not reach smallholder farmers who would otherwise benefit the most by having affordable and sustainable innovative inputs. Financial incentives, microloans, and inclusion in public procurement programs, investment in rural distribution systems, and quality certification systems are important in removing these barriers to output and need to be undertaken by the governments concerned.

19.6 CONCLUSION

The microbial technology is a revolutionary area in food security and sustainable agriculture. The overall result of the basic functions of raising soil fertility, holding pests and diseases at bay, boosting crop supplies and quality, and diminishing the deterioration of the environment is an attractive alternative to chemical-oriented farming. Microbial applications play a major role in achieving the United Nations' SDG 2: Zero Hunger by enhancing resource efficiency and, at the same time, SDG 12 (Responsible Consumption and Production) and SDG 13 (Climate Action).

Microbial technologies have their weaknesses in spite of their enormous potential. Constraints due to the economy include poor funding (subsidies), market access, and so on, preventing extensive adoption, in particular among smallholders. Other issues, such as social blockages, namely, misconceptions of the population and regulation bottlenecks, slow the pace as well. There are also technological obstacles that are in need of dire attention, such as disparate product efficiency and a lack of research on interactions between microbes and soil, and plants. The multi-sectoral cooperation is required to address these problems. Governments should develop the necessary policy settings, incentive frameworks; scientists should develop research levels into microbial consortia and adaptation to varying agro-climatic settings; the private sector should invest in delivery vehicles that they can scale, and educational institutes should integrate microbial literacy into the education agro-system. The prospects are even more thrilling when it comes to integrating synthetic biology and artificial intelligence (AI) in developing the next-generation microbial solutions, which are designed to be resilient, specific, and more effective. Due to the possibility of AI-driven formulation tool optimization of microbe selection to various soils and crops, innovation and precision in the agricultural field will be accelerated.

To conclude, there is no silver bullet in microbial technologies, yet they are a potent engine of change toward a more sustainable, resilient, and equitable international food system. Their tactical combination of policy, science, and practice can open a door to the future where the problem of hunger is not eliminated at the cost of chemical dependency, but is given life through living systems that coevolve with nature.

REFERENCES

Batista, B. D. and Singh, B. K., 2021. Realities and hopes in the application of microbial tools in agriculture. *Microbial Biotechnology*, *14*(4), pp. 1258–1268.

Crowther, T. W., Rappuoli, R., Corinaldesi, C., Danovaro, R., Donohue, T. J., Huisman, J., Stein, L. Y., Timmis, J. K., Timmis, K., Anderson, M. Z. and Bakken, L. R., 2024. Scientists' call to action: Microbes, planetary health, and the sustainable development goals. *Cell*, *187*(19), pp. 5195–5216.

Elias, A. M., Favaro, C. P., Bettiol, W., Ribeiro, C. and Farinas, C. S., 2022. Economic aspects on the production of biofertilizers as granular composites. *ACS Sustainable Chemistry & Engineering*, *10*(31), pp. 10426–10438.

Gao, J., Bambrah, C. K., Parihar, N., Kshirsagar, S., Mallarapu, S., Yu, H., Wu, J. and Yang, Y., 2024. Analysis of various machine learning algorithms for using drone images in livestock farms. *Agriculture*, *14*(4), p. 522.

Garcia, E. D. S., Quaresma, N., Aemro, Y. B., Coimbra, A. P. and De Almeida, A. T., 2024. Cooling with the sun: Empowering off-grid communities in developing countries with solar-powered cold storage systems. *Energy Research & Social Science*, *117*, p. 103686.

Hossain, M. E., Shahrukh, S. and Hossain, S. A., 2022. Chemical fertilizers and pesticides: Impacts on soil degradation, groundwater, and human health in Bangladesh. In V.P. Singh, S. Yadav, K.K Yadav, and R.N Yadava (Eds.), *Environmental Degradation: Challenges and Strategies for Mitigation* (pp. 63–92). Cham: Springer International Publishing.

Juncheed, K., Tanunchai, B., Wahdan, S. F. M., Thongsuk, K., Schädler, M., Noll, M. and Purahong, W., 2022. Dark side of a bio-based and biodegradable plastic? Assessment of pathogenic microbes associated with poly (butylene succinate-co-adipate) under ambient and future climates using next-generation sequencing. *Frontiers in Plant Science*, *13*, p. 966363.

Kumar, A., Saharan, B. S., Parshad, J., Gera, R., Choudhary, J. and Yadav, R., 2024. Revolutionizing Indian agriculture: The imperative of advanced biofertilizer technologies for sustainability. *Discover Agriculture*, *2*(1), p. 24.

Marzouk, S. H., Kwaslema, D. R., Omar, M. M. and Mohamed, S. H., 2025. Harnessing the power of soil microbes: Their dual impact in integrated nutrient management and mediating climate stress for sustainable rice crop production" A systematic review. *Heliyon*, *11*(1), e41158.

Mikiciuk, G., Miller, T., Kisiel, A., Cembrowska-Lech, D., Mikiciuk, M., Łobodzińska, A. and Bokszczanin, K., 2024. Harnessing beneficial microbes for drought tolerance: A review of ecological and agricultural innovations. *Agriculture*, *14*(12), p. 2228.

Mukherjee, R. and Sen, S., 2021. Agricultural sustainability through nitrogen fixation: Approaches and techniques. *Harvest*, *6*(1), pp. 48–55.

O'Callaghan, M., Ballard, R. A. and Wright, D., 2022. Soil microbial inoculants for sustainable agriculture: Limitations and opportunities. *Soil Use and Management*, *38*(3), pp. 1340–1369.

Qu, M., Merzendorfer, H., Moussian, B. and Yang, Q., 2022. Bioinsecticides as future mainstream pest control agents: Opportunities and challenges. *Frontiers of Agricultural Science and Engineering*, *9*(1), pp. 82–97.

Ravisankar, N., Ansari, M. A., Panwar, A. S., Aulakh, C. S., Sharma, S. K., Suganthy, M., Suja, G. and Jaganathan, D., 2021. Organic farming research in India: Potential technologies and way forward. *Indian J. Agron*, *66*(5th IAC Special issue), pp. S142–S162.

Rawat, P., Sharma, A., Shankhdhar, D. and Shankhdhar, S. C., 2022. Improvement of phosphorus uptake, phosphorus use efficiency, and grain yield of upland rice (*Oryza sativa* L.) in response to phosphate-solubilizing bacteria blended with phosphorus fertilizer. *Pedosphere*, *32*(5), pp. 752–763.

Ren, Z., Cai, T., Wan, Y., Zeng, Q., Li, C., Zhang, J., Ma, K., He, S., Li, J. and Wan, H., 2023. Unintended consequences: Disrupting microbial communities of *Nilaparvata lugens* with non-target pesticides. *Pesticide Biochemistry and Physiology*, *194*, p. 105522.

Sharma, N., Kang, D. K., Paik, H. D. and Park, Y. S., 2023. Beyond probiotics: A narrative review on an era of revolution. *Food Science and Biotechnology*, *32*(4), pp. 413–421.

Singh, S., Ashoka, P., Ahlawat, U., Changdeo, W. B., Rehsawla, R., Naruka, A. and Sharma, D., 2024. Mechanisms and applications of microbial biotechnology in soil health and agricultural productivity: A review. *Journal of Advances in Biology & Biotechnology*, *27*, pp. 1420–1438.

Sivaram, A. K., Abinandan, S., Chen, C., Venkateswartlu, K. and Megharaj, M., 2023. Microbial inoculant carriers: Soil health improvement and moisture retention in sustainable agriculture. *Advances in Agronomy*, *180*, pp. 35–91.

Suri, S., 2025. Building resilience: Sustainable agriculture for food security in a changing climate. *International Journal of Nutrition and Food Sciences*, *14*(2), pp. 98–104.

Tandon, A. and Singh, K., 2023. Advancements, Challenges, and Policy Interventions in Agriculture-Related Manufacturing in India: A Critical Analysis.

Teixidó, N., Usall, J. and Torres, R., 2022. Insight into a successful development of biocontrol agents: Production, formulation, packaging, and shelf life as key aspects. *Horticulturae*, *8*(4), p. 305.

Tensi, A. F., Ang, F. and Van der Fels-Klerx, H. J., 2022. Behavioural drivers and barriers for adopting microbial applications in arable farms: Evidence from the Netherlands and Germany. *Technological Forecasting and Social Change*, *182*, p. 121825.

Timofeeva, A. M., Galyamova, M. R. and Sedykh, S. E., 2023. Plant growth-promoting bacteria of soil: Designing of consortia beneficial for crop production. *Microorganisms*, *11*(12), p. 2864.

Vermelho, A. B., Moreira, J. V., Akamine, I. T., Cardoso, V. S. and Mansoldo, F. R., 2024. Agricultural pest management: The role of microorganisms in biopesticides and soil bioremediation. *Plants*, *13*(19), p. 2762.

Verstraete, W., Yanuka-Golub, K., Driesen, N. and De Vrieze, J., 2022. Engineering microbial technologies for environmental sustainability: Choices to make. *Microbial Biotechnology*, *15*(1), pp. 215–227.

Wuepper, D., Bukchin-Peles, S., Just, D. and Zilberman, D., 2023. Behavioral agricultural economics. *Applied Economic Perspectives and Policy*, *45*(4), pp. 2094–2105.

Zapaśnik, A., Sokołowska, B. and Bryła, M., 2022. Role of lactic acid bacteria in food preservation and safety. *Foods*, *11*(9), p. 1283.

20 Future Vision
The Path to Zero Hunger Using Microbial Technologies

Maria M. Adeyemi, Bunmi C. Kotun, and Oluwademilade Opesanwo

20.1 INTRODUCTION

Hunger is a major challenge affecting the world, affecting over 820 million people worldwide. With food security facing significant challenges like population growth, climate change, poverty, conflict such as civil war, terrorism, and inadequate food distribution, striving to end world hunger and achieve food security therefore became a goal for the United Nations (UN). All these challenges have effect on the food and nutrition of people around the world. One method being looked at, hoping to end this global hunger, and creating a hunger free future, is the use of microbial technologies which involve the use of microorganisms such as bacteria, fungi, and yeast to develop ideas to combat against human and environmental problems as they have the potential of transforming the food system, improving crop yields, enhancing nutrition, and improving the environment. Microbial technologies refer to the strategic use of microorganisms such as bacteria, fungi, and yeasts in various industrial, agricultural, and environmental applications to achieve beneficial outcomes. These technologies harness the natural metabolic capabilities of microorganisms to catalyze biochemical reactions that lead to the breakdown, conversion, or transformation of substances. In the specific context of food waste reduction, microbial technologies play a vital role by enabling the degradation of organic material, the recovery of nutrients, and the conversion of waste into useful bio-products. The scope of microbial technologies extends across numerous sectors, including agriculture, where they enhance soil fertility and support sustainable farming practices; bioenergy, where they are utilized to produce renewable energy sources such as biogas; pharmaceuticals, where microbial fermentation is instrumental in the synthesis of antibiotics and enzymes; and waste management, where they provide environmentally friendly alternatives to traditional disposal methods. Their versatility and adaptability make microbial technologies essential tools in the global effort to promote sustainability and mitigate environmental degradation which will alleviate hunger.

Some ways of achieving this zero hunger using microbial technologies includes Improved crop yield and nutrition and microbial-based fertilizers:

20.1.1 Improve Crop Yields and Nutrition

Improving crop yields and nutrition will help in eradicating world hunger, as increase in crop yields will allow for more distribution and less scarcity. It has been known that microbes can trigger the immune system of plants especially in the presence of plant pathogen (Adeyemi et al., 2025). These microbes help plants to suppress pathogens and indirectly promote plant growth.

Identified ways to achieve this improvement in crop yields and nutrition include:

20.1.1.1 Biofortification of Plants

Biofortification improves the quality and quantity of plant crops. Naturally occurring microbes maybe used as inoculants to maintain crop yields despite decreased resources such as water and

DOI: 10.1201/9781003650881-20

nutrient. Combining different classes of soil organisms within one inoculant has an advantage of creating multiple plant-growth-promoting mechanisms. Plant symbiotic relationships with *mycorrhizal* fungi alter root aquaporin gene expression and greatly increase the surface area over which plant root systems take up water and nutrients. Also, Soil bacteria on the root surface alter root phytohormone status thereby increasing growth and can make nutrients more available to the plant (Dodd & Ruiz-Lozano, 2012; Naik et al., 2024; Sagar et al., 2021).

20.1.1.2 Microbial-Based Fertilizers

Microbial fertilizers, also known as biofertilizers, are mixtures that constitute microorganisms such as actinomycetes, fungi, and bacteria that boost soil fertility by promoting plant development and the formation of new microbial community structures (Wang et al., 2024). Furthermore, crops are assisted by microbial fertilizers in overcoming biotic and abiotic stresses. Sustainable agriculture is now made possible by microbial fertilizers, which are thought to be renewable and environmentally safe. They are considered environmentally friendly organic and biological fertilizers made by fermenting beneficial microbial bacterial colonies (Wu et al., 2024).

20.2 TYPES OF MICROBIAL-BASED FERTILIZERS

Four categories can be used to organize the many kinds of biofertilizers that aid plant's growth at different stages of growth and all-round fertility of the soil as stated below.

20.2.1 N-Fixing Biofertilizers

About 80% of the nitrogen in the atmosphere in its free state is unable for plants to use, hence certain kinds of microorganisms are required to fix nitrogen, and therefore these microbes are referred to as biological nitrogen fixers (BNFs). They convert the inert N_2 into an organic form that plants may use. Up to 25% of the nitrogen in plants is obtained through N-fixation (Nosheen et al., 2021). They are divided into blue-green algae, symbionts like *Rhizobium, Frankia,* and *Azolla*, and free-living bacteria like *Azotobacter* and *Azospirillium. Rhizobium, Mesorhizobium, Azorhizobium, Bradyrhizobium, Sinorhizobium,* and *Allorhizobium* are among the N_2-fixing bacteria that are linked to legumes, while *Achromobacter, Alcaligenes, Arthrobacter, Acetobacter, Azomonas, Beijerinckia, Clostridium, Bacillus, Enterobacter, Erwinia, Desulfovibrio, Derxia, Corynebacterium, Campylobacterium, Herbaspirillum, Klebsiella, Lignobacter, Mycobacterium, Rhodospirillum, Rhodo-pseudomonas, Xanthobacter, Mycobacterium,* and *Methylosinus* are the bacteria that are linked to legumes (Nosheen et al., 2021; Daniel et al., 2022). They act by increasing the amount of N_2 in the soil through the process of fixing atmospheric nitrogen and making it available to plants.

20.2.2 Phosphate-Solubilizing Biofertilizers (PSB)

The insoluble forms of phosphate must be changed into soluble forms by microorganisms (Prabhu et al., 2019), the process of phosphate solubilization involves several bacterial and a few fungal species. Phosphate-solubilizing bacteria (PSB) use a variety of processes, such as ion exchange reactions, chelation, and the synthesis of organic acids, to turn insoluble phosphates, like HPO_4 and H_2PO_4, into soluble forms. Organic acids with hydroxyl (gluconic) and carboxyl (ketogluconic) groups that chelate the cation attached to the phosphate and change it into the soluble form that the plants can use are among the metabolites that the PSB can release. The PSB improves plant growth by supplying phosphate and other trace elements like Fe and Zn. Additionally, they produce the enzyme that eliminates the infections, shielding the plant from diseases. *Bacillus edaphicus, Arthrobacter* spp., *Bacillus, Mucilaginosus,* and *Bacillus circulans* can produce organic acids that degrade silicates and aid in the removal of metals to solubilize potassium (silicates) ions and make it available to plants. *Pseudomonas striata, B. circulans, Bacillus subtilis, Penicilium* spp., *Bacillus polymyxa,*

Agrobacterium, Microccocus, Flavobacterium, Aereobacterium. Aspergillus awamori, Penicillum spp., and *Trichoderma* spp. help dissolve bound phosphates, they secrete organic acids and lower soil pH by converting insoluble forms of P in the soil into soluble forms (Daniel et al., 2022). Other examples of phosphate solubilizers include strains of the bacterial genera *Pseudomonas, Bacillus, Rhizobium,* and *Enterobacter* as well as the fungus *Penicillium* and *Aspergillus.*

20.2.3 Plant-Growth-Promoting Rhizobacteria

Plant-growth-promoting rhizobacteria are a group of free-living symbiotic rhizosphere bacteria that colonize plant roots and promote plant growth. They benefit the host plant in a variety of ways, including nitrogen fixation and nodulation (Gouda et al., 2018). They function as biofertilizers by encouraging plant growth and development, enabling biotic and abiotic stress tolerance, and aiding in the mineralization of the soil through the breakdown of organic matter, also generate hormones that promote crop yields, enhance nutrient availability, and stimulate root growth. By enhancing resource acquisition or altering the plant's hormone levels, or by reducing the inhibitory effects of different pathogenic agents on plant growth and development, the bacteria can either directly or indirectly stimulate plant growth (Oleńska et al., 2020). Beneficial bacteria like *Pseudomonas fluorescens* and *B. subtilis* enhance plant growth by producing growth hormones and protecting against pathogens.

20.3 MECHANISMS OF ACTION OF MICROBIAL-BASED FERTILIZERS

Microbial-based fertilizers enhance soil fertility and plant growth through several mechanisms:

- *Nutrient Mobilization*: microbes convert atmospheric nitrogen, solubilize phosphorus and potassium, and decompose organic matter, making essential nutrients available to plants (Bhattacharyya and Jha, 2012; Glick, 2012).
- *Improved Soil Structure*: beneficial microbes, for example, *Bacillus, Pseud*omonas, and arbuscular mycorrhizal fungi promote soil aggregation and aeration, enhancing water retention and root penetration (Rashid et al., 2016).
- *Biocontrol of Plant Diseases*: some microbial fertilizers contain antagonistic microbes that suppress plant pathogens, reducing the need for chemical pesticides (Solórzano et al., 2025).
- *Enhanced Root Growth*: certain bacteria and fungi produce phytohormones that stimulate root elongation and branching, leading to better nutrient uptake (Egamberdieva et al., 2017).

20.3.1 Microbial Pest Control: Biological Control of Pests and Diseases Using Microorganisms

Microbial pest control is a biological approach to managing pests and plant diseases that utilize beneficial microorganisms, including bacteria, fungi, viruses, and protozoa. These organic substances are specifically designed to combat agricultural pests while minimizing environmental damage and protecting valuable produce (Sawangproh et al., 2025).

20.4 TYPES OF MICROBIAL AGENTS USED IN PEST CONTROL AND THEIR MODES OF ACTION

- *Bacterial Biopesticides*
  - *Bacillus thuringiensis (Bt)*: produces toxins that kill insect larvae, widely used against caterpillars, mosquitoes, and beetles.
  - *P. fluorescens*: suppresses plant pathogens and enhances plant defense mechanisms.

- *Fungal Biopesticides*
  - *Beauveria bassiana*: infects and kills insects by penetrating their cuticle.
  - *Metarhizium anisopliae*: used against soil-dwelling insects such as termites and beetles.
  - *Trichoderma* spp.: acts as a biocontrol agent against fungal plant pathogens.
- Viral Biopesticides
  - *Nucleopolyhedroviruses (NPVs)* and *Granuloviruses (GVs)*: specifically infect and kill insect pests, particularly caterpillars.
- Protozoan Biopesticides
  - *Nosema locustae*: a protozoan pathogen used against locusts and grasshoppers.

20.5 MECHANISM OF ACTION OF BIOPESTICIDES

Biopesticides work through various mechanisms, such as

a. *Biochemical Disruption*: where they interfere with the biochemical processes of pests, affecting their growth, development, or reproduction; an example is *Bt*. This bacterium produces delta-endotoxins (Cry and Cyt toxins), which disrupt the midgut epithelial cells of susceptible insect larvae when ingested, leading to their death (Sanahuja et al., 2011; Palma et al., 2014).
b. *Microbial Activity-Infecting or Parasitizing Pests*: *M. anisopliae*: this entomopathogenic fungus infects and kills a wide range of insect pests through direct infection and parasitism, not just by toxins or secreted metabolites. *M. anisopliae* germinates on the insect cuticle, penetrating it via enzymatic and mechanical action. It multiplies and consumes host tissues as it grows inside the hemocoel. The insect eventually perishes as a result of tissue damage and potentially secondary metabolite synthesis. The fungus releases conidia to infect new victims after sporulating on the cadaver after death (Pandey, 2013; Tadesse et al., 2025).
c. *Predation and Competition*: acting as natural predators or competitors of pests to limit their population growth, although these mechanisms are less commonly exploited than infection or biochemical disruption. Examples include the soil-dwelling bacteria *Myxococcus xanthus* actively feeds on other microorganisms, particularly plant pathogens, and demonstrates coordinated swarming activity. It releases antibiotics and lytic enzymes that degrade the cell walls of its prey. As a biocontrol agent, it can lower the numbers of bacterial and fungal phytopathogens (Muñoz-Dorado et al., 2016). By colonizing the rhizosphere and outcompeting pathogens for nutrients and space, *P. fluorescens* works through competitive exclusion. It produces siderophores, enzymes, and antibiotics (such as 2,4-diacetylphloroglucinol) that inhibit fungal pathogens like *Pythium* and *Fusarium* spp., which lower disease incidence and promote plant growth (Santoyo et al., 2016).
d. *Repellency-Deterring Pests from Attacking Crops*: *M. anisopliae* works by producing secondary metabolites and volatile organic compounds (VOCs) that deter termites and mosquitoes and locusts which are the target pests. Natural repellent properties are exhibited by VOCs like methyl isovalerate and 1-octen-3-ol (Moosa-Kazemi et al., 2021). The mechanism by which *B. bassiana* repels insects from treated plants is by the production of metabolites, such as beauvericin, which target pests such as thrips, whiteflies, and aphids and help to decrease their attraction to treated surfaces. *P. fluorescens*: acts by generating VOCs that discourage insect pests and causes systemic resistance in plants. The repellent action: *Diabrotica virgifera* (corn rootworm) a nematode and other pests are repelled by the VOCs from roots treated with *P. fluorescens* (Pieterse et al., 2014).

20.6 BENEFITS OF MICROBIAL PEST CONTROL

Minimizes environmental pollution, lessens dependency on chemical pesticides, only affects the targeted pest, protects biodiversity and beneficial insects, and does not pose a harmful risk to organisms that are not the intended target.

20.6.1 Microbial Food Production: Fermentation and Other Microbial Processes for Food Production

Microbial food production involves the use of beneficial microorganisms such as bacteria, yeasts, and molds in food processing. These microbes play a key role in fermentation, food preservation and sustainability, and the enhancement of nutritional value, texture, and flavor. Microbial-based food production has been practiced for centuries and is widely used in modern food biotechnology.

20.7 SCOPE IN MICROBIAL FOOD PRODUCTION

Fermented foods: In order to produce fermented foods like bread, cheese, yogurt, soy sauce, and traditional drinks, microorganisms are essential in the process. Dairy fermentation frequently uses lactic acid bacteria (LAB), such as *Lactobacillus* species, *Streptococcus thermophilus,* and *Leuconostoc* species. The manufacture of bread and alcoholic beverages depends heavily on yeasts, particularly *Saccharomyces cerevisiae*. Miso and soy sauce are examples of soy-based fermentations that involve molds like *Aspergillus oryzae* (Tamang et al., 2016).

Microbiological enzymes in the production of food: Enzymes such as lipases, amylases, and proteases are produced by microbes and are employed in food processing to improve digestibility, flavor, and texture. Meat tenderization, cheese manufacture, and starch conversion all involve the use of industrial enzymes derived from *Aspergillus niger* and *Bacillus subtilis* (Song et al., 2023).

Production of probiotics: Probiotics are good bacteria that are added to dairy products and dietary supplements for their health-promoting qualities. *Bifidobacterium bifidum* and *Lactobacillus acidophilus* are two species that are used for nutrient absorption, immunity, and gastrointestinal health (Hill et al., 2014).

Food Additives made by microbes: Microbes are designed to make organic acids (like citric acid), vitamins (e.g., B_{12}), and amino acids (such as glutamate). Glutamic acid, utilized in MSG, is produced by *Corynebacterium glutamicum*. The main organism used in the commercial synthesis of citric acid is *A. niger* (Hirasawa and Shimizu, 2016; Show et al., 2015).

Single-cell protein (SCP) production: Single-cell protein microorganisms are cultivated to produce high-protein biomass used as food or feed. Microbes like *Spirulina* (cyanobacteria), *Candida utilis*, and *Chlorella* (algae) are rich in protein and used to supplement human or animal diets (Nasseri et al., 2011).

Production of novel proteins by microbes: Alternative proteins are synthesized by microorganisms such as *Fusarium venenatum*, mycoprotein-producing fungi, *Spirulina* (cyanobacteria), genetically engineered *Escherichia coli*, or yeast (Finnigan et al., 2025).

Microbial flavorings and sweeteners: Using modified *S. cerevisiae* or *Corynebacterium*, microbes can biosynthesize natural sweeteners like steviol glycosides (e.g., Reb M) or taste enhancers like vanillin and methylbutenol and erythritol in yeast (Awofe et al., 2025) compared to plant extraction, this is more economical and environmentally friendly.

20.8 ADVANTAGES OF FOOD PRODUCTION BY MICROBES

Improved food preservation: Fermentation prevents the growth of spoilage microorganisms and extends food shelf life. Increased nutritional value: Microbial activity increases the bioavailability of vitamins, amino acids, and other nutrients. Sustainability: Produces food on a big scale using renewable microbiological resources. Health benefits: Bioactive substances and probiotics support immunological and digestive health.

20.8.1 Conversion of Organic Waste to Nutritious Food

A promising application of microbial technologies is the conversion of food waste into high-protein animal feed. This approach addresses two major challenges simultaneously: the reduction of food waste and the need for sustainable livestock nutrition. Certain microbial fermentation processes enhance the nutritional quality and safety of food waste by reducing pathogens and anti-nutritional factors. For instance, LAB are used to improve the palatability and shelf life of fermented food products. Additionally, insects such as black soldier fly larvae are cultivated on fermented food waste, converting it into a protein-rich biomass suitable for feeding poultry, fish, and livestock. This method significantly reduces the demand for conventional feed ingredients like soy and fishmeal and contributes to circular bio-economic practices. By diverting organic waste from landfills and transforming it into valuable animal feed, this technology presents a sustainable alternative within agricultural and aquaculture systems.

20.8.2 Supporting Sustainable Agriculture

This involves ensuring agricultural products are not depleted or destroyed by microorganisms/microbial growth without damaging the environment. This may be achieved through:

20.8.2.1 Organic Farming and Regenerative Agriculture

In recent years, global agricultural food production has become increasingly dependent on the use of non-renewable and/or scarce resources, and fossil fuel (e.g., for the production of mineral N-fertilizer and pesticides), mined minerals used as P and K fertilizers, and water used for irrigation (Rempelos et al., 2023).

20.8.2.2 Microbial-Based Soil Health Management

This uses the beneficial importance of soil microbes in the improvement of soil fertility and nutrient cycling and overall soil health. This approach involves using microbial inoculants and promoting the growth of beneficial microbes to enhance soil structure, improve nutrient availability, and increase plant resilience to disease (Adewara et al., 2024).

20.8.3 Waste Reduction and Food Preservation

Food waste continues to present significant environmental, economic, and social challenges. With increasing awareness of the need for sustainable development, microbial technologies have emerged as innovative and essential tools in addressing this issue. By leveraging the metabolic potential of microorganisms, food waste can be transformed into valuable resources, thus contributing to the goals of sustainability and circular economy models. Microbial technologies for waste reduction and food preservation include the following.

20.8.3.1 Anaerobic Digestion for Biogas Production

Anaerobic digestion is a well-established microbial process that utilizes anaerobic bacteria to degrade organic food waste in the absence of oxygen. This biochemical reaction results in the generation of biogas, a mixture primarily composed of methane and carbon dioxide, which can be captured and used as a renewable source of energy for electricity, heating, or fuel. In addition to biogas, the process yields a nutrient-rich byproduct known as digestate, which is often applied as a biofertilizer in agricultural fields. The dual benefit of energy recovery and soil amendment makes anaerobic digestion a particularly attractive solution for food waste management in municipal treatment facilities and large-scale agricultural operations.

This technology not only reduces the environmental burden of organic waste disposal but also contributes to the reduction of greenhouse gas emissions and promotes energy independence.

20.8.3.2 Fermentation and Bioconversion Technologies

Microbial fermentation is a versatile bioconversion process that transforms food waste into a variety of commercially valuable biochemical products. Utilizing specific microbial strains, food waste materials rich in carbohydrates, sugars, and starches are converted into substances such as bioethanol, lactic acid, and enzymes. Bioethanol production involves the fermentation of sugar-rich food waste, such as fruit peels and leftover bakery products, using yeast strains like *S. cerevisiae*. The resulting ethanol serves as an alternative fuel and a key industrial solvent. Lactic acid is produced through the fermentation of carbohydrate-dense waste materials by LAB. This compound is a precursor for polylactic acid, a biodegradable plastic that is increasingly used in sustainable packaging solutions. Enzyme production involves cultivating microbes on food waste substrates to extract industrially significant enzymes, including amylase, protease, and cellulase. These enzymes are essential in food processing, textiles, pharmaceuticals, and biofuel production. Through these processes, food waste is effectively repurposed, minimizing environmental impact and generating economic value.

20.8.3.3 Microbial Fuel Cells

Recent advancements have introduced the concept of microbial fuel cells (MFCs) (Kotun et al., 2024), an innovative technology that harnesses the ability of electrogenic bacteria to generate electricity from organic matter, including food waste. In MFCs, bacteria metabolize food waste and release electrons, which are captured through an anode and cathode system to produce an electric current. This form of bioelectrochemical energy generation offers a low-cost and environmentally friendly solution for small-scale electricity production, particularly in remote or off-grid locations.

Although still in the early stages of development, MFCs demonstrate the immense potential of integrating waste management with decentralized energy systems.

20.8.3.4 Bioplastic and Biopolymer Production

Certain bacterial strains, such as *Ralstonia eutropha*, are capable of synthesizing polyhydroxyalkanoates (PHAs), a type of biopolymer, using food waste as a carbon source. These bioplastics are biodegradable and provide a sustainable alternative to petroleum-based plastics.

The use of food waste as a substrate for bioplastic production reduces environmental pollution from both plastic and organic waste, contributing to a cleaner and more circular economy.

With increasing global interest in reducing plastic pollution, microbial biopolymer production represents a dual-purpose solution for sustainable packaging and food waste reduction.

20.8.3.5 Wastewater Treatment and Odor Control

Microbial technologies also play a critical role in the treatment of wastewater generated by food processing industries. Microorganisms are employed to break down organic contaminants in liquid waste streams, thereby reducing biological oxygen demand, controlling odors, and minimizing environmental impact. Effective microbial treatment of wastewater enhances the sustainability of food production and processing facilities by promoting water reuse and maintaining compliance with environmental regulations. In addition, biofilters containing specific microbial consortia are used to manage odors associated with food waste storage and treatment, improving air quality and reducing nuisances in surrounding communities.

Preservation of food will lead to a greater increase in available food in the world and reduce global hunger. Globally, an estimated 13% of food is lost between harvest and retail, while 17% of total food available to consumers is wasted at household, food service, and retail levels. This translates to a staggering 931 million metric tons of food wasted annually, with about 60% of that waste occurring in households. The combined toll of food loss and waste on the global economy is estimated to be around USD 1 trillion annually (FAO, 2021). If this can be curtailed and food preserved or even undergo rechauffe (the act of turning left over food into new dishes by adding new ingredients) instead of throwing away, this will greatly improve and help in the aim to end world hunger.

20.9 LIMITATIONS AND CHALLENGES TO UNDERSTANDING MICROBIAL TECHNOLOGIES

Despite their clear advantages, microbial technologies face several limitations and implementation challenges that can hinder their widespread adoption. Some of the most pressing issues include:

a. The need for highly controlled environmental conditions such as temperature, pH, and moisture to optimize microbial activity.
b. High initial capital investment for the setup of processing facilities and supporting infrastructure.
c. Regulatory hurdles and policy gaps that can slow the approval and scaling of microbial technologies.
d. Limited public awareness and understanding of microbial processes, leading to resistance or lack of support for their adoption.
e. Variability in the composition of food waste, which may affect the consistency and efficiency of microbial treatments. Addressing these challenges requires coordinated efforts among researchers, policymakers, and industry stakeholders to develop adaptive technologies, supportive regulations, and educational initiatives.

20.10 OBSTACLES AND UPCOMING DEVELOPMENTS TO MICROBIAL TECHNOLOGIES

a. *Microbial Contamination*: The possibility of harmful microorganisms compromising the safety of food.
b. *Regulatory Approval*: Verifying that microbiological food products fulfill quality and safety requirements.
c. *Biotechnological Developments*: Microbial food applications are growing thanks to genetic engineering and synthetic biology.

20.11 FUTURE OUTLOOK AND ROADMAP TO ZERO HUNGER

The aim and idea for global zero hunger has been around for a while with international bodies like the UNICEF and UN trying to put an end to world hunger in relation to the SDG 2, to prevent all forms of malnutrition and allow for proper nutrition in all children. The 2016 Global Hunger Index (GHI) presents a multidimensional measure of national, regional, and global hunger, focusing on how the world can get to zero hunger by 2030. The developing world has made substantial progress in reducing hunger since 2000. Therefore, the race to ending world hunger maybe by using microbes and microbial technologies. Rather than the idea of planting more or producing more, we can plant smarter with the help of microbes, thereby giving us an actual chance in achieving zero hunger. Through fermentation, precision biology, and waste conversion, and other methods, microbes can help rethink the very foundations of food production. Instead of relying on massive stretches of farmland, overworked soils, and resource-heavy livestock, we can grow protein in tanks, produce nutrients in labs, and turn what once was waste into nourishing food.

REFERENCES

Adewara, O. A., Adebayo-Olajide, T. C., Ayedun, J. S., Kotun, B. C., Adeleke, A. J., Brown, A. D., Alabi, O. J., & Ogunbanwo, S. T. (2024). Soil formation, soil health, and soil biodiversity. In S. A. Aransiola, B. R. Babaniyi, A. B. Aransiola, & N. R. Maddela (Eds.), *Prospects for Soil Regeneration and Its Impact on Environmental Protection. Earth and Environmental Sciences Library.* Springer, Cham. https://doi.org/10.1007/978-3-031-53270-2_5

Adeyemi, M. M., Olayemi, I. O., Adeboye A., & Thera-Sekgweng, S. N. (2025). Plant-based biotoxins: Applications and controls. In S. A. Aransiola, B. C. Kotun, & N. R. Maddela (Eds.), *Biotoxins in Food Threats and Benefits* (pp. 159–164). CRC Press.

Awofe, A. H., Monioluwa, L. A., Wuraola, A. B., & Oluyemisi, O. F. (2025). Microbial production of sweeteners and their industrial applications: Current status and future prospects. *Science World Journal*, *20*(1), 186–196. https://doi.org/10.4314/swj.v20i1.25

Bhattacharyya, P. N., & Jha, D. K. (2012). Plant growth-promoting rhizobacteria (PGPR): Emergence in agriculture. *World Journal of Microbiology and Biotechnology*, *28*(4), 1327–1350. https://doi.org/10.1007/s11274-011-0979-9

Daniel, A. I., Fadaka, A. O., Gokul, A., Bakare, O. O., Aina, O., Fisher, S., Burt, A. F., Mavumengwana, V., Keyster, M., & Klein, A. (2022). Biofertilizer: The future of food security and food safety. *Microorganisms*, *10*(6), 1220. https://doi.org/10.3390/microorganisms10061220

Dodd, I. C., & Ruiz-Lozano J. M. (2012). Microbial enhancement of crop resource use efficiency. *Current Opinion in Biotechnology*, *23*(2), 236–242. https://doi.org/10.1016/j.copbio.2011.09.005

Egamberdieva, D., Wirth, S. J., Alqarawi, A. A., Abd_Allah, E. F., & Hashem, A. (2017). Phytohormones and beneficial microbes: Essential components for plants to balance stress and fitness. *Frontiers in Microbiology*, *8*. https://www.frontiersin.org/journals/microbiology/articles/10.3389/fmicb.2017.02104

FAO, (2021). FAO Policy Series: Food Loss and Food Waste. FAO

Finnigan, T. J. A., Theobald, H. E., Bajka, B. (2025). Mycoprotein: A healthy and sustainable source of alternative protein-based foods. *Annual Review Food Science and Technology* *16*(1):105–125. https://doi.org/10.1146/annurev-food-111523-121802..

Glick, B. R. (2012). Plant growth-promoting bacteria: Mechanisms and applications. *Scientifica*, Article ID 963401. https://doi.org/10.6064/2012/963401

Gouda, S., Kerry, R. G., Das, G., Paramithiotis, S., Shin, H. S., & Patra, J. K. (2018). Revitalization of plant growth promoting rhizobacteria for sustainable development in agriculture. *Microbiological Research*, *206*, 131–140. https://doi.org/10.1016/j.micres.2017.08.016.

Hill, C., Guarner, F., Reid, G., et al. (2014). Expert consensus document: The international scientific association for probiotics and prebiotics consensus statement on the scope and appropriate use of the term probiotic. *Nature Reviews Gastroenterology & Hepatology*, *11*(8), 506–514. https://doi.org/10.1038/nrgastro.2014.66

Hirasawa, T., & Shimizu, H. (2016). Recent advances in amino acid production by microbial cells. *Current Opinion in Biotechnology*, *42*, 133–146. https://doi.org/10.1016/j.copbio.2016.04.017.

Kotun, B. C., Adewara, O. A., Oyetunji, O. E., Adebayo-Olajide, T. C., Ayedun, J. S., Adeoye, R., Egbeagu, U. U., & Ogunbanwo, S. T. (2024). Sustainability of microbial fuel cells, marine energy, and hydrogen. In N. R. Maddela, S. A. Aransiola, C. I. Ezugwu, L. K. W. Eller, L. Scalvenzi, & F. Meng, (Eds.), *Microbial Biotechnology for Bioenergy* (pp. 237–262). Elsevier. https://doi.org/10.1016/B978-0-443-14112-6.00012-2

Moosa-Kazemi, S. H., Asgarian, T. S., Sedaghat, M. M., & Javar, S. (2021). Pathogenic fungi infection attributes of malarial vectors *Anopheles maculipennis* and Anopheles superpictus in central Iran. *Malaria journal*, *20*(1), 393. https://doi.org/10.1186/s12936-021-03927-4

Muñoz-Dorado, J. et al (2016). Myxobacteria: moving, killing, feeding, and surviving together. *Frontiers in Microbiology*, *7*, 781. https://:doi.org/10.3389/fmicb.2016.00781

Naik, B., Kumar, V., Rizwanuddin, S., Mishra, S., Kumar, V., Saris, P. E. J., Khanduri N., Kumar, A., Pandey, P., Gupta, A. K., Khan, J. M., & Rustagi, S. (2024). Biofortification as a solution for addressing nutrient deficiencies and malnutrition, *Heliyon*, *10*(9), e30595.

Nasseri, A. T., Rasoul-Amini, S., Morowvat, M. H., & Ghasemi, Y. (2011). Single cell protein: Production and process. *American Journal of Food Technology*, *6*(2), 103–116. https://doi.org/10.3923/ajft.2011.103.116

Nosheen, S., Ajmal, I., & Song, Y. (2021). Microbes as biofertilizers, a potential approach for sustainable crop production. *Sustainability*, *13*(4), 1868. https://doi.org/10.3390/su13041868

Oleńska, E., Małek, W., Wójcik, M., Swiecicka, I., Thijs, S., & Vangronsveld, J. (2020). Beneficial features of plant growth-promoting rhizobacteria for improving plant growth and health in challenging conditions: A methodical review. *Science of the Total Environment, 743*, 140682. https://doi.org/10.1016/j.scitotenv.2020.140682

Palma, L., Muñoz, D., Berry, C., Murillo, J., & Caballero, P. (2014). *Bacillus thuringiensis* toxins: An overview of their biocidal activity. *Toxins, 6*(12), 3296–3325. https://doi.org/10.3390/toxins6123296

Pandey, A. (2013). Field evaluation of *Beauveria bassiana* and *Metarhizium anisopliae* against the cutworm, *Agrotis ipsilon* (Hufnagel) damaging potato in Uttarakhand Hills. *Journal of Biological Control, 27*, 293–297.

Pieterse, C. M. J. et al. (2014). Induced systemic resistance by beneficial microbes. *Annual Review of Phytopathology, 52*, 347–375. https://doi.org/10.1146/annurev-phyto-082712-102340

Prabhu, N., Borkar, S., & Garg, S. (2019). Phosphate solubilization by microorganisms: Overview, mechanisms, applications and advances. *Advances in Biological Science Research*, 161–176.

Rashid, M. I., Mujawar, L. H., Shahzad, T., Almeelbi, T., Ismail, I. M. I., & Oves, M. (2016). Bacteria and fungi can contribute to nutrients bioavailability and aggregate formation in degraded soils. *Microbiological Research, 183*, 26–41. https://doi.org/10.1016/j.micres.2015.11.007

Rempelos, L., Kabourakis, E., & Leifert, C. (2023). Innovative organic and regenerative agricultural production. *Agronomy, 13*(5), 1344. https://doi.org/10.3390/agronomy13051344

Sagar, A., Rathore, P., Ramteke, P. W., Ramakrishna, W., Reddy, M. S., & Pecoraro, L (2021). Plant growth promoting *rhizobacteria*, arbuscular mycorrhizal fungi and their synergistic interactions to counteract the negative effects of saline soil on agriculture: Key macromolecules and mechanisms. *Microorganisms, 9*(7), 1491. https://doi.org/10.3390/microorganisms9071491

Sanahuja, G., Banakar, R., Twyman, R. M., Capell, T., & Christou, P. (2011). *Bacillus thuringiensis*: A century of research, development and commercial applications. *Plant Biotechnology Journal, 9*(3), 283–300. https://doi.org/10.1111/j.1467-7652.2011.00595.x

Santoyo, G., Moreno-Hagelsieb, G., Orozco-Mosqueda M. dC. & Glick B. R. (2016). Plant growth-promoting bacterial endophytes. *Microbiological Research, 183*, 92–99. https://doi.org/10.1016/j.micres.2015.11.008.

Sawangproh, W., Paejaroen, P., Afifah, L. et al. (2025) Microbial pesticides: A bibliometric analysis of global research trends (1973–2024). *Egypt Journal of Biology and Pest Control, 35*(2). https://doi.org/10.1186/s41938-025-00840-9

Show, P. L., Oladele, K. O., Siew, Q. Y., et al. (2015). Overview of citric acid production from *Aspergillus niger. Frontiers in Life Science, 8*(3), 271–283. https://doi.org/10.1080/21553769.2015.1033653

Solórzano, R., Ramírez Maguiña, H. A., Johnson, L., Ureta Sierra, C., & Cruz, J. (2025). Current progress in microbial biocontrol of banana *Fusarium* wilt: A systematic review. *Agronomy, 15*(3), 619. https://doi.org/10.3390/agronomy15030619

Song, P., Zhang, X., Wang, S., Xu, W., Wang, F., Fu, R., & Wei, F. (2023). Microbial proteases and their applications. *Frontiers in Microbiology, 14*. https://www.frontiersin.org/journals/microbiology/articles/10.3389/fmicb.2023.1236368

Tadesse Mawcha, K., Malinga, L., Muir, D., Ge, J., & Ndolo, D. (2025). Recent advances in biopesticide research and development with a focus on microbials. *F1000Research*, 13, 1071. https://doi.org/10.12688/f1000research.154392.2

Tamang, J. P., Watanabe, K., & Holzapfel, W. H. (2016). Review: Diversity of microorganisms in global fermented foods and beverages. *Frontiers in Microbiology, 7*, 377. htpps://doi.org/10.3389/fmicb.2016.0037

Wang, T., Xu, J., Chen, J., Liu, P., Hou, X., Yang, L., & Zhang, L. (2024). Progress in microbial fertilizer regulation of crop growth and soil remediation research. *Plants (Basel, Switzerland), 13*(3), 346. https://doi.org/10.3390/plants13030346

Wu, W., Chen, Y., Dou, X., Liao, D., Li, K., An, C., Li, G., &Dong, Z. (2024). Microbial fertilizers improve soil quality and crop yield in coastal saline soils by regulating soil bacterial and fungal community structure. *Science of the Total Environment, 949*, 175127.

21 Microorganisms in the Context of Global Trends
From Precision Agriculture to Functional Foods

Mukul Machhindra Barwant, Balwant Singh, Nur Koç Koyun, and Odangowei Inetiminebi Ogidi

21.1 INTRODUCTION

Reducing the yield losses from harmful pests and disease activities to the harvestable sections of the crops is one way to help ensure that food supplies keep up with urbanization and population growth for sustainable food security. By 2050 or later, the world's food consumption is expected to double, posing several difficulties to agricultural sustainability (Tijjani & Khairulmazmi, 2021). The development of food microbiology-based solutions that may combine environmental responsibility, safety, and usefulness has been fueled by the growing desire for more sustainable food systems. The most recent and significant developments in the nexus of food innovation, sustainability, and microbiology are reviewed in this study. In the twenty-first century, the search for more sustainable food systems has emerged as a top worldwide concern. The agro-food industry is currently thought to be responsible for up to 34% of worldwide greenhouse gas emissions. In this transformation, microbial fermentation is a fundamental component. In addition to producing popular fermented foods like yogurt, sauerkraut, kefir, and tempeh, fermentation techniques are being used more and more to generate value-added products from agro-industrial waste, supporting circular bioeconomy models with smaller environmental impacts (Nascimento & Barros, 2025).

Culture and tradition shine through in the wide range of fermented foods and beverages from the various locations they originate from. While our tropical forebears weathered droughts quite fine, our temperate and colder brethren weathered winters with the help of fermentation. Fermentation, in its most basic form, is the gradual decomposition of organic compounds into their component acids or alcohols by means of bacteria or enzymes. For a very long time, fermented foods and alcoholic drinks were staples in the diets of nearly every human culture on every continent. Proteins, vitamins, minerals, and other nutrients can be reliably and significantly supplied by these foods, and they are often in good preservation (Swain et al., 2014; Tamang et al., 2020).

The use of agrochemicals, which include nutrients for plants and growth regulators (PGRs), is on the rise in order to feed the world's growing population (Tijjani & Khairulmazmi, 2021; Ogidi & Akpan, 2023). In addition to driving up production costs, overusing chemicals increases the likelihood of land degradation and deteriorating environmental quality. To enhance crop output, maximize income, and safeguard both people and the environment, one of the greatest and most appropriate solutions is to use microorganisms for sustainability. Plant growth-promoting rhizobacteria (PGPR) are just one of many microbes that can increase plant development. Others include actinomycetes, fungi, and bacteria.

Our existing food system is based on unsustainable methods that frequently fall short of feeding a rising population a healthy diet. New sustainable supplies and methods of nourishment are therefore desperately needed. Because of their low carbon footprint, minimal dependence

DOI: 10.1201/9781003650881-21

on land, water, and seasonal fluctuations, as well as their advantageous nutritional profile, microorganisms have drawn interest as a potential new food source. The use of microbes has also increased with the development and application of new techniques, particularly in synthetic biology, which shows tremendous promise for meeting many of our nutritional demands (Graham & Ledesma-Amaro, 2023; Iyiola et al., 2023). Microorganisms' impact on agricultural yield, soil fertility, and food preservation makes them essential to attaining global food security. Rhizosphere-associated fungi and bacteria, including nitrogen-fixing bacteria (*Rhizobium, Azospirillum*, and *Frankia*) and arbuscular mycorrhizal fungi, improve nutrient availability and promote sustainable crop yields (Glick, 2012). Fermentation is a historic technique in food production that has attracted fresh scientific attention. Nowadays, microbes are used to improve the functional and nutritional value of food (Ogidi & Tawariowei, 2023). Numerous health advantages, such as altered gut microbiota, better digestion, and increased immunity, have been associated with probiotics (*Lactobacillus*, *Bifidobacterium*, and *Saccharomyces boulardii*) (Hill et al., 2014).

Sharif et al. (2021) have been proposed that the use of microbes in food has long been known, recent studies have shown the wide range of health advantages and environmental savings that may be obtained by doing so. Single cell protein (SCP), commonly known as biomass, microbial protein, or bioprotein, is a bulk of dried cells. Although bacteria, fungi, yeast, and algae can all make SCP, fungi and bacteria are the main producers of this protein. Their rapid growth rate and comparatively greater protein level in their chemical structure were the primary causes of the high-protein output from these sources.

Meals and drinks that were fermented were among the first processed foods that humans consumed. Yoghurt, cultured milk, wine, beer, sauerkraut, kimchi, and fermented sausage were once considered beneficial to produce due to their extended shelf life, safety, and organoleptic attributes. It is now widely acknowledged that fermented foods can also have enhanced nutritional and functional properties as a result of substrate change and the formation of bioactive or bioavailable end products. To add to that, many fermented foods include living bacteria, some of which share genes with probiotics. There may be health benefits to eating fermented foods beyond what is initially thought, even if there have been little clinical studies on the topic (Marco et al., 2017). Bacteria play an important role in material recycling, which improves plant health by promoting nutrient uptake and protecting plants from phytopathogen attacks via biocontrol agents and the negative consequences of nutrient hyper-accumulation. Not putting too much emphasis on these bacteria's use is crucial (Tijjani & Khairulmazmi, 2021). Nowadays, it's not uncommon for the most popular probiotics to come from a limited number of species. Because of the organisms' varied evolutionary histories and unknown characteristics that make them potential novel therapeutics, our expanding knowledge of the gut microbiota is challenging this paradigm (Ogwu et al., 2025). Because their development is likely more suited to a pharmaceutical delivery method than a food delivery method, these organisms are often operationally referred to as next-generation probiotics. This term overlaps with the newly developing idea of live biotherapeutic medicines (O'Toole et al., 2017). Protein is an essential source of nitrogen and amino acids for all known living things. Amino acids are critical building blocks for many proteins, including hormones, enzymes, and structural proteins. Proteins can also be used as a source of energy in extreme conditions. The nutritional value of a protein is based on its amino acid composition. Dietary protein usually contains 20 amino acids. Some of these, like arginine, cysteine, glycine, glutamine, proline, and tyrosine, are beneficial but cannot be synthesized by humans or animals. Dietary protein also contains leucine, isoleucine, tryptophan, methionine, prosine, and histidine. Protein can also be supplied by cultivating various microbes and algae, preferably ones with a protein biomass of 30% or more and the ability to provide an adequate supply of all the essential amino acids. Proteins produced by microbes are usually known as SCPs, yet filamentous fungi and filamentous algae are examples of microbes that can be multicellular. In addition to their direct use as SCP, microbes can boost protein demand by enhancing the protein content or quality of fermented foods. Bacterial SCP usually has a protein level of 50–80% by dry weight, and the amount of essential amino acids should be at least as high as what is recommended by the FAO (Ritala et al., 2017).

21.2 MICROBIAL TECHNOLOGIES IN PRECISION AGRICULTURE

Precision agriculture is a farming paradigm change that maximizes inputs and outputs by fusing ecological principles with contemporary technologies. Microorganisms, especially helpful soil microorganisms, are essential for increasing crop sustainability and productivity. By using microbial technology in precision agriculture, microbial interventions are guaranteed to be targeted both geographically and temporally, increasing productivity and reducing environmental effects. The monitoring and deployment of microbial consortia can be customized to particular crop and soil conditions with the help of digital technologies, artificial intelligence (AI), and metagenomics, bringing in a new era of intelligent, microbe-driven agriculture. A comprehensive strategy, the circular economy seeks to establish a regenerative and sustainable economic structure that reduces waste, encourages resource efficiency, and protects the environment. The use of advantageous bacteria in precision biotechnology has become a key strategy for promoting the circular economy (Singh et al., 2024). In the quest for long-term answers to urgent global issues, the symbiotic interaction between microbiological life and the environment has gained attention. Chemicals such as fertilizers and insecticides have become increasingly prevalent in order to meet the needs of the fast growing population. The health of the soil suffers when these chemicals are used carelessly. One promising approach to balancing environmental conservation and agricultural output is the judicious use of beneficial microorganisms. The strategic use of advantageous microorganisms for environmentally friendly pest management that is sustainable is emphasized by the vanguard of biopesticide innovation (Pandey et al., 2024).

21.2.1 Biofertilizers for Soil Fertility and Crop Nutrition

Biofertilizers are natural mixtures of living or dormant microbial strains that, when applied to soil, plant surfaces, or seeds, improve nutrient availability and encourage plant development. In contrast to chemical fertilizers that supply nutrients directly, biofertilizers fix nitrogen, solubilize potassium and phosphates, produce phytohormones, and enhance soil structure, all of which benefit the soil ecosystem.

Nitrogen (N_2): Nitrogen (N) is the fourth most abundant element in all biomass, which includes all forms of life, behind oxygen, carbon, and hydrogen. It is a necessary component of life. Adenosine triphosphate (ATP), an energy transfer molecule, amino acids, nucleic acids, and chlorophyll, for instance, all require nitrogen (Singh et al., 2024). Organic N is a significant source of nitrogen in soils that must be transformed into plant-available inorganic N through microbial mineralization, which combines nitrification and ammonification (Paul & Frey, 2023). Symbiotic bacteria, free-living bacteria, and archaea are all members of the tiny and varied group of microorganisms known as diazotrophs (N_2-fixing), which contains nitrogenases. *Rhizobia* are symbiotic N_2-fixing bacteria that are most commonly researched in agriculture (Mitter et al., 2021). The majority of these organisms belong to the *Rhizobiaceae* family, which also includes *Bradyrhizobium, Sinorhizobium, Azorhizobium, Mesorhizobium,* and *Rhizobium* (Ensifer). Legumes (family *Fabaceae*) are able to build symbiotic relationships with *Rhizobia* by growing nodules on their roots or stems. *Bradyrhizobium, Sinorhizobium* (Ensifer), and *Rhizobium* are three of the *Rhizobia* commonly used as biofertilizers in farming (Werner & Newton, 2005).

Over 113 different plant species have shown an improvement in N availability and acquisition when exposed to the most famous free-living diazotrophs, the *Azospirillum* species (Pandey et al., 2024). Their PGP capabilities are extensive as well. *Cyanobacteria, Azotobacter* sp., and *Azospirillum* sp. are only a few examples of the heterotrophic, free-living diazotrophic microorganisms that can fix atmospheric N_2 in both bulk soil and the rhizosphere, in contrast to symbiotic N_2-fixing bacteria.

According to Moreira-Coello et al. (2019) and Dakora et al. (2008), free-living diazotrophs play an especially important role in nitrogen acquisition in crops that do not contain legumes. It explains Microorganisms Involved in Nitrogen, Phosphorus, and Potassium Solubilization (Table 21.1).

Phosphorus (P): As one of the plant kingdom's most important nutrients, phosphorus has an indirect or direct impact on every biological process. As an example, in order for plants to carry out respiration, energy transfer, signal transduction, photosynthesis, and molecular biosynthesis, P is an absolute must. It is one of the main factors limiting plant development in many ecosystems across the globe, despite the fact that soils contain substantial levels of P in both organic and inorganic forms (Raghothama, 2015). Soil bacteria are able to convert insoluble soil P into plant-usable forms via a web of solubilization and mineralization activities. In order to make inorganic P (such as tricalcium phosphate, hydroxyapatite, and rock phosphate) more soluble, phosphate-solubilizing microorganisms (PSMs) produce and secrete a variety of compounds (Pradhan et al., 2018). In addition, microorganisms can secrete chelating compounds that release associated soluble phosphates by mobilizing and capturing cations from insoluble phosphates, such as Ca^{+2}, Al^{+3}, and Fe^{+3}. The use of mineral P fertilizers can have negative consequences for the environment, such as the eutrophication of freshwater bodies, whereas PSM improves P bioavailability and hence decreases the need for these inputs (Alori et al., 2017). The P solubilizers that have been studied the most belong to the genera *Aspergillus, Pseudomonas, Bacillus, Rhizobium, Enterobacter,* and *Penicillium* (Anand et al., 2016). Soil microorganisms may increase P bioavailability in a number of ways, one of the most important being the mineralization of organic phosphate molecules. In order for plants to

TABLE 21.1

Key Microorganisms Involved in Nitrogen, Phosphorus, and Potassium Solubilization

Nutrient	Microbial Function	Representative Genera	Mechanism	Citation
Nitrogen (N)	Nitrogen fixation (symbiotic)	*Rhizobium, Bradyrhizobium, Mesorhizobium, Sinorhizobium, Azorhizobium*	Symbiotic root or stem nodulation and atmospheric N_2 fixation via nitrogenase enzymes	Mitter et al. (2021), Werner and Newton (2005)
	Nitrogen fixation (free-living)	*Azospirillum, Azotobacter, Beijerinckia, Clostridium, Anabaena, Nostoc*	Non-symbiotic atmospheric N_2 fixation in rhizosphere or soil	Singh et al. (2024), Moreira-Coello et al. (2019)
Phosphorus (P)	Phosphate solubilization	*Bacillus, Pseudomonas, Rhizobium, Penicillium, Aspergillus, Enterobacter*	Secretion of organic acids (e.g., gluconic, citric) to convert insoluble P into bioavailable form	Alori et al. (2017), Pradhan et al. (2018)
	Organic P mineralization	*Bacillus, Streptomyces, Pseudomonas, Rhizobium*	Production of phosphatases and phytases to release P from organic compounds (e.g., phytate)	Mitter et al. (2021), Raghothama (2015)
Potassium (K)	Potassium solubilization	Bacillus mucilaginosus, Frateuria aurantia, Pseudomonas, Burkholderia, Aspergillus	Organic acid production (e.g., tartaric, oxalic); chelation; mineral weathering	Sharma et al. (2016), Srivastava et al. (2020)
	K mineral weathering (fungi + bacteria)	*Trichoderma, Cladosporium, Glomus, Acidithiobacillus, Paenibacillus*	Solubilize K from feldspar/mica using acidification and enzymatic weathering processes	Kour et al. (2020), Sattar et al. (2019)

absorb phosphate, enzymes like phytases and phosphatases (such as phosphodiesterases and phosphomonoesterases) break down organic P molecules (Mitter et al., 2021; Table 21.1).

Potassium (K): Potassium, a vital macronutrient, is an essential inorganic cation found in plant cytoplasm and essential for proper cell structure and function. Among its many important metabolic roles, it plays a role in photosynthesis and protein synthesis (Srivastava et al., 2020). Soil potassium, which is abundant both on Earth and in plants, ranks second in soil nutrients, behind only nitrogen (Kour et al., 2020). Feldspar and mica minerals (e.g., biotite and muscovite) comprise nearly all of the soil's non-exchangeable potassium, at roughly 98%. Soil solutions contain only 0.1–0.2%, which plants easily absorb, while organic matter and clay particles adsorb an additional 1–2%. Microbes can increase plant availability of potassium by, for example, solubilizing the element, a key stage in the potassium cycle (Sharma et al., 2016). Similar to how organic acids like tartaric, citric, oxalic, gluconic, lactic, and malic acid are produced and released, the most well-known way for microbes to solubilize potassium is the same as it is for potassium. Organic acid exchange reactions and chelation are two more important mechanisms for K release (Sattar et al., 2019). According to Mitter et al. (2021) (Table 21.1), several soil bacterial and fungal species can dissolve K minerals. These include: *Aspergillus*, *Cladosporium*, *Macrophomina*, *Sclerotinia*, *Trichoderma*, *Glomus*, *Penicillium*, *Rhizobium*, *Acidithiobacillus*, *Paenibacillus*, *Pseudomonas*, and *Burkholderia*.

21.2.2 Biopesticides and Disease Management Strategies

Increasing the production of nutrient-dense food while reducing the negative effects of crop production systems on the land, water, and climate are the two problems facing the agricultural community (Barwant et al., 2025b, 2025c, 2025e). Since plant diseases result in enormous production and financial losses to crops that feed a large section of the world's population, controlling plant pathogens will play a key role in addressing these difficulties. It is necessary to develop and implement new methods and technology to improve the sustainability of crop production systems and, more crucially, plant disease control. Utilizing cutting-edge geoinformatic methods, technological advancements in computing and sensing infrastructure (such as cloud-based, big data-driven applications) will help monitor and manage biologicals, like cover crops and beneficial microbes, and pesticides to lessen the environmental impact of cropping systems and plant disease control. In order to help farmers manage cropping systems and disease management measures, which are becoming more complex yet more sustainable, geospatial tools are being developed. Moreover, cloud-based, big data-driven technologies and geoinformatics are being used to accelerate the improvement of agricultural germplasm; improved resistance to abiotic stress and infections, as well as compatibility with various cropping methods and environmental circumstances, are required (Roberts et al., 2021).

In order to control pests and infections, biopesticides are bioformulations made from naturally existing bacteria or microbial metabolites. Particularly in precision agriculture systems that depend on focused treatments, they are increasingly seen as a safer and more sustainable substitute for synthetic pesticides.

Food losses affect food safety and security since they are characterized as a decrease in the amount and quality of food produced and stored (Ogidi et al., 2025b). Plant pathogens are responsible for some of the most important losses (Barwant et al., 2025a, 2025d, 2025f). In order to avoid microbiological infections, chemical pesticides have been employed widely. However, because of their toxicity and decreased effectiveness, researchers are now working to create substitutes. Microbial biopesticide (MBP)-based substitutes are typically safer and less harmful to the environment than traditional pesticides. Formulations based on biopesticides have grown in variety and quantity over the past few years, and they have drawn interest from the commercial community. Achieving optimal disease control requires an understanding of the mechanisms by which biopesticides regulate the disease. As a preventative approach to control diseases, MBPs are frequently employed in augmentative biocontrol techniques. Since the populations of the microbial antagonists that make up those biopesticide compounds are often self-sustaining for a short period of time (partial or one growing season), they must be treated multiple times. These products are mass-produced through industrial fermentation. Using a variety of methods, biopesticides provide protection. Certain of these mechanisms, like biotrophic mycoparasitism, are quite specialized, while others, like certain antibiotics, have a wider range of action and can even affect bacteria that may be advantageous to host plants. Comprehending the ways in which biopesticides work is essential for evaluating their range of action and forecasting how they can affect beneficial bacteria (Vero et al., 2023).

A variety of chemicals used in agriculture for insect control are referred to as pesticides. One major issue with the use of these synthetic pesticides is their lack of selective toxicity, biodegradability, and environmental buildup. The most prevalent ones, like toxaphene and dichlorodiphenyltrichloroethane (DDT), can affect soil for a long time and can seep into aquatic and groundwater environments. Due to their danger to people, certain pesticides have been outlawed in numerous nations, including dichlorvos, anthraquinone, carbofuran, carbendazim, dinocap, paraquat, methomyl, aldicarb, and diuron. In both terrestrial and aquatic settings, pesticides have the potential to cause pollutants to bioaccumulate in food chains. The term "biopesticide" refers to a broad category of compounds that come from natural sources, such as microbes, plants, and biochemicals. The incorporation of genetic material into agricultural products to provide protection from insect damage is an additional strategy. Biopesticides are considered the best alternative to synthetic pesticides because to their minimal environmental dangers, specificity, effectiveness, and sustainability. Microbial pesticides are an advancement in pest control above conventional agrochemical pesticides due to their efficacy and biodegradability (Ogidi et al., 2023). The word "biopesticide" encompasses a wide range of substances derived from biological sources, including plants, microorganisms, and biochemical (Makgoba et al., 2024). The incorporation of genetic material into agricultural products to provide protection from insect damage is an additional strategy. Biopesticides are considered the best alternative to synthetic pesticides because of their minimal environmental dangers, specificity, effectiveness, and sustainability. Microbial pesticides are an advancement in pest control over conventional agrochemical pesticides due to their efficacy and biodegradability (Ayilara et al., 2023; Vermelho et al., 2024). Up to 40% of food spoilage occurs during production, postharvest storage, transportation, and marketing due to microbial contamination. A number of environmental conditions, including temperature, humidity, and exposure to ultraviolet (UV) rays, influence their development and reproduction in the wild. Most of the active components in microbial insecticides are bioactive compounds that bacteria create. Since microbial populations tend to be transient, biopesticides derived from microbial byproducts must be reapplied. In industrial settings, fermentation is used to mass-produce them. The microorganisms are not easy to work with due to their high hydrophobicity and tendency to aggregate in solutions. MBPs consist of microorganisms that occur naturally, such as bacteria, fungus, baculoviruses, or microbes linked with nematodes. They aren't dangerous to humans or non-target animals and can be cheap. Issues like target customization and insect resistance are effectively tackled by MBPs. More than 3000 distinct bacteria have recently been identified as potential insect killers (Karnwal & Kapoor, 2020).

With a market share of 75–95% for MBPs, *Bacillus thuringiensis* is one of the most widely used and effective bioinsecticides for biologically managing pests. The leaf-feeding polyphagous pest *Ectropis obliqua* (*Geometridae, Lepidoptera*) attacks a wide variety of host plants, including flowers, sesame, tea, and soybeans. The Baculoviridae family includes two subgroups of dsDNA viruses, the granuloviruses (GVs) and the nuclear polyhedroviruses (NPVs). They are both effective biological control agents (BCAs), but the bug has to ingest one of them to make it sick. Additionally, they enhance plant immunity by stimulating more antioxidant molecule formation, which in turn makes plants more resistant to disease. Researchers have found that certain microalgae, such as *Anabaena laxa*, *Calothrix* sp., *Chlorella vulgaris*, and *Fischerella* sp., create secondary metabolites with antibacterial, larvicidal, and insecticidal effects (Karnwal & Kapoor, 2020; Vermelho et al., 2024).

21.2.2.1 Metagenomics and AI in Microbial Agriculture

In contemporary microbial agriculture, metagenomics and AI are revolutionary instruments. When combined, they make it possible to thoroughly profile the microbiomes associated with plants and soil and to create data-driven microbial solutions for crop management at specific sites. Precision agriculture, bioinformatics, big data, AI, metagenomics, and machine learning (ML) are just a few examples of how technological advances in the last few decades have revolutionized farming. These innovations have revolutionized crop management, making it more efficient, productive, and environmentally friendly. AI has become indispensable for real-time decision-making and sophisticated data evaluation.

It comes in many forms, including deep neural networks and ML algorithms. The development of crop varieties more suited to environmental factors like drought or soil salinity, as well as the identification of genes causing desirable traits like disease resistance and stress tolerance, have resulted from advances in bioinformatics and genomic technologies like highthroughput sequencing (HTS) and next-generation sequencing (NGS). Metagenomics and bioinformatics have been used in practical agricultural settings to create sustainable farming methods (Ogidi & Izah, 2024). To create efficient microbial inoculants used as biofertilizers, for instance, metagenomic investigations have shown advantageous microbial communities that support plant growth and inhibit soilborne diseases (Villamar-Torres et al., 2025).

Though its vast, intricate datasets have presented many difficulties, metagenomics—the study of genetic material from environmental samples—has shed light on the complexities of microbial diversity and function. The perfect answer is offered by AI, which has strong data analysis capabilities. Comprehensive investigation of microbial communities has been made possible by the crucial field of metagenomics in environmental research. However, its full potential has been impeded by issues with data processing, such as taxonomic classification, functional annotation, and data quality. In order to facilitate data preparation, taxonomic profiling, and functional prediction, AI is a game-changer. ML algorithms and deep neural networks are skilled at identifying significant patterns in these intricate metagenomic datasets (Rani et al., 2024).

Agroecosystem research can benefit greatly from the use of metagenomics. For the development of new compounds with therapeutic value, biotechnological breakthroughs, and sustainable agriculture, metagenomics can help comprehend the vast array of microbial diversity found in the soil environment. Although this strategy has had some success, the great majority of soil microorganisms still play a part. The advent of metagenomics has opened up new possibilities for understanding the microbiome composition in a certain microniche. Importantly, this approach reveals not only the microbial composition (i.e., the identities and relative abundances of all taxa) but also the functional gene composition of microbes (e.g., bacteria, archaea, and eukaryotes) within a community (Nwachukwu & Babalola, 2022).

21.2.2.2 Microorganisms in Functional Foods

The idea of functional foods was initially presented in Japan in 1984 as a result of extensive, methodical research into the newly discovered functions of foods. The purpose of this study, which was funded by Japan's Ministry of Education, Science, and Culture, was to develop a regulatory framework for approving specific foods with health benefits claims. This would enable the public to select healthier options and, in turn, lower the high healthcare expenses incurred by the government (Hilachuk et al., 2021). Foods that offer health advantages beyond simple sustenance are known as functional foods. The range of functional foods has been greatly increased by the addition of microorganisms, particularly in the form of probiotics, prebiotics, postbiotics, and fermentation processes. These microbial therapies have the potential to create bioactive chemicals, improve nutritional bioavailability, alter gut microbiota, and aid in the management or prevention of chronic illnesses (Tamang et al., 2016).

21.2.2.3 Probiotics, Prebiotics, and Postbiotics in Human Health

Probiotics selectively digest prebiotic food fibers, which facilitate their colonization in the gut. Probiotic development results in the formation of fermentation byproducts, or postbiotics, which reduce pH and produce bacteriocins that prevent the growth of enteropathogens (Ogidi & Felagha, 2024). Once their life cycle is complete, their dead residues parabiotics, such as exopolysaccharides and cell wall glycoproteins) also prevent pathogens from adhering to the gut epithelium and forming biofilms. A systemic response is observed along many gut-organ axes, demonstrating that these positive effects are not limited to the gut (Kango & Nath, 2024). The word "probiotics" was first used by Lilly and Stillwell (1965) to refer to compounds secreted by one bacterium that promoted the growth of another. According to current definitions, probiotics are "live microorganisms which upon ingestion impart health promotion effects which are not limited to the nutritional effects." Live, nonpathogenic microorganisms known as probiotics are given to enhance microbial balance, especially in the gastrointestinal system. A number of new strains, including *S. boulardii*, *Bacillus clausii*, *Bacillus coagulans*, *Streptococcus thermophilus*, and others, are among the most common members of the genera *Lactobacillus* and *Bifidobacterium*. Particularly when comparing microorganisms that live in mucus or connected to the intestinal wall (parietal microbiota) with those that coexist with food in transit and feces, the microbiota arranges itself as an ecosystem and varies from one organ to another.

In the human gut, probiotics preferentially use prebiotics, which are oligosaccharides that are not broken down and absorbed by the small intestine. The idea of prebiotics as non-digestible oligosaccharides was first proposed by Gibson et al. (2017), who also stated that fructo-oligosaccharides (FOS) were the only prebiotic that was accessible. Numerous studies have been conducted since then, and a range of indigestible oligosaccharides are being investigated for the good of humanity. Prebiotics are defined as "substrates selectively utilized by probiotics to confer health benefits" in order to support the idea of prebiotics in light of future scientific discoveries. Prebiotic oligosaccharides show the antimicrobial action either through direct or indirect mechanism. Prebiotics also preferentially boost the probiotic population, which in turn produces SCFA and many antimicrobial peptides, so indirectly inhibiting the growth of harmful bacteria. Lately, probiotic-derived exopolysaccharides (EPS) from *Lactobacillus delbrueckii* and *Lacticaseibacillus rhamnosus* demonstrated prebiotic potential with other probiotics, including *Faecalibacterium prausnitzii* and *Ruminococcus bromii*, and enhanced the production of SCFA in addition to having antiproliferative and antioxidant effects on cancer cell lines. Additionally, it employed EPS from *S. thermophilus* and *Enterococcus faecium*, which demonstrated antiproliferative effects on cancer cell lines, antioxidant capability, and ACE inhibition in addition to promoting the proliferation of *F. prausnitzii* and *R. bromii* (Tarique et al., 2022).

Postbiotics, sometimes called metabiotics, are metabolites or cell-free supernatants (CFS) and soluble components that probiotics release. Defensins, bacteriocins, short-chain fatty acids, branched-chain amino acids, and a few others are recognized to have health-promoting qualities. A new definition of postbiotics has been put forth by the International Scientific Association for Probiotics and Prebiotics (ISAPP): "a preparation of inanimate microorganisms and/or their components that confers a health benefit on the host." Which means the microbe that's good for the host's health will be either dead or partially dead (enzymes, peptides, metabolites, cell wall, etc.). Anvar et al. (2023) distinguish postbiotics from probiotics, which incorporate the benefits of live bacteria. A spray-dried encapsulated *Ferulago angulata* extract nano-emulsion and a postbiotic metabolite from *Lactiplantibacillus plantarum* subsp. *plantarum* were recently added to the butter to improve its physical, chemical, and sensory qualities (Ogidi et al., 2025a). Research has demonstrated that ducks fed probiotics like *Bacillus subtilis* and *Clostridium butyricum* produce more organic acids, such as benzoic acid, fumaric acid, phosphoric acid, and formic acid. The results showed that the ducks' weight, feed consumption, and egg production rate all went up (Salminen et al., 2021).

21.2.2.4 Microbial Fermentation for Nutrient-Enriched Foods

Microbial biotransformation alters the biochemical makeup of foods through fermentation, greatly increasing their nutritional value. This technique modifies the overall nutritional content and profile of food products, decreases anti-nutritional factors, and increases nutrient bioavailability. Knowing how to use microbial activity to create functional foods with more health benefits requires an awareness of these factors. Bioavailability is the percentage of nutrients that the body absorbs and uses after consuming them. In the course of fermentation, microorganisms like molds, yeast, and bacteria develop enzymes that simplify complicated compounds so the human digestive system can absorb them more readily. According to Samtiya et al. (2021), fermentation can convert lipids into free fatty acids, proteins into amino acids, and carbs into simple sugars.

Phytic acid is broken down by phytases produced by bacteria like LAB during fermentation, which releases the bound minerals and increases their accessibility for absorption in the human digestive system. In Asian cuisines, for instance, fermenting soybeans is a traditional procedure that raises the bioavailability of iron and zinc, which are normally poor due to the high phytic acid content of raw soybeans. Because they bind to indigestible substances and form complexes that are difficult for the human digestive system to break down, minerals in plant-based diets frequently have limited bioavailability. By breaking down these complexes and releasing minerals, fermentation provides a solution and increases their accessibility (Sawant et al., 2025). On a global scale, dietary diversity is greatly influenced by fermented food products, especially those derived from legumes and cereals. Important for maintaining a healthy diet, these goods are a must-have for any pantry. Nutritionally, food undergoes a transformation during fermentation, which improves its palatability, organoleptic attributes, and bioavailability. This food processing procedure is also famous for improving the sensory aspects of the dish by adding different flavors, textures, and odors. In addition to these uses, it can reduce harmful bacteria, lectins, phytic acid, proteinase inhibitors, oxalic and tannic acids, and other factors that reduce nutrients' bioavailability and bioaccessibility, as well as increase the economic worth of food goods by preserving them (Adebo et al., 2022).

21.2.2.5 Microbiome-Based Dietary Interventions

A complex community of trillions of microorganisms, the human gut microbiome is essential to host metabolism, immune regulation, and general health. The development of microbiome-based dietary treatments, which seek to alter the makeup and activity of gut microbes for better health outcomes, is a result of advances in microbiome research.

These interventions usually include the use of particular dietary fibers, fermented foods, and live beneficial microbes (probiotics), which selectively stimulate the growth of beneficial microbiota or restore microbial balance (Zmora et al., 2019). Using customized nutrition plans based on each person's unique microbiome profile is one promising strategy. Studies such as those conducted by Zeevi et al. (2015) serve as examples of this, showing that individualized dietary interventions informed by gut microbiota analysis greatly enhanced glycemic control in people. Furthermore, it has been demonstrated that foods high in polyphenols, like berries, cocoa, and green tea, have a positive impact on the diversity and function of gut microbes, which help to reduce inflammation and promote cardioprotection (Cardona et al., 2013).

Combinations of probiotics and prebiotics, known as synbiotics, are another type of microbiome-targeted dietary intervention that work together to increase the activity and survival of good bacteria. For therapeutic purposes in gastrointestinal illnesses and metabolic syndrome, such combinations are being investigated more and more (Swanson et al., 2020). Further demonstrating the possibility of microbiome alteration in severe gut-related disorders like *Clostridium difficile* infection is fecal microbiota transplantation (FMT), however more clinical in nature (Cammarota et al., 2019).

According to Hitch et al. (2022), the gut microbiome is located where the environment and the host meet, and it has the power to alter how the host reacts to stimuli and exposures that are relevant to the disease. This is clear from the way intestinal bacteria interact with the immune system, such as promoting immunological maturation in infancy, regulating the development of immune cell populations and their mediators, or impacting therapeutic efficacy through immune response regulation. Throughout daily life, a variety of factors influence the dynamics of the gut environment, and the therapeutic and preventative potential of microbiome-based therapies is only now becoming apparent. The purpose, implementation, and modes of operation of these strategies differ. Phage therapy, probiotics, and prebiotics are examples of treatments that target certain taxa or strains, while nutritional methods like FMT alter the overall community.

21.3 CHALLENGES AND OPPORTUNITIES AND FUTURE DIRECTIONS IN MICROBIAL APPLICATIONS

21.3.1 Challenges in Microbial Applications

Although promising, the use of microbes in functional meals is hindered by a number of issues that prevent widespread use and commercialization. Some of the most important considerations are the viability and stability of probiotic strains as they go through the food production, storage, and digestion processes.

Before they reach the gut, where they start to work, the majority of probiotic bacteria, including *Lactobacillus* and *Bifidobacterium*, lose viability due to their sensitivity to heat, oxygen, moisture, and acidic environments (Hill et al., 2014). Technically complicated, the industrial formulation of these strains into various food matrices such as dairy, drinks, or baked goods may necessitate the use of protective delivery technologies, such as microencapsulation, in order to preserve functionality (Chávarri et al., 2010). Furthermore, product development is complicated by the strain-specificity of probiotic effects. Each microbiological candidate must undergo extensive clinical validation since functional characteristics like cholesterol reduction, immunomodulation, or anti-inflammatory effect are not consistent across species or even strains (Sanders et al., 2019). This increases the regulatory burden because many jurisdictions demand that probiotic and prebiotic health claims be rigorously substantiated, which delays market approval even more. Additionally, regulations are not harmonized globally. For instance, other nations might permit health claims with little proof, despite the EU's strict EFSA regulations. This could result in inconsistent product quality and erode consumer confidence (Kleerebezem et al., 2019). Host variability presents another difficulty; depending on variables such as age, nutrition, immunological condition, and microbiota makeup, individual reactions to microbial functional foods can vary greatly (Zmora et al., 2019). It is challenging to extrapolate the advantages of a microbial strain to the larger population due to this inter-individual heterogeneity. Last, concerns regarding horizontal gene transfer, allergenicity, and long-term health effects are raised by the safety of introducing novel strains, particularly genetically engineered or next-generation probiotics (O'Toole et al., 2017). Post-market surveillance and thorough safety evaluations are crucial but undeveloped, especially in environments with limited resources, as seen in Figure 21.1.

21.3.2 Opportunity Microbial Applications

Notwithstanding the difficulties, there are a ton of chances to enhance food systems and human health through the use of microbes in functional foods. Probiotic, prebiotic, and postbiotic innovations have been sparked by the increased consumer desire for natural, health-promoting products across a range of food categories. Researchers can now create targeted microbial therapies that alter human metabolism, immunity, and even mood by identifying health-associated microbial signatures thanks to developments in metagenomics and microbiome profiling (Suez et al., 2018). By customizing microbial functional foods to each person's unique microbiome type and health requirements, these techniques enable the creation of individualized nutrition strategies (Kho &

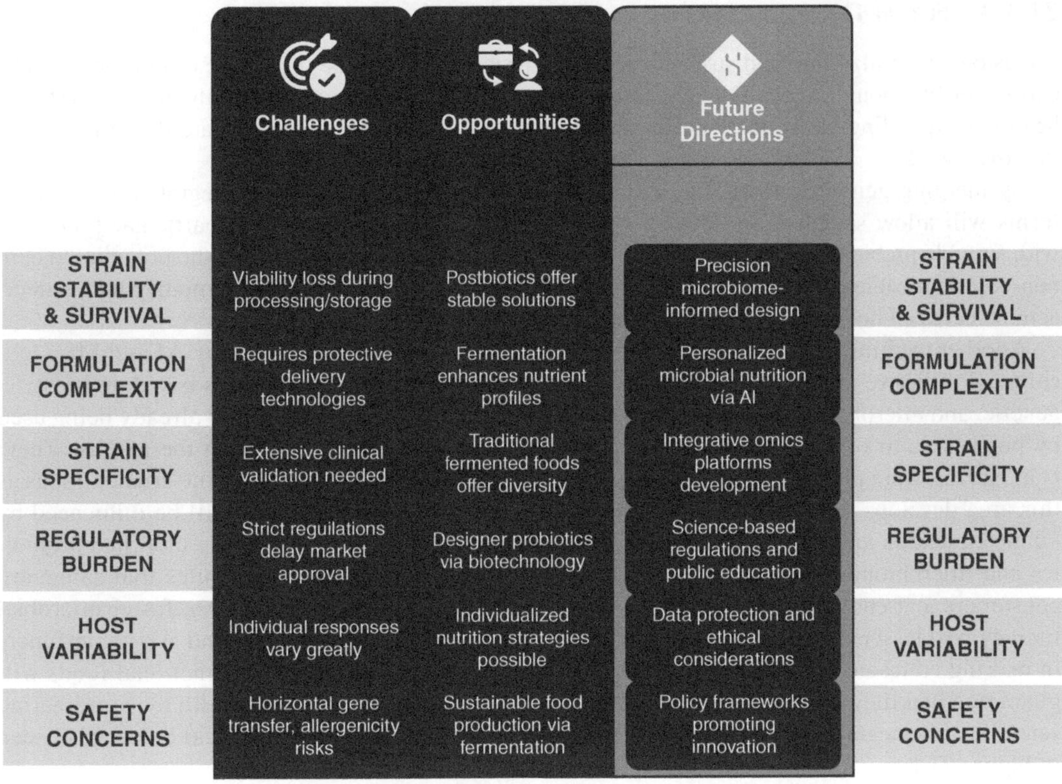

FIGURE 21.1 Microbial applications: Challenges, opportunities, and future directions.

Lal, 2018). Additionally, the variety of microbiological sources offers unrealized possibilities. From kimchi and kefir to kanji and idli, traditional fermented foods from many cultures contain a variety of microbial consortia with special metabolic and health-promoting qualities. New probiotics and ethnically tailored formulations for local markets may result from the methodical identification and characterization of these strains (Tamang et al., 2020). Furthermore, by boosting the bioavailability of vitamins (like folate and B12), eliminating anti-nutritional elements (like phytates and oxalates), and generating bioactive peptides with antihypertensive and antioxidant qualities, microbial fermentation can improve the nutritional profiles of food (Pessione, 2012). Postbiotics, a relatively new field that encompasses peptides, SCFAs, and cell wall components as well as non-viable microbial metabolites, is another promising area. According to Aguilar-Toalá et al. (2018), postbiotics offer functional benefits without limiting microbial viability, making them a safer and more stable option for food inclusion, especially in foods that have been heat-processed.

Furthermore, the combination of microbial applications with biotechnology, including synthetic biology and CRISPR-Cas9, creates opportunities for the creation of designer probiotics that can withstand stomach acid, deliver medicinal compounds, or detect intestinal inflammation (Charbonneau et al., 2020). Last but not least, microbial applications can support the production of sustainable food. Compared to animal agriculture, fermentation-based food systems are less resource-intensive and produce less waste, which is consistent with the circular economy and food security tenets (Marco et al., 2017). Microbial enzymes are employed to transform agro-industrial waste into useful nutrients, supporting the creation of zero-waste functional foods, while microbial biomass from fungi and algae is being investigated as an alternate source of protein, as seen in Figure 21.1.

21.3.3 Future Direction Microbial Applications

Precision, customization, and sustainability will probably determine the future of microbial applications in functional meals. Our understanding of the gut microbiome's significance in systemic health is expanding, and functional food designers are beginning to incorporate this information into their work.

By merging genomes, transcriptomics, proteomics, and metabolomics, integrative omics platforms will allow scientists to develop next-generation synbiotics, matching particular probiotics with suitable prebiotics for desired health effects. These specific methods have the potential to cure non-communicable diseases like diabetes, obesity, and mental health issues by modifying the interactions between hosts and microbes through food.

Another exciting area is personalized microbial nutrition, which is aided by AI and ML algorithms. Predictive models can suggest or create personalized functional foods based on lifestyle, genetic, and microbiome data. Direct-to-consumer microbiome testing kits are already being used by businesses to offer customized probiotic diets and supplements based on the results (Zeevi et al., 2015). Data protection, regulatory monitoring, and fair access will become crucial issues as this area develops. Finally, the development of microbial functional foods will be influenced by public education and policy frameworks. In light of new goods like genetically modified probiotics and microbiome modulators, governments must develop science-based rules that safeguard consumers and encourage innovation. Concurrently, educational programs that foster microbial literacy will lessen consumer skepticism about fermented or probiotic foods and assist consumers in making educated decisions (Peterson et al., 2020). In the end, microbial functional foods will guarantee that they make a significant contribution to world nutrition and health by coordinating scientific advancements with consumer demands, ethical norms, and ecological balance, as seen in Figure 21.1.

21.4 CONCLUSION

Once thought of only as disease-causing or spoiling agents, microorganisms are now important contributors to solving the problems of agriculture, nutrition, and global sustainability in the twenty-first century. Through the use of biofertilizers, biopesticides, and microorganisms that promote plant growth, microbes have become an increasingly important component of precision agriculture, facilitating nutrient management, stress tolerance, and soil health. At the same time, their contribution to food innovation, particularly in the area of functional foods, has shown that probiotics, prebiotics, postbiotics, and fermented nutrient-enriched foods have enormous potential to enhance human health. Farming systems and nutritional methods are being revolutionized by the convergence of microbial science with new fields like metagenomics, synthetic biology, and microbiome-based dietary interventions. By improving the nutritional profile of foods, these microbial advances not only provide sustainable substitutes for chemical inputs in agriculture, but they also aid in the fight against chronic diseases and worldwide starvation. Furthermore, the incorporation of microbiome-based insights into personalized health and precision nutrition suggests that dietary recommendations will eventually be customized to each person's unique microbial composition. Microorganisms provide long-term, economical, and environmentally beneficial solutions as the world's population rises and environmental problems worsen. It will be crucial to maximize their potential through ethical research, public education, and supportive legislative frameworks. In the end, microbes—nature's tiniest engineers—are at the forefront of a bio-revolution that supports the global objectives of human welfare, climate resilience, and food security. Only through interdisciplinary cooperation, strong regulatory frameworks, and ongoing innovation that upholds ecological balance and public confidence will their full potential be achieved.

REFERENCES

Adebo, J. A., Njobeh, P. B., Gbashi, S., Oyedeji, A. B., Ogundele, O. M., Oyeyinka, S. A., & Adebo, O. A. (2022). Fermentation of cereals and legumes: Impact on nutritional constituents and nutrient bioavailability. *Fermentation*, *8*(2), 63.

Aguilar-Toalá, J. E., Garcia-Varela, R., Garcia, H. S., Mata-Haro, V., González-Córdova, A. F., Vallejo-Cordoba, B., & Hernández-Mendoza, A. (2018). Postbiotics: An evolving term within the functional foods field. *Trends in Food Science & Technology*, *75*, 105–114.

Alori, E. T., Glick, B. R., & Babalola, O. O. (2017). Microbial phosphorus solubilization and its potential for use in sustainable agriculture. *Frontiers in Microbiology*, *8*, 971.

Anand, K. U. M. A. R., Kumari, B. A. B. Y., & Mallick, M. A. (2016). Phosphate solubilizing microbes: An effective and alternative approach as biofertilizers. *Int. J. Pharm. Pharm. Sci*, *8*(2), 37–40.

Anvar, S. A., Rahimyan, D., Golestan, L., Shojaee, A., & Pourahmad, R. (2023). Butter fortified with spray-dried encapsulated *Ferulago angulata* extract nanoemulsion and postbiotic metabolite of *Lactiplantibacillus plantarum* subsp. Plantarum improves its physicochemical, microbiological and sensory properties. *International Journal of Dairy Technology*, *76*(2), 381–392.

Ayilara, M. S., Adeleke, B. S., Akinola, S. A., Fayose, C. A., Adeyemi, U. T., Gbadegesin, L. A., ... & Babalola, O. O. (2023). Biopesticides as a promising alternative to synthetic pesticides: A case for microbial pesticides, phytopesticides, and nanobiopesticides. *Frontiers in Microbiology*, *14*, 1040901.

Barwant, M. M., Ogidi, O. I., Singh, B., Patil, V. S., & Sharma, S. A. B. (2025a). Climate change and pesticide dynamics. In B. R. Babaniyi & E. E. Babaniyi (Eds.), *The Interplay of Pesticides and Climate Change*. Springer, Cham. https://doi.org/10.1007/978-3-031-81669-7_7

Barwant, M. M., Ogidi, O. I., Yogita, C., & Munje, R. (2025b). Study of consumers choices on pesticides use and sustainability. In B. R. Babaniyi & E. E. Babaniyi (Eds.), *The Interplay of Pesticides and Climate Change*. Springer, Cham. https://doi.org/10.1007/978-3-031-81669-7_18

Barwant, M. M., Sharma, S. A. B., Lodh, D., Sil, D. C., & Ogidi, O. I. (2025c). Immedicable effects of organic pesticides on aquatic environment. In B. R. Babaniyi & E. E. Babaniyi (Eds.), *The Interplay of Pesticides and Climate Change*. Springer, Cham. https://doi.org/10.1007/978-3-031-81669-7_16

Barwant, M. M., Singh, B., Ogidi, O. I., Dar, M. A., & Muhammad, M. (2025d). Integrative physiology of food safety management. In T. Sarkar, A. Hamad, & A. Chatterjee (Eds.), *Physiological Perspectives on Food Safety: Exploring the Intersection of Health and Nutrition*. Springer, Cham. https://doi.org/10.1007/978-3-031-94582-3_17

Barwant, M. M., Singh, S., Ogidi, O. I., Singh, S., & Naik, B. S. S. S. (2025e). Economic implications of climate change and pesticide use on agriculture. In B. R. Babaniyi & E. E. Babaniyi (Eds.), *The Interplay of Pesticides and Climate Change*. Springer, Cham. https://doi.org/10.1007/978-3-031-81669-7_19

Barwant, M. M., Singh, S., Singh, S., Naik, B. S. S.S., & Ogidi, O. I. (2025f). Resilience of ecosystems: Navigating the challenges of pesticide exposure and climate change. In B. R. Babaniyi & E. E. Babaniyi (Eds.), *The Interplay of Pesticides and Climate Change*. Springer, Cham. https://doi.org/10.1007/978-3-031-81669-7_3

Cammarota, G., Ianiro, G., Tilg, H., Rajilić-Stojanović, M., Kump, P., Satokari, R., ... & Gasbarrini, A. (2017). European consensus conference on faecal microbiota transplantation in clinical practice. *Gut*, *66*(4), 569–580.

Cardona, F., Andrés-Lacueva, C., Tulipani, S., Tinahones, F. J., & Queipo-Ortuño, M. I. (2013). Benefits of polyphenols on gut microbiota and implications in human health. *The Journal of Nutritional Biochemistry*, *24*(8), 1415–1422.

Charbonneau, M. R., Isabella, V. M., Li, N., & Kurtz, C. B. (2020). Developing a new class of engineered live bacterial therapeutics to treat human diseases. *Nature Communications*, *11*(1), 1738.

Chávarri, María et al (2010). Microencapsulation of a probiotic and prebiotic in alginate-chitosan capsules improves survival in simulated gastro-intestinal conditions. *International Journal of Food Microbiology*, *142*(1–2), 185–189.

Dakora, F. D., Chimphango, S. B., Valentine, A. J., Elmerich, C., & Newton, W. E. (Eds.). (2008). *Biological Nitrogen Fixation: Towards Poverty Alleviation through Sustainable Agriculture: Proceedings of the 15th International Nitrogen Fixation Congress and the 12th International Conference of the African Association for Biological Nitrogen Fixation* (Vol. 42). Springer Science & Business Media.

Gibson, G. R., Hutkins, R., Sanders, M. E., Prescott, S. L., Reimer, R. A., Salminen, S. J., ... & Reid, G. (2017). Expert Consensus Document: The International Scientific Association for Probiotics and Prebiotics (ISAPP) Consensus Statement on the definition and scope of prebiotics. *Nature Reviews Gastroenterology & Hepatology*, *14*(8), 491–502.

Glick, B. R. (2012). Plant growth-promoting bacteria: Mechanisms and applications. *Scientifica, 2012*(1), 963401.

Graham, A. E., & Ledesma-Amaro, R. (2023). The microbial food revolution. *Nature Communications, 14*(1), 2231.

Hilachuk, D., Crisóstimo, C., Silva, A. J. A. D., & Paula, D. D. (2021). Patenting activity on functional foods: A Brazilian scenario. *Journal of Technology Management & Innovation, 16*(2), 70–81.

Hill, C., Guarner, F., Reid, G., Gibson, G. R., Merenstein, D. J., Pot, B., ... & Sanders, M. E. (2014). The international scientific association for probiotics and prebiotics consensus statement on the scope and appropriate use of the term probiotic. *Nature Reviews Gastroenterology & Hepatology, 11*(8), 506–514.

Hitch, T. C., Hall, L. J., Walsh, S. K., Leventhal, G. E., Slack, E., de Wouters, T., ... & Clavel, T. (2022). Microbiome-based interventions to modulate gut ecology and the immune system. *Mucosal Immunology, 15*(6), 1095–1113.

Iyiola, A. O., Kolawole, A. S., Ajayi, F. O., Ogidi, O. I., & Ogwu, M. C. (2023). Chapter 22: Sustainable water use and management for agricultural transformation in Africa. In *Water Resources Management for Rural Development: Challenges and Mitigation* (pp. 287–300). Elsevier. https://doi.org/10.1016/B978-0-443-18778-0.00001-5

Kango, N., & Nath, S. (2024). Prebiotics, probiotics and postbiotics: The changing paradigm of functional foods. *Journal of Dietary Supplements, 21*(5), 709–735.

Karnwal, A., & Kapoor, D. (2020). Soil microbes as biopesticides: Agricultural applications and future prospects. In *Current Trends in Microbial Biotechnology for Sustainable Agriculture* (pp. 499–524). Singapore: Springer Singapore.

Kho, Z. Y., & Lal, S. K. (2018). The human gut microbiome–a potential controller of wellness and disease. *Frontiers in Microbiology, 9*, 356589.

Kleerebezem, M., Binda, S., Bron, P. A., Gross, G., Hill, C., van Hylckama Vlieg, J. E., ... & Ouwehand, A. C. (2019). Understanding mode of action can drive the translational pipeline towards more reliable health benefits for probiotics. *Current Opinion in Biotechnology, 56*, 55–60.

Kour, D., Rana, K. L., Kaur, T., Yadav, N., Halder, S. K., Yadav, A. N., ... & Saxena, A. K. (2020). Potassium solubilizing and mobilizing microbes: Biodiversity, mechanisms of solubilization, and biotechnological implication for alleviations of abiotic stress. In *New and Future Developments in Microbial Biotechnology and Bioengineering* (pp. 177–202). Elsevier.

Lilly, D. M., & Stillwell, R. H. (1965). Probiotics: growth-promoting factors produced by microorganisms. *Science, 147147.*

Makgoba, L., Abrams, A., Röösli, M., Cissé, G., & Dalvie, M. A. (2024). DDT contamination in water resources of some African countries and its impact on water quality and human health. *Heliyon, 10*(7).

Marco, M. L., Heeney, D., Binda, S., Cifelli, C. J., Cotter, P. D., Foligné, B., ... & Hutkins, R. (2017). Health benefits of fermented foods: Microbiota and beyond. *Current Opinion in Biotechnology, 44*, 94–102.

Mitter, E. K., Tosi, M., Obregón, D., Dunfield, K. E., & Germida, J. J. (2021). Rethinking crop nutrition in times of modern microbiology: Innovative biofertilizer technologies. *Frontiers in Sustainable Food Systems, 5*, 606815.

Moreira-Coello, V., Mouriño-Carballido, B., Marañón, E., Fernández-Carrera, A., Bode, A., Sintes, E., ... & Varela, M. M. (2019). Temporal variability of diazotroph community composition in the upwelling region off NW Iberia. *Scientific Reports, 9*(1), 3737.

Nascimento, A. P. S., & Barros, A. N. (2025). Sustainable innovations in food microbiology: Fermentation, biocontrol, and functional foods. *Foods, 14*(13), 2320.

Nwachukwu, B. C., & Babalola, O. O. (2022). Metagenomics: A tool for exploring key microbiome with the potentials for improving sustainable agriculture. *Frontiers in Sustainable Food Systems, 6*, 886987.

O'Toole, P. W., Marchesi, J. R., & Hill, C. (2017). Next-generation probiotics: The spectrum from probiotics to live biotherapeutics. *Nature Microbiology, 2*(5), 1–6.

Ogidi, O. I., & Akpan, U. M. (2023). Impacts of chemical use in agricultural practices: Perspectives of soil microorganism and vegetation. In *One Health Implications of Agrochemicals and Their Sustainable Alternative*. Springer Nature Singapore Pte Ltd.

Ogidi, O. I., & Felagha, I. (2024). Gut microbiome and diet interaction. In *Nutrition Controversies and Advances in Autoimmune Disease*. IGI Global.

Ogidi, O. I., & Izah, S. C. (2024). Emerging technologies in the global South food industry: Prospects and challenges. In *Food Safety and Quality in Global South*. Springer Nature Singapore Pte Ltd. https://doi.org/10.1007/978-981-97-2428-4_7

Ogidi, O. I., Izah, S. C., & Akpan, U. M. (2023). Food quality and agrochemical use: Integrated monitoring, assessment, and management policies. In *One Health Implications of Agrochemicals and Their Sustainable Alternative*. Springer Nature Singapore Pte Ltd.

Ogidi, O. I., Okara, E., & Barwant, M. M. (2025a). Nanocarriers in plant genetic engineering. In J. Chen (Ed.), *Nanocarriers in Plant Science and Agriculture* (pp. 89–124). IGI Global Scientific Publishing. https://doi.org/10.4018/979-8-3693-9869-2.ch004

Ogidi, O. I., & Tawariowei, A. M. (2023). Multi-elemental analysis and potential human health risk assessment of selected canned foods sold in Nigeria. *Food and Environment Safety Journal*, *22*(4). http://dx.doi.org/10.4316/fens.2023.025

Ogidi, O. I., Wenapere, C. M., & Chukwunonso, O. A. (2025b). Enhancing food safety and quality control with computer vision systems. In D. Bora & R. Bania (Eds.), *Computer Vision Techniques for Agricultural Advancements* (pp. 51–88). IGI Global Scientific Publishing. https://doi.org/10.4018/979-8-3693-8019-2.ch003

Ogwu, M. C., Ogidi, O. I., Izah, S. C., & Odubo, T. C. (2025). Therapeutic and preventive effects of natural compounds derived from microbes. In K. G. Ramawat & J. M. Mérillon (Eds.), *Natural Products*. Springer, Berlin, Heidelberg. https://doi.org/10.1007/978-3-642-36202-6_287-1

Pandey, K. P., Priyadarsini, M., Kushwaha, J., Jurel, A., & Dhoble, A. S. (2024). Microbial resurgence: A new era in agriculture and environmental management. In *Industrial Microbiology and Biotechnology: A New Horizon of the Microbial World* (pp. 763–782). Singapore: Springer Nature Singapore.

Paul, E., & Frey, S. (Eds.). (2023). *Soil Microbiology, Ecology and Biochemistry*. Elsevier.

Peterson, S. B., Bertolli, S. K., & Mougous, J. D. (2020). The central role of interbacterial antagonism in bacterial life. *Current Biology*, *30*(19), R1203–R1214.

Pessione, E. (2012). Lactic acid bacteria contribution to gut microbiota complexity: Lights and shadows. *Frontiers in Cellular and Infection Microbiology*, *2*, 86.

Pradhan, A., Pahari, A., Mohapatra, S., & Mishra, B. B. (2018). Phosphate-solubilizing microorganisms in sustainable agriculture: Genetic mechanism and application. In *Advances in Soil Microbiology: Recent Trends and Future Prospects: Volume 2: Soil-Microbe-Plant Interaction* (pp. 81–97). Singapore: Springer Singapore.

Raghothama, K. G. (2015). Phosphorus and plant nutrition: An overview. In *Agronomy Monographs. American Society of Agronomy, Crop Science Society of America, and Soil Science Society of America* (pp. 353–378). Madison, WI. https://doi.org/10.2134/agronmonogr46.c11

Rani, J., Kumar, D., Sharma, V., & Handique, M. (2024). Metagenomics and AI: A synergistic approach to data-driven insights. In *Genomic Intelligence* (pp. 21–39). CRC Press.

Ritala, A., Häkkinen, S. T., Toivari, M., & Wiebe, M. G. (2017). Single cell protein—State-of-the-art, industrial landscape and patents 2001–2016. *Frontiers in Microbiology*, *8*, 2009.

Roberts, D. P., Short Jr, N. M., Sill, J., Lakshman, D. K., Hu, X., & Buser, M. (2021). Precision agriculture and geospatial techniques for sustainable disease control. *Indian Phytopathology*, *74*(2), 287–305.

Salminen, S., Collado, M. C., Endo, A., Hill, C., Lebeer, S., Quigley, E. M., … & Vinderola, G. (2021). The international scientific association of probiotics and prebiotics (ISAPP) consensus statement on the definition and scope of postbiotics. *Nature Reviews Gastroenterology & Hepatology*, *18*(9), 649–667.

Samtiya, M., Aluko, R. E., Puniya, A. K., & Dhewa, T. (2021). Enhancing micronutrients bioavailability through fermentation of plant-based foods: A concise review. *Fermentation*, *7*(2), 63.

Sanders, M. E., Merenstein, D. J., Reid, G., Gibson, G. R., & Rastall, R. A. (2019). Probiotics and prebiotics in intestinal health and disease: From biology to the clinic. *Nature Reviews Gastroenterology & Hepatology*, *16*(10), 605–616.

Sattar, A., Naveed, M., Ali, M., Zahir, Z. A., Nadeem, S. M., Yaseen, M., … & Meena, H. N. (2019). Perspectives of potassium solubilizing microbes in sustainable food production system: A review. *Applied Soil Ecology*, *133*, 146–159.

Sawant, S. S., Park, H. Y., Sim, E. Y., Kim, H. S., & Choi, H. S. (2025). Microbial fermentation in food: Impact on functional properties and nutritional enhancement—A review of recent developments. *Fermentation (Basel)*, *11*(1).

Sharif, M., Zafar, M. H., Aqib, A. I., Saeed, M., Farag, M. R., & Alagawany, M. (2021). Single cell protein: Sources, mechanism of production, nutritional value and its uses in aquaculture nutrition. *Aquaculture*, *531*, 735885.

Sharma, A., Shankhdhar, D., & Shankhdhar, S. C. (2016). Potassium-solubilizing microorganisms: Mechanism and their role in potassium solubilization and uptake. In *Potassium Solubilizing Microorganisms for Sustainable Agriculture* (pp. 203–219). New Delhi: Springer India.

Singh, R. P., Yadav, P., Kumar, I., Kumar, A., & Gupta, R. K. (2024). Precision biotechnology using beneficial microbes as a fundamental approach to the circular economy. In *The Potential of Microbes for a Circular Economy* (pp. 73–103). Academic Press.

Srivastava, A. K., Shankar, A., Nalini Chandran, A. K., Sharma, M., Jung, K. H., Suprasanna, P., & Pandey, G. K. (2020). Emerging concepts of potassium homeostasis in plants. *Journal of Experimental Botany*, *71*(2), 608–619.

Suez, J., Zmora, N., Zilberman-Schapira, G., Mor, U., Dori-Bachash, M., Bashiardes, S., … & Elinav, E. (2018). Post-antibiotic gut mucosal microbiome reconstitution is impaired by probiotics and improved by autologous FMT. *Cell*, *174*(6), 1406–1423.

Swain, M. R., Anandharaj, M., Ray, R. C., & Parveen Rani, R. (2014). Fermented fruits and vegetables of Asia: A potential source of probiotics. *Biotechnology Research International*, *2014*(1), 250424.

Swanson, K. S., Gibson, G. R., Hutkins, R., Reimer, R. A., Reid, G., Verbeke, K., … & Sanders, M. E. (2020). The international scientific association for probiotics and prebiotics (ISAPP) consensus statement on the definition and scope of synbiotics. *Nature Reviews Gastroenterology & Hepatology*, *17*(11), 687–701.

Tamang, J. P., Cotter, P. D., Endo, A., Han, N. S., Kort, R., Liu, S. Q., … & Hutkins, R. (2020). Fermented foods in a global age: East meets west. *Comprehensive Reviews in Food Science and Food Safety*, *19*(1), 184–217.

Tamang, J. P., Watanabe, K., & Holzapfel, W. H. (2016). Diversity of microorganisms in global fermented foods and beverages. *Frontiers in Microbiology*, *7*, 377.

Tarique, M., Abdalla, A., Masad, R., Al-Sbiei, A., Kizhakkayil, J., Osaili, T., … & Ayyash, M. (2022). Potential probiotics and postbiotic characteristics including immunomodulatory effects of lactic acid bacteria isolated from traditional yogurt-like products. *LWT*, *159*, 113207.

Tijjani, A., & Khairulmazmi, A. (2021). Global food demand and the roles of microbial communities in sustainable crop protection and food security: An overview. In G. Seneviratne, & J. S. Zavahir (Eds.), *Role of Microbial Communities for Sustainability* (pp. 81–107). Springer.

Vermelho, A. B., Moreira, J. V., Akamine, I. T., Cardoso, V. S., & Mansoldo, F. R. (2024). Agricultural pest management: The role of microorganisms in biopesticides and soil bioremediation. *Plants*, *13*(19), 2762.

Vero, S., Garmendia, G., Allori, E., Sanz, J. M., Gonda, M., Alconada, T., … & Wisniewski, M. (2023). Microbial biopesticides: Diversity, scope, and mechanisms involved in plant disease control. *Diversity*, *15*(3), 457.

Villamar-Torres, R. O., Factos-Laiño, K. N., Yánez-Cajo, D., Mayorga-Morejon, K. R., & Jazayeri, S. M. (2025). An overview to the new era in efficient crop management: Artificial intelligence, machine learning, big data, bioinformatics, metagenomics and precision agriculture. *Journal of Animal & Plant Sciences*, *35*(3), 638–659.

Werner, D., & Newton, W. E. (Eds.). (2005). *Nitrogen Fixation in Agriculture, Forestry, Ecology, and the Environment* (Vol. 4). Springer Science & Business Media.

Zeevi, D., Korem, T., Zmora, N., Israeli, D., Rothschild, D., Weinberger, A., … & Segal, E. (2015). Personalized nutrition by prediction of glycemic responses. *Cell*, *163*(5), 1079–1094.

Zmora, N., Suez, J., & Elinav, E. (2019). You are what you eat: Diet, health and the gut microbiota. *Nature Reviews Gastroenterology & Hepatology*, *16*(1), 35–56.

22 Microbial Biotechnologies and the Future of Agriculture in a Changing Climate

Balwant Singh, Mukul Machhindra Barwant, Amjid Khan, and Kingsley Erhons Enerijiofi

22.1 INTRODUCTION

Climate change is a major threat to world agriculture, threatening the stability of food systems and jeopardizing the attainment of food security objectives (Barwant et al., 2025a). Increasing temperatures, changed patterns of precipitation, frequent flooding, droughts, and the increased frequency and severity of extreme weather events are negatively impacting crop yields, soil conditions, and the supply of natural resources (Barwant et al., 2025a, 2025e). These environmental stresses undermine agricultural productivity and increase the exposure of farming populations, especially in developing countries whose livelihood depends significantly on agriculture (Wheeler & von Braun, 2013; IPCC, 2021). Apart from abiotic stresses, climate change also exacerbates biotic stresses like pest infestations and the occurrence of diseases, thus further undermining the resilience of agroecosystems and aggravating the food insecurity situation (Lesk et al., 2016). For the Food and Agriculture Organization (FAO), more than 820 million individuals globally are undernourished, with climate variability being a primary factor in this phenomenon (FAO, 2021a).

In an effort to tackle these intricate and interconnected issues, a revolution in agricultural production is needed—a shift toward sustainable, resilient, and innovative technologies. Among them, microbial biotechnologies have proved to be a revolutionary instrument in contemporary agriculture. Nutrient cycling, soil fertility, disease suppression, and stress reduction are all significantly impacted by microorganisms like cyanobacteria, fungus, actinomycetes, and bacteria. According to Liu et al. (2020) and Vurukonda et al. (2016), the rhizosphere microbiome promotes plant health by modulating nutrient uptake, improving water usage efficiency, and increasing resistance to drought, salt, and heavy metals. More and more, sustainable agriculture is using these phyto-symbionts—which include rhizobia that fix nitrogen, bacteria that solubilize phosphate, and arbuscular mycorrhizal fungi (AMF)—as biofertilizers, biostimulants, and biopesticides.

Sustainable development goal (SDG) 2 of the United Nations is to end world hunger by 2030, and one natural way to achieve this is through the incorporation of microbial biotechnologies into farming systems. Breakthroughs in metagenomics, systems biology, and synthetic biology have made it possible for scientists to decipher intricacies in plant-microbe interactions and design microbial consortia specific to respective agro-climatic niches (Mitter et al., 2019). These breakthroughs have led to the creation of next-generation bioinoculants that can operate effectively in unfavorable environmental conditions, lowering the use of chemical inputs and fostering ecological harmony (Nadeem et al., 2014).

In addition, microbial-based technologies also contribute to environmental sustainability by augmenting carbon sequestration, lowering greenhouse gases, and restoring degraded soils (Pascault et al., 2013). As agriculture evolves in response to the increasing effects of climate change, microbial biotechnologies provide a viable platform for strengthening crop productivity, providing healthy soils, and guaranteeing food systems for coming generations. The strategic utilization of microbial

DOI: 10.1201/9781003650881-22

innovations not only contributes to the ecological intensification of farming but also opens avenues for inclusive and resilient agriculture-based food production systems that are indispensable for realizing global food security under conditions of climate uncertainty (Ogidi & Izah, 2024; Ogidi et al., 2025b).

22.1.1 Climate Change and Farming Problems

Farming is among the sectors most exposed to the diversified impacts of climate change. Increasing global temperatures, unpredictable precipitation patterns, weather extremes, and heightened frequency of floods and droughts are directly affecting crop productivity, soil quality, and water supply (IPCC, 2021). Climatic stresses have increased biotic and abiotic stresses in agricultural systems, which has disturbed food production and resulted in food insecurity, especially in the developing world (Wheeler & von Braun, 2013). The reduction in agricultural productivity caused by climate-induced stressors is projected to lower global crop yields 10–25% by 2050 based on today's emission path (Lesk et al., 2016).

In addition, soil degradation, loss of biological diversity, and lower nutrient use efficiency exacerbate these issues, compromising the resilience of agroecosystems (FAO, 2017). Conventional methods of agriculture tend to fail to tackle these intricate problems, prompting the need for novel, sustainable, and adaptive solutions. The United Nations SDG 2 (Zero Hunger) demands revolutionary approaches to promoting agricultural sustainability and resilience against the backdrop of a changing climate (UN, 2015).

22.1.2 Contribution of Microbial Biotechnologies to Future Agriculture

The development of climate-smart agriculture (CSA) has a bright future thanks to microbial biotechnologies. Liu et al. (2020) and Vurukonda et al. (2016) found that soil-associated microbiomes, which include plant growth-promoting bacteria (PGPB), endophytes, AMF, and rhizobia, play an essential role in enhancing plant health, nutrient cycling, and resilience to both biotic and abiotic challenges. Chemical fertilizers and pesticides are rendered unnecessary by these bacteria that contribute to plant growth (Ogidi et al., 2023). They fix nitrogen, solubilize phosphorus, make plants drought resistant, and inhibit diseases.

Modern advances in microbial biotechnology, such as microbial consortia engineering, synthetic biology, and metagenomics, have opened up new possibilities for sustainable agriculture through the manipulation and optimization of microbial traits (Mitter et al., 2019). When it comes to creating biofertilizers, biopesticides, and biostimulants that are compatible with farming and can withstand climate change, these technologies are crucial (Ogidi & Akpan, 2023; Ogidi et al., 2024). In situations where water is scarce, for instance, research has demonstrated that bioinoculants derived from microbes that can withstand drought can improve water use efficiency and agricultural yields (Nadeem et al., 2014).

The incorporation of microbial biotechnologies in crop cultivation systems also promotes carbon sequestration and greenhouse gas mitigation by enhancing soil health and plant-microbe association (Pascault et al., 2013). It is therefore a nature-based solution that addresses climate risks, food security, and realizes the dream of the 21st-century sustainable agriculture.

22.2 MICROBIAL BIOTECHNOLOGIES FOR CLIMATE RESILIENCE

Microbial biotechnologies have come forward as a viable option for improving climate resilience of agricultural production by exploiting the functional capabilities of beneficial soil microbes. The long-term viability of agricultural production relies on plant growth-promoting microorganisms (PGPMs), including PGPB and AMF. These microbes stimulate plant development by a number of processes, such as biological nitrogen fixation (BNF), phosphate solubilization, phytohormone synthesis, siderophore synthesis, and systemic resistance induction (Bhattacharyya & Jha, 2012). In challenging environmental settings, PGPB taxa such as *Azospirillum*, *Bacillus*, *Pseudomonas*, and *Rhizobium* have shown great promise for enhancing root architecture, nutrient uptake, and stress tolerance (Vurukonda et al., 2016). The majority of land plants also have AMF in their roots. These symbioses improve water and nutrient acquisition, especially micronutrients and phosphorus. Additionally, AMF modulate antioxidant mechanisms and osmotic adjustment, which increases plant resistance to drought, salinity, and heavy metal toxicity (Smith & Read, 2008; Begum et al., 2019).

Nitrogen-fixing bacteria, particularly symbiotic rhizobia and free-living diazotrophs like *Azotobacter* and *Azospirillum*, are central to maintaining soil fertility and reducing reliance on synthetic nitrogen fertilizers. These microorganisms convert atmospheric nitrogen (N_2) into bioavailable ammonium, thereby enhancing soil nitrogen content and promoting plant growth in nitrogen-deficient soils (Mus et al., 2016). Nitrogen-fixing microbes' integration into cropping systems not only enhances the nutrition of plants but also reduces greenhouse gas emissions linked to chemical fertilizers (Smercina et al., 2021). In addition, microbial strategies that promote soil organic matter (SOM) accumulation and aggregate stability help to enhance long-term soil health improvement, which is key to climate adaptation and productivity under resource-constrained conditions.

Under rising climatic stresses like drought and salinity, microbial inoculants offer an environmentally friendly option for stress alleviation. Microbes possessing stress-management characteristics—like the production of 1-aminocyclopropane-1-carboxylate (ACC) deaminase, exopolysaccharides (EPS), and compatible solutes—assist the plant in regulating its physiological processes under water scarcity and salt stress conditions (Nadeem et al., 2014). ACC deaminase-producing bacteria reduce ethylene production in stressed plants, stimulating root growth and sustaining growth (Glick, 2014). EPS-forming microbes enhance the physical structure and water retention capacity of the soil and hence drought tolerance. Microbial inoculants also affect ion transport and osmotic adjustment by modulating the Na^+/K^+ homeostasis and antioxidant enzyme activity in salt stress plants (Upadhyay et al., 2012).

Table 22.1 describes the together, microbial biotechnologies provide multi-purposed means to increase agricultural resilience in the context of climate change. Through support of plant growth, enhanced soil fertility, and abiotic stress tolerance, these biological interventions are a pillar of sustainable agriculture and make a significant contribution toward realizing the targets of food security and Zero Hunger in an uncertain climate.

22.2.1 Plant Growth-Promoting Microbes (PGPB and AMF)

PGPMs, particularly PGPB and AMF, play a pivotal role in fostering sustainable and climate-resilient agriculture. Nitrogen fixation, phosphate solubilization, siderophore production, phytohormone synthesis (e.g., indole-3-acetic acid), and suppression of phytopathogens are some of the ways in which PGPB amplify plant growth (Bhattacharyya & Jha, 2012). The capacity of prominent genera including *Azospirillum*, *Pseudomonas*, *Bacillus*, and *Rhizobium* to increase nutrient uptake, root development, and stress tolerance in various crops has been the subject of much research (Vurukonda et al., 2016).

Mutualistic symbioses are formed by the vast majority of terrestrial plants with AMF, a phylum of *Glomeromycota*. These fungi significantly enhance plant nutrient acquisition, particularly phosphorus, and improve water absorption through extensive hyphal networks. Furthermore, AMF modulate plant physiological responses under stress by enhancing antioxidant defense systems, regulating osmolyte production, and improving membrane stability (Smith & Read, 2008; Begum et al., 2019; Barwant et al., 2025a). These traits enable AMF to support plant productivity under climate-induced stress conditions such as drought, salinity, and nutrient limitations, contributing to overall agroecosystem resilience.

TABLE 22.1
Microbial Biotechnologies for Climate Resilience in Agriculture

Microbial Strategy	Microbial Group	Mechanism of Action	Impact on Plant and Soil Health	References
Plant Growth-Promoting Bacteria (PGPB)	*Azospirillum*, *Bacillus*, *Pseudomonas*, *Rhizobium*	Production of IAA, siderophores, ACC deaminase; nutrient solubilization	Enhance root growth, nutrient uptake, and abiotic stress tolerance	Bhardwaj et al. (2014), Backer et al. (2018)
Arbuscular Mycorrhizal Fungi (AMF)	*Glomus*, *Rhizophagus*, *Funneliformis*	Improved phosphorus uptake, enhanced root colonization, osmolyte accumulation	Increase drought resistance, soil aggregation, and carbon sequestration	Smith and Read (2008), Begum et al. (2019)
Nitrogen-Fixing Bacteria	*Rhizobium*, *Azotobacter*, *Bradyrhizobium*, *Frankia*	Biological nitrogen fixation through nitrogenase enzyme	Reduce need for chemical fertilizers, improve soil fertility and plant nitrogen content	Bhattacharyya and Jha (2012)
Drought Stress Mitigation	*Bacillus subtilis*, *Enterobacter cloacae*, AMF	Osmoregulation, antioxidant enzyme activation, ABA modulation	Enhance plant water use efficiency and root system architecture under water stress	Ngumbi and Kloepper (2016), Hashem et al. (2016)
Salinity Stress Alleviation	*Pseudomonas fluorescens*, *Halotolerant Bacillus*, AMF	Ion homeostasis, osmolyte synthesis, increased proline and soluble sugars	Improve salt tolerance, membrane stability, and chlorophyll content	Egamberdieva et al. (2017), Evelin et al. (2009)

22.2.2 Nitrogen-Fixing Bacteria and Enhancement of Soil Health

Nitrogen-fixing bacteria are essential components of sustainable agriculture due to their role in converting atmospheric nitrogen (N_2) into bioavailable forms usable by plants. These include symbiotic bacteria such as *Rhizobium*, *Bradyrhizobium*, and *Frankia*, which associate with legumes, and free-living or associative diazotrophs like *Azotobacter* and *Azospirillum*. BNF not only reduces the need for synthetic nitrogen fertilizers but also promotes long-term soil fertility and structure (Mus et al., 2016).

Incorporating nitrogen-fixing bacteria into farming systems can mitigate the environmental footprint of agriculture by decreasing nitrous oxide emissions and preventing nitrogen leaching into water bodies. Moreover, these microbes contribute to SOM buildup and influence soil microbial diversity, which are key indicators of healthy and resilient soils (Smercina et al., 2021; Singh et al., 2024). The integration of such microbial technologies thus aligns with agroecological principles and helps maintain productivity under nutrient-deficient and degraded soil conditions.

22.2.3 Microbial Approaches for Tolerance to Drought and Salt Stress

Abiotic stressors, such as drought and salinity, are expected to worsen due to climate change, although microbial inoculants have recently shown great promise in mitigating their negative impacts. In stressed plants, reducing ethylene levels improves root elongation and growth vigor;

this is because some PGPB strains express 1-aminocyclopropane-1-carboxylate (ACC) deaminase. In order to help plants deal with drought, some microorganisms create EPS, which increase the quantity of water available to them by improving soil aggregation and water retention (Nadeem et al., 2014).

In saline soils, microbial consortia are capable of alleviating ion toxicity through the regulation of ion uptake and maintaining Na^+/K^+ homeostasis inside the plant cells. Furthermore, inoculation with halotolerant microbes enhances antioxidant activity and the accumulation of osmolytes in plants, which would strengthen their physiological adaptation to salt stress (Upadhyay et al., 2012). AMF also contribute to stress mitigation by maintaining hydraulic conductivity and improving osmotic adjustment, allowing the plants to bear productivity under harsh environmental conditions (Begum et al., 2019). Stress-tolerant microbial strains as bioformulations can be effectively used for supporting plant performance in degraded and marginal lands in an economically viable, environmentally friendly, and scalable way. Such strategies not only improve crop resilience and yield stability but also reduce synthetic inputs, which hence promote climate-smart and sustainable agricultural practices (Singh et al., 2024).

22.3 ADVANCES IN MICROBIAL-BASED AGRICULTURAL INPUTS

Among these innovations, biofertilizers are a critical strategy aimed toward improving the availability of nutrients while concurrently diminishing the reliance on chemical inputs. Biofertilizers are preparations that are composed of living microorganisms which may be either applied directly to colonize the rhizosphere or introduced into the interior of the plant, thereby facilitating growth by enhancing the supply or availability of primary nutrients to the host plant. They operate by nitrogen fixation, phosphate solubilization, and growth-promoting substance production, hence adding to soil fertility and sustainability over the long term (Mahanty et al., 2017). *Rhizobium*, *Azospirillum*, *Azotobacter*, *Bacillus*, and phosphate-solubilizing bacteria *Pseudomonas* and *Bacillus* species are some of the most commonly employed microbial strains in biofertilizer preparation. These microbial inoculants increase nutrient uptake, root growth, and plant growth, particularly under conditions of stress. Their integration into farming systems supports regenerative agriculture by enhancing soil microbial diversity, reducing greenhouse gas emissions from fertilizer overuse, and restoring degraded soils (Bhardwaj et al., 2014).

Alongside biofertilizers, biopesticides derived from beneficial microorganisms have gained prominence for eco-friendly pest and disease management. Microbial biopesticides offer a sustainable alternative to chemical pesticides by utilizing antagonistic microbes or their bioactive metabolites to suppress pathogens and pests (Barwant et al., 2025d, 2025f). Fungal species like *Trichoderma*, *Beauveria*, and *Metarhizium*, and bacterial species like *Bacillus thuringiensis* and *Pseudomonas fluorescens* are widely used as biocontrol agents (Glare et al., 2012). These microorganisms operate through mechanisms like antibiosis, parasitism, competition, and the induction of systemic resistance in the host plant. Biopesticides are very specific to cause harm to target organisms, biodegradable, and environmentally friendly, thus making them effective as part of integrated pest management (IPM) strategy. Furthermore, their use avoids pesticide resistance, decreases chemical residues in foodstuffs, and protects beneficial insect species, thus contributing to biodiversity and agroecosystem balance (Pieterse et al., 2014).

Recent advances have been achieved through the use of microbial consortia, such as the utilization of compatible functionally diverse microbial strains that work synergistically to further enhance plant and soil health. In contrast to single-strain inoculants, microbial consortia are engineered to conduct several processes simultaneously, including nitrogen fixation, phosphorus solubilization, suppression of diseases, and alleviation of abiotic stress (Bashan et al., 2014). Multifunctional consortia, composed mostly of bacteria, fungi, and occasionally actinomycetes, provide greater persistence and flexibility under changing environmental conditions. They improve soil structure, organic matter content, and microbial enzymatic activities, and enhance crop productivity as well as ecosystem services. Application of microbial consortia is especially efficient in marginal or degraded soils, where the stress factor complexity calls for a holistic and dynamic biological approach (Compant et al., 2019). Together with improved omics technologies and microbial formulation science, engineered synthetic microbial communities adapted to particular crops and environmental settings are now being designed to optimize agricultural advantages while reducing environmental implications.

Overall, such advancements in microbial-based crop inputs represent a promising pathway to CSA. They help toward sustainable intensification by enhancing input use efficiency, rebuilding soil and ecological health, and aiding global attempts at attaining food security and Zero Hunger in the face of a changing climate (Table 22.2).

22.3.1 Biofertilizers and Their Role in Sustainable Farming

Biofertilizers, which are made up of living microorganisms that increase the availability and uptake of nutrients by plants, are a pillar of sustainable agriculture. In contrast to synthetic fertilizers, biofertilizers enhance soil fertility through biological mechanisms like nitrogen fixation, phosphate solubilization, and potassium mobilization, in the process cutting down the environmental impact of agriculture (Malusá & Vassilev, 2014). Nitrogen-fixing bacteria *Rhizobium*, *Azotobacter*, and *Azospirillum*, phosphate-dissolving bacteria *Bacillus* and *Pseudomonas*, and potassium-dissolving strains *Frateuria aurantia* (Mahanty et al., 2017) are some of the popular biofertilizers. In addition to improving nutrient content, these microorganisms also lead to better plant vigor, yield, and tolerance to environmental stressors (Pathak et al., 2025).

With the growing international concerns for environmental degradation and resource depletion, biofertilizers are being highly advocated as green alternatives that replenish soil microbial diversity, enhance organic matter content, and minimize the dependency on chemical fertilizers (Bhardwaj et al., 2014; Singh & Singh, 2023). Biofertilizer application is in harmony with regenerative agriculture and climate-smart farming by enhancing nutrient use efficiency and mitigating greenhouse

TABLE 22.2
Advances in Microbial-Based Agricultural Inputs

Input Type	Microbial Agents	Mechanism of Action	Agronomic/Environmental Benefit	References
Biofertilizers	*Rhizobium, Azospirillum, Azotobacter, Phosphobacteria, Frankia*	Nitrogen fixation, phosphate solubilization, potassium mobilization, phytohormone production	Enhance nutrient availability, reduce chemical fertilizer use, improve soil fertility	Bhardwaj et al. (2014), Vessey (2003)
AMF-based Biofertilizers	*Glomus, Funneliformis, Rhizophagus*	Enhance phosphorus uptake, improve water and micronutrient acquisition	Promote plant growth under abiotic stress; boost soil aggregation	Smith and Read (2008), Begum et al. (2019)
Biopesticides (Bacterial)	*B. thuringiensis, P. fluorescens, B. subtilis*	Produce toxins, antibiotics, lytic enzymes; induces systemic resistance in plants	Control insect pests and bacterial/fungal pathogens; safe for non-target organisms	Glare et al. (2012), Kumar et al. (2021)
Biopesticides (Fungal)	*Trichoderma harzianum, Beauveria bassiana, Metarhizium anisopliae*	Mycoparasitism, competition, antibiosis, and plant defense induction	Effective against soil-borne pathogens and insect pests; promote root health	Mukherjee et al. (2020), Sharma et al. (2014)

gas emissions from agriculture. In addition, the use of high-performance carrier materials, nano-encapsulation, and formulation technologies has greatly enhanced the shelf life and field efficacy of microbial inoculants in various agro-ecological zones (Adesemoye & Kloepper, 2009; Pathak et al., 2025).

22.3.2 Biopesticides for Eco-Friendly Pest and Disease Management

Biopesticides, obtained from microbial organisms or their bioactive metabolites, represent a sustainable tool for the management of agricultural pests and diseases. The agents, especially bacterial strains (*B. thuringiensis*), fungal antagonists (*Trichoderma* spp., *B. bassiana*), and insect-pathogenic nematodes, exert pest control through different modes of action such as parasitism, antibiosis, competition, and induction of systemic resistance in plants (Glare et al., 2012). In contrast to traditional pesticides, biopesticides are target-specific, biodegradable, and have minimal harm to non-target organisms and the environment.

The use of biopesticides has become essential in controlling pest resistance and minimizing the pesticide load in agroecosystems. IPM programs depend more and more on microbial biopesticides to control diseases brought about by soilborne pathogens like *Fusarium*, *Pythium*, and *Rhizoctonia* (Jensen et al., 2020; Barwant et al., 2025a). Moreover, some strains of PGPR not only inhibit pathogens but also induce plant immune systems, which improve the plant to fight various biotic stresses (Pieterse et al., 2014). Production of recombinant and metabolite-formulated products is also broadening the efficacy and application areas of microbial biopesticides in agricultural fields.

22.3.3 Microbial Consortia for Holistic Soil and Plant Health

Microbial consortia, composed of functionally compatible and synergistic microbial strains, are gaining attention as next-generation agricultural inputs that support holistic plant and soil health. Unlike single-strain inoculants, microbial consortia are designed to perform multiple functions simultaneously—such as nutrient solubilization, stress alleviation, and disease suppression—thus enhancing their adaptability and efficacy in field conditions (Frey-Klett et al., 2011). These consortia usually consist of a mixture of nitrogen-fixing bacteria, phosphate solubilizers, AMF, and biocontrol agents to form a strong microbial community which interacts in a positive way with the plant rhizosphere.

Application of microbial consortia has been promising in enhancing crop yields, soil organic carbon, and microbial diversity in degraded and low-input agricultural systems (Bashan et al., 2014). In addition, consortia-based bioinoculants promote soil aggregation, water holding capacity, and microbial enzymatic processes, leading to more efficient ecosystem functioning and climate resilience (Compant et al., 2019). New genomic and metabolomic technologies are now making it possible to screen and design synthetic microbial communities that can be optimized for particular crops, soils, and weather conditions, further opening their potential for agricultural sustainability.

22.4 TECHNOLOGICAL INNOVATIONS IN MICROBIAL RESEARCH

Microbial research technological innovations (Figure 22.1) have far-reaching implications for enhancing understanding and utilization of microbial communities in agriculture to facilitate accurate, effective, and directed interventions toward climate-resilient agriculture. Of these, metagenomics and microbiome engineering have transformed microbial ecology through the revelation of the composition, diversity, and functional potential of complex microbial community symbionts of plants and soils. Metagenomics circumvents culturing requirements by applying high-throughput sequencing technologies to characterize the combined genomes of microbial communities, which disclose previously unculturable and cryptic taxonomic groups

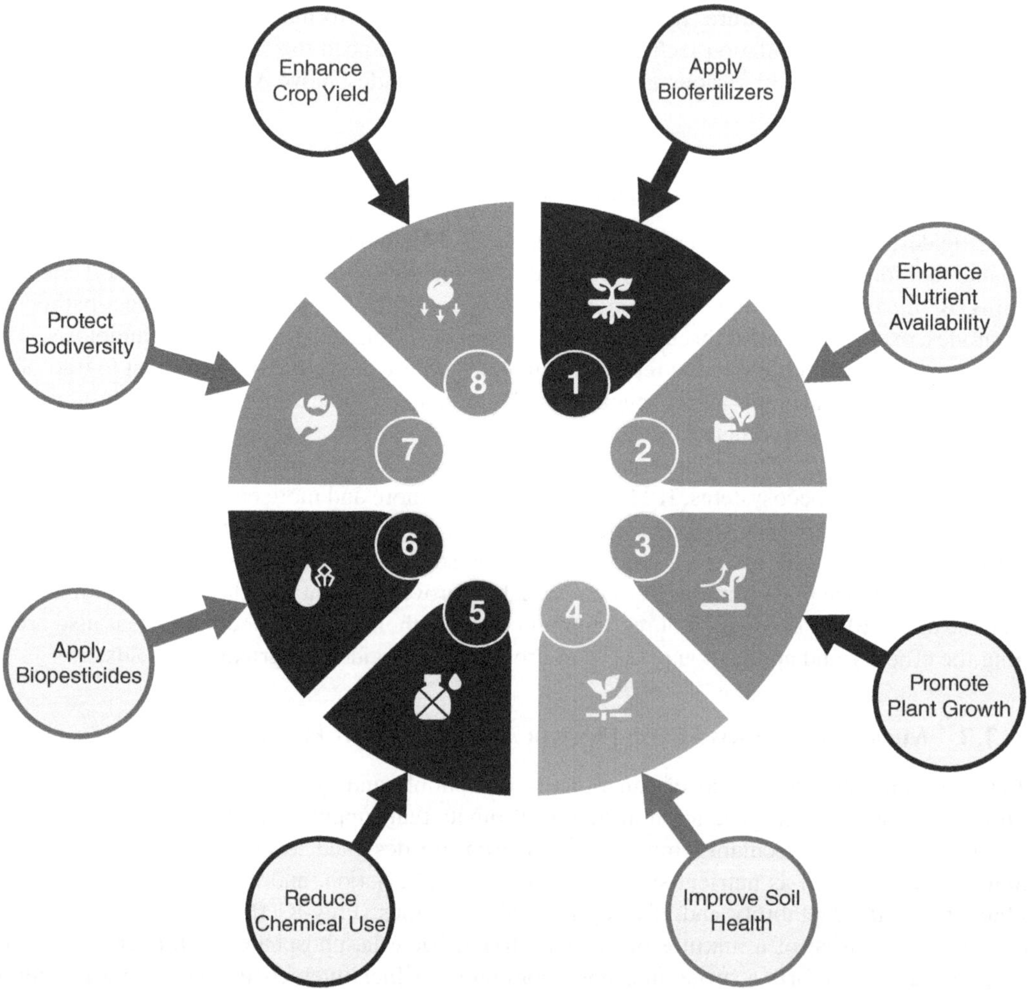

FIGURE 22.1 Microbial-based agricultural inputs.

with fundamental roles in plant health, nutrient cycling, and stress management (Zarraonaindia et al., 2015). This has enabled the establishment of core microbiomes and keystone species determining the performance of crops under abiotic stress, and the formulation of microbiome-driven approaches to engineer robust agroecosystems. In addition, microbiome engineering, informed by systems biology, machine learning, and ecological modeling, allows the manipulation of microbial populations to boost desirable functions like nitrogen fixation, phosphorus solubilization, and disease resistance (Busby et al., 2017).

Synthetic biology provides a second forceful set of tools to enhance the performance of microbes in agriculture through the rational design and construction of microbes with new or improved properties. With genome editing tools such as CRISPR-Cas9, scientists can edit microbial genomes to improve metabolic pathways for increased production of phytohormones, tolerance to stress, or biocontrol activities (Sun et al., 2022). For example, microbial chassis can be designed to produce particular genes for drought resistance or salinity tolerance, thus addressing customized needs of crop growth in marginal environments. Synthetic biology also facilitates the creation of "designer consortia," where communities of multi-strain microbes are composed from complementary functions to supply synergistic benefits to plants. These engineered systems not only enhance efficiency and specificity but also provide environmental safety features, like self-limiting circuits or kill switches, to avoid unwanted ecological effects (Peyraud et al., 2018).

No less significant are advances in intelligent delivery systems that ensure the efficient application and viability of microbial formulations under varied field conditions. Conventional inoculant delivery systems are generally plagued with an unsatisfactory shelf life, low viability, and compromised efficacy caused by environmental stress. Recent developments in encapsulation technologies like alginate beads, chitosan matrices, and nanomaterial-based carriers provide shielding to microbial cells and allow for controlled release at the root zone, hence improving colonization and functional expression (Mitter et al., 2019). In addition, intelligent delivery systems can be engineered to recognize unique environmental stimuli (such as moisture or pH) to trigger microbial release just when and where required. Such innovations greatly enhance the consistency and scalability of microbial-based crop inputs as well as ensure consistent performance in diverse agro-climatic zones.

Complementing each other, these advances are redefining microbial biotechnology from a trial-and-error art form to a precision science. Combining metagenomics, synthetic biology, and smart formulation technologies has the potential to unleash next-generation solutions for sustainable agriculture. By increasing our capacity to describe, design, and deploy useful microbes in a targeted and reproducible fashion, these tools provide a strong route toward climate-resilient food production systems that meet the ideals of Zero Hunger and environmental care (Figure 22.2).

22.4.1 Metagenomics and Microbiome Engineering

Metagenomics has transformed our perception of microbial diversity and function within plant production systems through the culture-independent analysis of total microbial populations. Metagenomics, through next-generation sequencing (NGS), gives insights into the taxonomic composition, metabolic capabilities, and ecological relationships of microbial assemblages that colonize the rhizosphere, phyllosphere, and endosphere (Zarraonaindia et al., 2015). This data enables researchers to discern valuable microbial characteristics relating to plant growth promotion, nutrient cycling, and abiotic stress tolerance even in unculturable or low-abundance taxa. Consequently, metagenomics is a basis for assembling microbial databases and developing functionally optimized microbial inoculants specific to a given crop and environment.

Microbiome engineering is an extension of metagenomic knowledge whereby the function and composition of microbial communities associated with plants are deliberately manipulated to bring about beneficial effects, like improved tolerance to drought, pests, or salt (Busby et al., 2017). Through methods like selective plant breeding for pleasant microbiome characteristics, synthetic community assembly, and host-microbe co-selection, microbiome engineering seeks to direct the rhizosphere microbiota in a manner conducive to sustainable agriculture. These approaches usher in new frontiers in agroecology by considering the microbiome as a dynamic and engineerable entity.

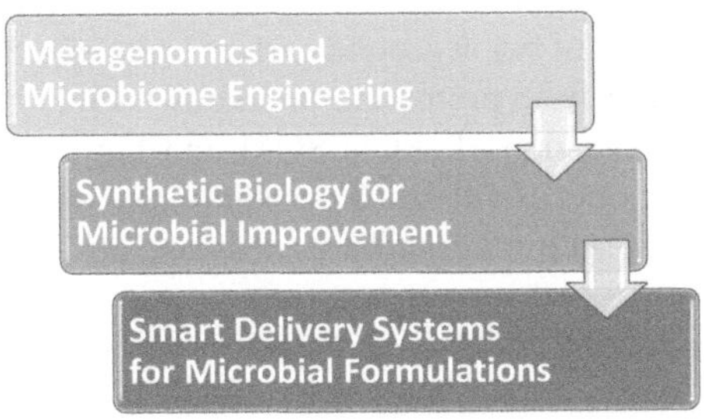

FIGURE 22.2 Technological innovations in microbial research.

22.4.2 Synthetic Biology for Microbial Improvement

Synthetic biology allows for microbes with customized functionalities to be designed and constructed using genome editing, pathway improvement, and chassis design. With the help of tools such as CRISPR-Cas9 and other gene-editing tools, scientists can accurately control microbial genomes to improve characteristics including nitrogen fixation, phosphate solubilization, ACC deaminase production, siderophore release, and the production of plant hormones (Sun et al., 2022). These genetic improvements enable improved survival in adverse soil environments, more efficient interaction with plant roots, and the provision of selected agronomic benefits.

Synthetic biology also makes it possible to design "designer microbial consortia" where engineered microbes are paired according to complementary functions to create synthetic communities with solid ecological function. This ensures that microbes not only accomplish single activities but also work well in a community to achieve entire plant and soil wellness. Notably, synthetic biology allows for the creation of genetic controls like kill-switches or quorum-sensing regulation systems to avoid environmental spillage or runaway growth (Peyraud et al., 2018). Such advances promise next-generation bioinoculants that are safer, more stable, and very effective under field conditions.

22.4.3 Smart Delivery Systems for Microbial Formulations

Notwithstanding the increasing success of microbial-based crop inputs, field-level performance is frequently contingent upon the survival, colonization, and functioning of the microbial inoculants upon application. Smart delivery systems intend to address these limitations by enhancing the stability, targeted delivery, and functional expression of the microbial formulations. New technologies in encapsulation, including alginate, chitosan, starch polymers, and nanomaterials, shield microbes against environmental stress factors (e.g., UV light, drying, and temperature fluctuations) and facilitate controlled release within the rhizosphere (Mitter et al., 2019).

These systems can also be created to be triggered by environmental signals like soil pH, temperature, or moisture content, so microbes are engaged exactly when they are required. The application of biodegradable and bio-compatible material in formulation also maximizes environmental safety and sustainability. Moreover, improvements in spray-dried, lyophilized, and nano-emulsified inoculants have boosted shelf life and compatibility with mechanized farming practices, increasing the scalability and commercial applicability of microbial technologies (Malusá & Vassilev, 2014).

22.5 ENVIRONMENTAL AND POLICY CONSIDERATIONS

Microbial technologies provide exciting opportunities not just for enhancing agricultural productivity but also for resolving environmental issues such as climate change mitigation. Among the major environmental contributions of soil microbes are carbon sequestration. Soil microorganisms, that is, bacteria, fungi, and actinomycetes, are primarily responsible for decomposition and processing of organic matter, thus leading to carbon stabilization in microbial biomass and necromass. AMF, especially, promote carbon allocation to the soil by plants from their symbiotic relationships with roots, resulting in an enhancement of belowground carbon input and the facilitation of soil aggregation, which shields organic matter from premature breakdown (Sokol & Bradford, 2019). Microbial communities also assist in the creation of stable SOM and humus, which act as long-term sinks for carbon (Liang et al., 2019). Successful soil microbiota management therefore offers a chance to enhance carbon sequestration within agriculture, decrease atmospheric CO_2 concentration, and improve soil health and resilience.

In spite of the increasing scientific evidence of the advantages of microbial biotechnologies, various factors complicate their adoption on a large scale in agricultural systems. One of the primary limitations is the variability in field-level efficacy of microbial inoculants, which usually occurs as a result of variability in soil characteristics, climatic factors, and indigenous microbial flora (Compant et al., 2019). Survival and performance of applied microbials are also influenced by formulation problems like low shelf life, inefficient delivery systems, and insufficient microbial compatibility with particular agroecosystems. In addition, insufficient standardized protocols for quality control, product testing for efficacy, and certification demolish user confidence and retard commercialization. From a socio-economic perspective, poor farmer awareness, technical illiteracy, poor extension services, and poor defined economic incentives are other inhibitory factors preventing large-scale adoption of microbial technologies, especially among small-scale and poor resource farmers (Backer et al., 2018).

In order to transcend these constraints and realize the full potential of microbial biotechnologies, enabling policy approaches are needed. Most agricultural policies and subsidy regimes are still very much oriented toward chemical fertilizers and pesticides, with little regulatory acceptance of bio-based inputs. Embedding microbial solutions within mainstream agricultural policy involves setting legal definitions in place, guidelines for product registration, and safety regulations for biofertilizers, biopesticides, and microbial soil conditioners (Malusá & Vassilev, 2014). Policymakers further need to encourage research and innovation through funding instruments, green technology tax relief, and public–private collaboration. At a global level, microbial technologies can be integrated into CSA plans, nationally determined contributions (NDCs), and SDGs and Paris Agreement programs. Farmer training schemes, demonstration schemes, and participatory research can bring the gap between laboratory innovation and on-farm adoption. Essentially, a multi-level policy framework that promotes innovation, guarantees product quality, promotes extension services, and complements climate objectives is necessary for mainstreaming microbial biotechnologies as climate-resilient and sustainable agricultural tools.

22.5.1 Microbial Role in Carbon Sequestration

Microorganisms are an integral part of soil carbon dynamics and have tremendous potential to help with climate change mitigation via carbon sequestration. Soil microorganisms such as bacteria, fungi, and archaea stabilize organic matter through decomposition, the alteration of plant residues, and the creation of microbial necromass, which forms the bulk of soil organic carbon (Liang et al., 2019). Mycorrhizal fungi, especially arbuscular and ectomycorrhizal fungi, increase carbon input into the soil by enhancing root biomass and enabling the transfer of photosynthetically fixed carbon to stable soil aggregates (Sokol & Bradford, 2019). Microbial processes also affect the formation of humus and stabilization of soil aggregates, hence preventing loss of carbon from erosion and leaching.

Evidence has been demonstrated through current research that microbial communities can be manipulated for optimal carbon sequestration potential. For instance, inoculation with plant growth-promoting rhizobacteria (PGPR) and mycorrhizal fungi have been associated with augmented belowground biomass and carbon contribution, particularly in degraded or marginal soils (Lal, 2020). In conservation agriculture and low-tillage systems, microbial amendment also enhances carbon sequestration by improving soil structure, water holding capacity, and microbial respiration efficiency. Microbes are therefore being increasingly regarded as the vital partners in counteracting greenhouse gas emissions from agricultural ecosystems.

22.5.2 Challenges to Large-Scale Microbial Technology Adoption

Notwithstanding their established advantages, a number of technical, environmental, and socio-economic challenges limit the large-scale adoption of microbial technologies in agriculture. Chief among these is the variability in field performance resulting from discrepancies in soil properties, climatic conditions, crop species, and indigenous microbial communities (Compant et al., 2019). The performance of microbial inoculants is also affected by low formulation stability, limited shelf life, and insufficient delivery systems under adverse field environments. In addition, the absence of standardized laboratory protocols for quality control, storage, and application prevents many of the commercial microbial products from being reliable.

Socio-economically, low awareness among farmers, insufficient extension services, and inadequate demonstration of cost-benefit ratios in field experiments can hinder adoption. Growers will also be reluctant to switch chemical inputs with bio-based options because they will fear the short-term yield effects. Further, the intellectual property rights (IPRs), regulation barriers, and scattered markets for microbial products are sources of unequal access and uneven product quality (Backer et al., 2018). These gaps need to be addressed through joint action by scientific confirmation, private sector innovation, training for farmers, and supportive institutional arrangements.

22.5.3 Policy Guidelines for Implementing Microbial Solutions Into Agriculture

Implementing microbial technologies into conventional agricultural practices requires all-encompassing policy intervention. Existing agricultural policies in most nations are mainly geared toward chemical inputs and intensive production systems, with little scope for biologically derived alternatives. There is an urgent need for regulatory systems that acknowledge biofertilizers, biopesticides, and microbial soil conditioners as key components of sustainable agriculture (Malusá & Vassilev, 2014). Policies must aim at setting requirements for product registration, quality control, effectiveness testing, and environmental protection to boost confidence among stakeholders.

At global level, microbial technologies can be integrated into CSA policies as well as national pledges under the SDGs and the Paris Agreement. Incentive schemes like subsidies on microbial inputs, provision of indigenous bioresource use, and public–private collaboration for research and development can promote acceleration and adoption (FAO, 2021b). In addition, capacity building initiatives and farmer field schools must be established to enhance the knowledge and on-farm practical skills needed to utilize microbial solutions effectively. Finally, a harmonious integration of environmental sustainability, technological development, and sound policy frameworks is necessary to realize the full potential of microbial biotechnologies in reshaping agriculture and securing food in the climate change era.

22.6 FUTURE DIRECTIONS AND CONCLUSION

The future of agriculture in the context of climate change and food insecurity will further depend on the incorporation of microbial biotechnologies into sustainable agriculture systems. As international agriculture struggles to deal with increasing heat, unpredictable rainfall, soil fertility loss, and excessive chemical applications, microbes present a natural, environmentally friendly, and cost-effective solution to boosting crop yields, soil quality, and climatic adaptability. Continued developments in metagenomics, systems biology, synthetic biology, and artificial intelligence will further transform microbial use in agriculture by allowing for the specific identification, design, and application of positive microbes for specific environmental and crop requirements (Busby et al., 2017; Sun et al., 2022).

In the future, microbial consortia—precisely engineered communities of actinomycetes, fungi, and bacteria—will become substitutes for single-strain bioinoculants because of their increased functional diversity and compatibility with ecosystems. These consortia will enhance several functions of plants at once, including nutrient uptake, promotion of plant growth, resistance to disease, and tolerance of abiotic stress. In addition, the establishment of future-generation bioformulations with nanocarriers, intelligent polymers, and environmentally responsive encapsulation systems will significantly enhance the viability, targeted delivery, and field performance of microbiome inputs (Mitter et al., 2019; Ogidi et al., 2025a). In combination with progress in microbiome engineering and host-microbe interactions, scientists will be capable of navigating plant-associated microbiomes toward desirable trajectories under altered climatic conditions.

Still, for the maximum utilization of microbial technologies, a number of scientific, economic, and policy challenges would need to be overcome. Uniform regulatory systems, sound quality control measures, and public–private collaborations will be instrumental to guarantee the safe, efficient, and extensive use of microbial inputs in agricultural production (Malusá & Vassilev, 2014). In addition, greater investment in the education of farmers, capacity development, and participatory research will be essential to closing the gap between innovation and application, particularly in low- and middle-income nations where hunger is most severe. Special focus should also be given to incorporating microbial technologies into climate-resilient agriculture plans and national food security policy, thus solidifying the position of microbial biotechnology as a foundation of secure food systems.

In summary, microbial biotechnologies have revolutionary power to revolutionize contemporary agriculture for sustainability, better soil and plant health, and alleviation of the harmful effects of climate change. SDG 2 (Ending Hunger) and SDG 13 (Climate Action) are particularly aligned with their strategic application inside the United Nations. Microbiological solutions will play a crucial role in future scientific and technological developments that aim to secure food supplies around the world, strengthen agroecosystems, and pave the way for agriculture to adapt to climate change.

REFERENCES

Adesemoye, A. O., & Kloepper, J. W. (2009). Plant–microbes interactions in enhanced fertilizer-use efficiency. *Applied Microbiology and Biotechnology*, 85(1), 1–12. https://doi.org/10.1007/s00253-009-2196-0

Backer, R., Rokem, J. S., Ilangumaran, G., Lamont, J., Praslickova, D., Ricci, E., … Smith, D. L. (2018). Plant growth-promoting rhizobacteria: Context, mechanisms of action, and roadmap to commercialization of biostimulants for sustainable agriculture. *Frontiers in Plant Science*, 9, 1473. https://doi.org/10.3389/fpls.2018.01473

Barwant, M. M., Singh, B., Ogidi, O. I., Dar, M. A., & Muhammad, M. (2025a). Integrative physiology of food safety management. In T. Sarkar, A. Hamad, & A. Chatterjee (Eds.), *Physiological Perspectives on Food Safety: Exploring the Intersection of Health and Nutrition*. Springer, Cham. https://doi.org/10.1007/978-3-031-94582-3_17

Barwant, M. M., Ogidi, O. I., Singh, B., Patil, V. S., Sharma, S. A. B. (2025b). Climate change and pesticide dynamics. In B. R. Babaniyi & E. E. Babaniyi (Eds.), *The Interplay of Pesticides and Climate Change*. Springer, Cham. https://doi.org/10.1007/978-3-031-81669-7_7

Barwant, M. M., Ogidi, O. I., Yogita, C., Munje, R. (2025d). Study of consumers choices on pesticides use and sustainability. In: B. R. Babaniyi & E. E. Babaniyi (Eds.), *The Interplay of Pesticides and Climate Change*. Springer, Cham. https://doi.org/10.1007/978-3-031-81669-7_18

Barwant, M. M., Sharma, S. A. B., Lodh, D., Sil, D. C., Ogidi, O. I. (2025f). Immedicable effects of organic pesticides on aquatic environment. In B. R. Babaniyi & E. E. Babaniyi (Eds.), *The Interplay of Pesticides and Climate Change*. Springer, Cham. https://doi.org/10.1007/978-3-031-81669-7_16

Barwant, M. M., Singh, S., Ogidi, O. I., Singh, S., Naik, B. S. S. S. (2025e). Economic implications of climate change and pesticide use on agriculture. In B. R. Babaniyi & E. E. Babaniyi (Eds.), *The Interplay of Pesticides and Climate Change*. Springer, Cham. https://doi.org/10.1007/978-3-031-81669-7_19

Bashan, Y., de-Bashan, L. E., Prabhu, S. R., & Hernandez, J. P. (2014). Advances in plant growth-promoting bacterial inoculant technology: Formulations and practical perspectives (1998–2013). *Plant and Soil*, 378(1–2), 1–33. https://doi.org/10.1007/s11104-013-1956-x

Begum, N., Qin, C., Ahanger, M. A., Raza, S., Khan, M. I., Ashraf, M., & Ahmed, N. (2019). Role of arbuscular mycorrhizal fungi in plant growth regulation: Implications in abiotic stress tolerance. *Frontiers in Plant Science*, 10, 1068. https://doi.org/10.3389/fpls.2019.01068

Bhardwaj, D., Ansari, M. W., Sahoo, R. K., & Tuteja, N. (2014). Biofertilizers function as key player in sustainable agriculture by improving soil fertility, plant tolerance and crop productivity. *Microbial Cell Factories*, 13(1), 66. https://doi.org/10.1186/1475-2859-13-66

Bhattacharyya, P. N., & Jha, D. K. (2012). Plant growth-promoting rhizobacteria (PGPR): Emergence in agriculture. *World Journal of Microbiology and Biotechnology*, 28(4), 1327–1350. https://doi.org/10.1007/s11274-011-0979-9

Busby, P. E., Soman, C., Wagner, M. R., Friesen, M. L., Kremer, J., Bennett, A., … & Dangl, J. L. (2017). Research priorities for harnessing plant microbiomes in sustainable agriculture. *PLoS Biology*, 15(3), e2001793. https://doi.org/10.1371/journal.pbio.2001793

Compant, S., Samad, A., Faist, H., & Sessitsch, A. (2019). A review on the plant microbiome: Ecology, functions, and emerging trends in microbial application. *Journal of Advanced Research*, 19, 29–37. https://doi.org/10.1016/j.jare.2019.03.004

Egamberdieva, D., Wirth, S. J., Alqarawi, A. A., Abd_Allah, E. F., & Hashem, A. (2017). Phytohormones and beneficial microbes: Essential components for plants to balance stress and fitness. *Frontiers in Microbiology*, 8, 2104. https://doi.org/10.3389/fmicb.2017.02104

Evelin, H., Kapoor, R., & Giri, B. (2009). Arbuscular mycorrhizal fungi in alleviation of salt stress: A review. *Annals of Botany*, 104(7), 1263–1280. https://doi.org/10.1093/aob/mcp251

FAO (2017). *The Future of Food and Agriculture – Trends and Challenges*. Food and Agriculture Organization of the United Nations. https://www.fao.org/3/i6583e/i6583e.pdf

FAO (2021a). *Harnessing Biotechnologies for Sustainable Agrifood Systems*. Food and Agriculture Organization of the United Nations. /https://digitallibrary.in.one.un.org/TempPdfFiles/6444_1.pdf

FAO (2021b). *The State of Food Security and Nutrition in the World 2021: Transforming Food Systems for Food Security, Improved Nutrition and Affordable Healthy Diets for All*. Food and Agriculture Organization of the United Nations. https://www.fao.org/3/cb4474en/cb4474en.pdf

Frey-Klett, P., Burlinson, P., Deveau, A., Barret, M., Tarkka, M., & Sarniguet, A. (2011). Bacterial–fungal interactions: Hyphens between agricultural, clinical, environmental, and food microbiologists. *Microbiology and Molecular Biology Reviews*, 75(4), 583–609. https://doi.org/10.1128/MMBR.00020-11

Glare, T. R., Caradus, J. R., Gelernter, W. D., Jackson, T. A., Keyhani, N. O., Köhl, J., … & Stewart, A. (2012). Have biopesticides come of age? *Trends in Biotechnology*, 30(5), 250–258. https://doi.org/10.1016/j.tibtech.2012.01.003

Glick, B. R. (2014). Bacteria with ACC deaminase can promote plant growth and help to feed the world. *Microbiological Research*, 169(1), 30–39. https://doi.org/10.1016/j.micres.2013.09.009

Hashem, A., Abd_Allah, E. F., Alqarawi, A. A., & Egamberdieva, D. (2016). Arbuscular mycorrhizal fungi enhance salinity tolerance of *Trifolium alexandrinum* by altering photosynthetic and antioxidant pathways. *Journal of Plant Interactions*, 11(1), 1–9. https://doi.org/10.1080/17429145.2015.1129574

IPCC (2021). *Climate Change 2021: The Physical Science Basis*. Intergovernmental Panel on Climate Change. Cambridge University Press. https://www.ipcc.ch/report/ar6/wg1/

Jensen, B., Knudsen, I. M., & Andersen, B. (2020). Microbial-based biopesticides in IPM: Current status and future prospects. In J. L. Capinera (Ed.), *Encyclopedia of Entomology* (pp. 245–260). Springer.

Kumar, S., Kaushik, N., & Saini, P. (2021). Biopesticides: Current status and prospects in pest management. *Journal of Plant Diseases and Protection*, 128, 11–26. https://doi.org/10.1007/s41348-020-00360-0

Lal, R. (2020). Soil organic matter content and crop yield. *Journal of Soil and Water Conservation*, 75(2), 27A–32A. https://doi.org/10.2489/jswc.75.2.27A

Lesk, C., Rowhani, P., & Ramankutty, N. (2016). Influence of extreme weather disasters on global crop production. *Nature*, 529(7584), 84–87. https://doi.org/10.1038/nature16467

Liang, C., Schimel, J. P., & Jastrow, J. D. (2019). The importance of anabolism in microbial control over soil carbon storage. *Nature Microbiology*, 4(6), 738–747. https://doi.org/10.1038/nmicrobiol.2017.105

Liu, H., Brettell, L. E., Qiu, Z., & Singh, B. K. (2020). Microbiome-mediated stress resistance in plants. *Trends in Plant Science*, 25(8), 733–743. https://doi.org/10.1016/j.tplants.2020.03.014

Mahanty, T., Bhattacharjee, S., Goswami, M., Bhattacharyya, P., Das, B., Ghosh, A., & Tribedi, P. (2017). Biofertilizers: A potential approach for sustainable agriculture development. *Environmental Science and Pollution Research*, 24(4), 3315–3335. https://doi.org/10.1007/s11356-016-8104-0

Malusá, E., & Vassilev, N. (2014). A contribution to set a legal framework for biofertilisers. *Applied Microbiology and Biotechnology*, 98(15), 6599–6607. https://doi.org/10.1007/s00253-014-5828-y

Mitter, B., Brader, G., Pfaffenbichler, N., Sessitsch, A., & Compant, S. (2019). Next generation microbiome applications for crop production—Limitations and the need of knowledge-based solutions. *Current Opinion in Microbiology*, 49, 59–65. https://doi.org/10.1016/j.mib.2019.10.006

Mukherjee, P. K., Horwitz, B. A., Herrera-Estrella, A., Schmoll, M., & Kenerley, C. M. (2020). Trichoderma research in the genome era. *Annual Review of Phytopathology*, 60(1), 47–72. https://doi.org/10.1146/annurev-phyto-082712-102353

Mus, F., Crook, M. B., Garcia, K., Costas, A. G., Geddes, B. A., Kouri, E. D., … Ané, J. M. (2016). Symbiotic nitrogen fixation and the challenges to its extension to nonlegumes. *Applied and Environmental Microbiology*, 82(13), 3698–3710. https://doi.org/10.1128/AEM.01055-16

Nadeem, S. M., Ahmad, M., Zahir, Z. A., Javaid, A., & Ashraf, M. (2014). The role of mycorrhizae and plant growth promoting rhizobacteria (PGPR) in improving crop productivity under stressful environments. *Biotechnology Advances*, 32(2), 429–448. https://doi.org/10.1016/j.biotechadv.2013.12.005

Nadeem, S. M., Zahir, Z. A., Naveed, M., & Arshad, M. (2014). Rhizobacteria containing ACC-deaminase confer salt tolerance in maize grown on salt-affected fields. *Canadian Journal of Microbiology*, 60(5), 379–384. https://doi.org/10.1139/w09-092

Ngumbi, E., & Kloepper, J. (2016). Bacterial-mediated drought tolerance: Current and future prospects. *Applied Soil Ecology*, 105, 109–125. https://doi.org/10.1016/j.apsoil.2016.04.009

Ogidi, O. I., & Akpan, U. M. (2023). Impacts of chemical use in agricultural practices: Perspectives of soil microorganisms and vegetation. In M. C. Ogwu, & S. Chibueze Izah (Eds.), *One Health Implications of Agrochemicals and their Sustainable Alternatives* (vol. 34). Sustainable Development and Biodiversity. Springer. https://doi.org/10.1007/978-981-99-3439-3_27

Ogidi, O. I., & Izah, S. C. (2024). Emerging technologies in the Global South Food Industry: Prospects and challenges. In M. C. Ogwu, S. C. Izah, & N. R. Ntuli (Eds.), *Food Safety and Quality in the Global South*. Springer. https://doi.org/10.1007/978-981-97-2428-4_7

Ogidi, O. I., Izah, S. C., & Akpan, U. M. (2023). Food quality and agrochemical use: Integrated monitoring, assessment, and management policies. In *One Health Implications of Agrochemicals and Their Sustainable Alternative*. Springer Nature Singapore Pte Ltd.

Ogidi, O. I., Izah, S. C., & Akpan, U. M. (2023). Food quality and agrochemical use: Integrated monitoring, assessment, and management policies. In M. C. Ogwu, & S. Chibueze Izah (Eds.), *One Health Implications of Agrochemicals and their Sustainable Alternatives* (vol. 34). Sustainable Development and Biodiversity. Springer. https://doi.org/10.1007/978-981-99-3439-3_15

Ogidi, O. I., Izah, S. C., & Tobia, P. S. (2024). Biosurfactant production from cassava processing wastes. In M. C. Ogwu, S. Chibueze Izah, A. A. Cunha Alves, & S. Chandra Babu (Eds.), *Sustainable Cassava: Strategies from Production Through Waste Management*. Elsevier.

Ogidi, O. I., Okara, E., & Barwant, M. M. (2025a). Nanocarriers in plant genetic engineering. In J. Chen (Ed.), *Nanocarriers in Plant Science and Agriculture* (pp. 89–124). IGI Global Scientific Publishing. https://doi.org/10.4018/979-8-3693-9869-2.ch004

Ogidi, O. I., Wenapere, C. M., & Chukwunonso, O. A. (2025b). Enhancing food safety and quality control with computer vision systems. In D. Bora & R. Bania (Eds.), *Computer Vision Techniques for Agricultural Advancements* (pp. 51–88). IGI Global Scientific Publishing. https://doi.org/10.4018/979-8-3693-8019-2.ch003

Pascault, N., Ranjard, L., Kaisermann, A., & Maron, P. A. (2013). Stimulation of different functional groups of bacteria by various plant residues as a driver of soil priming effect. *Ecosystems*, 16(5), 810–822. https://doi.org/10.1007/s10021-013-9650-7

Pathak, A., Kumar, A., Singh, B., Saha, V., Barwant, M. M. (2025). Microbial diversity and their functional roles in agriculture. In B. Singh & M. M. Mukul (Eds.), *Symbiotic Partners in Agriculture: PGPF and PGPB for Sustainable Crop Production* (Vol 1). Akinik Publication. https://doi.org/10.22271/ed.book.3178

Peyraud, R., Cottret, L., Marmiesse, L., & Genin, S. (2018). Control of plant diseases by engineering of microbial synthetic communities. *Pathogens*, 7(3), 63. https://doi.org/10.3390/pathogens7030063

Pieterse, C. M., Zamioudis, C., Berendsen, R. L., Weller, D. M., van Wees, S. C., & Bakker, P. A. (2014). Induced systemic resistance by beneficial microbes. *Annual Review of Phytopathology*, 52, 347–375. https://doi.org/10.1146/annurev-phyto-082712-102340

Sharma, P., Singh, A., Pandey, A., & Gupta, V. K. (2014). Role of *Trichoderma* spp. in biocontrol and plant growth promotion. *Plant Microbes Symbiosis: Applied Facets*, 259–282. https://doi.org/10.1007/978-81-322-2068-8_12

Singh, B., & Singh, V. K. (2023). Diversity of wood-inhabiting macrofungi from District Ayodhya, Uttar Pradesh, India. *Kavak*, 59(3), 01–11. https://doi.org/10.36460/Kavaka/59/3/2023/51-61

Singh, B., Tripathi, S., Tyagi, S., Tripathi, A., Dubey, P., Kumar, S. (2024). Phytostimulation and induction of systemic resistance by plant growth promoting fungi. In P. Mathur & S. Roy (Eds.), *Plant Microbiome and Biological Control. Sustainability in Plant and Crop Protection* (Vol 20). Springer, Cham. https://doi.org/10.1007/978-3-031-75845-4_10

Smercina, D. N., Evans, S. E., Friesen, M. L., & Tiemann, L. K. (2021). To fix or not to fix: Controls on free-living nitrogen fixation in the rhizosphere. *Applied and Environmental Microbiology*, 87(3), e02096–20. https://doi.org/10.1128/AEM.02546-18

Smith, S. E., & Read, D. J. (2008). *Mycorrhizal Symbiosis* (3rd ed.). Academic Press.

Sokol, N. W., & Bradford, M. A. (2019). Microbial formation of stable soil carbon is more efficient from belowground than aboveground input. *Nature Geoscience*, 12(1), 46–53. https://doi.org/10.1038/s41561-018-0258-6

Sun, Y., Cheng, Z., Ma, Y., Chen, Y., & Chen, J. (2022). Synthetic biology for microbial formulation and precision agriculture. *Biotechnology Advances*, 55, 107885. https://doi.org/10.1016/j.biotechadv.2021.107885

UN (2015). *Transforming Our World: The 2030 Agenda for Sustainable Development*. United Nations. https://sdgs.un.org/2030agenda

Upadhyay, S. K., Singh, J. S., Saxena, A. K., & Singh, D. P. (2012). Impact of PGPR inoculation on growth and antioxidant status of wheat under saline conditions. *Plant Biology*, 14(5), 605–611. https://doi.org/10.1111/j.1438-8677.2011.00549.x

Vessey, J. K. (2003). Plant growth promoting rhizobacteria as biofertilizers. *Plant and Soil*, 255(2), 571–586. https://doi.org/10.1023/A:1026037216893

Vurukonda, S. S. K. P., Vardharajula, S., Shrivastava, M., & SkZ, A. (2016). Enhancement of drought stress tolerance in crops by plant growth-promoting rhizobacteria. *Microbiological Research*, 184, 13–24. https://doi.org/10.1016/j.micres.2015.12.003

Wheeler, T., & von Braun, J. (2013). Climate change impacts on global food security. *Science*, 341(6145), 508–513. https://doi.org/10.1126/science.1239402

Zarraonaindia, I., Owens, S. M., Weisenhorn, P., West, K., Hampton-Marcell, J., Lax, S., … Gilbert, J. A. (2015). The soil microbiome influences grapevine-associated microbiota. *mBio*, 6(2), e02527–14. https://doi.org/10.1128/mBio.02527-14

23 Sustainability and Future Agricultural Policies Supporting Microbial Technologies

Balwant Singh, Mudasir A. Dar, Shailendra Kumar, Kingsley Erhons Enerijiofi, Mukul Machhindra Barwant, and Odangowei Inetiminebi Ogidi

23.1 INTRODUCTION

Agricultural sustainability is now a multidimensional idea that considers not just productivity but ecological integrity, social equity, and economic viability. The growing frequency of climate-associated stressors like drought, floods, heat waves, and soil salinization is negatively affecting global crop yields, particularly in vulnerable parts of the Global South. Here, microbial technologies are on the move as sustainable options that fit into climate-resilient agriculture practices and agro-ecological strategies. They provide a biologically derived alternative to synthetic inputs and play key roles in supporting natural processes critical to plant growth and soil recovery (Barwant et al., 2025b, 2025c, 2025e).

The rhizosphere microbiome is a dynamic interface in which advantageous microbial populations interact with plant roots, impacting nutrient acquisition, disease suppression, and mitigation of environmental stress. These microbe-plant interactions are not only symbiotic but also crucial to attaining stability of the agroecosystem. For example, nitrogen-fixing bacteria such as *Rhizobium*, *Azospirillum*, and *Frankia* fix atmospheric nitrogen into plant-accessible forms, which downplays the use of chemical nitrogen fertilizers, which are energy-dependent and lead to nitrous oxide emissions (Rosenblueth et al., 2018). In a similar manner, phosphate-solubilizing bacteria increase phosphorus availability, a nutrient usually trapped in insoluble forms in most types of soils.

With agricultural policies moving toward green and circular economies, incorporating microbial technologies into conventional farming is turning into a policy imperative. Governments and global institutions are beginning to identify the potential of microbiome-based interventions in realizing food security goals. The Food and Agriculture Organization (FAO) promotes sustainable bioinputs, understanding their contribution to lowering reliance on fossil fuel-based agrochemicals and enhancing long-term soil quality (FAO, 2021a). In addition, microbial products formulated for particular crops and weather conditions are now being designed using the tools of omics technologies to make precision agriculture high-yielding and efficient in resource use.

The issue of sustainability in agriculture is not complete without the discussion of the microbial aspect of soil fertility and plant productivity. Microbial consortia not only facilitate plant nutrition but also act as ecological engineers by enhancing soil structure, carbon sequestration, and stabilization of organic matter. For instance, the use of arbuscular mycorrhizal fungi (AMF) has been found to significantly enhance water retention capacity and drought resistance in cereal crops (Barwant et al., 2025f), which is essential in arid and semi-arid areas where there is scarce irrigation resource (Berruti et al., 2016).

Recent trends in agriculture innovation highlight a call for systems-based, integrative strategies that place microbiome science within policy agendas. Sustainable agriculture's future is to transition away from high-input chemical-based farming systems to biologically intensive

DOI: 10.1201/9781003650881-23

systems where microbes are key to crop health, nutrient cycling, and yield stability. This paradigm change not only assures ecological equilibrium but also serves the pressing worldwide call for more food production with less resource input and thus contributes to the fulfillment of SDG 2 (Zero Hunger).

23.1.1 Role of Sustainability in Future Agriculture

Global food security, environmental protection, and socio-economic stability in rural areas can all be achieved through sustainable agriculture (Ogidi et al., 2025). The projected global population of 9.7 billion in 2050 (UN DESA, 2019) means that food demand is expected to increase by 60%, placing a greater strain on the environment and increasing the impact of agriculture. According to Foley et al. (2011), conventional intensive agricultural practices have a tendency to lead to a number of negative outcomes, including environmental degradation, biodiversity loss, soil fertility loss, and water shortages. Thus, we must move quickly to develop and implement agricultural practices that are adaptable to climate change and ecological sustainability.

Sustainable agriculture combines natural ecological principles with advanced technology to sustain productivity at low environmental costs. It focuses on resource conservation, ecosystem services, climate change adaptation, and soil health over the long term (Pretty et al., 2018). Soil aggregation, nutrient cycling, disease control, and plant growth are all greatly aided by microbial technologies such biofertilizers, biopesticides, and microbial consortia (Bhattacharyya & Jha, 2012). In line with regenerative agriculture and the circular economy, these microbial technologies promote sustainability goals such as restoring agroecosystems and reducing dependence on synthetic agrochemicals. Furthermore, according to the UN SDG Report (2023), sustainable methods must be employed to end hunger, provide food security, and improve nutrition in order to achieve SDG 2 (Zero Hunger). Microbial-based strategies are becoming more acknowledged as key drivers of sustainable intensification, thus creating a resilient and future-capable agricultural system.

23.1.2 Significance of Microbial Technologies in Food Security

Microbial technologies have been instrumental in mitigating food security challenges by enhancing agricultural yield, nutrient use efficiency, and stress tolerance. Beneficial microbes, including nitrogen-fixing bacteria, mycorrhizal fungi, and plant growth-promoting rhizobacteria (PGPR), significantly influence soil fertility and plant health (Vessey, 2003; Smith & Read, 2008). As per Lugtenberg and Kamilova (2008), these microorganisms are essential for the regulation of critical physiological processes, including the establishment of systemic resistance, phosphate solubilization, nitrogen fixation, and phytohormone synthesis. Biofertilizers are produced by microbial inoculants, which enhance crop yields, mitigate environmental damage, and reduce dependence on chemical fertilizers. The inoculation of legumes with *Rhizobium* spp. and cereals with *Azospirillum* spp. both enhance nitrogen supply and biomass production (Bashan et al., 2014). In the same vein, AMF enhance the absorption of immobile minerals such as zinc and phosphorus by plants in abiotic stress situations, such as drought and salinity (Smith & Smith, 2011; Barwant et al., 2025f).

In addition, microbial biopesticides provide a green alternative to chemical pesticides in that they specifically target pathogens and pests while having low off-target consequences (Ogidi & Akpan, 2023). Microorganisms like *Bacillus thuringiensis* (Bt), *Trichoderma harzianum*, and *Pseudomonas fluorescens* have been used effectively to control pests in integrated pest management (IPM) programs, which has helped in lowering pesticide residues in food products (Keswani et al., 2014). Improvements in genomics, metagenomics, and synthetic biology are further speeding the design of personalized microbial consortia optimized for crop- and soil-specific purposes (Mitter et al., 2019a). These technologies can potentially transform future agriculture systems to become more productive, resilient, and sustainable—key drivers of global food security (Ogidi & Izah, 2024).

23.2 MICROBIAL TECHNOLOGIES FOR SUSTAINABLE AGRICULTURE

Biofertilizers from plant growth-promoting microorganisms (PGPMs) provide eco-friendly alternatives to enhance crop productivity, soil health, and plant pest management. Microbial technologies, especially biofertilizers, biopesticides, and microbial consortia, form the core for realizing the objectives of sustainable agriculture and climate-resilient agriculture and food production systems (Barwant et al., 2025d, 2025g). Table 23.1 explains about the adoption of these not only curbs dependency on chemical inputs but also improves ecosystem services, thus complementing agroecological change.

23.2.1 Biofertilizers and Soil Health Improvement

Biofertilizers are products made of live or latent microbial strains that infect the rhizosphere or plant tissues and establish forages to make nutrients available by biological means such as fixation of nitrogen, solubilization of phosphate, and mobilization of potassium (Bhattacharyya & Jha, 2012). These microbial inoculants are crucial for ensuring soil fertility and porosity, enhancing

TABLE 23.1
Microbial Technologies for Sustainable Agriculture

Microbial Technology	Microorganisms Involved	Key Functions in Agriculture	Applications/Benefits	References
Biofertilizers	*Rhizobium*, *Azotobacter*, *Azospirillum*, *Phosphate-Solubilizing Bacteria* (PSB), *Mycorrhiza*	Nitrogen fixation, phosphorus solubilization, nutrient mobilization	Enhance soil fertility, reduce synthetic fertilizer use	Bhardwaj et al. (2014), Singh et al. (2021)
Biopesticides	B. thuringiensis, P. fluorescens, Trichoderma spp., Beauveria bassiana	Antagonistic activity against pests and pathogens	Eco-friendly pest control	Kumar et al. (2022), Jat et al. (2021)
Plant Growth-Promoting Rhizobacteria (PGPR)	*Pseudomonas* spp., *Bacillus* spp., *Enterobacter* spp.	Hormone production (IAA, gibberellins), nutrient uptake, disease suppression	Improve plant growth and stress tolerance	Glick (2012), Vimal et al. (2017)
Mycorrhizal Technology	Arbuscular mycorrhizal fungi (*Glomus* spp.)	Symbiotic nutrient exchange, enhanced water and phosphorus uptake	Improve drought resistance and soil health	Smith and Read (2008), Sharma et al. (2020)
Composting and Microbial Decomposition	Cellulolytic and ligninolytic fungi and bacteria (*Aspergillus*, *Bacillus*, *Pseudomonas*)	Organic matter decomposition, nutrient recycling	Waste reduction and compost production	Singh and Prabha (2023), Meena et al. (2021)
Bioremediation	Pseudomonas, Bacillus, Phanerochaete chrysosporium, Trametes versicolor	Degradation of pesticides, heavy metals, and pollutants	Soil and water detoxification	Ahemad (2015), Kumar and Verma (2020)
Microbial Consortia (Co-inoculants)	Combination of PGPR, AMF, N-fixers	Synergistic effects on plant growth and nutrient cycling	Broad-spectrum benefits in various crops	Verma et al. (2022), Sharma et al. (2020)
Microalgae-based Biofertilizers	*Anabaena*, *Nostoc*, *Chlorella*, *Spirulina*	Nitrogen fixation, organic acid secretion, photosynthesis	Soil enrichment and carbon sequestration	Rahi et al. (2020), Singh et al. (2023)

decompositions of organic matter, and promoting plant growth traits. *Rhizobium*, *Azospirillum*, *Azotobacter*, *Phosphobacteria*, and AMF are typical biofertilizers (Singh et al., 2024).

For instance, nitrogen-fixing bacteria like *Rhizobium* in legumes and *Azotobacter* in cereals increase nitrogen availability, minimize the use of synthetic fertilizer, and enhance root and shoot biomass (Bashan et al., 2014). Likewise, phosphate-solubilizing microorganisms (PSMs) such as *Bacillus megaterium* and *Pseudomonas putida* transform insoluble phosphorus into bioavailable forms and hence increase phosphorus uptake while minimizing the application of rock phosphate fertilizers (Rodríguez & Fraga, 1999). AMF establish symbiotic relationships with plant roots and enhance nutrient uptake, particularly phosphorus and micronutrients such as zinc and copper, in addition to their ability to tolerate drought and aggregate soil (Smith & Read, 2008).

Biofertilizer use not only enhances crop yield but also enhances soil microbial diversity, enzyme activity, and carbon sequestration capacity (Adesemoye & Kloepper, 2009). Therefore, the use of biofertilizers in agriculture is an environmentally friendly and cost-effective approach to rehabilitating degraded soils and advancing regenerative agriculture (Singh & Singh, 2023; Barwant et al., 2025f).

23.2.2 Biopesticides and Eco-Friendly Pest Management

Glare et al. (2012) found that biopesticides, which are biocontrol agents derived from microorganisms like bacteria, fungi, viruses, and actinomycetes, successfully manage a variety of pest insects and plant diseases. They provide environmentally friendly substitutes for chemical pesticides by specifically acting against the target pests while having minimal impacts on non-target organisms, lowering environmental pollution and pesticide resistance (Glare et al., 2012).

Bacterial biopesticides like *Bt* yield insecticidal toxins that kill many insect larvae, especially *Lepidoptera*, *Diptera*, and *Coleoptera* species (Bravo et al., 2011). Fungal biopesticides like *T. harzianum* and *B. bassiana* inhibit plant pathogens and insect pests through pathways like mycoparasitism, nutrient and space competition, and the release of secondary metabolites and enzymes (Harman et al., 2004). Besides, *P. fluorescens* and *Bacillus subtilis* are also capable of triggering systemic resistance in host plants, thereby lowering the incidence of disease caused by pathogens such as *Fusarium*, *Rhizoctonia*, and *Pythium*.

The use of biopesticides is an additive to IPM strategies by lowering pesticide residues in food webs, maintaining natural predators, and balancing agroecosystem behavior. In addition, biopesticides meet organic farming requirements and are appropriate for smallholder and resource-poor farmers, offering a sustainable means of pest control without compromising biodiversity (Ogidi et al., 2023).

23.2.3 Microbial Consortia for Climate Resilience

Microbial consortia, which are mixtures of several desirable microbes, provide synergistic advantages in promoting plant growth and tolerance to various environmental stresses. In contrast to single-strain inoculants, consortia include multiple functional microbes that can work together to enhance nutrient uptake, reduce abiotic stress (e.g., drought, salinity, heat), and defend against pathogens (Mitter et al., 2019a; Barwant et al., 2025a).

According to Grover et al. (2011), when plants are subjected to climate stress, microbial consortia have a role in regulating the production of phytohormones including gibberellins, cytokinins, and indole-3-acetic acid (IAA). They also increase the activity of antioxidant enzymes. When crops such as wheat, corn, and soybeans are subjected to drought stress, it has been found that co-inoculation of *Azospirillum*, *Pseudomonas*, and AMF improves their photosynthetic ability, water-use efficiency, and nutrient uptake (Timmusk et al., 2014). Also, through functional redundancy and cross-feeding relationships, microbial consortia maintain rhizospheric resilience, aggregate soil, and increase microbial biomass carbon (Finkel et al., 2017). Central to sustaining productivity under fluctuating climatic conditions are the interactions between plants, microbes, and the environment, which are fostered by these complex microbial communities.

Synthetic microbial consortia tailored to specific agroecosystems are becoming more feasible thanks to developments in systems biology, metagenomics, and microbial ecology. Future agricultural policies aiming at climate resilience and sustainable intensification may be bolstered by these developments, which could revolutionize adaptable farming techniques.

23.3 AGRICULTURAL POLICIES ENABLING MICROBIAL TECHNOLOGIES

The shift to sustainable agriculture is not just a technological necessity but a policy imperative as well. Microbial technologies like biofertilizers, biopesticides, and microbial consortia have the potential for providing solutions to address the twin hurdles of enhanced food production without compromising environmental integrity. Their large-scale deployment, however, hinges considerably on the availability of enabling policy environments, regulation, and strategic alliances. At the global and national levels, conducive agricultural policies are now being harmonized with sustainability objectives to mainstream microbial inputs into farm systems (Figure 23.1).

23.3.1 Global Policy Frameworks for Sustainable Farming

International organizations have come to appreciate the pivotal function of microbial and biological inputs in reshaping farming practices for sustainability. Sustainable intensification of agriculture

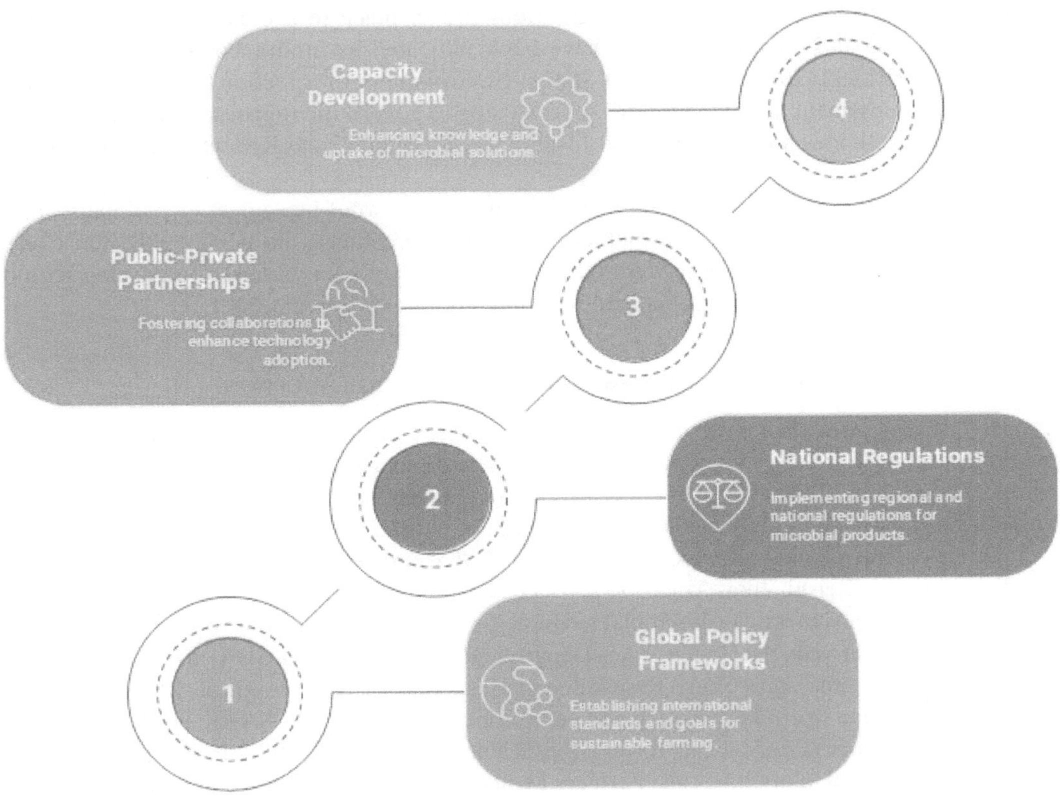

FIGURE 23.1 Acheving suctainable agriculture.

through the application of nature-based solutions and ecosystem services is explicitly promoted by the United Nations SDGs—SDG 2 (Zero Hunger), SDG 13 (Climate Action), and SDG 15 (Life on Land)—in 2023. To achieve these goals, the FAO has long advocated for sustainable agriculture based on agroecological principles that prioritize soil health, biodiversity, and microbiome management (FAO, 2021b).

In order to enhance nutrient cycling, soil fertility, and resilience to climate extremes, microbial technologies are being promoted by global forums such as the Global Soil Partnership and the International Initiative for the Conservation and Sustainable Use of Soil Biodiversity (FAO, 2020). In addition, the protocols and Convention on Biological Diversity (CBD) stress the need of protecting important microbes and encouraging their sustainable usage, especially in agricultural settings (CBD, 2022). The Global Bioeconomy Summit has also encouraged governments to consider microbial technologies in their bioeconomy plan, explaining that they hold considerable potential for green innovation and circular agriculture (Global Bioeconomy Summit, 2020) (Figure 23.1).

23.3.2 National and Regional Regulations for Microbial Products

At the regional and national levels, microbial product development and commercialization are regulated by different regulatory systems, which tend to affect their availability and acceptability to farmers. In India, guidelines for the quality control, labeling, and standards of efficacy of biofertilizers and biopesticides are provided by the Bureau of Indian Standards (BIS) and the Fertilizer Control Order. These regulations are vital to the establishment of microbial product credibility and for the safeguarding of farmers from inferior formulation (Choudhary & Johri, 2009).

Microbial biopesticides in the European Union (EU) are regulated under Regulation (EC) No 1107/2009, where microbial strains are subjected to stringent testing for safety, environmental stability, and efficacy. Although these regulations are in place to provide safety, the lengthy approval periods and compliance costs have been indicated as hindrances to innovation and market entry (Ehlers, 2011). The United States' Environmental Protection Agency (EPA) uses a tiered risk assessment approach for microorganisms to govern the registration and application of microbial pesticides under the Federal Insecticide, Fungicide, and Rodenticide Act (FIFRA) (USEPA, 2020).

In order to align international trade and facilitate cross-border movement of bioinputs, international regulatory convergence and mutual recognition agreements are being called for. Codex Alimentarius and OECD guidelines on microbial products are being referred to develop harmonized standards on registration and trade of microbial inoculants (OECD, 2021).

23.3.3 Role of Public-Private Partnerships in Technology Adoption

The large-scale successful uptake of microbial technologies calls for synergistic partnerships among public institutions, private companies, academic research, and agricultural communities. Public-private partnerships (PPPs) have proven to be effective means of filling gaps in research, funding, knowledge sharing, and supply chain logistics for microbial products (Spielman et al., 2007).

Programs at the national level like India's National Mission on Sustainable Agriculture (NMSA) and Paramparagat Krishi Vikas Yojana (PKVY) have entailed collaborations with private biotech companies and NGOs to upscale microbial technologies in organic agriculture and climate-resilient farming. In a similar vein, the United States Department of Agriculture (USDA) also promotes PPPs through its Sustainable Agriculture Research and Education (SARE) program, where it finances microbial innovation in crop production and environmental care (USDA-SARE, 2021).

Multinational companies such as Bayer Crop Science, Novozymes, and UPL have made investments in R&D related to microbes and partnered with start-ups, universities, and agricultural cooperatives. These partnerships offer both finances and technical capabilities and generate innovations specific to local agroecological needs and marketplace requirements.

Capacity development activities, farm training programs, and online extension platforms are being co-designed through PPPs to enhance knowledge and uptake of microbial solutions. Such activities are vital in ensuring that the benefits of microbial technologies are fairly shared and make significant contributions to food security, sustainable development, and environmental recovery (Figure 23.1).

23.4 CHALLENGES IN POLICY IMPLEMENTATION

Even though microbial technologies have been widely accepted to support sustainable agriculture and food security, there are a number of policy-level and institutional bottlenecks that are inhibiting their proper implementation and extensive use. These are regulatory barriers, commercialization blockades, and weak linkage between science innovation and policy formulation. Overcoming these hindrances is critical to take full advantage of microbial technologies in realizing sustainability and Zero Hunger goals.

In spite of increasing recognition of microbial technologies as sustainable substitutes for conventional synthetic agrochemicals, many serious challenges discourage their effective integration into policy frameworks for agriculture. Among the most significant of these challenges is the complexity and lack of consistency in regulatory systems that manage the approval and commercialization of microbial products. In most nations, regulatory systems apply the same level of scrutiny and long approval time for microbial bioinputs as chemical insecticides, even though they tend to have lower environmental risk and higher specificity. Such regulation creates barriers that not only lengthen and expense product registration but also discourage innovation, especially for small- and medium-scale enterprises (Ehlers, 2011; Chandler et al., 2011). For instance, the EU's Regulation (EC) No 1107/2009 demands extensive toxicological, environmental, and efficacy testing for microbial pesticides, which can take up to seven years, making market entry challenging for newer, ecologically safe formulations (Kabaluk et al., 2010). In developing nations, including India, while regulatory guidelines like the FCO and Insecticides Act establish a basis for biofertilizer and biopesticide quality control, strain-level specificity deficiency, outdated procedures, and weak enforcement of standards complicate further variability in quality and lower farmer confidence (Bhattacharyya & Jha, 2012).

Commercialization and market access to microbial technologies do not receive strong support from weak infrastructure, lack of investment, fragmented supply chains, and poor farmer awareness. The prevalence of synthetic inputs in agricultural markets, supported by subsidies and well-established distribution systems, poses a major entry barrier to competition from bio-based alternatives. Most microbial products demand special storage conditions and possess shorter shelf lives relative to their chemical analogs, making handling and delivery more complicated, particularly in the tropics and in developing countries (Herrmann & Lesueur, 2013). Further, variation in product performance depending on agroecological zones, in addition to unstandardized formulations and inadequate extension services, has led to low rates of adoption by farmers. Small-scale producers of microbial inputs, responsible for a lot of innovation in this area, tend to suffer from scaling up production, dealing with intricate regulatory frameworks, and accessing large markets because of small financial and technical capacities (Jansson & Hofmockel, 2020).

Another main challenge is the gap between scientific research, technological development, and agricultural policymaking. While scientific research in plant-microbe interactions, soil microbiomes, and bioinoculants has progressed at a fast pace, their translation into field-relevant solutions and evidence-based policy interventions has been limited (Finkel et al., 2017). Policymakers do not have access to strong field data and participatory research that capture the agronomic and socio-economic local context, resulting in poorly target policies to farmer requirements and microbial product performance in actual environments. Further, the lack of open-access microbial strain databases, unaddressed intellectual property rights (IPR) issues, and limited interdisciplinary collaborations between regulators, researchers, and extension agencies also restrict the scaling of microbial technologies (Chandra et al., 2020). Closing this gap demands stronger science-policy interfaces, participatory governance arrangements, and support for translational research platforms that connect laboratory innovation with on-the-ground implementation. Strengthening institutional frameworks for multi-stakeholder interaction will be essential to overcoming these system-level challenges and allowing microbial technologies to play a significant role in sustainable agriculture and food security.

23.4.1 Regulatory Barriers and Approval Processes

One of the principal hindrances to microbial technology adoption is the complexity and variability of regulatory frameworks across nations. Biofertilizer and biopesticide regulatory approval procedures are frequently long, fragmented, and expensive, deterring innovation and delaying market access for novel microbial products (Ehlers, 2011). Microbial products are subject to identical strict regulations as chemical pesticides in many areas, even though they pose lower environmental risk and specificity, imposing an unreasonable regulatory burden (Chandler et al., 2011).

For instance, in the EU, microbial biocontrol products fall under Regulation (EC) No 1107/2009, which requires extensive dossiers for evaluations of safety, efficacy, and environmental impact. These steps can require approval in five to seven years, impairing the competitiveness of microbial technologies (Kabaluk et al., 2010). Likewise, in India, while biofertilizers are identified under the FCO and biopesticides under Insecticide Act, respectively, they are subjected to limited regulation by the absence of microbial strain specificity, archaic testing protocols, and weak quality standard enforcement (Bhattacharyya & Jha, 2012).

In addition, border regulation harmonization is still minimal, making international trade and distribution of microbial products difficult. Science-based, risk-related, and harmonized regulatory frameworks reflecting the distinctive nature of microbial inputs and allowing for their safe yet timely commercialization are urgently needed.

23.4.2 Market Accessibility and Commercialization Challenges

Commercialization of microbial technologies, especially in low- and middle-income nations, encounters major hurdles involving infrastructure, investment, supply chain logistics, and market awareness. Small and medium enterprises (SMEs), who are more likely to be the prime drivers of microbial innovation, do not possess the financial and technical capabilities to address intricate regulatory requirements as well as expand production (Jansson & Hofmockel, 2020).

Market fragmentation, unstandardized product formulations, and inconsistent efficacy across various agroclimatic conditions are among the limitations for farmer confidence and adoption. In addition, poor cold chain and storage infrastructure influence shelf-life and viability of microbial inoculants, particularly in the tropics (Herrmann & Lesueur, 2013). These are exacerbated by constrained distribution networks and limited numbers of trained extension staff to inform farmers about appropriate use, benefits, and application methods.

Furthermore, conventional agricultural input markets are still dominated by chemical products because of prevailing subsidies, embedded supply chains, and familiarity of farmers. Absent proper incentives, policy backing, and public promotion campaigns, microbial technologies cannot match traditional agrochemicals in terms of access and perceived effectiveness.

23.4.3 Bridging the Gap Between Research and Policy

One key challenge in achieving the full potential of microbial technologies is the disjunction among academic research, on-the-ground practices, and policy design. Although scientific breakthroughs on microbiome research, metagenomics, and synthetic biology have resulted in the identification of new strains and mechanisms, their translation into market-ready products and policy proposals is narrow (Finkel et al., 2017).

Policymaking is frequently limited by the absence of live data, field-validation, and stakeholder engagement. Numerous policies are prepared without complete integration of farmer expertise, microbial ecology knowledge, or system-based assessment, resulting in implementation gaps. Intellectual property concerns, the unavailability of open-access microbial databases, and poor interdisciplinary coordination also slow down knowledge transfer (Chandra et al., 2020).

To close the gap, more effective science-policy interfaces are needed, such as multi-stakeholder platforms that engage researchers, policymakers, private sector players, and farming communities. Adaptive regulation, participatory policy design, and investment in translational research have the potential to enhance the adoption of microbial technologies and their integration into national and international agriculture systems.

23.5 FUTURE DIRECTIONS IN POLICY AND INNOVATION

The future policy directions in agriculture need to give utmost importance to microbial technologies as drivers of change toward sustainable food systems. The essential policy development among the critical ones is to provide monetary incentives and subsidies to encourage microbial inputs like biofertilizers, biopesticides, and microbial soil conditioners. These biological options tend to suffer from commercial uptake due to costly production, poor farmer awareness, and market volatility. Subsidies that lower the cost pressure on both producers and final users can drive adoption at scale. For instance, assistance by the Indian government under programs such as the PKVY has been instrumental in enhancing the application of organic and microbial inputs by smallholder farmers (Ministry of Agriculture & Farmers Welfare, Government of India, 2020). In a parallel manner, international structures like the EU's Common Agricultural Policy (CAP) have implemented eco-schemes that indirectly subsidize microbial inputs as part of wider soil health and climate objectives (European Commission, 2021a).

Climate-smart agriculture and regenerative approaches are increasingly being recognized as key pillars in countering the impacts of climate change on the food system. Microbial solutions naturally align with these objectives through enhanced carbon sequestration, improved nutrient-use efficiency, and increased soil biodiversity. The integration of microbial technologies into national climate-smart agriculture action plans and regenerative agriculture plans could extend their coverage and environmental effect. Kenya and Brazil are among countries that have started working to couple microbial inoculants with carbon farming and sustainable intensification programs (FAO, 2021c). The focus on circularity, resource use efficiency, and ecosystem services also favors microbial technologies as core elements of agroecological innovation (Pretty et al., 2018).

In order to ensure long-term adoption and innovation, strengthening the research-education-extension continuum is critical. Investments in microbial genomics, product formulation, and field trials need to be supported with the construction of region-specific microbial consortia that are specific to local agroecologies. PPPs and global collaborations can support developing research infrastructure, aligning regulatory protocols, and facilitating data sharing. Simultaneously, extension education for farmers through demonstration plots, mobile advisory, and participatory learning systems is essential for establishing confidence in microbial products and enhancing application skills. Research indicates that awareness and on-field training enormously raise the rates of adoption of biocontrol products and biofertilizers (Meena et al., 2017). Agricultural university and vocational institute curriculum reform should also involve microbial biotechnology courses to empower future professionals with the scientific and practical know-how (Mitter et al., 2019a). Educating farmers and researchers and providing them with access is the basis to ensure that microbial technologies live up to their promise in producing food for an expanding global populace.

23.5.1 Financial Incentives and Subsidies for Microbial Solutions

To accelerate the adoption of microbial technologies in agriculture, targeted financial incentives and subsidies are essential. Unlike synthetic inputs, biofertilizers and biopesticides often face initial market entry challenges due to limited awareness, perceived uncertainty in efficacy, and underdeveloped supply chains. Governments can promote these eco-friendly alternatives by offering input subsidies, tax exemptions, and credit support for microbial product manufacturers and farmers using them (Bhardwaj et al., 2014). For instance, India's PKVY and NMSA have included financial support for bioinput use and organic, including microbial inoculants (Ministry of Agriculture & Farmers Welfare, Government of India, 2020). Likewise, the EU's CAP permits eco-schemes that promote the application of biocontrol agents and organic soil improvers, indirectly favoring microbial innovations (European Commission, 2021b). Financial support structures can reduce entry barriers, stimulate private sector investments, and facilitate scale-up operations of microbial product development and supply.

23.5.2 Climate-Smart Policies and Regenerative Agriculture Initiatives

Incorporation of microbial technologies into climate-smart and regenerative agriculture approaches has great potential to meet the interconnected challenges of food security, climate change, and soil degradation. Microbial products like PGPR, nitrogen-fixing bacteria, and mycorrhizal fungi help in carbon sequestration, synthetic input reduction, and water-use efficiency—principles paramount in climate-resilient agriculture (Singh et al., 2016; Delgado-Baquerizo et al., 2017). Policies supportive of agroecological intensification and regenerative agriculture can involve mandates or incentives for the application of microbial-based biofertilizers, biopesticides, and compost inoculants. Countries like Brazil and Kenya are piloting programs that link microbial technology deployment with climate-smart agriculture goals through farmer cooperatives and national soil health strategies (FAO, 2021a). Additionally, the One Health approach, which links ecosystem, plant, and human health, can further reinforce the adoption of environmentally benign microbial inputs within sustainable food systems.

23.5.3 Strengthening Research, Education, and Farmer Awareness

One of the pillars for scaling microbial technology is the strengthening of research infrastructure and extension education. Dedicated public-private research centers addressing microbial biodiversity, functional genomics, and product formulation for agroclimatically diverse regions need to be created. They must interact with national agricultural research systems (NARS) and international centers in order to develop standardized protocols, bioresource repositories, and field evaluation networks (Mitter et al., 2019b). Besides, incorporation of microbial technology modules into curricula in agricultural universities and vocational training will lead to the development of the next generation of sustainable farming practitioners and biotechnologists. Farmer education is still crucial—participatory learning, demonstration trials, and digital extension platforms need to be utilized to create awareness regarding the usage of microbial products, benefits, and method of application (Meena et al., 2017). In the absence of farmer trust and knowledge, policy and innovation are insufficient to bring microbial technologies into widespread use.

23.6 CONCLUSION

Embracing microbial technologies within sustainable agriculture platforms is a critical strategy for global food security with reduced ecological footprints. As conventional agriculture continues to struggle with soil loss, climate uncertainty, and over-reliance on chemical inputs, microbial alternatives like biofertilizers, biopesticides, and microbial consortia can be ecologically appropriate and economically accessible options. These technologies not only increase crop productivity and nutrient availability but also play an important role in contributing to soil health, carbon sequestration, and resilience to abiotic stresses (Vessey, 2003; Bhattacharyya et al., 2020).

Their extensive use depends on facilitating policies, regulatory harmonization, and effective institutional support. Present-day obstacles—vitiated from uncertain regulatory regimes, sluggish processes of approval, entry hurdles in the market, and farmers' ignorance—emphasize the importance of visionary policy reforms and intelligent investments in research and education (Mitter et al., 2019b; Rouphael & Colla, 2020). Standardization across countries, encouraging the private sector to participate, implementing simplified product registration processes will be vital for scaling microbial technologies in agriculture.

Looking ahead, policies that integrate microbial technologies with climate-smart and regenerative agriculture agendas can provide dual benefits of productivity enhancement and environmental stewardship. Financial subsidies, farmer training programs, and PPPs should be prioritized to bridge the gap between laboratory innovation and field-level application. Strengthening the research-extension interface and incorporating microbial science into agricultural curricula will further ensure that future stakeholders are equipped to harness these innovations effectively.

Finally, a multisectoral and multidisciplinary approach will be vital in integrating microbial technologies into mainstream agricultural policies. This will not only promote the objectives of Zero Hunger and sustainable development, but also create resilient agri-food systems that are able to endure future climatic and economic stresses (Pretty et al., 2018; FAO, 2021a). By putting microbes at the core of agricultural change, we can unlock a more sustainable, inclusive, and food-secure future.

REFERENCES

Adesemoye, A. O., & Kloepper, J. W. (2009). Plant–microbes interactions in enhanced fertilizer-use efficiency. *Applied Microbiology and Biotechnology*, 85(1), 1–12. https://doi.org/10.1007/s00253-009-2196-0

Ahemad, M. (2015). Enhancing phytoremediation of heavy metals by soil microbes. *Ecological Engineering*, 74, 213–223. https://doi.org/10.1016/j.ecoleng.2014.10.027

Barwant, M. M., Ogidi, O. I., Singh, B., Patil, V. S., Sharma, S. A. B. (2025a). Climate change and pesticide dynamics. In B. R. Babaniyi & E. E. Babaniyi (Eds.), *The Interplay of Pesticides and Climate Change*. Springer, Cham. https://doi.org/10.1007/978-3-031-81669-7_7

Barwant, M. M., Ogidi, O. I., Yogita, C., Munje, R. (2025b). Study of consumers choices on pesticides use and sustainability. In B. R. Babaniyi & E. E. Babaniyi (Eds.), *The Interplay of Pesticides and Climate Change*. Springer, Cham. https://doi.org/10.1007/978-3-031-81669-7_18

Barwant, M. M., Sharma, S. A. B., Lodh, D., Sil, D. C., Ogidi, O. I. (2025c). Immedicable effects of organic pesticides on aquatic environment. In B. R. Babaniyi & E. E. Babaniyi (Eds.), *The Interplay of Pesticides and Climate Change*. Springer, Cham. https://doi.org/10.1007/978-3-031-81669-7_16

Barwant, M. M., Singh, B., Ogidi, O. I., Dar, M. A., Muhammad, M. (2025d). Integrative physiology of food safety management. In T. Sarkar, A. Hamad & A. Chatterjee (Eds.), *Physiological Perspectives on Food Safety: Exploring the Intersection of Health and Nutrition*. Springer, Cham. https://doi.org/10.1007/978-3-031-94582-3_17

Barwant, M. M., Singh, S., Ogidi, O. I., Singh, S., Naik, B. S. S. S. (2025e). Economic implications of climate change and pesticide use on agriculture. In B. R. Babaniyi & E. E. Babaniyi (Eds.), *The Interplay of Pesticides and Climate Change*. Springer, Cham. https://doi.org/10.1007/978-3-031-81669-7_19

Barwant, M. M., Singh, B., Sharma, S., Gore, N. T., Mohammad, A. (2025f). Role of mycorrhizal association in leguminous plant growth. In P. Mathur & A. Gupta (Eds.), *Recent Trends and Applications of Leguminous Microgreens as Functional Foods*. Springer, Cham. https://doi.org/10.1007/978-3-031-75678-8_13

Barwant, M. M., Singh, S., Singh, S., Naik, B. S. S. S., Ogidi, O. I. (2025g). Resilience of ecosystems: Navigating the challenges of pesticide exposure and climate change. In B. R. Babaniyi & E. E. Babaniyi (Eds.), *The Interplay of Pesticides and Climate Change*. Springer, Cham. https://doi.org/10.1007/978-3-031-81669-7_3

Bashan, Y., de-Bashan, L. E., Prabhu, S. R., & Hernandez, J. P. (2014). Advances in plant growth-promoting bacterial inoculant technology: Formulations and practical perspectives. *Plant and Soil*, 378(1), 1–33. https://doi.org/10.1007/s11104-013-1956-x

Berruti, A., Lumini, E., Balestrini, R., & Bianciotto, V. (2016). Arbuscular mycorrhizal fungi as natural biofertilizers: Let's benefit from past successes. *Frontiers in Microbiology*, 6, 1559. https://doi.org/10.3389/fmicb.2015.01559

Bhardwaj, D., Ansari, M. W., Sahoo, R. K., & Tuteja, N. (2014). Biofertilizers function as key player in sustainable agriculture by improving soil fertility, plant tolerance and crop productivity. *Microbial Cell Factories*, 13(1), 66. https://doi.org/10.1186/1475-2859-13-66

Bhattacharyya, P. N., & Jha, D. K. (2012). Plant growth-promoting rhizobacteria (PGPR): Emergence in agriculture. *World Journal of Microbiology and Biotechnology*, 28, 1327–1350. https://doi.org/10.1007/s11274-011-0979-9

Bhattacharyya, P. N., Pal, R., & Jain, R. K. (2020). Biofertilizers and biopesticides in sustainable agriculture. In B. Giri, R. Prasad, & A. Varma (Eds.), *Biofertilizers for Sustainable Agriculture and Environment* (pp. 1–23). Springer. https://doi.org/10.1007/978-3-030-18933-4_1

Bravo, A., Likitvivatanavong, S., Gill, S. S., & Soberón, M. (2011). *Bacillus thuringiensis*: A story of a successful bioinsecticide. *Insect Biochemistry and Molecular Biology*, 41(7), 423–431. https://doi.org/10.1016/j.ibmb.2011.02.006

CBD (Convention on Biological Diversity). (2022). *Sustainable Use of Soil Biodiversity in Agriculture*. https://www.cbd.int/agro/soil

Chandler, D., Bailey, A. S., Tatchell, G. M., Davidson, G., Greaves, J., & Grant, W. P. (2011). The development, regulation and use of biopesticides for integrated pest management. *Philosophical Transactions of the Royal Society B: Biological Sciences*, 366(1573), 1987–1998. https://doi.org/10.1098/rstb.2010.0390

Chandra, S., Askari, M. F., Kumari, M., & Srivastava, R. (2020). Bridging the gap between agricultural microbiology research and policy. *Journal of Applied and Natural Science*, 12(2), 181–187.

Choudhary, D. K., & Johri, B. N. (2009). Interactions of *Bacillus* spp. and plants—with special reference to induced systemic resistance (ISR). *Microbiological Research*, 164(5), 493–513. https://doi.org/10.1016/j.micres.2008.08.007

Delgado-Baquerizo, M., Maestre, F. T., Reich, P. B., Jeffries, T. C., Gaitan, J. J., Encinar, D., … & Singh, B. K. (2017). Microbial diversity drives multifunctionality in terrestrial ecosystems. *Nature Communications*, 7, 10541. https://doi.org/10.1038/ncomms10541

Ehlers, R. U. (2011). Regulation of biological control agents and the EU policy support action REBECA. *IOBC-WPRS Bulletin*, 55, 47–50.

European Commission. (2021a). *The Common Agricultural Policy at a Glance*. https://ec.europa.eu/info/food-farming-fisheries/key-policies/common-agricultural-policy_en

European Commission. (2021b). *The Common Agricultural Policy: 2023–2027*. https://ec.europa.eu

FAO. (2020). *State of Knowledge of Soil Biodiversity – Status, Challenges and Potentialities*. Food and Agriculture Organization of the United Nations. https://doi.org/10.4060/cb1928en

FAO. (2021a). *Microbial Solutions for Climate-Smart Agriculture: Lessons from Developing Countries*. Rome: Food and Agriculture Organization of the United Nations.

FAO. (2021b). *Microbiome: The Missing Link in Sustainable Agri-Food Systems*. Food and Agriculture Organization of the United Nations. https://www.fao.org/documents/card/en/c/cb4443en

FAO. (2021c). *The State of the world's Land and Water Resources for Food and Agriculture – Systems at Breaking Point*. Rome: Food and Agriculture Organization of the United Nations.

Finkel, O. M., Castrillo, G., Herrera Paredes, S., Salas González, I., & Dangl, J. L. (2017). Understanding and exploiting plant beneficial microbes. *Current Opinion in Plant Biology*, 38, 155–163. https://doi.org/10.1016/j.pbi.2017.04.018

Foley, J. A., Ramankutty, N., Brauman, K. A., Cassidy, E. S., Gerber, J. S., Johnston, M., … & Zaks, D. P. M. (2011). Solutions for a cultivated planet. *Nature*, 478(7369), 337–342. https://doi.org/10.1038/nature10452

Glare, T. R., Caradus, J. R., Gelernter, W. D., Jackson, T. A., Keyhani, N. O., Köhl, J., … & Stewart, A. (2012). Have biopesticides come of age? *Trends in Biotechnology*, 30(5), 250–258. https://doi.org/10.1016/j.tibtech.2012.01.003

Glick, B. R. (2012). Plant growth-promoting bacteria: Mechanisms and applications. *Scientifica*, 2012, 963401.

Global Bioeconomy Summit. (2020). *Bioeconomy Policy Frameworks around the World*. https://gbs2020.net

Grover, M., Ali, S. Z., Sandhya, V., Rasul, A., & Venkateswarlu, B. (2011). Role of microorganisms in adaptation of agriculture crops to abiotic stresses. *World Journal of Microbiology and Biotechnology*, 27, 1231–1240. https://doi.org/10.1007/s11274-010-0572-7

Harman, G. E., Howell, C. R., Viterbo, A., Chet, I., & Lorito, M. (2004). Trichoderma species—Opportunistic, avirulent plant symbionts. *Nature Reviews Microbiology*, 2(1), 43–56. https://doi.org/10.1038/nrmicro797

Herrmann, L., & Lesueur, D. (2013). Challenges of formulation and quality of biofertilizers for successful inoculation. *Applied Microbiology and Biotechnology*, 97(20), 8859–8873. https://doi.org/10.1007/s00253-013-5228-8

Jansson, J. K., & Hofmockel, K. S. (2020). Soil microbiomes and climate change. *Nature Reviews Microbiology*, 18(1), 35–46. https://doi.org/10.1038/s41579-019-0265-7

Jat, R. S., Meena, R. H., & Sharma, S. K. (2021). Biopesticides for sustainable pest management in agriculture. *Indian Journal of Agricultural Sciences*, 91(7), 1019–1026.

Kabaluk, T., Svircev, A., Goettel, M., & Woo, S. (Eds.). (2010). *The Use and Regulation of Microbial Pesticides in Representative Jurisdictions Worldwide*. IOBC Global.

Keswani, C., Singh, H. B., Hermosa, R., & García-Estrada, C. (2014). Myconanotechnology: New opportunities for development of eco-friendly fungal based nanobiopesticides. *Frontiers in Plant Science*, 5, 579. https://doi.org/10.3389/fpls.2014.00579

Kumar, S., Sharma, R., & Verma, P. (2022). Role of biopesticides in sustainable agriculture. *Journal of Applied and Natural Science*, 14(1), 97–104.

Kumar, A., & Verma, J. P. (2020). Phytoremediation: Role of rhizospheric microbes and microbe-plant interactions. In *Rhizosphere Engineering* (pp. 283–306). Academic Press.

Lugtenberg, B., & Kamilova, F. (2008). Beneficial rhizosphere pseudomonads. *Экологическая генетика*, *6*(2), 4–12.

Lugtenberg, B., & Kamilova, F. (2009). Plant-growth-promoting rhizobacteria. *Annual Review of Microbiology*, 63, 541–556. https://doi.org/10.1146/annurev.micro.62.081307.162918

Meena, R. K., Bohra, J. S., Singh, R. K., & Meena, V. S. (2017). Response of microbial inoculants and phosphorus levels on yield and quality of chickpea under rainfed condition. *Journal of Soil Science and Plant Nutrition*, 17(1), 45–59. https://doi.org/10.4067/S0718-95162017005000004

Meena, V. S., Maurya, B. R., Verma, J. P., Meena, R. S., & Jatav, G. K. (2017). Potassium-solubilizing microbes for sustainable agriculture. In V. S. Meena (Ed.), *Potassium Solubilizing Microorganisms for Sustainable Agriculture* (pp. 1–17). Springer.

Meena, R. K., Sharma, A., & Saini, A (2021). Role of microbial decomposition in nutrient cycling and organic farming. *International Journal of Environmental Sciences*, 6(2), 45–54.

Ministry of Agriculture & Farmers Welfare, Government of India. (2020). *Operational Guidelines for Paramparagat Krishi Vikas Yojana (PKVY)*. https://agricoop.gov.in

Mitter, B., Brader, G., Afzal, M., Compant, S., Naveed, M., & Sessitsch, A. (2019a). Advances in elucidating beneficial interactions between plants, soil, and bacteria. *Frontiers in Plant Science*, 10, 1601. https://doi.org/10.3389/fpls.2019.01601

Mitter, B., Brader, G., Pfaffenbichler, N., & Sessitsch, A. (2019b). Next generation microbiome applications for crop production—limitations and the need of knowledge-based solutions. *Current Opinion in Microbiology*, 49, 59–65. https://doi.org/10.1016/j.mib.2019.10.005

OECD. (2021). *Safety Assessment of Microbial Inoculants Used in Agriculture and Food*. Organisation for Economic Co-operation and Development. https://www.oecd.org/env/ehs/biotrack/

Ogidi, O. I., & Akpan, U. M. (2023). Impacts of chemical use in agricultural practices: Perspectives of soil microorganism and vegetation. In M.C. Ogwu, & S. Chibueze Izah, (Eds.), *One Health Implications of Agrochemicals and Their Sustainable Alternative*. Springer Nature Singapore Pte Ltd.

Ogidi, O. I., & Izah, S. C. (2024). Emerging technologies in the global South food industry: Prospects and challenges. In M.C. Ogwu, S.C. Izah, & N.R. Ntuli (Eds.), *Food Safety and Quality in Global South*. Springer Nature Singapore Pte Ltd. https://doi.org/10.1007/978-981-97-2428-4_7

Ogidi, O. I., Izah, S. C., & Akpan, U. M. (2023). Food quality and agrochemical use: Integrated monitoring, assessment, and management policies. In M.C. Ogwu, & S. Chibueze Izah, (Eds.), *One Health Implications of Agrochemicals and Their Sustainable Alternative. Springer Nature Singapore Pte Ltd.*

Ogidi, O. I., Wenapere, C. M., & Chukwunonso, O. A. (2025). Enhancing food safety and quality control with computer vision systems. In D. Bora & R. Bania (Eds.), *Computer Vision Techniques for Agricultural Advancements* (pp. 51–88). IGI Global Scientific Publishing. https://doi.org/10.4018/979-8-3693-8019-2.ch003

Pretty, J., Benton, T. G., Bharucha, Z. P., Dicks, L. V., Flora, C. B., Godfray, H. C. J., … & Wratten, S. (2018). Global assessment of agricultural system redesign for sustainable intensification. *Nature Sustainability*, 1(8), 441–446. https://doi.org/10.1038/s41893-018-0114-0

Rahi, D. K., Kumar, V., & Choudhury, A. R. (2020). Application of microalgae in agriculture. *Indian Journal of Agricultural Research*, 54(1), 12–19.

Rodríguez, H., & Fraga, R. (1999). Phosphate solubilizing bacteria and their role in plant growth promotion. *Biotechnology Advances*, 17(4–5), 319–339. https://doi.org/10.1016/S0734-9750(99)00014-2

Rosenblueth, M., Ormeño-Orrillo, E., López-López, A., Rogel, M. A., Reyes-Hernández, B. J., Martínez-Romero, J. C., & Martínez-Romero, E. (2018). Nitrogen fixation in cereals. *Frontiers in Microbiology*, 9, 1794. https://doi.org/10.3389/fmicb.2018.01794

Rouphael, Y., & Colla, G. (2020). Editorial: Biostimulants in agriculture. *Frontiers in Plant Science*, 11, 40. https://doi.org/10.3389/fpls.2020.00040

Sharma, M. P., Saini, R. P., & Kumar, A. (2020). Mycorrhizal fungi: Role in improving crop productivity. *Indian Phytopathology*, 73(2), 199–207.

Singh, R. P., Kumar, A., & Kumar, S. (2021). Biofertilizer and their application in agriculture: A review. *Indian Journal of Agricultural Sciences*, 91(6), 815–820.

Singh, J. S., Kumar, A., Rai, A. N., & Singh, D. P (2016). Cyanobacteria: A precious bio-resource in agriculture, ecosystem, and environmental sustainability. *Frontiers in Microbiology*, 7, 529. https://doi.org/10.3389/fmicb.2016.00529

Singh, D. P., & Prabha, R (2023). Microbial approaches in composting and sustainable solid waste management. *Waste Management & Research*, 41(1), 7–17.

Singh, B., & Singh, V. K. (2023). Diversity of wood-inhabiting macrofungi from District Ayodhya, Uttar Pradesh, India. *Kavak*, 59(3), 1–11. https://doi.org/10.36460/Kavaka/59/3/2023/51-61

Singh, B., Tripathi, S., Tyagi, S., Tripathi, A., Dubey, P., Kumar, S. (2024). Phytostimulation and induction of systemic resistance by plant growth promoting fungi. In P. Mathur & S. Roy (Eds.), *Plant Microbiome and Biological Control. Sustainability in Plant and Crop Protection* (vol 20). Springer, Cham. https://doi.org/10.1007/978-3-031-75845-4_10

Singh, S., Yadav, S., & Rathi, N. (2023). Role of cyanobacteria in sustainable agriculture. *Asian Journal of Microbiology, Biotechnology & Environmental Sciences*, 25(2), 327–335.

Smith, S. E., & Read, D. J. (2008). *Mycorrhizal Symbiosis* (3rd ed.). Academic Press.

Smith, F. A., & Smith, S. E. (2011). Roles of arbuscular mycorrhizas in plant nutrition and growth: New paradigms from cellular to ecosystem scales. *Annual Review of Plant Biology*, 62, 227–250. https://doi.org/10.1146/annurev-arplant-042110-103846

Spielman, D. J., Hartwich, F., & Grebmer, K. V. (2007). Public–private partnerships and developing-country agriculture: Evidence from the international agricultural research system. *Public Administration and Development*, 27(5), 261–276. https://doi.org/10.1002/pad.464

Timmusk, S., Behers, L., Muthoni, J., Muraya, A., & Aronsson, A. C. (2014). Perspectives and challenges of microbial application for crop improvement. *Frontiers in Plant Science*, 5, 713. https://doi.org/10.3389/fpls.2014.00713

UN DESA (United Nations Department of Economic and Social Affairs). (2019). *World Population Prospects 2019*. https://population.un.org/wpp

UN SDG Report. (2023). *The Sustainable Development Goals Report 2023*. United Nations. https://unstats.un.org/sdgs/report/2023/

USDA-SARE. (2021). *Sustainable Agriculture Research and Education: Projects and Impacts*. United States Department of Agriculture. https://www.sare.org/

USEPA. (2020). *Biopesticides Regulation*. U.S. Environmental Protection Agency. https://www.epa.gov/ingredients-used-pesticide-products/what-are-biopesticides

Verma, J. P., Yadav, J., Tiwari, K. N., & Kumar, A. (2022). Application of microbial consortia for sustainable agriculture. *Biocatalysis and Agricultural Biotechnology*, 42, 102365.

Vessey, J. K. (2003). Plant growth promoting rhizobacteria as biofertilizers. *Plant and Soil*, 255(2), 571–586. https://doi.org/10.1023/A:1026037216893

Vimal, S. R., Singh, J. S., Arora, N. K., & Singh, S. (2017). Soil-plant-microbe interactions in stressed agriculture management: a review. *Pedosphere*, *27*(2), 177–192.

Index

Note: Locators in *italics* represent figures and **bold** indicate tables in the text.

A

D

E

G

H

Q

R

S

T

U

V

W

X

Y

Z

For Product Safety Concerns and Information please contact our
EU representative GPSR@taylorandfrancis.com Taylor & Francis
Verlag GmbH, Kaufingerstraße 24, 80331 München, Germany